Synthesis Series in Biomedical Engineering

Volume 1

Sensory Organ Replacement and Repair
Gerald E. Miller
2006

Artificial Organs
Gerald E. Miller
2006

Signal Processing of Random Physiological Signals
Charles S. Lessard
2006

Image and Signal Processing for Networked E-Health Applications
Ilias G. Maglogiannis, Kostas Karpouzis, Manolis Wallace
2006

Volume 2

Fundamentals of Respiratory Sounds and Analysis
Zahra Moussavi
2006

Basic Probability Theory for Biomedical Engineers
John D. Enderle, David C. Farden, Daniel J. Krause
2006

Intermediate Probability Theory for Biomedical Engineers
John D. Enderle, David C. Farden, Daniel J. Krause
2006

Advanced Probability Theory for Biomedical Engineers
John D. Enderle, David C. Farden, Daniel J. Krause
2006

Volume 3

Capstone Design Courses: Producing Industry-Ready Biomedical Engineers
Jay R. Goldberg
2007

BioNanotechnology
Elisabeth S. Papazoglou, Aravind Parthasarathy
2007

Bioinstrumentation
John D. Enderle
2006

Volume 4

Multimodal Imaging in Neurology: Special Focus on MRI Applications and MEG
Hans-Peter Müller, Jan Kassubek
2007

Estimation of Cortical Connectivity in Humans: Advanced Signal Processing Techniques
Laura Astolfi, Fabio Babiloni
2007

Brain–Machine Interface Engineering
Justin C. Sanchez , José C. Principe
2007

Introduction to Statistics for Biomedical Engineers
Kristina M. Ropella
2007

Volume 5

Introduction to Continuum Biomechanics
Kyriacos A. Athanasiou, Roman M. Natoli
2008

The Effects of Hypergravity and Microgravity on Biomedical Experiments
Thais Russomano, Gustavo Dalmarco, Felipe Prehn Falcão
2008

A Biosystems Approach to Industrial Patient Monitoring and Diagnostic Devices
Gail Baura
2008

Volume 6

Neural Interfacing: Forging the Human-Machine Connection
Thomas D. Coates Jr.
2008

Quantitative Neurophysiology
Joseph V. Tranquillo
2008

Tremor: From Pathogenesis to Treatment
Giuliana Grimaldi, Mario Manto
2008

Volume 7

Introductory Medical Imaging
A. A. Bharath
2008

Lung Sounds: An Advanced Signal Processing Perspective
Leontios J. Hadjileontiadis
2008

An Outline of Informational Genetics
Gérard Battail
2008

Volume 8

Basic Feedback Controls in Biomedicine
Charles S. Lessard
2009

Understanding Atrial Fibrillation: The Signal Processing Contribution
Luca Mainardi , Leif Sörnmo , Sergio Cerutti
2008

Volume 9

Landmarking and Segmentation of 3D CT Images
Graham Boag
2009

Phonocardiography Signal Processing
Rasha Bassam
2009

Introduction to Biomedical Engineering: Biomechanics and Bioelectricity Vol 1
Christensen
2009

Introduction to Biomedical Engineering: Biomechanics and Bioelectricity Vol 2
Christensen
2009

Synthesis Series in Biomedical Engineering 4

Multimodal Imaging in Neurology: Special Focus on MRI Applications and MEG
Hans-Peter Müller, Jan Kassubek

Estimation of Cortical Connectivity in Humans: Advanced Signal Processing Techniques
Laura Astolfi, Fabio Babiloni

Brain–Machine Interface Engineering
Justin C. Sanchez , José C. Principe

Introduction to Statistics for Biomedical Engineers
Kristina M. Ropella

ISBN: 9781608451876

www.morganclaypool.com

A PublicationintheMorgan&ClaypoolPublishers

SYNTHESISSERIESINBIOMEDICALENGINEERING

SeriesEditors:JohnD.Enderle

Synthesis Series in Biomedical Engineering 4

This volume includes:

Multimodal Imaging in Neurology: Special Focus on MRI Applications and MEG

Hans-Peter Müller, Jan Kassubek

Estimation of Cortical Connectivity in Humans: Advanced Signal Processing Techniques

Laura Astolfi, Fabio Babiloni

Brain–Machine Interface Engineering

Justin C. Sanchez, José C. Principe

Introduction to Statistics for Biomedical Engineers

Kristina M. Ropella

Multimodal Imaging in Neurology: Special focus on MRI applications and MEG

Hans-Peter Müller and Jan Kassubek
University of Ulm, Germany

MORGAN & CLAYPOOL PUBLISHERS

ABSTRACT

The field of brain imaging is developing at a rapid pace and has greatly advanced the areas of cognitive and clinical neuroscience. The availability of neuroimaging techniques, especially magnetic resonance imaging (MRI), functional MRI (fMRI), diffusion tensor imaging (DTI) and magnetoencephalography (MEG) and magnetic source imaging (MSI) has brought about breakthroughs in neuroscience.

To obtain comprehensive information about the activity of the human brain, different analytical approaches should be complemented. Thus, in "intermodal multimodality" imaging, great efforts have been made to combine the highest spatial resolution (MRI, fMRI) with the best temporal resolution (MEG or EEG). "Intramodal multimodality" imaging combines various functional MRI techniques (e.g., fMRI, DTI, and/or morphometric/volumetric analysis). The multimodal approach is conceptually based on the combination of different noninvasive functional neuroimaging tools, their registration and cointegration. In particular, the combination of imaging applications that map different functional systems is useful, such as fMRI as a technique for the localization of cortical function and DTI as a technique for mapping of white matter fiber bundles/tracts.

This booklet gives an insight into the wide field of multimodal imaging with respect to concepts, data acquisition, and postprocessing. Examples for intermodal and intramodal multimodality imaging are also demonstrated.

KEYWORDS

diffusion tensor imaging, functional magnetic resonance imaging, intermodal multimodality, intramodal multimodality, magnetic resonance imaging, magnetoencephalography, multimodal imaging

Preface

The field of brain imaging is developing at a rapid pace and has greatly advanced the areas of cognitive and clinical neuroscience. Cognitive neuroscience aims at understanding the functional and neural architecture of cognitive functions. The availability of neuroimaging techniques, especially magnetic resonance imaging (MRI), functional MRI (fMRI), diffusion tensor imaging (DTI), and magnetoencephalography (MEG)/magnetic source imaging (MSI) has brought many new insights in cognitive neuroscience as well as in the understanding, and diagnosis of brain diseases.

To obtain comprehensive information about the activity of the human brain, various analytical approaches and different methods of activation should be complemented. Thus, in "intermodal multimodality" imaging, great efforts have been performed to combine the highest spatial resolution (MRI, fMRI) with the best temporal resolution (MEG or EEG). Additionally, "intramodal multimodality" imaging combines various functional MRI techniques (DTI, fMRI, and/or morphometry). Comprehensive information can be provided by radionuclide techniques (PET or SPECT) or by MR spectroscopy (MRS).

The multimodal approach is conceptually based on the combination of different noninvasive functional neuroimaging tools, their registration and cointegration. In particular, intramodal multimodality is useful, for example, the combination of imaging applications on the same technical approach, such as MRI that maps different functional systems, fMRI as a technique for the localization of cortical function, and DTI as a technique for mapping of white matter fiber bundles/tracts. MSI as an additional imaging technique of cortical function is of particular advantage in epilepsy diagnostics.

This booklet provides first insights for engineers or medical doctors who deal with multimodal imaging. Basic measurement techniques as well as procedures for the overlay of results from different measurement techniques with a special focus on MRI-based techniques and MSI are described in detail. Examples for intermodal and intramodal multimodality imaging are demonstrated. The authors have tried to do away with the basic details and formulas where possible, nevertheless have given formulas and mathematics where the understanding is assisted.

Acknowledgments

Dr. Alexander Unrath, University of Ulm, Germany, is thankfully acknowledged for his engagement in course of the DTI experiments and intramodal integration.

Dr. Sergio Nicola Erné, Friedrich Schiller University, Jena, Germany, is thankfully acknowledged for his engagement in course of the MEG experiments and the intermodal integration.

Prof. Dr. Siegfried Stapf, Technical University Illmenau, Germany, is thankfully acknowledged for assistance in the MRI chapters.

Contents

CHAPTER 1

INTRODUCTION

At the beginning of the 20th century, neuroimaging began with a technique called pneumoencephalography [1]. This process basically consisted of draining the cerebrospinal fluid and replacing it with air, altering the relative density of the brain and its surroundings. In this way, X-ray contrasts were improved. Computed tomography (CT) and magnetic resonance imaging (MRI) were discovered between 1970 and 1980. These technologies were considerably less harmful. The next neuroimaging techniques to be developed were single photon emission computed tomography (SPECT) and positron emission tomography (PET) that allowed the mapping of the brain functionalities providing more than just static images of the brain's structure. The end of the last century showed the great breakthrough in mapping human brain functionalities by developing functional MRI (fMRI) with abilities that opened the door to direct observation of cognitive activities. This technique includes, among other options, the study of emotions, social functions (social neuroscience), decision-making (neuroeconomics) and executive functions. Clinical neuroscience aims at understanding, diagnosing and treating diseases of the central nervous system such as neurodegenerative diseases, stroke, epilepsy, and psychiatric disorders. In today's aging society, brain-related diseases are an increasing cost factor. Therefore, both from a scientific and a social point of view, brain imaging is one of the most challenging research areas of the 21st century.

In addition to these localization techniques, the temporal evolution of brain signals may be of interest too. Techniques with a high temporal resolution are electroencephalography (EEG) and magnetoencephalography (MEG). Both techniques are completely noninvasive. The technical equipment needed for EEG is in principle restricted to electrodes, an amplifier, and a detection device, whereas MEG is a much more complicated technique (for details refer to Section 2.2). The high temporal resolution of EEG/MEG can complement the spatial resolution of CT/MRT for comprehensive information – called intermodal multimodal imaging.

Multimodality imaging is widely considered to involve the incorporation of two or more imaging modalities. The localization of signals detected by MEG or EEG, PET or SPECT has an anatomical background with limited spatial resolution. The combination of these modalities with more morphologically oriented ones such as MRI (or CT) offers great opportunities. These methodologies constrain and complement each other. Combined, they can therefore improve our interpretation of functional neural organization. Therefore, it is useful to investigate the brain functions in a multimodal noninvasive way by combining different neuroimaging techniques, thus making use of their individual advantages. Furthermore, the acquisition of individual volume-rendering three-dimensional (3-D) MRI data allows improved models of the anatomical basis or of the head volume conductor (see Chapter 2). For example, one approach to analyze MEG data with particular respect to the inverse problem is to impose anatomic and physiologic constraints for magnetic source imaging (MSI) derived from other imaging modalities such as fMRI in order to reduce the solution space. Combinations of MEG and fMRI in the same paradigm were described in several studies [2,3,4,5], both by comparing the localization of the results as well as applying fMRI constraints to MEG data analysis. Efforts have also been made for the combination of electroencephalography and fMRI [6]. Another field of research is magnetic resonance spectroscopy (MRS) and its intramodality multimodal integration [7].

The main topic of the following chapters is the multimodal combination presented in the form of short introductions to the various measurement techniques. The focus is on the principles of multimodal imaging. Basically, two ways of multimodal imaging can be distinguished. The first way is intermodal multimodality – results of different measurement techniques have to be combined and transformed to a common coordinate frame. To perform this task modality specific markers have to be used to obtain reference points in each specific modality/measurement technique. As an example the combination of MRI, fMRI, and MEG is explained in detail in Section 4.1.

The second way is called intramodal multimodality. Here, results obtained from one single modality such as MRI are combined in a common coordinate frame. Intersubject comparison, group averaging, and comparison between subject groups with different clinical categorizations can be performed by using, for example, the Montreal Neurological Institute (MNI) coordinate frame as a common basis. Transformation rules have to be set up that allows the transformation of results of any single measurement into this coordinate frame. The focus in Section 4.2 is the comprehensive information of diffusion tensor imaging (DTI) results along with morphometry results, while fMRI results complements the data.

Examples of clinical applications and an outlook of future applications of multimodal imaging will complete the walk through different aspects of multimodal imaging.

CHAPTER 2

Neurological Measurement Techniques and First Steps of Postprocessing

2.1 MAGNETIC RESONANCE IMAGING (MRI)

2.1.1 Principles of Magnetic Resonance

MRI was introduced in clinical routine in the early 1980s. It has become a preferred neuroimaging modality because of its feasibility to generate images of slices in arbitrary orientation with different tissue contrast [8,9,10,11]. Nuclei [protons are preferred as $I=1/2$, and they possess the largest gyromagnetic ratio (γ, see below)] possess an angular momentum. This is coupled with a magnetic dipole moment, which is proportional to the angular momentum. In the presence of a strong magnetic field a nucleus with quantum number I has $2I+1$ discrete energy states

$$E_m = -\gamma\hbar B_z m \tag{2.1}$$

where γ is the gyromagnetic ratio, $\hbar$ is the Planck's constant, B_z is the flux density of magnetic field, m is the magnetic quantum number. A change of energy levels means emission or absorption of a photon with

$$\hbar\omega = E_{m-1} - E_m = \gamma\hbar B_z \quad m = \pm 1 \tag{2.2}$$

where ω is the angular frequency.

In an external magnetic field, the occupation of energy states [Eq. (2.1)] is in accordance to Boltzmann's statistics

$$\frac{N_{m-1}}{N_m} = e^{-\hbar\omega/kT} \tag{2.3}$$

where k is the Boltzmann's constant, and T is the temperature.

In this way, a surplus of nuclear spins leads to a macroscopic magnetization align parallel to the magnetic field resulting in an angular momentum and an associated macroscopic magnetic moment. If this parallel alignment is disturbed, a torque acts on the magnetic moment of the sample, resulting in a precession of the magnetization with the Larmor frequency

$$\omega = \gamma B_z \tag{2.4}$$

In MR, precession of the nuclear magnetization is stimulated by disturbing the alignment of the nuclear magnetization parallel to the main magnetic field by a radio frequency (RF) field according to Eq. (2.4). This is realized by a coil around the sample with its field axis orthogonal to the static magnetic field.

The RF magnetic flux density B_1 of a pulse of duration t leads to a rotation of the macroscopic magnetization about an angle α between the macroscopic magnetization and the static magnetic field

$$\alpha = \gamma B_1 t \tag{2.5}$$

resulting in a precessing nuclear magnetization $\vec{M}$ with components

$$\begin{aligned} M_x &= -M_0 \sin(\alpha)\sin(\phi) \\ M_y &= M_0 \sin(\alpha)\cos(\phi) \\ M_z &= M_0 \cos(\alpha) \end{aligned} \tag{2.6}$$

where ϕ is the angle of the precession.

The same coil can also be used for signal detection. By switching off the pulse, the signal induced is proportional to the precessing nuclear magnetization and decays with time, and the original state of equilibrium is reestablished, with the magnetization aligned parallel to the main magnetic field described by two separate relaxation time constants, T_1 and T_2 (Bloch's equations [12]).

$$\begin{aligned} \frac{dM_z}{dt} &= \gamma(\vec{M}x\vec{B})_z + \frac{M_0 - M_z}{T_1} \\ \frac{dM_{xy}}{dt} &= \gamma(\vec{M}x\vec{B})_{xy} + \frac{M_{xy}}{T_2} \end{aligned} \tag{2.7}$$

The measured signal is called free induction decay (FID).

Stimulation of the magnetic moments with $\alpha = 90^\circ$ [Eq. (2.5)] leads to a precession in the x/y plane decaying with time. The inhomogeneous dephasing precessing magnetic moments can be refocused with a second pulse ($\alpha = 180^\circ$) at time τ causes a spin echo (SE) at time 2τ (Fig. 2.1).

The mobility of the molecules that contain the nucleus causes relaxation. Each nucleus is surrounded by other magnetic moments that are in constant Brownian motion and therefore produce a continuously changing magnetic perturbation field. The part of the magnetization vector that is

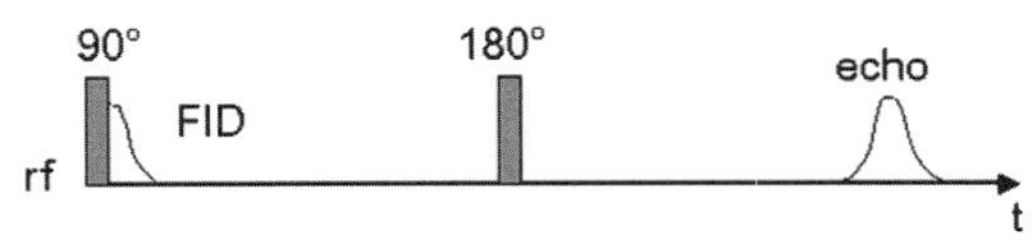

FIGURE 2.1: Spin echo pulse sequence.

parallel to the main magnetic field relaxes with the spin-lattice relaxation time T_1. This is caused by interactions between the observed nucleus and unexcited nuclei in the environment as well as by electric fields in the environment. T_1 is the time required for the longitudinal component of the magnetization vector to be restored to 63% of its original magnitude.

The perpendicular part of the magnetization relaxes with the spin–spin relaxation time T_2. This is caused by interactions between the already excited spinning nuclei. T_2 is the time required for the transverse component of the magnetization vector to drop to 37% of its original magnitude after initial excitation. Nevertheless, T_1 and T_2 are coupled variables. The solution of Bloch's equations is

$$\begin{aligned} M_{xy}(t) &= M_x(t) + iM_y(t) = iM_0 e^{-i\omega t} e^{-t/T_2} \\ M_z(t) &= M_0(1 - e^{-t/T_1}) \end{aligned} \tag{2.8}$$

The measurement of T_2 is only possible in highly homogeneous magnetic fields. In practice, the effective relaxation time T_2^* which is directly related to the main magnetic field deviations ΔB_z is valid

$$\frac{1}{T_2^*} = \frac{1}{T_2} + \frac{1}{2}\gamma \Delta B_z \tag{2.9}$$

Therefore, T_2^* directly measures the local magnetic field inhomogeneities.

2.1.2 Magnetic Resonance Imaging

The required static magnetic field is provided by a superconducting magnet. Typical magnetic field strength for medical applications and human brain research is 1.5 or 3.0 T. The homogeneity is assured by additional field gradients called shim system. The RF sender provides pulses with a frequency $\omega/2\pi$ of about 64 or 128 MHz (proton resonance), respectively. When the gradient system applies gradient pulses in the order of 100 mT/m, the detection system acquires the data (Fig. 2.2).

In order to analyze the signal with respect to its spatial origin, at least one of the two fields required for MR measurement has to vary over space. The simplest way for a two-dimensional (2-D) MR imaging is a standard spin echo sequence (Fig. 2.3).

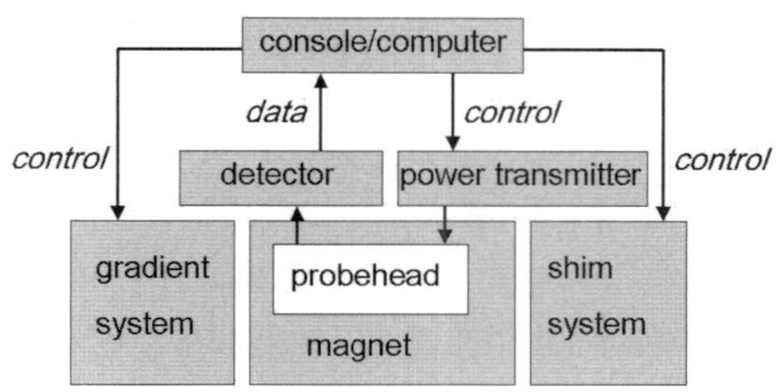

FIGURE 2.2: Block diagram of an MR tomograph.

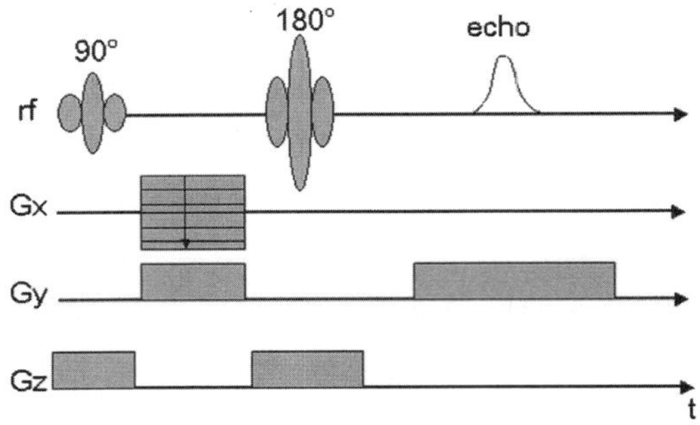

FIGURE 2.3: Typical 2-D slice imaging pulse sequence with spin echo.

Applying a gradient in z-direction during stimulation with an amplitude-modulated RF-pulse causes a slice selection perpendicular to z. The spatial information in x- and y-direction is encoded by a phase-encoding gradient in x-direction and a read-out gradient in y-direction. The pulse sequence has to be repeated for each step of the phase-encoding gradient. The read-out gradient has to be "refocused" before the 180°-pulse. The information is sampled in the reciprocal k-space and is then Fourier transformed into r-space, thereby reproducing the spin-density distribution of the measured object. However, the information consists of complex numbers and thereby the phase of the signal as function of k can be derived, the so-called k-space imaging. This information is used, e.g., in chemical shift imaging [11]. A 3-D image can be recorded by repeating the 2-D sequence for several slices. The respective slice selection is achieved by changing the stimulating RF-frequency.

Two parameters are available for signal manipulation: the sequence repetition time T_{R} and the echo time T_{E}. The use of a long T_{E} allows transverse relaxation of the system to become effective before signal acquisition, whereas, rapid repetition of the pulse sequence prevents longitudinal magnetization from reestablishing. The signal intensity in a picture element (2-D: pixel, 3-D: voxel) is given by

$$M_{xy}(x, y, z) = M_0(x, y, z)(1 - e^{-T_{\mathrm{R}}/T_1(x,y,z)})e^{-T_{\mathrm{E}}/T_2(x,y,z)} \qquad (2.10)$$

The repetition time and the echo time can be adjusted so that the image contrast due to different types of tissue is determined by M_0, T_1 or T_2. Short values of T_{R} and T_{E} give T_1 weighted images whereas long T_{E} and short T_{R} give spin-density weighted images. Long values of T_{R} as well as T_{E} give T_2 weighted images. An example is shown in Fig. 2.4.

Improvement and simplification is done by a gradient echo sequence (Fig. 2.5). Here, short repetition times are combined with small flip angles (α) [13]. Thus, it is possible to record multiple gradient echoes. The possibility of recording images within seconds occurs [echo-planar imaging (EPI)]. As gradient echo sequences are fast, they are well suited for 3-D data acquisition. The sensitivity of gradient echo sequences to magnetic field inhomogeneities can be utilized for image effects in the human body, which responds sensitively to changes in magnetic susceptibility.

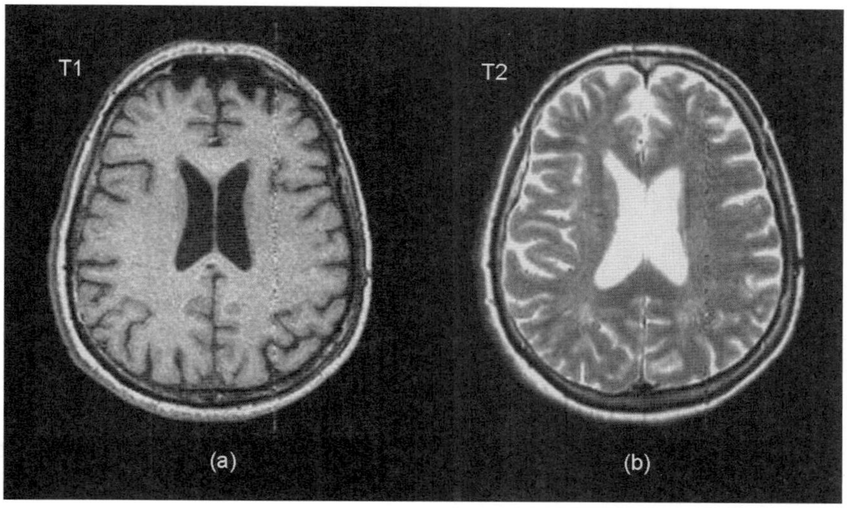

FIGURE 2.4: Example for different image contrasts of appreciatively the same slice of a human brain. (a) T_1-weighted image. (b) T_2-weighted image.

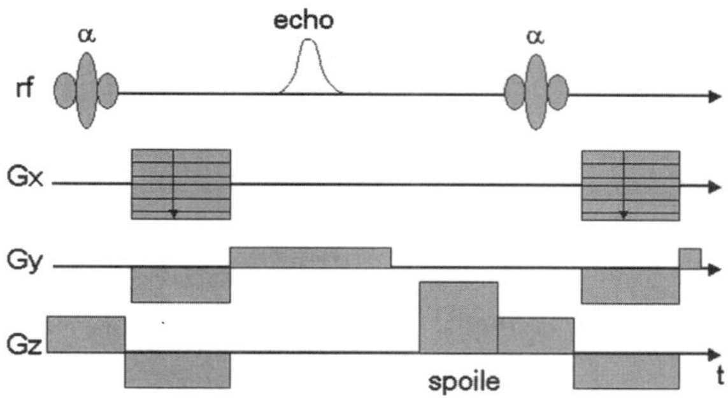

FIGURE 2.5: Gradient echo sequences for fast image acquisition.

For more details about magnetic resonance instrumentation and pulse sequences the reader is referred to [11,9,8,10].

2.1.3 Mathematical Excursion 1: Smoothing, Interpolation and Normalization During Postprocessing

In order to improve the image quality and to unify the image or recording dimensions, several postprocessing procedures can be applied.

2.1.3.1 Smoothing and Interpolation. As different MRI scans can be recorded with different voxel resolutions and often the recording technique provides voxels with nonisotropic size (e.g., the slice thickness is larger than the in-plane voxel resolution or pixel size), the transformation to an isotropic voxel size may be of interest. The standard transformation procedure is a linear nearest neighbor transformation with

$$I_{\text{target}}(i, j, k) = \sum a_v I_v(l, m, n) \tag{2.11}$$

where $I_{\text{target}}(i, j, k)$ is the voxel intensity at the new grid coordinates i,j,k, and l,m,n are the original recorded voxel coordinates in x,y,z direction. $I_v(l, m, n)$ are the voxel intensities of the eight recorded voxels that contribute to the target intensity with factors a_v, the eight weighting factors for the interpolation.

Interpolation and smoothing are classical image postprocessing procedures for which a large variety of methods exist. Mishra et al. [14] developed the idea of an anisotropic image interpolation method. Here, the kernel for interpolation is weighted by a factor depending on the local gradients. Thus, the sharpness of the image is preserved. For smoothing, a Gaussian kernel can be chosen, e.g., a kernel sphere around each voxel with Gaussian shape is used. This kernel is then weighted with the local gradients by

$$I_{\text{target}}(i, j, k) = \frac{\sum I_v(l, m, n)/(r_v + g_{\text{target}})}{\sum 1/(r_v + g_{\text{target}})} \tag{2.12}$$

where r_v is the distance between I_{target} and I_v. g_{target} is the absolute value of the gradient at position (i,j,k). In this way, the local gradients weigh the interpolation kernel with a sharpness dependency.

2.1.3.2 Normalization. The measured intensity of T_1- or T_2-weighted images depends on acquisition parameters and thus does not reflect a measurable value. For comparison of recordings from different subjects or from different scanners, intensity normalization may be helpful. For this task, the average $\bar{\mu}_1$, $\bar{\mu}_2$ and standard deviation σ_1, σ_2 of the signal intensities have to be calculated.

$$\bar{\mu}_i = \frac{1}{N}\sum_{v=1}^{N} I_v \quad \sigma_i^2 = \frac{1}{N}\sum (I_v - \bar{\mu}_i)^2 \tag{2.13}$$

Intensity transformation of data set 1 to the parameters of data set 2 is performed by

$$I_{1,\text{new}}(i, j, k) = \frac{\sigma_2}{\sigma_1}(I_1(i, j, k) + \bar{\mu}_2 - \bar{\mu}_1) \tag{2.14}$$

Note that the normalization can be performed only after transformation to isotropic voxels. It has to be assured that the recorded data contain the same volume – otherwise errors due to different scanning regions or (in case of different subjects) due to different head sizes can occur. It has to be assured that only voxels inside the head are taken into account. Different signal-to-noise (SNR) ratios can also contribute to an error. A detailed consideration of this topic will extend the focus of this booklet and is left to the study of the interested reader.

2.1.4 Functional Magnetic Resonance Imaging

MRI for mapping human brain functionality has rapidly increased during the 1990s. Now, cerebral physiological responses during neural activation can be measured noninvasively [15,16]. fMRI uses the blood oxygenation-level-dependent (BOLD) contrast [17,18,19]. Here, slight physiological alterations, such as neuronal activation resulting in changes of blood flow and blood oxygenation are detected. These signal changes are related to changes in the concentration of deoxyhemoglobin which acts as an intravascular contrast agent for fMRI [20]. The vast majority of fMRI examinations is performed with BOLD-based methods using T_2^*-weighted gradient echo pulse sequences sensitive to local distortions in the magnetic field (susceptibility sensitive techniques) [21,22,23].

Various fMRI methods have good spatial and sufficient temporal resolution, limited by the precision with which the autoregulatory mechanisms of the brain adjust blood flow in space according to the metabolic demands of neuronal activity. Since these methods are completely noninvasive without the need for contrast agents or ionizing radiation, repeated single-subject studies are feasible.

The fMRI experiment consists of a functional template or protocol (e.g., alternating activation and rest for a certain time as the classical "block design"), which induces a functional response in the brain. The aim of an fMRI experiment is to detect this stimulus response, resulting from the BOLD effect, in a defined volume element. During the evaluation of the resulting fMRI image series the physiologically induced signals have to be separated from noise or from artifacts resulting from patient movement or MRI detection techniques. The functional information of a voxel has to be extracted from its functional time course. Therefore, for each functional time point one fMRI volume is recorded. The complete 4-D data set (three dimensions in space, one dimension in time) consists of subsequently recorded 3-D volumes, and thus a functional time course exists for each voxel of a volume. The acquisition of these functional volumes runs over periods lasting up to several minutes. Therefore, it is essential for the following functional analysis to eliminate even very small head movements. Thus, prior to the analysis of a voxel time course, the sequentially acquired volumes have to be corrected for these head motions. The majority of motion correction algorithms found in the appropriate literature is voxel-based matching algorithms [24,25,26].

As a further fMRI specific problem, in most cases, the signal intensity difference between neighbored voxels in a slice is in the range of the expected BOLD effect. Therefore, complex algorithms have to be developed to extract the functional regions. Common techniques for the analysis of a voxel time course are deterministic techniques, i.e., the similarity to the stimulus function is calculated [27,28], statistically analytical (the differences between activation and rest of two samples are calculated by performing Student's t-test, Kolmogorov-Smirnov test, etc.) [27,29], and have an application of clustering techniques, to improve the SNR of the results [27,30,31].

2.1.4.1 Motion Correction. The functional information is contained in a series of volumes, which are acquired over a certain time period. Thus, prior to the analysis of a voxel time course, the sequentially acquired functional volumes should be corrected for head motion caused by translation and rotation. Motion correction can be performed under the following assumptions:

- The brain is considered as a rigid body that can move slightly with six degrees of freedom.
- There are no relative movements of single parts (the brain is not pulsating).
- The central slice of a 3-D fMRI volume acts as reference slice for the motion correction.
- The first volume of a series of volumes contributing to an fMRI experiment acts as reference coordinate system.

With these prerequisites, the motion correction is performed in the following way:

First, the rotation and shifting matrices $\vec{\vec{R}}$ and $\vec{T}$ are determined for a reference slice of the volume to be corrected

$$\vec{x}^{(2)} = \vec{\vec{R}}\vec{x}^{(1)} + \vec{T} \tag{2.15}$$

where $\vec{x}^{(1)}$ and $\vec{x}^{(2)}$ are the position vectors of voxels before and after transformation, respectively. The Euler angles (ϕ, θ, ψ) define the rotation matrix

$$\vec{\vec{R}} = \begin{pmatrix} \cos\phi\cos\psi - \sin\phi\sin\psi\cos\theta & -\sin\phi\cos\psi - \cos\phi\sin\psi\cos\theta & \sin\psi\sin\theta \\ \cos\phi\sin\psi + \sin\phi\cos\psi\cos\theta & -\sin\phi\sin\psi + \cos\phi\cos\psi\cos\theta & -\cos\psi\sin\theta \\ \sin\phi\sin\theta & \cos\phi\sin\theta & \cos\theta \end{pmatrix} \tag{2.16}$$

The fitting can be performed, e.g., either by Levenberg–Marquardt method or by Conjugated Simplex method [32]. Second, the resulting matrices are used for the transformation of the whole 3-D volume using Eq. (2.15).

2.1.4.2 Functional Analysis. The task of functional analysis of block designs is to measure the difference of activation and rest intensities. In Fig. 2.6 a typical fMRI stimulus–response curve in a block design is displayed.

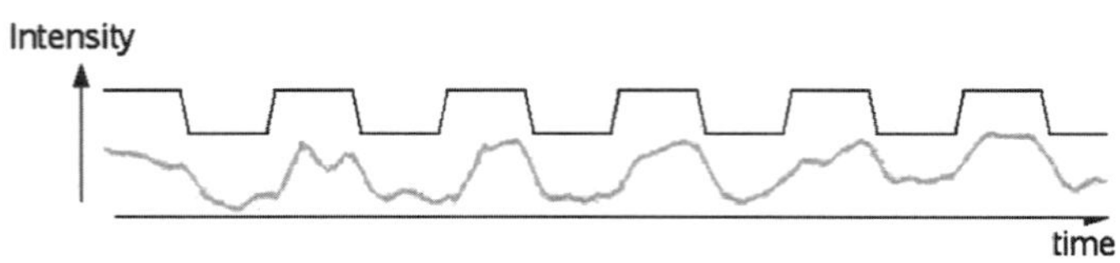

FIGURE 2.6: Stimulus function (dark blue) and response (light blue) in a voxel for a typical fMRI experiment.

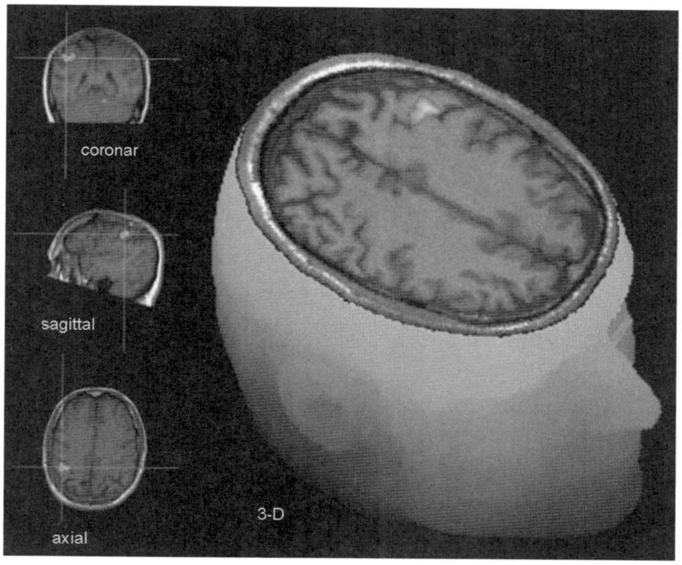

FIGURE 2.7: fMRI detected response from a right hand second finger movement. Activation result is displayed in color. The results are overlaid to a T_1-weighted 3-D image.

In the following excursion, examples for analysis methods are proposed. Using these methods, the resulting activation for a right hand second finger movement is obtained (Fig. 2.7).

Different approaches are event-related designs that associate brain processes with discrete events occurring at any point during the scanning session. The statistical analysis methods are identical for block designs. The deterministic methods can only be applied after a re-sorting of the data.

2.1.5 Mathematical Excursion 2: Are Two Distributions Different?

The decision that two distributions are different can be performed either by deterministic methods or by statistical methods.

2.1.5.1 Deterministic Methods. The deterministic tools reveal the correlation between the time course and the stimulus function. In fMRI experiments, the measurement protocol requires that the stimulus function be periodic with a time period τ and a frequency $\omega/2\pi$. The total measurement time is then $T = Mt$, and L is the number of time points in the time course, i.e., the number of volumes contributing to the fMRI experiment. $S(t_i)$ is the transformed time course function with average $\langle S(t_i)\rangle = 0$. This is reached by

$$S(t_i) = I(t_i) - \langle I(t_i)\rangle_T \qquad (2.17)$$

where $I(t_i)$ is the functional time course intensity of the appropriate voxel. It is possible to express the transformed time course function in Fourier terms (phase sensitive detection).

$$S(t_i) = \sum S_n \sin(n\omega t_i + \phi_n) \tag{2.18}$$

2.1.5.2 Correlation Analysis with a Sinus Wave Stimulus Function. The correlation of a functional time course to a sinus wave stimulus function is the analogy to the two-phase sensitive detection of a monochromatic signal. This is performed by the multiplication of the functional time course signal with a sinusoidal in-phase reference function

$$R_1(t_i) = \sin(\omega t_i) \tag{2.19}$$

and a sinusoidal out-of-phase reference function (shifted by $\pi/2$):

$$R_2(t_i) = \cos(\omega t_i) \tag{2.20}$$

$$K_{\text{inphase}} = \frac{1}{L}\sum S(t_i)R_1(t_i) \tag{2.21}$$

and

$$K_{\text{outphase}} = \frac{1}{L}\sum S(t_i)R_2(t_i). \tag{2.22}$$

By use of the Fourier expression of $S(t_i)$ (2.18) and the orthogonality of the sinus harmonics

$$\sum_{i=0}^{L} \sin(n\omega t_i)\sin(m\omega t_i) = 0 \quad \forall n \neq m \tag{2.23}$$

two terms are obtained

$$K_{\text{inphase}} = \frac{S_1}{2}\cos(\phi_1) \quad \text{and} \quad K_{\text{outphase}} = \frac{S_1}{2}\sin(\phi_1). \tag{2.24}$$

Now, C_{si} is the result of the correlation analysis component of the amplitude of the spectrum of a sinusoidal time course function.

$$C_{si} = \sqrt{K^2_{\text{inphase}} + K^2_{\text{outphase}}} = \frac{S_1}{\sqrt{2}} \tag{2.25}$$

or in terms of the functional time course intensity

$$C_{si} = \frac{1}{L}\sqrt{\left(\sum (I(t_i) - \langle I(t_i)\rangle_T)\sin(\omega t_i)\right)^2 + \left(\sum (I(t_i) - \langle I(t_i)\rangle_T)\cos(\omega t_i)\right)^2}. \tag{2.26}$$

C_{si} extracts the monochromatic contribution of the functional time course function. The result is weighted with the functional time course intensity of the appropriate voxel. Due to the analogy of phase sensitive detection, phase shifts between time course and stimulus function are automatically resolved. It is obvious from these calculations that noise contribution can only occur from noise components that fit into the monochromatic contribution, i.e., in the first harmonic.

The calculation of the corresponding significance used in Eq. (2.31) (see below) needs

$$S_{si} = \frac{C_{si}}{\sqrt{\frac{1}{L}\sum\left((I(t_i) - \langle I(t_i)\rangle_T)T(t_i) - \frac{1}{L}\sum(I(t_i) - \langle I(t_i)\rangle_T)T(t_i)\right)^2}} \tag{2.27}$$

where $T(t_i) = 1$ is the activation, $T(t_i) = -1$ is the rest is the stimulus function with average $\langle T(t_i)\rangle_T = 0$ over the time course. Note that the significance reveals the degree of correlation between time course and stimulus and does not take the functional time course intensity into account.

2.1.5.3 Correlation Analysis with a Square Wave Stimulus Function. A square wave stimulus function ($T(t_i) = 1$: activation, $T(t_i) = -1$: rest, $\langle T(t_i)\rangle_T = 0$) in Fourier components is

$$R(t_i) = \frac{4}{\pi}\left(\sin(\omega t_i) + \frac{1}{3}\sin(3\omega t_i) + \frac{1}{5}\sin(5\omega t_i) + \dots\right) \tag{2.28}$$

A correlation analysis reveals

$$C_{sq} = \frac{1}{L}\sum S(t_i)R(t_i) = \sum \frac{S_n}{n}\cos(\phi_n), \quad (n \text{ odd}) \tag{2.29}$$

or in terms of the functional time course intensity

$$C_{sq} = \frac{1}{L}\sum(I(t_i) - \langle I(t_i)\rangle_T)T(t_i). \tag{2.30}$$

C_{sq} is the mathematical analogy to the phase sensitive detection of a signal with a square wave reference function. This function extracts the correlation of the voxel time course function to the stimulus function. The result is weighted with the signal intensity of the time course function. It is obvious that Eq. (2.29) extracts only the sinus components of $S(t_i)$, i.e. due to the fact that the phase of the voxel time course versus the stimulus function is unknown. C_{sq} has to be calculated as the maximum of the function obtained by progressively shifting the stimulus function for each time point of a period. Thus, if the input signal of interest is not a pure sinusoidal signal but has odd harmonics of the frequency $1/\tau$, the output has additional contributions from these odd harmonics.

To test the statistical significance of signals based on this correlation coefficient, s_{sq}, is calculated according to the rules given in [32]

$$S_{\mathrm{sq}} = B_{inc}\left(\frac{L-2}{2}, \frac{1}{2}, s_{sql}\right) \tag{2.31}$$

using the abbreviations

$$s_{sql} = \frac{L-2}{L-2+s_{sqn}^2}, \quad s_{sqn} = s_{sq}\sqrt{\frac{L-2}{(1+s_{sq})(1-s_{sq})}}, \quad s_{sq} = \frac{C_{sq}}{\sqrt{\frac{1}{L}\sum(I(t_i) - \langle I(t_i)\rangle_{t_i})T(t_i))^2}}.$$

Here, B_{inc} is the incomplete beta function which returns a number between zero and one, and represents the probability that s_{sq} could be this large or larger just by chance.

Equation (2.31) reveals the significance of the degree of correlation between time course and stimulus and does not take the functional time course intensity into account, whereas Eq. (2.27) calculates the correlation weighted by the intensity of the time course function.

As a complete understanding of all aspects of the BOLD effect could not be reached yet [27], the theoretical functional time course is also not known exactly. Now, C_{sq} extracts all odd harmonics of a measured sinusoidal time course function, including noise that fits into one of the higher harmonics, whereas C_{si} only cares about the ($n=1$)-component of the time course function and of the noise. Therefore, one has to note carefully if C_{sq} or C_{si} is being calculated.

2.1.5.4 Statistical Analysis Methods. The Student's t-test S_t calculates the significance of a difference of means of two ensembles (ensemble M – for activated time points (stimulus "on") and ensemble N – rest time points (stimulus "off"), m and n are the corresponding ensemble members).

$$S_t = 1 - B_{inc}\left(\frac{m+n}{2} - 2, \frac{1}{2}, s_{st}\right) \tag{2.32}$$

with the abbreviations

$$s_{st} = \frac{m+n-2}{m+n-2+t^2}, \quad t = \frac{\overline{M} - \overline{N}}{\sqrt{(\sigma^2(M)/m + \sigma^2(N)/n)}}$$

where $\overline{M}$ and $\overline{N}$ are the average values for each ensemble and σ^2 the variances [32].

Another statistical test is the Kolmogorov–Smirnov test that calculates the significance of the maximum value of the absolute difference between two ensembles, for details see [32]. Another group of statistical tests are the so-called nonparametric tests that replace the measured values by their ranking when they are in a sorted order. For details see [32]. Many other statistical tests can be thought of and it is left to the reader to study them in detail. For detailed information about fMRI experiments, the reader is referred to [33,34, 35,36,37,38,39].

2.1.6 Diffusion Weighted Imaging

The diffusion of water molecules in the presence of a strong magnetic gradient results in an MRI signal loss as a result of the dephasing of spin coherence. The application of a pair of strong gradients

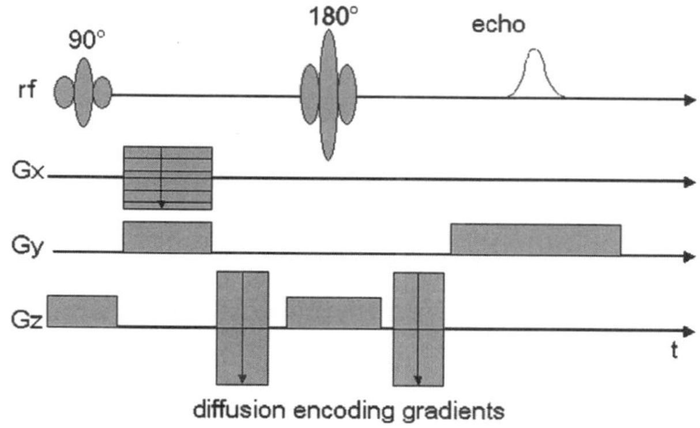

FIGURE 2.8: Typical diffusion encoding pulse sequence.

to detect differences in the diffusivity of water molecules among various biological tissues is known as diffusion weighting [40,23,41]. In diffusion-weighted imaging (DWI), the signal intensity in each voxel S is influenced by the b-value (diffusion sensitization) which is related to the gradient strength G. Diffusion is described by using a single scalar parameter, the diffusion coefficient D. The effect of diffusion on the MRI signal is an attenuation A that depends on D and the b-value

$$A = e^{-Db} \tag{2.33}$$

Thus, the apparent diffusion coefficient (ADC) for each image voxel can be calculated.

A typical pulse sequence is displayed in Fig. 2.8. 3-D images are recorded for different diffusion encoding gradients (typical b-values are 0, 500 s/mm^2, 1000 s/mm^2). An ADC map of a patient with a lesion in the right hemisphere is displayed in Fig. 2.9. The area of the lesion has reduced diffusion and thus appears dark in the ADC map.

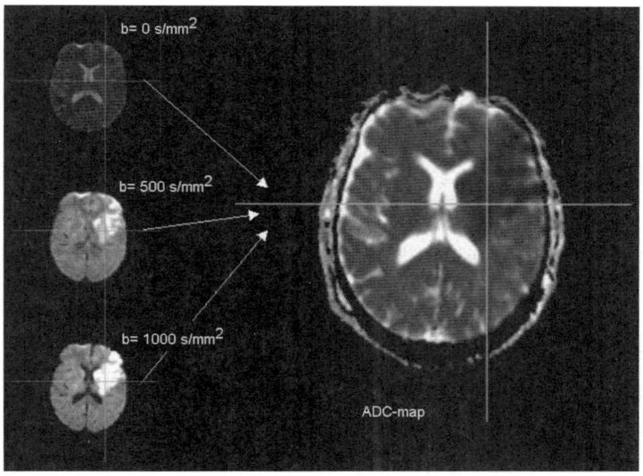

FIGURE 2.9: ADC-map (axial slice) for a patient with a vascular lesion in the right hemisphere.

2.1.7 Diffusion Tensor Imaging (DTI)

The advanced extension of DWI that addresses the visualization of white matter networks is DTI. In DTI at least six gradient directions are recorded to scan a diffusion matrix for each voxel. In this way, not only the diffusion amplitude (as in DWI) is mapped, but also information about the orientational dependence can be obtained.

DTI can be used to characterize the orientational dependence of ^{1}H diffusion in human brain white matter [42,43]. The measured diffusion is apparently maximal along the fiber direction and restricted in the perpendicular direction. ^{1}H diffusion can be mapped spatially, and the observation that this diffusion is reduced in acutely ischemic brain tissue as well as in degenerated brain tissue, is responsible for the increasingly widespread use of these techniques in clinical neuroimaging [44,45, Nimski, 2005c, 46,47].

However, in the presence of anisotropy in white matter, diffusion can no longer be characterized by a single scalar coefficient [40,48], but requires a tensor $\vec{\vec{D}}$, which fully describes molecular mobility along each direction and correlation between these directions [49,50]. Diffusion anisotropy is mainly caused by the orientation of fiber tracts in white matter and is influenced by its micro- and macrostructural features. Of the microstructural features, intraaxonal organization appears to be of major influence on diffusion anisotropy. Other features include the density of fiber and cell packing, degree of myelination, and individual fiber diameter. On a macroscopic scale, the variability in the orientation of all white matter tracts in an imaging voxel influences the degree of anisotropy assigned to the respective voxel [51,52].

The elements of the symmetric tensor $\vec{\vec{D}}$ can be measured by diffusion gradients along at least six noncollinear directions so that b [Eq. (2.33)] has become a tensor, resulting in signal attenuation

$$\ln(A) = -(b_{xx}D_{xx} + 2b_{xy}D_{xy} + 2b_{xz}D_{xz} + 2b_{yz}D_{yz} + b_{yy}D_{yy} + b_{zz}D_{zz}) \tag{2.34}$$

This requires taking into account possible interactions between imaging and diffusion gradients and even between imaging gradients that are applied in orthogonal directions (cross terms) [53,54].

The second-rank diffusion tensor $\vec{\vec{D}}$ can be diagonalized, leaving only three nonzero elements along the main diagonal of the tensor, the eigenvalues (λ_1, λ_2, λ_3). The eigenvalues reflect the shape or configuration of the spheroid. The mathematical relationship between the principal coordinates of the spheroid and the laboratory frame is described by the eigenvectors ($\vec{v}_1$, $\vec{v}_2$, $\vec{v}_3$) (Fig. 2.10).

2.1.7.1 Eddy Current Correction. DTI data acquired with echo-planar imaging technique are highly sensitive to eddy current-induced geometric distortions that vary with the magnitude and direction of the diffusion sensitizing gradients. This distortion has a greater effect on the images than the subject head movement and has to be corrected prior to a motion correction as described earlier. For the correction of this distortion, several methods were proposed. The technique of Shen et al.

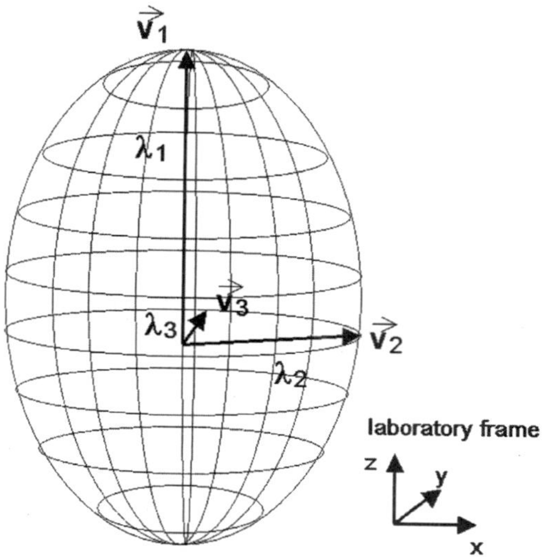

FIGURE 2.10: Diffusion spheroid (schematically) with three eigenvectors ($\vec{v}_1$, $\vec{v}_2$, $\vec{v}_3$) and three eigenvalues (λ_1, λ_2, λ_3). The eigenvectors usually do not coincide with the Cartesian frame (x,y,z).

[55] relies on collecting pairs of images with diffusion sensitizing gradients reversed – these paired images are distorted with eddy currents in opposite directions. A columnwise correction in the image domain along the phase encoding direction (anterior–posterior) was applied. This was performed by searching for the maximum value of the cross-correlation between two corresponding columns (of two paired volumes) while one is shifted and scaled (fitting routine – Simplex method [32]). Each column was then corrected by applying opposite shifts and scales equal to half of the correction.

As the recording technique provides voxels with nonisotropic size, the DTI data sets are transformed as a first analysis step into an isotropic grid with voxel size 1.0 mm × 1.0 mm × 1.0 mm. The transformation was performed according to Eq. (2.11).

2.1.7.2 Fractional Anisotropy (FA) Mapping. The diffusion anisotropy can be quantified by several indices [56], each of which distinguishes between a prolonged spheroid (which would be expected in oriented fiber bundles of white matter) and a sphere (which represents isotropic diffusion). Out of these, the fractional anisotropy (FA) is calculated

$$F_f = \sqrt{\frac{(\lambda_1 - \bar{\lambda})^2 + (\lambda_2 - \bar{\lambda})^2 + (\lambda_3 - \bar{\lambda})^2}{\lambda_1^2 + \lambda_2^2 + \lambda_3^2}} \qquad (2.35)$$

where $\bar{\lambda}$ is the arithmetic average of the three eigenvalues. After the eddy current correction and an optional rigid brain motion correction, the FA maps can be calculated (Fig. 2.11).

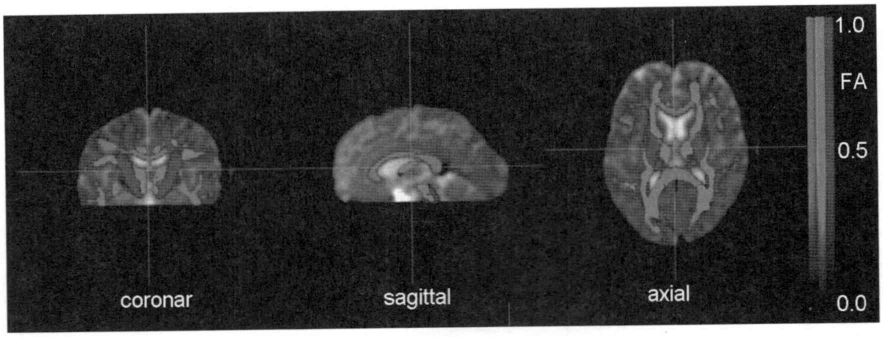

FIGURE 2.11: FA maps of a healthy subject, aged 36 years. The FA maps were overlaid on the b_0 image (for the display a threshold on the FA-values of 0.2 was applied). The color coding was as follows: red for major eigenvector mainly in left-right direction, blue for major eigenvector mainly in inferior-superior direction, and green for major eigenvector mainly in posterior-anterior direction.

2.1.7.3 Fiber Tracking (FT). Insights into the link between functional brain regions and anatomical fiber connections are essential for an integrated understanding of the organization of the human brain and central nervous system. For MRI-based fiber tracking (FT) anisotropic diffusion is characterized to determine the preferred diffusion direction. By calculation of the diffusion spheroid, the eigenvector corresponding to the major eigenvalue is seen as the direction of fastest diffusion and indicates the fiber direction in white matter regions (see scheme in Fig. 2.12).

Based on this directional information, different methods and algorithms have been proposed to estimate white matter connectivity [57,58,59,60,61]:

- *Streamline tracking technique (STT)*: It models the propagation in the major Eigenvector field of the brain.
- *Tensor deflection (TEND)*: It uses the entire diffusion tensor $\vec{\vec{D}}$ to deflect the incoming vector [57] that results in the outgoing vector (2.36) $\vec{v}_{\text{out}} = \vec{\vec{D}}\vec{v}_{\text{in}}$

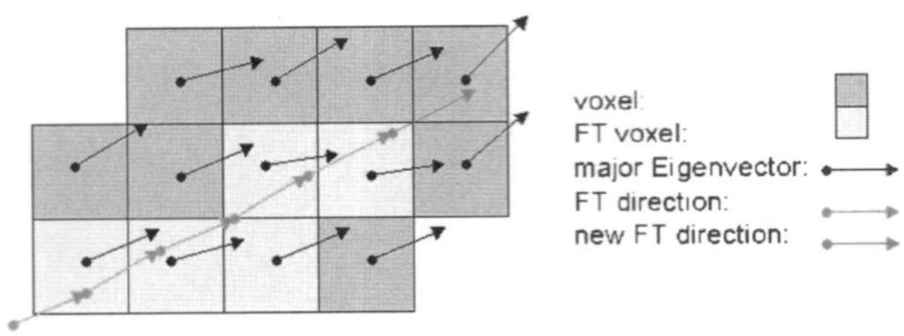

FIGURE 2.12: Scheme for FT. Following the average of surrounding major eigenvectors (black arrows) directions step by step, a voxel is classified as belonging to FT (light yellow) or not (light blue), depending on whether the seed point of the FT direction (green arrow or red arrow, respectively) is at the voxel site.

- *Tensorline algorithm*: It uses STT for highly prolate diffusion tensors and TEND for the diffusion tensors which have a more oblate or spherical tensor shape.

$$\vec{v}_{\text{out}} = f\vec{v}_1 + (1-f)((1-g)\vec{v}_{\text{in}} + \bar{\bar{D}}\vec{v}_{\text{in}}) \tag{2.37}$$

where f and g are user-defined weighting factors that vary between 0 and 1. The algorithm has three terms: (a) an STT term (the direction of the major Eigenvector is weighted by f), (b) a TEND term weighted by $(1-f)g$, and (c) an undeviated term weighted by $(1-f)(1-g)$. All vectors have to be normalized to unity before they can be used in Eq. (2.37).

Probabilistic methods – the global connectivity is estimated by the calculation of the uncertainty in the direction of the eigenvectors at each voxel. The best estimate of probability density function at all voxels of the track is selected as the fiber track [62].

In our observations, a set of parameters had to be adjusted for a reasonable FT, and the validation of the result was the plausibility judged by an experienced analyst. Parameters to be adjusted were as follows:

- *Threshold for the scalar product*: The threshold for the scalar product of the first eigenvectors (angle between directions) of two consecutive FT positions: not necessarily voxels, since FT positions result from float numbers depending on the step width (i.e., the distance between two FT positions) and the first eigenvector directions is usually about 0.9.
- *Choice between STT, TEND, or the combination of both*: The choice between STT, TEND, or the combination of both – in the latter, the parameters f and g had to be adjusted.
- *The step width*: The step width (distance between two consecutive FT positions) for a reasonable FT, variations of the step width between 0.5 mm and 1.0 mm are possible.

FT results are displayed in Fig. 2.13 (TEND, step width 0.5 mm, threshold 0.9).

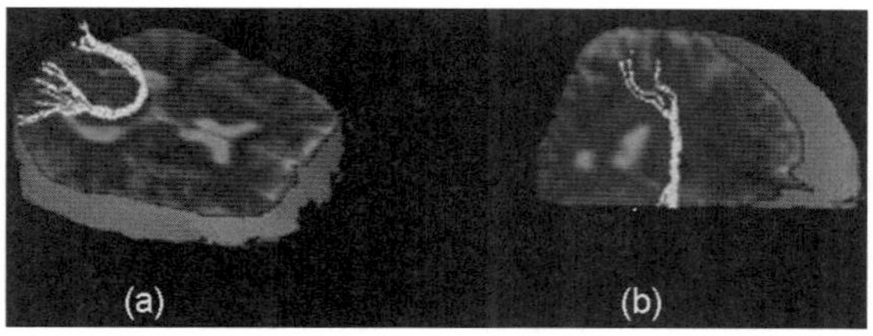

FIGURE 2.13: Fiber tracking results for a healthy subject – the images are glass-brain plots of the FTs overlaid on 3-D views of ($b = 0$)-data: (a) starting point in the dorsal corpus callosum, (b) starting point in the left internal capsule.

2.1.8 Magnetic Resonance Spectroscopy (MRS)

MRS takes advantage of the fact that protons reside in different chemical environments, depending on the molecular structure. The distribution of these protons can be displayed as a spectrum with the main peak composed of water and other peaks depending on the molecular structure. Nevertheless, the multimodal integration MRS is mathematically similar to the integration of fMRI or DTI and thus the focus of this synopsis is not set to MRS.

2.2 MAGNETOENCEPHALOGRAPHY

2.2.1 Principles of Magnetoencephalography

Biomagnetic signals from the human brain are extremely weak in comparison to the earth's magnetic field (5×10^{-5} T) or magnetic urban noise: they are in the order of femtotesla (10^{-15} T) with frequency between 1 Hz and 1 kHz. Thus, highly sensitive sensors are needed and the influence of disturbing magnetic fields has to be reduced.

This sensor is the superconducting quantum interference device (SQUID). A SQUID sensor uses the phenomenon of superconductivity, i.e., a superconductor loses its electric dc resistance below a certain temperature and a magnetic field is expelled from the interior of a superconductor (Meissner–Ochsenfeld effect). The magnetic flux Φ_{sc} caused by the external magnetic field B enclosed by a superconducting loop with the area A is kept constant:

$$\Phi_{sc} = BA = const. \tag{2.38}$$

The magnetic flux is quantified

$$\Phi_{sc} = n\Phi_0 = n\hbar/2e \tag{2.39}$$

where n is an integer, $\hbar$ is the Planck's constant, e is the elementary charge, and Φ_0 is the magnetic flux quantum.

To keep the magnetic flux constant, a screening current I_{sc} is kept flowing through the superconducting circuit when the external magnetic field has changed. Two main classes of SQUIDs depending on the operating principle exist: dc SQUIDs consists of two weak links in the circuit whereas RF SQUIDs have only one. For dc SQUIDs, if a dc bias current is applied to the SQUID loop, then a voltage U can be measured. This voltage is sensitive to the external magnetic field and thus a SQUID represents a flux to voltage converter.

The SQUID operates below 4.2 K, i.e., is usually placed in liquid helium. The typical area covered by a SQUID chip is $4 \times 4\,\text{mm}^2$ and the field sensitivity of a biomagnetic measurement system can be expressed by flux density noise in $fT/\sqrt{Hz}$. The best values achieved were $2.5fT/\sqrt{Hz}$. To measure the spatial magnetic field distribution/dependence, an array of SQUIDs is placed in a Dewar under superconducting conditions. To obtain a better SNR by reducing the ambient disturbances, the recording system is placed in a magnetically shielded room. The shielding is based

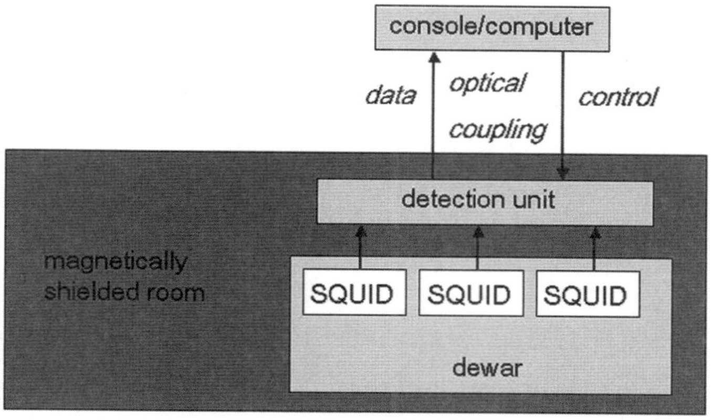

FIGURE 2.14: Schematical MEG recording device. An array of SQUIDs is placed in a Dewar. The whole detector system is operating inside a magnetically shielded room. The coupling to the console is optically controlled.

on diversion of the flux by means of layers of metal with low magnetic permeability. Special sensor configurations are additionally used to reduce signal contributions from homogeneous magnetic disturbances. Figure 2.14 shows a schematic overview of an MEG device.

The MEG signals are derived from the net effect of ionic currents flowing in the dendrites of neurons during synaptic transmission. In accordance with Maxwell's equations, any electrical current will produce an orthogonally oriented magnetic field, which can be measured by MEG. The net currents can be thought of as current dipoles, which are currents defined to have an associated position, orientation, and magnitude, but no spatial extent. According to the right-hand rule, a current dipole gives rise to a magnetic field that flows around the axis of its vector component.

In order to generate a signal that is detectable, approximately 50,000 active neurons are needed. Since current dipoles must have similar orientations to generate magnetic fields that reinforce each other, it is often the layer of pyramidal cells in the cortex, which are generally perpendicular to its surface, that give rise to measurable magnetic fields. Furthermore, it is often bundles of these neurons located in the sulci of the cortex with orientations parallel to the surface of the head that project measurable portions of their magnetic fields outside of the head. For more details about biomagnetic instrumentation refer to [63,27].

Examples of MEG recordings are shown in Fig. 2.15. The task performed was a right hand second finger movement. Due to the low SNR, single-event display is not useful, so (see Chapter 4) the events were detected and averaged according to trigger myogram. Other examples for MEG are given in [64,65,66].

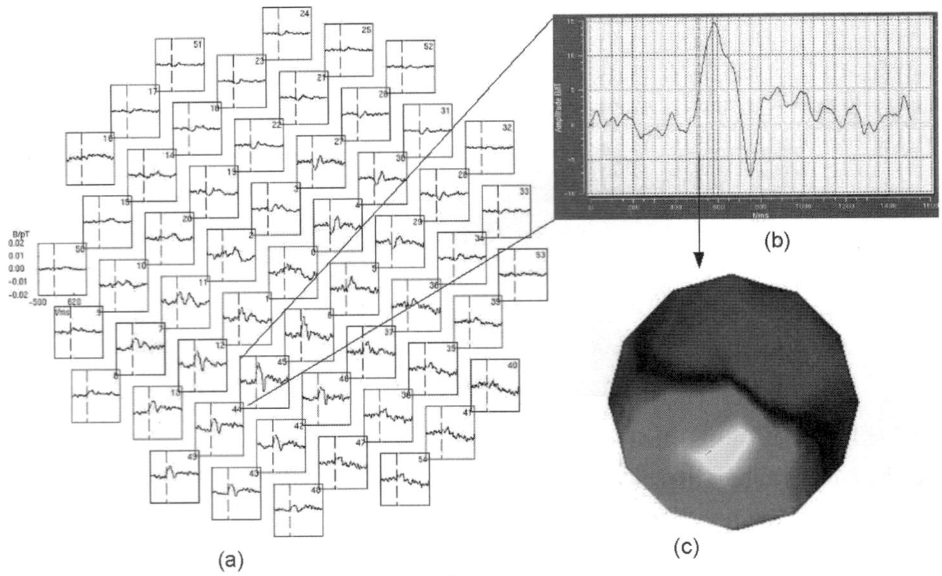

FIGURE 2.15: Averaged MEG event. (a) Sensor distribution plot – the averaged event is displayed at the respective sensor position. (b) Single channel recording of a brain event. (c) Magnetic field distribution for a certain time point marked as line in (b) (yellow-red: positive magnetic field, blue: negative magnetic field).

2.2.2 Source Reconstruction

In order to determine the location of the activity within the brain, advanced signal processing techniques that use the magnetic fields measured outside the head to estimate the location of the source of that activity are used. This is referred to as the inverse problem.

To represent the simplest kind of source, a current dipole is often used to explain the relationship between neuronal activity and the measured electric and magnetic fields. The current dipole $\vec{q}$ is a point source, a mathematical model to represent a short element of current. A current density within a volume conductor Ω induces a magnetic field outside. When the conductivity σ within Ω and the electric current are known, Maxwell's equations and the continuity equation

$$\nabla \cdot \vec{J} = -\frac{\partial \rho}{\partial t} \tag{2.40}$$

where $\vec{J}$ is the current density and ρ is the charge density can be used to calculate the electric field $\vec{E}$ and the magnetic field $\vec{B}$. The current dipole is related to the primary current density $\vec{J}_p$ in a differential volume element dv.

$$\vec{q} = \int \vec{J}_p dv \quad \text{or} \quad \vec{J}_p(r) = \vec{q}\delta \tag{2.41}$$

The total current density can be divided into two components

$$\vec{J} = \vec{J}_p + \vec{J}_\Omega \qquad (2.42)$$

with $\vec{J}_\Omega = \sigma\vec{E} = -\sigma\nabla V$ is the volume current which results from the effect of electric field on charge carriers in the conducting medium. Solving Eqs. (2.40–2.42) together with Maxwell's relations, a relation between the measured magnetic field and the current density is obtained

$$\nabla^2\vec{B} = -\mu_0\nabla\times\vec{J} \qquad (2.43)$$

with $\vec{B} = 0$ at ∞.

Equation (2.43) is of Poisson's type. Analytical solutions can be found only for simple geometries and conductivity distributions, i.e., volume conductors like half space, sphere, or ellipsoid. The reader interested in the theory of solution of the forward and inverse problems is referred to [63,67].

In the example presented here, the MEG source reconstruction process is performed using single dipole representations of the underlying source by the Boundary Element Method: for a piecewise homogeneous volume conductor, consisting of N homogeneous and isotropic compartments, the magnetic field generated by an active source can be calculated by the Geselowitz's formula [68,63]. The global field is given by the contribution of the infinite medium magnetic field and by the secondary sources represented by the electric potential $V(\vec{r})$ at the conductivity interfaces s_0 and s. An integral equation describes the potential on each surface between compartments. The homogeneous single-compartment conductor model, usually sufficient for MEG [67] is used in the following. A tessellation of the surface into an appropriate number of triangles is needed, and the potential values at each of these triangles are approximated using a collocation method with linear basis functions [69]. This means that the integral equation for the potential must be satisfied only at the M node points represented by the vertices of the triangles where the potential V is calculated from

$$\left(I - \frac{1}{2\pi}\Omega\right)V = \frac{2\sigma_0}{\sigma}V_0 \qquad (2.44)$$

where V_0 is the vector of the infinite medium potential at the M nodes and W the matrix of the solid angles.

Once the numerical expression of the secondary source is obtained, the global magnetic field is analytically calculated as a linear combination of vector functions $a_i(r)$ [70,71]

$$B(r) = B_0(r) + \sum_{i=1}^{M} V_i a_i(r) \qquad (2.45)$$

A dipole localization at a given time instant in the averaged data can be performed using either a linear iterative algorithm [72] or a simplex conjugate algorithm for double check. The cost function is based on the match of the measured map with the predicted map for the trial solution according to a least squares approach or a correlation-based criterion

$$J_{LS} = \left\| \vec{B} - \overrightarrow{LQ^T} \right\|^2 \tag{2.46}$$

where $\vec{B}$ is the measured signal map and $\vec{B}_{\mathrm{cal}} = \overrightarrow{LQ^T}$ is the predicted map calculated from the estimated dipole parameters.

A linear iterative algorithm for inverse calculation [72] consists of a least squares parameters estimation combined with a first-order approximation of the field variation corresponding to a given parameter increment. This method linearizes locally, the problem to calculate increments for position Dr and momentum Dd thus transforming the standard nonlinear least square parameter estimation into a linear iteration scheme (computationally similar to a Levenberg–Marquardt approach).

The simplex conjugate algorithm is a standard nonlinear least squares parameter estimation routine, an algorithm with randomly distributed starting points for each iteration is used to estimate the position of the dipole.

Generally, the localization procedure inside the whole localization algorithm is performed twice: first, an analytically represented conductor is used as volume conductor model (spherical model for total head MEG) and in this way an approximation of the final dipole position is achieved. The second run of the localization algorithm is performed using the actual volume conductor (realistic shaped volume conductor or its ellipsoid approximation), and a refinement of the previous solution is found. The use of the ellipsoid approximation can be used as an alternative to the realistic shaped volume conductor; nevertheless, for the dipole localizations presented here no significant differences could be observed.

Both of the algorithms (nonlinear minimization or its linearized version) provide a solution for the dipole location. The three parameters given by the position of the dipole have, in fact, a nonlinear dependency on the field due to the lead field matrix relating the field to the dipole position [68]. The strength $\vec{Q}$ of the localized source is, instead, linearly related to the field and can thus be obtained through direct pseudoinverse solution, given the location of the dipole. For both of the inverse algorithms, the linear system, derived from the maximum likelihood approach, is solved to get the momentum components $\vec{Q}$ once the position has been determined:

$$\left(\vec{\vec{L}}^T \vec{\vec{L}} \right) \vec{Q} = \vec{\vec{L}}^T \vec{B} \tag{2.47}$$

where $\vec{\vec{L}}$ is the lead field matrix and $\vec{B}$ is the matrix $3 \times N_s$ of the components of the field at every measurement site (N_i is the number of channels of the MEG system).

2.2.3 Modeling

To obtain the structural background for MEG source reconstruction, 3-D models from MRI data are generated and used as a morphological basis for the MSI reconstruction. Tissue segmentation is performed using the voxel intensity above or below respective thresholds.

Tissue segmentation is a topic already covered extensively in scientific literature. Thus, in the framework of this study, the strategy to obtain the inner skull is explained in short. The outer skin can be obtained from MP-RAGE data sets by setting up ray traces from the image borders to the center of the head, and subsequently (starting from the image borders and going to the center of the head) defining the outer skin as the first point where the image voxel intensity is higher than the noise threshold.

Alternatively, edge detection algorithms can be used. The simplest one is to use the derivative instead of the original data.

$$D(i,j,k) = \frac{1}{2}|I_{(i+1,j,k)} - I_{(i-1,j,k)}| + \frac{1}{2}|I_{(i,j+1,k)} - I_{(i,j-1,k)}| + \frac{1}{2}|_{(i,j,k+1)} - I_{(i,j,k-1)}| \qquad (2.48)$$

Instead of using the derivative distance of one, two or three voxels or a combination of them can be used.

Traveling further along a ray trace, the inner skull then can be derived as the minimum of the derivative of the voxel intensity values. This definition properly works for regions that are relevant for MEG (e.g., parietal cortex of the hemisphere). For example, eye holes, mouth, etc. are not relevant for the brain source reconstruction, and there is no need for a high-resolution segmentation of those head regions. Finally, the brain tissue can be segmented by a region-growing algorithm working with simple thresholds that need to be adjusted.

A threshold-dependent region grow algorithm follows the following iteration process:

1. Mark a starting point.
2. Test if the intensities of the neighbored voxels to a marked voxel are in the range adjusted by the thresholds.
3. Mark the voxels.
4. Go to 1 till no further voxel is marked.

The triangulation of the resulting tissue components is performed by expanding an already triangulated sphere positioned in the center of the respective tissue until the tissue boundaries are reached for each triangle edge separately, thus rendering a realistic volume conductor model. Additionally, for use in realistic models which have a complex boundary and in the case of a very superficially located source – in order to avoid error due to tessellation – an ellipsoid could be used which is fitted with its center, length, and orientation to the axis to the realistic model. The realistic volume conductor used for source reconstruction is the triangulated inner skull.

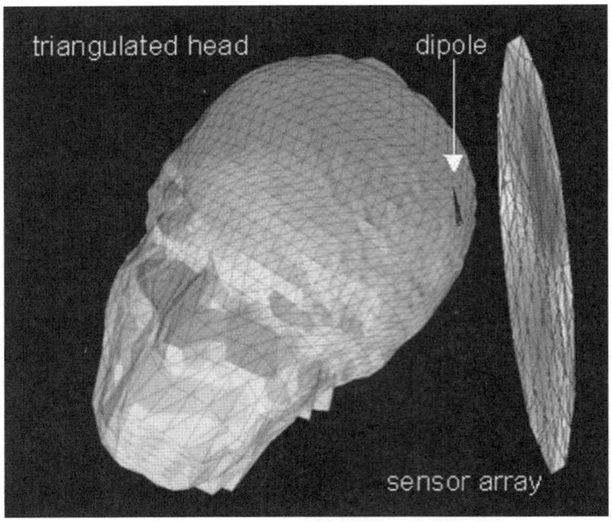

FIGURE 2.16: 3-D visualization of a triangulated head surface (about 1700 triangles) with dipole localization result displayed as a pin. On the left side, a sensor array with magnetic field distribution for a certain time point is displayed schematically.

2.2.4 Dipole Localization

Using the model described in Section 2.3 and with the fitting procedure described in Section 2.2 an example of a single dipole localization for the magnetic field distribution at a defined time point is calculated (Fig. 2.16).

2.3 OTHER TECHNIQUES

To add quantitative information of regionally disturbed brain function or regional tumor metabolism into the common framework of image analysis and display, the integration of PET and SPECT into multimodal functional neuroimaging can be used. Generally, PET as well as SPECT generates images of the regional and temporal distribution of a radio-labeled pharmaceutical after injection. For SPECT, pharmaceuticals are usually labeled with the photon emitters ^{99m}Tc or ^{123}I, which due to their radiochemical properties and their half-life of 6 hours and 13 hours, respectively, allow either a daily on-site preparation of the radiopharmaceuticals or a delivery from a regional provider. The positron emitters needed for PET, however, have half-lives ranging from 2 hours (^{18}F) to 2 minutes (^{15}O) so that for all preparations of radiopharmaceuticals apart from ^{18}F labeled compounds, a local cyclotron facility is mandatory.

The image resolution, depending on the technique and instrumentation, is ranging from a spatial resolution of 10–15 mm for SPECT to 4 mm for PET [7].

PET and SPECT as techniques for scientific investigation in humans are limited by the need for clearance by ethics committee to inject radioactive material into participants. The minimization of radiation dose to the subject is an attractive feature of the use of short-lived radionuclides.

In short, the raw data collected by a PET scanner are a list of "coincidence events" representing near-simultaneous detection of annihilation photons by a pair of detectors. Each coincidence event represents a line in space connecting the two detectors along which the positron emission occurred.

Coincidence events can be grouped into projections images called sinograms. The sinograms are sorted by the angle of each view and tilt, the latter in 3-D case images. A normal PET data set has millions of counts for the whole acquisition.

A detailed overview and technical details for PET and SPECT is beyond the scope of this synopsis, which puts special focus on different neuroimaging techniques.

Integration into multimodal imaging is performed in analogy to the other techniques. Using PET specific markers and detecting them by a marker detection system (Chapter 3) allows for easy integration of the results [7].

Another brain mapping technique performing the neurophysiologic measurement of the electrical activity of the brain is called EEG. EEGs are frequently used in experimentation because the process is noninvasive. EEG is performed by recordings from electrodes placed on the scalp, or subdurally, or in the cerebral cortex. The resulting traces represent an electrical signal (postsynaptic potentials) from a large number of neurons. Really, not electrical currents are measured, but voltage differences between different parts of the brain.

EEG is capable of detecting changes in electrical activity in the brain with a very high temporal resolution (of the order of milliseconds). In conventional scalp EEG, the recording is obtained by placing electrodes on the scalp after application of a conductive gel to reduce impedance.

To perform dipole or current source reconstruction as in MEG, a map of electrode signals has to be recorded. This cannot be performed by placing single electrodes, but the subject has to wear a plastic cap where the electrodes are inserted in small holes. Electrode placement is determined by measuring and marking the scalp using a system of placement that is reliable and reproducible (see Chapter 3).

The limitation in contrast to MEG is that MEG has fixed sensor positions but EEG sensor positions vary from each recording. Thus, for current source reconstruction, each electrode position has to be detected before the measurement which can be a long-lasting process or might even result in inaccuracy of the results.

From an economical point of view EEG should be preferred to MEG but the technical point of view tells us superiority of MEG versus EEG. In the following, EEG will not be presented in detail, as well as another brain mapping technique, i.e., transcranial magnetic stimulation.

CHAPTER 3

Coordinate Transformation

3.1 TRANSFORMATION IN INTERMODAL MULTIMODALITY

Multimodal integration requires the 3-D registration of all involved modalities in one common coordinate frame. As it is usually not possible to detect modality specific markers in the environment of any other modality, problems arise with multimodal mapping techniques. Moreover, specific measurement modalities often need different marker locations.

MRI results have a high spatial resolution and if no high temporal resolution is required, they can stand alone. On the other hand, MEG (and EEG) mapping ask for the combination of the resulting data with the anatomical data provided by MRI. Simple marker substitution for the appropriate modality at identical positions introduces an additional error source to the overall system. Simultaneous detection of different types of markers in an optical manner and registration in a common 3-D coordinate frame overcomes the outlined problem.

If the multimodality is based on two imaging modalities a different approach is possible. Instead of transforming from one coordinate frame to another, it is possible to transform all image information from each modality to a stereotactic standard space.

3.1.1 Tracking Systems

Different tracking systems based on different physical principles are available. Usually, tracking systems allow for detection of position and orientation. For multimodal imaging, the orientation detection is not necessary, the position of the modality specific markers is the only parameter needed. Tracking systems work with a sender and a receiver unit: the detector unit has to be placed at the appropriate position, and the sender detects the receiver position.

The electromagnetic tracking system contains three coils with perpendicular axes producing an electromagnetic ac field, which is inducing currents in the receiver. In this way, the actual position of the receiver can be calculated. The problem is of the metal in the surrounding of the system that might disturb the measurement because the electromagnetic fields cause currents in these metals. To overcome this problem, dc currents can be used. These and other disturbances such as other ferromagnetic materials and the earth's magnetic field have been overcome by shielding and earth field compensation (Polhemus, Colchester, Vermont, USA [73]).

An acoustical tracking system is based on ultrasound sender and receiver, which is different in the physical principle but has the same mathematical basis for calculating the positions.

The disadvantage of these tracking systems is that the recording time increases with an increasing number of markers and errors due to patient movements that are difficult to avoid. Improvement is possible by optical systems where the recording time is less than 1 s independent of the number of markers.

3.1.2 Optical Systems

Optical systems have no need for a sender unit, since the optical image recorded by a camera contains all information needed. To evaluate the position of an object, it has to be recorded from at least two cameras where the position and the orientation of the camera axes have to be known exactly. A good example is the OptiCoS [74]. In the following, the detection of marker positions and the appropriate coordinate transformations are explained on the basis of this system.

The system utilizes color video cameras, homogeneously distributed over a 1000 mm diameter circle line at equal distance (Fig. 3.1). The overlapping of the recording segments allows for a 3-D reconstruction by transforming the 2-D coordinates of the appropriate camera image into 3-D coordinates. This transformation is possible as the camera locations are defined and the optical axis of each camera is tilt against the circle plane by an angle of 45° thus focusing a common point 500 mm distant.

The cameras are working with a resolution of 768 × 576 pixels providing approximately 0.5 mm resolution within the investigation field. Detecting a marker means determining the position of its optical center within the corresponding camera image.

FIGURE 3.1: OptiCoS setup with superimposed coordinate frame and lines representing camera locations.

3.1.3 Mathematical Excursion 3: Calculation of 3-D Coordinates from 2-D Images

Each marker position detected by a camera can be expressed by a line connecting the center of the associated camera lens and the center of the marker. Therefore, the 3-D positions of the markers can be determined by establishing these line equations and calculating their intersection points.

In local 2-D coordinate frame of a camera, a marker vector $\vec{a}$ is defined by the pixel column address a_x and pixel row address a_y and the focal distance a_z between lens and image plane. Each camera coordinate frame can be represented globally by its origin vector $\vec{d}_i$ (center of camera lens) and its Euler angles (ϕ_i,θ_i,ψ_i). The latter describes the revolutions necessary for the local coordinate frames to be transformed to align with the global coordinate frame. The geometric conditions resulting from the chosen mechanical setup facilitate these transformations (Table 3.1). Local marker vectors $\vec{a}$ can be transformed to global coordinate vectors $\vec{a}'$ using the matrix $\vec{\vec{R}}_i$ (i: camera, k: marker).

$$\vec{a}'_{ik} = \vec{\vec{R}}_i \cdot \vec{a}_{ik} \tag{3.1}$$

$$\vec{\vec{R}}_i = \begin{pmatrix} \cos\phi_i\cos\psi_i - \sin\phi_i\sin\psi_i\cos\theta_i & -\sin\phi_i\cos\psi_i - \cos\phi_i\sin\psi_i\cos\theta_i & \sin\psi_i\sin\theta_i \\ \cos\phi_i\sin\psi_i + \sin\phi_i\cos\psi_i\cos\theta_i & -\sin\phi_i\sin\psi_i + \cos\phi_i\cos\psi_i\cos\theta_i & -\cos\psi_i\sin\theta_i \\ \sin\phi_i\sin\theta_i & \cos\phi_i\sin\theta_i & \cos\theta_i \end{pmatrix} \tag{3.2}$$

After this transformation, the lines between the cameras and the marker can be described generally

$$p = \vec{d}'_p + \kappa \cdot \vec{a}'_p \quad \text{and} \quad q = \vec{d}'_q + \nu \cdot \vec{a}'_q \tag{3.3}$$

The markers 3-D coordinates are provided by the intersection point ($\vec{p} = \vec{q}$). Due to the limited accuracy of the optical components two lines p and q associated with the same marker but with different cameras will not likely intersect at a common point but rather approach very close.

TABLE 3.1: Euler angles for transformation

	ψ_i Global z axis	θ_i Modified global x axis	ϕ_i Modified global z axis
Camera 0	$\pi/6$	$-\pi/4$	0
Camera 1	$\pi/2$	$-\pi/4$	0
Camera 2	$-\pi/6$	$\pi/4$	π
Camera 3	$\pi/6$	$\pi/4$	π
Camera 4	$\pi/2$	$\pi/4$	π
Camera 5	$-\pi/6$	$-\pi/4$	0

Mathematically, the intersection point is found at half the minimum distance between the lines. Let $\vec{a}_p$ and $\vec{a}_q$ be normalized, let

$$\vec{\Delta} = \vec{d}_p - \vec{d}_q \tag{3.4}$$

and

$$\cos(\alpha) = \vec{a}_q \cdot \vec{a}_p \tag{3.5}$$

then κ and ν can be expressed in closed form as

$$\begin{pmatrix} \kappa \\ \nu \end{pmatrix} = \frac{1}{\sin^2(\alpha)} \begin{pmatrix} 1 & \cos(\alpha) \\ \cos(\alpha) & 1 \end{pmatrix} \begin{pmatrix} -\vec{\Delta} \cdot \vec{a}_p \\ \vec{\Delta} \cdot \vec{a}_q \end{pmatrix} \tag{3.6}$$

The midpoint can be expressed along the closest connection between the two lines as

$$\vec{d}_p + \kappa \vec{a}_p - \frac{1}{2}\vec{s} \tag{3.7}$$

where

$$\vec{s} = \frac{\vec{\Delta} \cdot (\vec{a}_p \times \vec{a}_q)}{|\vec{a}_p \times \vec{a}_q|^2} \vec{a}_p \times \vec{a}_q \tag{3.8}$$

is the vector which connects the two lines with minimal distance. Experience has shown that an intersection point can be assumed if $|s| < 1$ mm.

3.1.4 Mathematical Excursion 4: Coordinate Transformation for Intermodal Multimodality

The following coordinate transformation is valid for all markers independent on the marker detection system (Polhemus, Opticos, etc.). Thus data of different modalities are transformed into one unique coordinate frame.

In this, the coordinates of the first modality markers are known as relative positions of the markers of the second modality. Subsequently, following transformations of three coordinate frames are to be performed

- The coordinate frame of the first modality measurement device (x, y, z).
- The coordinate frame of the marker detection system (x', y', z').
- The coordinate frame of the second modality measurement device (x'', y'', z'').

First, from the marker detection of the first modality measurement, the positions $(\vec{p})$ are known and rotation and shifting matrices $\vec{\vec{R}}_1$ and $\vec{T}_1$ are determined for the transformation from marker detection system coordinate frame ($\vec{p}'$) to first modality coordinate frame (suggested method – Conjugated Simplex [32])

$$\vec{p} = \vec{\vec{R}}_1 \vec{p}' + \vec{T}_1 \tag{3.9}$$

Second, from marker detection in the second modality ($\vec{m}''$) the rotation and shifting matrices $\vec{\vec{R}}_2$ and $\vec{T}_2$ are determined for the transformation from second modality coordinate frame to marker detection system coordinate frame ($\vec{m}'$) (suggested method: Conjugated Simplex [32])

$$\vec{m}' = \vec{\vec{R}}_2 \vec{m}'' + \vec{T}_2 \tag{3.10}$$

Then, any coordinate in the second modality coordinate frame ($\vec{a}''$) (e.g., position of functional active voxel) can be transformed to the coordinate frame of the first modality by

$$\vec{a} = \vec{\vec{R}}_1 (\vec{\vec{R}}_2 \vec{a}'' + \vec{T}_2) + \vec{T}_1 \tag{3.11}$$

3.2 TRANSFORMATION TO A STEREOTACTIC STANDARD SPACE – INTRAMODAL MULTIMODALITY

The fundamental advantage of using spatial normalization is that results obtained from different subjects can be averaged to finally perform a comparison of groups of patients with certain disorders and healthy subjects. Talairach and Tournoux [75] suggested a transformation algorithm to a standard atlas involving the identification of various brain landmarks and piecemeal scaling of brain quadrants. An alternative approach was to use automated brain registration algorithms [76, 26]. Here, a spatial normalization to the Montreal Neurological Institute (MNI) stereotactic standard space via a scanner- and sequence-specific template data set is presented. The MNI defined a new standard brain by using a large series of MRI scans of normal controls. This resulted in the MNI atlas [77].

An example for standard space normalization of MRI data is explained in the following using a template-specific normalization. The template used for the normalization was created in two iterative steps.

3.2.1 First Iteration Step

From each modality (T_1 weighted, T_2 weighted or functional) one volume is arranged with eight operator-defined landmarks [anterior commissure (AC), posterior commissure (PC), vertical PC plane (VCP) superior, VCP inferior, prolonged AC-PC front, prolonged AC-PC back, prolonged PC-left hemisphere border, prolonged PC-right hemisphere border, see Fig. 3.2]. According to these landmarks, an affine transformation (transformation, rotation, dilation) into the MNI space is performed. Template 1 ($A_{\text{templ},1}$) is then created by arithmetic averaging of the voxel intensities of the transformed volumes of single subjects. Thus, a template is created for each modality.

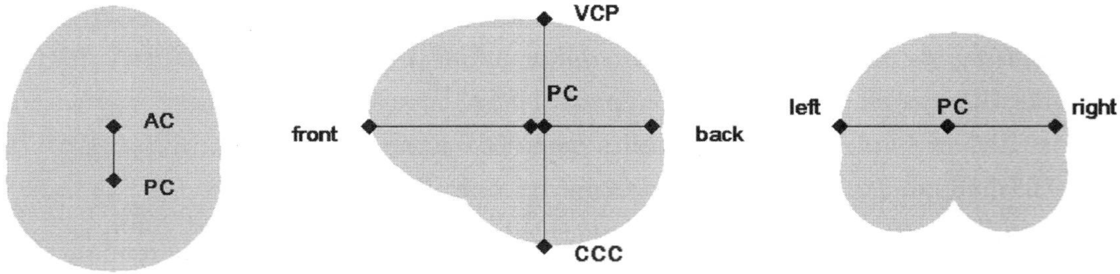

FIGURE 3.2: Anatomical landmarks as used for MNI transformation.

3.2.2 Second Iteration Step

In order to improve the quality of the template, template $A_{templ,1}$ is used for the template-dependent transformation of the original data. For this transformation, a spherical coordinate frame is set up, and dilation factors along about 400 arrows (center at PC) are calculated (Fig. 3.3).

The mismatch between the intensities along an arrow of the volume of the data set to be fitted (via edge detection as well as via correlation) and the volume of the template $A_{templ,1}$ is minimized according to the squared differences (χ^2). The dilation factors are then interpolated at the respective voxel position for each voxel to get the normalization on a voxelwise basis. Template 2 ($A_{templ,2}$) is created by arithmetic averaging of the voxel intensities. That way, a modality-specific

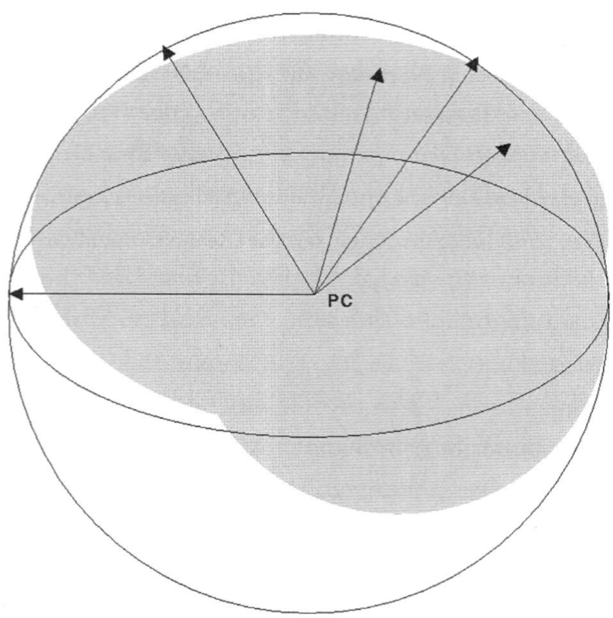

FIGURE 3.3: Template-dependent transformation – spherical coordinate frame with center at PC.

template is created. Single subject modality-specific data sets can be normalized according to MNI dimensions.

The MNI transformation for each individual data set is performed by minimizing the mismatch between intensity regions of the volume of the data set to be fitted (following the basic ideas of Ashburner and Friston [78]) and of the template $A_{\text{templ},2}$ according to the squared differences (χ^2). Using the resulting transformation matrix, the other volumes of the modality are then to be transformed with the identical transformation parameters.

CHAPTER 4
Examples for Multimodal Imaging

To increase the understanding regarding the functional organization of the human brain, different aspects of the functional neuroanatomy must be considered. This includes, but is not limited to, the specific localization of eloquent areas, the type of processing in the brain, the time course of processing, and the interactions between the functions. The correlation of anatomy and function is one of the most important parameters in modern brain imaging analysis techniques. With this approach, different imaging techniques offer tradeoffs between spatial and temporal resolution [79]. In the following sections different approaches to combine information from different modalities are presented.

4.1 INTERMODAL MULTIMODALITY: MAGNETOENCEPHALOGRAPHY AND FUNCTIONAL MAGNETIC RESONANCE IMAGING

As we have seen in Section 2.2, MEG is a technique that provides a very high temporal resolution (i.e., on the order of 1 ms), mainly restricted only by the sampling frequency of the recording system [80]. Since magnetic fields are not distorted by conduction inhomogeneities in the head, MEG is a potentially powerful localization tool of brain function, providing spatial discrimination below 5 mm. However, MEG has a fundamental source identification problem, the so-called inverse problem: determining the active site in the brain from the magnetic field pattern recorded outside the skull. Thus, some a priori assumptions must be made about the source [63]. Usually, signal generators in the brain are described as current dipoles, which are physiologically reasonable models for relatively small active cortical areas.

Functional magnetic resonance imaging (fMRI) is a different approach to noninvasive analysis of the brain function, which has shown of widespread use over the last decade. The fMRI, based upon the Blood Oxygenation Level Dependent (BOLD) contrast, measures changes in the local concentration of deoxyhemoglobin by susceptibility-weighted MRI and can provide excellent spatial sampling in the range of 1 mm [81]. Therefore, it is obvious that the two measurement techniques, MEG and fMRI, detect signals with completely different physical origins. The localization of this origin is not expected to coincide for both techniques. The temporal resolution, due to the intrinsic inertia of blood flow changes, has a lower limit of about 100 ms. Moreover, the biological mechanisms underlying the BOLD contrast are complex because of many variables such as blood flow, blood volume, and subsequent oxygen change including factors like age of the subject [82].

These methodologies constrain and complement each other. When combined, they can therefore improve our interpretation of functional neural organization. Consecutively, it is useful to investigate brain function in a multimodal noninvasive way by combining different imaging techniques, thus making use of their individual advantages. Furthermore, the acquisition of individual volume-rendering morphological MRI data allows improved models of the head volume conductor. With particular respect to the inverse problem, one approach to analyzing MEG data is to impose anatomical and physiological constraints for neural electromagnetic source modeling derived from other imaging modalities such as fMRI to reduce the solution space. Combinations of MEG and fMRI in the same paradigms were described in several studies [5,2,3,4], both as a comparison of the localization of the results and an application of fMRI constraints to MEG data analysis. One example of a bimodal study using the latter approach is explained here in detail [83].

4.1.1 Results from Magnetoencephalography

The MEG measurements were performed on the 55-channel magnetometer system in Ulm, Germany [84]. The sensor system operated inside a magnetically shielded room [85] and consisted of a planar Dewar, containing a complex architectural structure with sensors distributed over two levels. The first level, i.e., the primary measurement plane, held 55 SQUID sensors. The sensing elements were integrated magnetometers with a square shape of 12.7 mm in diagonal. The sensors were uniformly distributed over the inner surface of the dewar, according to a hexagonal grid, covering a circular surface of about 230 mm in diameter. The measurement plane was 18 mm from the outer dewar bottom. Nineteen additional SQUIDs were mounted on the second level and were used as reference channels 90 mm from the measurement plane. Operating in the shielded environment, the system showed on all channels a white noise level better than $10fT/\sqrt{Hz}$ at 10 Hz.

A software gradiometer setup was performed by subtracting online the background field sensed by selected reference channels from the signal of each primary channel. Three orthogonal electrocardiographic (ECG) leads for offline cancellation of the cardiac components, three bipolar EEG leads, and a respiration channel were used in addition. The recording sampling frequency was 8200 Hz. Signals were digitally filtered and decimated to a final sampling frequency of 1025 Hz and a bandwidth between DC and 250 Hz.

The standard measurement protocol of motor-related activity consisted of repeated flexion movements (FM) of the index finger of the right hand every 1.5 s. Five minutes of magnetic activity were recorded for the task. The self-paced movements were performed with a frequency of about 40 movements per min. The planar Dewar was positioned approximately 1 cm over the parietal cortex of the hemisphere contralateral to the movements (left hemisphere), covering the pericentral region. As a trigger signal for the finger movements, the surface myogram of the flexor digitorum muscle was recorded.

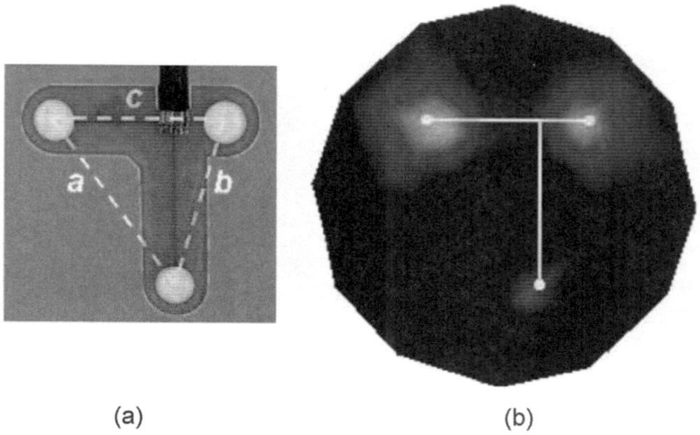

FIGURE 4.1: MEG recording determines the locations of the coils. (a) Schematic coil arrangement. (b) Magnetic field map reconstructed from a 15-s coil measurement with a 55-channel magnetometer system.

For the localization of the MEG marker coils, the coils were activated for a 15-s recording before the measurement started (Fig. 4.1).

4.1.2 Results from Functional Magnetic Resonance Imaging

The MRI and fMRI data were recorded with a 1.5 T scanner (Siemens, Magnetom, Erlangen, Germany) using echo planar imaging (EPI). For the acquisition of the volume-rendering MRI data set of the whole brain, a T_1-weighted magnetization-prepared rapid-acquisition gradient echo sequence was used (MP-RAGE, $T_R = 9.7$ ms, $T_E = 3.93$ ms, flip angle 15°, matrix size 256×256 mm^2, voxel size $1.0 \times 1.0 \times 1.0$ mm^3), consisting of 160–200 sagittal partitions depending on the head size.

Functional data sets consisted of 130 recorded volumes in a block design changing between 10 volumes of rest (each 3 s) and 10 volumes of activation (each 3 s), 32 axial slices with 64×64 voxel each (voxel size $3.6 \times 3.6 \times 4.0$ mm^3, field of view (FOV) $230 \times 230 \times 128$ mm^3, $T_R = 50$ ms, $T_R = 2820$ ms).

The fMRI activation maps were overlaid on each individual 3-D brain. The respective T_1-weighted morphology covering the identical area as the functional data sets consisted of 32 slices with 256×256 voxels (voxel size $0.9 \times 0.9 \times 4.0$ mm^3, FOV $230 \times 230 \times 128$ mm^3, $T_E = 12$ ms, $T_E = 468$ ms). This data set was used for the coregistration of the functional localization results to the whole head recordings.

The whole head volume-rendering structural MRI data set serves, additionally, as the morphological basis for the superposition of the MEG results (see below). The true-scale imaging accuracy was tested for this pulse sequence with a maximum error of 1 mm [74].

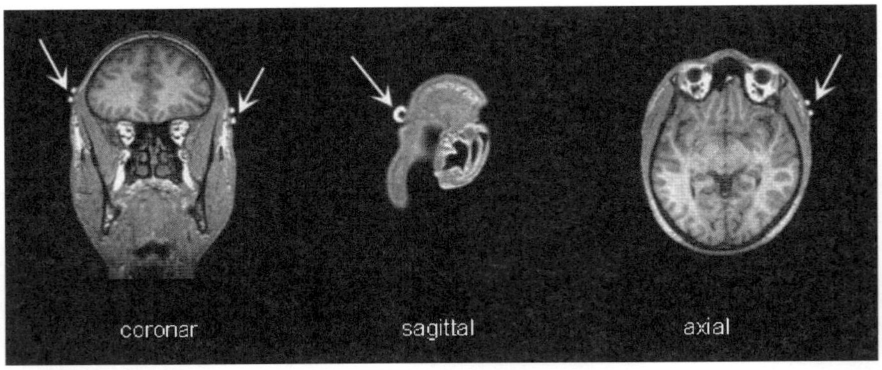

FIGURE 4.2: MRI marker detection. Hydrogel-based marker chips show up in MRI scans as ring structures (see arrows) and can be located manually by the operator.

4.1.3 Coregistering – Optical Marker Detection

For MSI, the combination of MEG data with the morphological data provided by the MRI is necessary. Coregistration may be based on distinct locations (marked by fiducials, electrodes or similar devices) or on surface matching algorithms using numerous intrinsic reference points. The adjustment of morphological MRI and MEG data in this example is accomplished by an optical localizer for the reasons outlined below. Optical systems can work with any well visible marker. For this reason, of course, the MRI markers and the MEG coils have to be prepared in a way that they produce striking optical images, e.g., color encoding. Combined fMRI/MEG recordings require different marker positions for the different modalities. These different marker positions are then transferred into one coordinate frame. Three MRI fiducial markers at suitable anatomical locations (Fig. 4.2) and three MEG markers (Fig. 4.3) were optically detected and coregistered by the optical coregistration system (OptiCoS) (Section 3.1) [86].

4.1.4 Multimodal Signal Processing and Results

The fMRI data during the motor activation paradigm were recorded according to the protocol described above. A detailed discussion of fMRI data processing and analysis algorithms is given in [87]. In that work, it has been shown by computer simulations that basically, deterministic analysis leads to an activation concept that is independent from the noise level, and vice versa, the statistical analysis leads to an activation concept where the activation is intended as deviation from noise and in such way that the activation becomes related to the noise level. The deterministic methods exclude true positive results with increasing noise and the statistical methods include false positive results with increasing noise. Nevertheless, a decision which of the analysis methods should be the method of choice for fMRI-analysis could not be made. In the example of data recorded from a right hand second finger movement, all the analysis methods listed here showed identical results (Fig. 4.4).

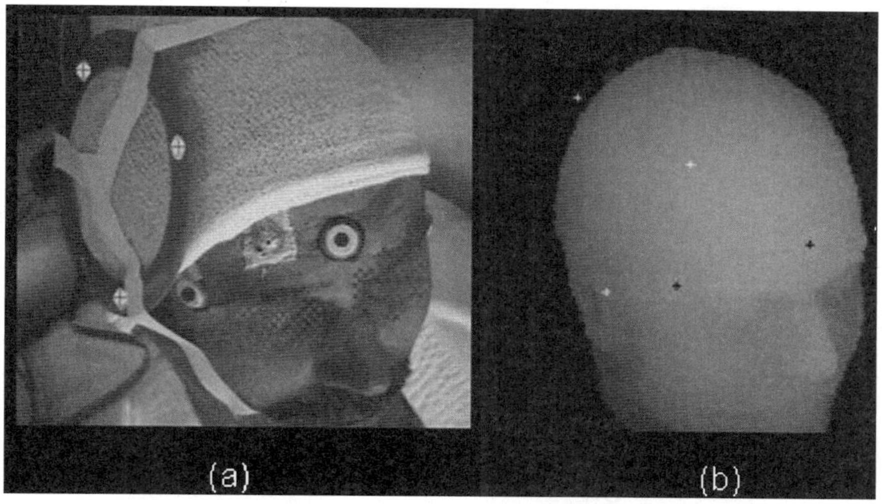

FIGURE 4.3: (a) Images of human head acquired by six cameras. Two marker types can be distinguished: MRI capsules marked with black dots and magnetic coils marked with light dots. (b) Artificial 3-D projection resulting from optical coregistration shows MEG markers as well as MRI markers.

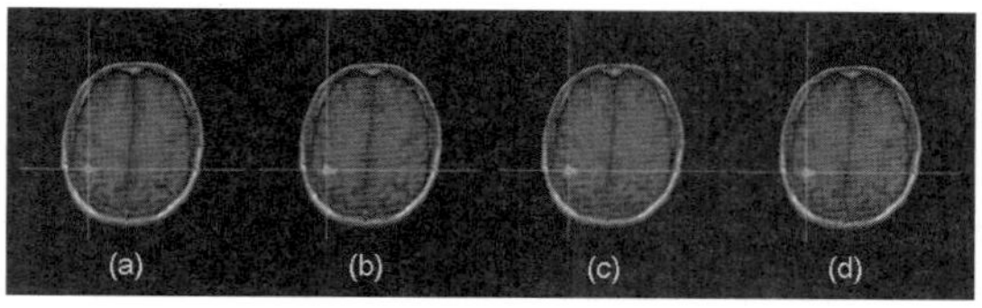

FIGURE 4.4: fMRI analysis of data recorded from a right hand second finger movement with different analysis methods (axial slice). (a) Deterministic correlation analysis. (b) Deterministic sinus wave analysis. (c) Statistical Student's t-test. (d) Statistical Kolmogorov–Smirnov test.

The functional regions in this course of experiments were identified by calculating the statistical test values [Eq. (2.32)].

In Fig. 4.5, functional activation in the primary motor cortex for FM is displayed. The T_1-weighted MRI data (covering the identical area as the fMRI data) are fitted to the T_1-weighted MP-RAGE data. Thus, the fMRI activation results can be superimposed onto the 3-D MRI data.

In addition, a segmented and triangulated surface is created from the MP-RAGE data. Using these surfaces as volume conductor models, the dipole localization can be performed (see below).

Epochs for averaging had to be selected from the MEG signal recordings. Fig. 4.6a shows examples for trigger signals from the specified FM myogram. A time-point in the trigger signal was defined automatically (occasionally having to be interactively corrected) for the trigger epochs that were defined by a similar selection process. Using these trigger time-points for each epoch, the series of signal epochs are selected for averaging. Fig. 4.6b–d show examples for averaged MEG signal epochs in different layout displays.

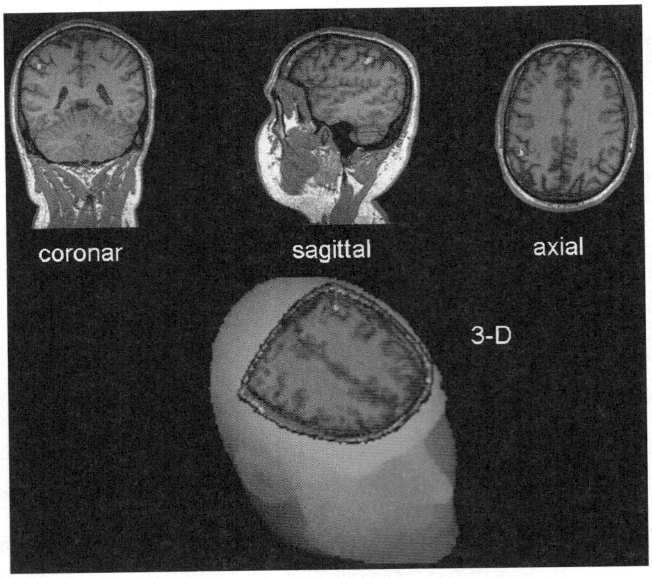

FIGURE 4.5: Functional activation in the left primary cortex (yellow/red) obtained by fMRI for a motor task (FM). The results are superimposed on the MP-RAGE.

Fig. 4.7 shows MEG source reconstruction results. To obtain the structural background for MEG source reconstruction, 3-D models from MRI data are generated and used as a morphological basis for the reconstruction. The triangulated surface is displayed in Fig. 4.7a. An ellipsoid that is fitted to the triangulated surface is shown in Fig. 4.7b. This ellipsoid can also act as volume conductor to simplify the source reconstruction.

Fig. 4.8a shows the correspondence between functional brain activation in the primary motor cortex identified by fMRI and the dipole localization obtained from MEG for a time-point 55 ms after myogram onset (Fig. 4.6a). The localization results for 20 subsequent time-points around the selected time-point revealed an average distance to the fMRI in the range of millimeters. To obtain interactivity between the modalities, the positions of fMRI active regions can be transferred to the MEG dipole localization procedure as seeds for the first guess or as possible dipole locations. The dipole localization result with the use of fMRI seeds (yellow) as first guess in the fitting procedure is shown in Fig. 4.8b.

It cannot be expected that the source of the magnetic field distribution for a certain time-point (resulting from postsynaptic potentials) exactly match with the activity revealed by the BOLD effect, which integrates over the entire block of movements, sometimes described as the "brain or vein" conflict. Localization differences in the range of several millimeters are therefore reasonable.

This accuracy is sufficient to use the regions identified as "active" in the fMRI analysis as possible sites of sources or as first guess for the fit for MEG source reconstruction.

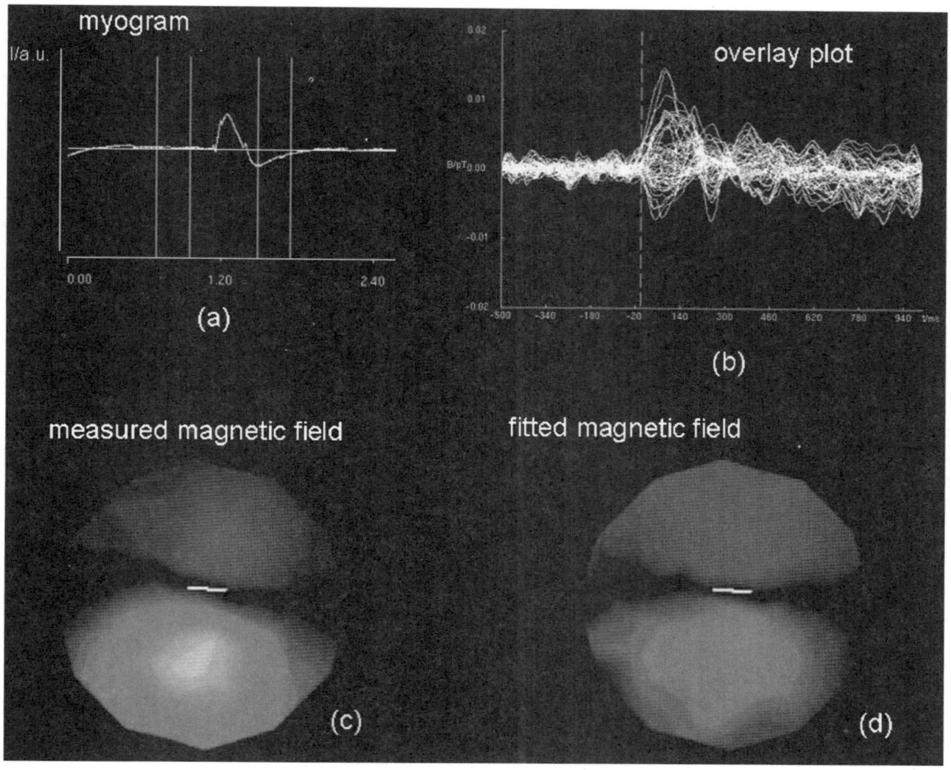

FIGURE 4.6: (a) Display of rectified electromyogram time course used as trigger for flexion movement producing MEG detectable brain activation. (b) Averaged MEG signal epochs of motor-evoked fields in overlay display, the dashed line corresponds to the trigger time-point in (a), the solid line marks the time-point of the dipole fit. (c) Measured magnetic field pattern for a defined time-point. (d) Fitted magnetic field pattern.

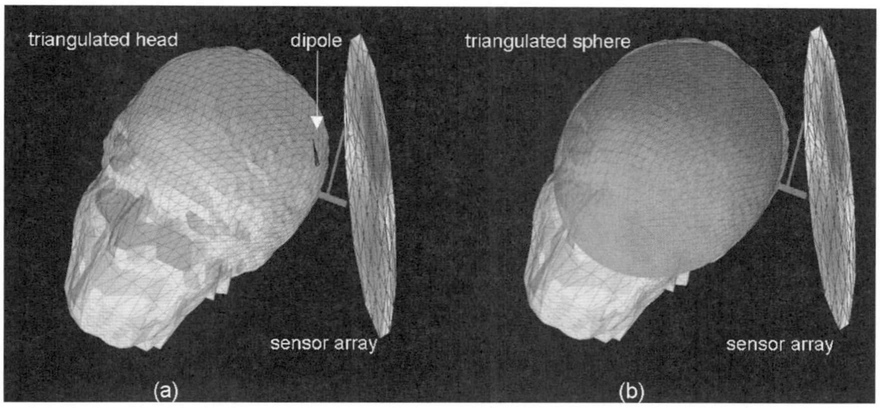

FIGURE 4.7: MEG results in 3-D visualization. (a) Triangulated head surface (about 1700 triangles) with dipole localization result displayed as a pin. (b) Idealized ellipsoid as an alternative conductor model.

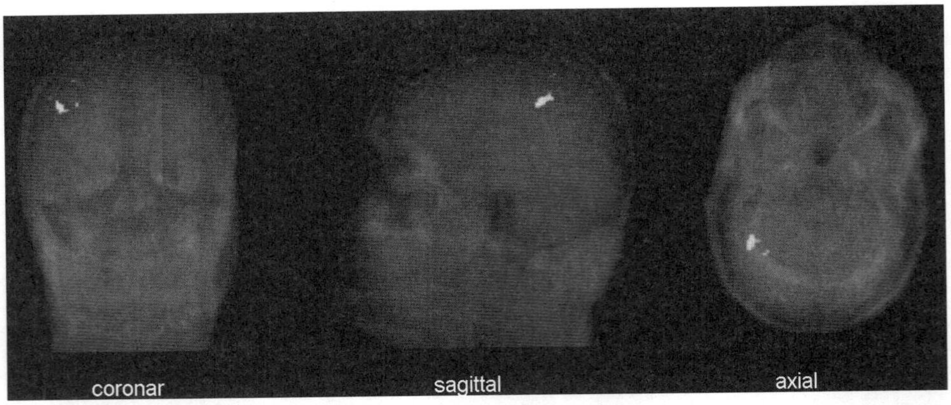

FIGURE 4.8: Functional activation in the left primary motor cortex (yellow) obtained by fMRI. The results are displayed on the morphological background of MRI 3-D recordings. The glass view represents the average over all respective slices. The single dipole localization is superimposed as a red dot.

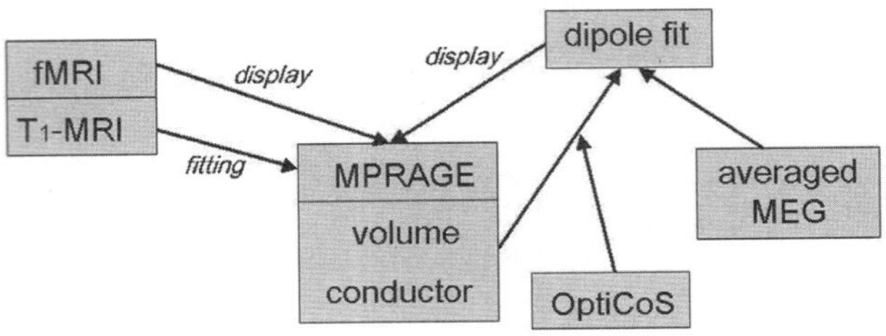

FIGURE 4.9: Intermodal multimodality scheme.

To pursue synergetic interactions between the results of the two modalities, MEG and fMRI, it is essential to perform the combined analyses in one single software environment. The optical coregistration system (OptiCoS) provides the coordinate transformations and from hereon all analysis work is performed in one unique coordinate frame bound to the human head. All morphological and nonmorphologic information are available in one unique coordinate frame. For an overview, Fig. 4.9 illustrates major parts of this intermodal multimodality analysis concept. With the help of the MP-RAGE data and the optical marker detection by OptiCoS, the results of fMRI localization and dipole localization are overlaid on the identical individual MP-RAGE. Thus, comprehensive information of fMRI localization with MEG becomes feasible.

4.2 INTRAMODAL MULTIMODALITY: DIFFUSION TENSOR IMAGING AND MORPHOMETRY

As an example for intramodal multimodality, differences between normal subject groups and patients with certain disorders can be quantified. Comparison of groups of patients and healthy subjects can only be made after transformation of the single brain data into normalized stereotactic space. To explain the concurrence between diffusion tensor imaging (DTI) and morphometry, a study on patients with thinning of the corpus callosum (CC) was conducted. The subjects with CC thinning (which was present in the course of a neurodegenerative disease named hereditary spastic paraplegia (HSP) [88]) seemed particularly prone to this kind of analysis, since the CC is one of the cerebral structures/areas with the strongest accumulation of oriented fibers connecting the two hemispheres. Assuming HSP to be a white-matter disease, DTI should be able to detect axonal loss or at least damage of oriented white-matter regions [89].

4.2.1 The TIFT-Software

The following results have been created by the tensor imaging and fiber tracking (TIFT) software [90]. This software allows the scenario of a complete intramodal multimodality analysis. TIFT provides various quantification and visualization possibilities for the analysis of DTI, DWI, fMRI, and 3-D morphology data. TIFT provides several analysis methods, quantification methods via region of interest (ROI) analysis, stereotactic normalization techniques, group averaging, several interactive fiber tracking algorithms, and various 2-D and 3-D visualization options. The analysis methods can be applied to single subject data before and after MNI normalization as well as to the patient groups. The structure of the software aims at minimization of operator dependency providing analysis in a fast, comprehensive, and reproducible way.

The TIFT software provides new features in several respects:

- Stereotactic normalization techniques to the MNI space as implemented in TIFT are the prerequisites for group analysis. The TIFT software includes a new normalization technique based on scanner and sequence-specific templates and therefore can be adapted or newly created by every user depending on the requirements.
- Single subject data sets can be individually normalized and thus, FA mapping, ROI analysis, and FT can be performed in MNI space. The averaging of FA maps allows for the comparison of healthy subjects and patients with a similar pattern of white-matter changes, e.g., certain neurodegenerative diseases, at group level.
- ROI analyses offer the possibility to quantify diffusion properties in defined brain regions demonstrating the differences between patients and controls. This feature can be applied to individual data (before or after MNI deformation) or to group-averaged data. The TIFT-software can combine the ROI analysis with fiber tracking. First studies of such techniques have been reported by other authors [91,92]. This feature is part of a new comprehensive technology for the analysis of disturbed interconnectivity.
- In the field of FT, the TIFT-software provides several techniques and a set of parameters in which the operator has a wide choice for optimizing the FT for the needs of the specific brain

region under inspection. The implemented normalization and averaging techniques allow for FT at group level for the first time.

- The fast and operator-independent implementation can be made use of in presurgical mapping of white matter. After MRI scanning the whole procedure including DICOM-conversion, motion correction, eddy-current correction, FA analysis, and FT takes no more than 30 min.

4.2.2 Study Protocol

DTI scanning protocols were performed on a 1.5 T scanner (Symphony, Siemens Medical, Erlangen, Germany). Six healthy controls (three men, three women, average age 32.7 ± 4.5 years) and six patients with CC thinning due to complicated HSP (three men, three women, average age 32.5 ± 12.1 years), as model data for degenerative diseases in the CC, underwent scanning.

The DTI study protocol consisted of 13 volumes (45 slices, 128×128 voxel, slice thickness 2.2 mm, in-plane voxel size 1.5 mm $\times$ 1.5 mm) representing 12 gradient directions and one scan with gradient 0 (b_0). T_E and T_R were 93 ms and 8000 ms, respectively; b was 800 s/mm^2; five scans were averaged. For the acquisition of the volume-rendering MRI data set of the whole brain used in morphometry (MP-RAGE), a T_1-weighted magnetization-prepared rapid-acquisition gradient echo sequence was used (MP-RAGE, $T_R = 9.7$ ms, $T_E = 3.93$ ms, flip angle 15°, matrix size 256×256 mm^2, voxel size $1.0 \times 0.96 \times 0.96$ mm^3), consisting of 160–200 sagittal partitions depending on the head size.

4.2.3 Template Creation

For the transformation into MNI space, the creation of study specific templates is necessary (see scheme in Fig. 4.10). The templates were created by the rules explained in Chapter 3.

The normalization of the MP-RAGE data sets is a straightforward procedure (for the created template see Fig. 4.11) whereas the normalization of DTI data sets is more sophisticated.

The normalization parameters are calculated for the b_0 volume and have to be applied consequently to all other volumes of the recording. The functional information also has to be transformed. The rotation parameters of the rigid brain transformation have to be stored so that the tensor $\vec{\vec{D}}$ of each voxel can be aligned by rotation to the main gradient directions when the MNI normalization of the DTI data sets is performed. Following the basic ideas of Alexander et al. [93], the dilation matrices are used for the alignment of the tensor $\vec{\vec{D}}$ of each voxel to the surrounding vox-

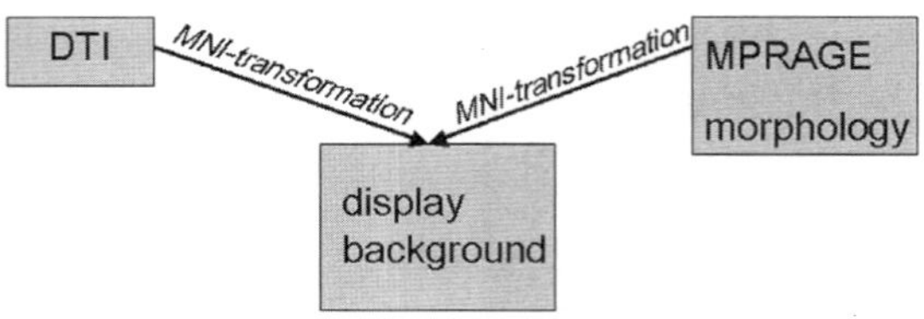

FIGURE 4.10: Intramodal multimodality scheme.

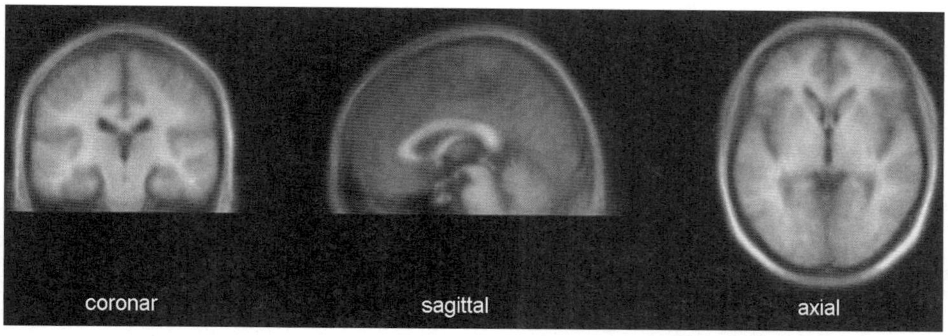

FIGURE 4.11: MP-RAGE template for normalization. Display focus is PC.

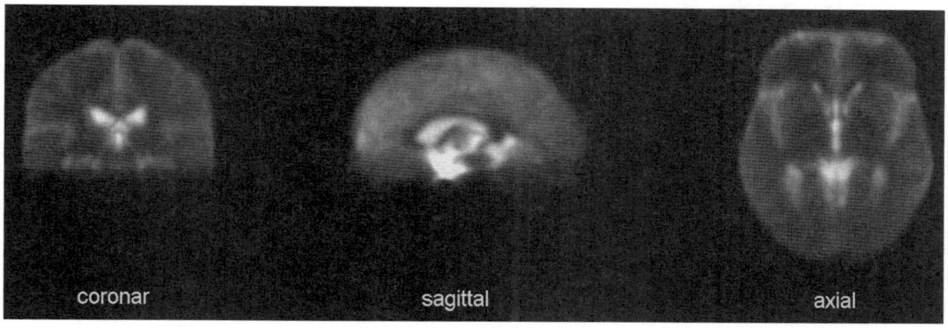

FIGURE 4.12: DTI template for normalization. Display focus is PC.

els resulting in a 3-D vector matrix containing three rotation angles at each voxel site. Strategies for spatial transformation of DTI data sets were also suggested by Ashburner et al. [78] and Park et al. [94,95].

Arithmetic averaging of all normalized single subject DTI data sets leads to an averaged DTI data set (Fig. 4.12). This approach is chosen in order to address two independent items. First, the voxel intensities of all subject's DTI data sets are arithmetically averaged. Second, the tensor fine-correction for each subject's data set (resulting from eddy current correction and normalization correction), i.e., a 3-D vector matrix for each data set, is arithmetically averaged. This averaged tensor fine-correction is applied to the calculation of the FA map of the averaged DTI data set.

In Fig. 4.13 a visual comparison between FA maps before and after MNI normalization is shown. ROI analyses have shown coincidence for the FA values before and after normalization [90].

4.2.4 Averaging for Different Subjects

Studies at group level may be important if the common clinical phenotype is supposed to be due to lesion of defined brain areas. Consequently, averaging of results for different subjects is necessary. Each individual brain has to be transferred into stereotactic space, and in a second step, the arithmetic averaging of the results is possible voxel by voxel. The averaging of the individual results can be performed in two ways:

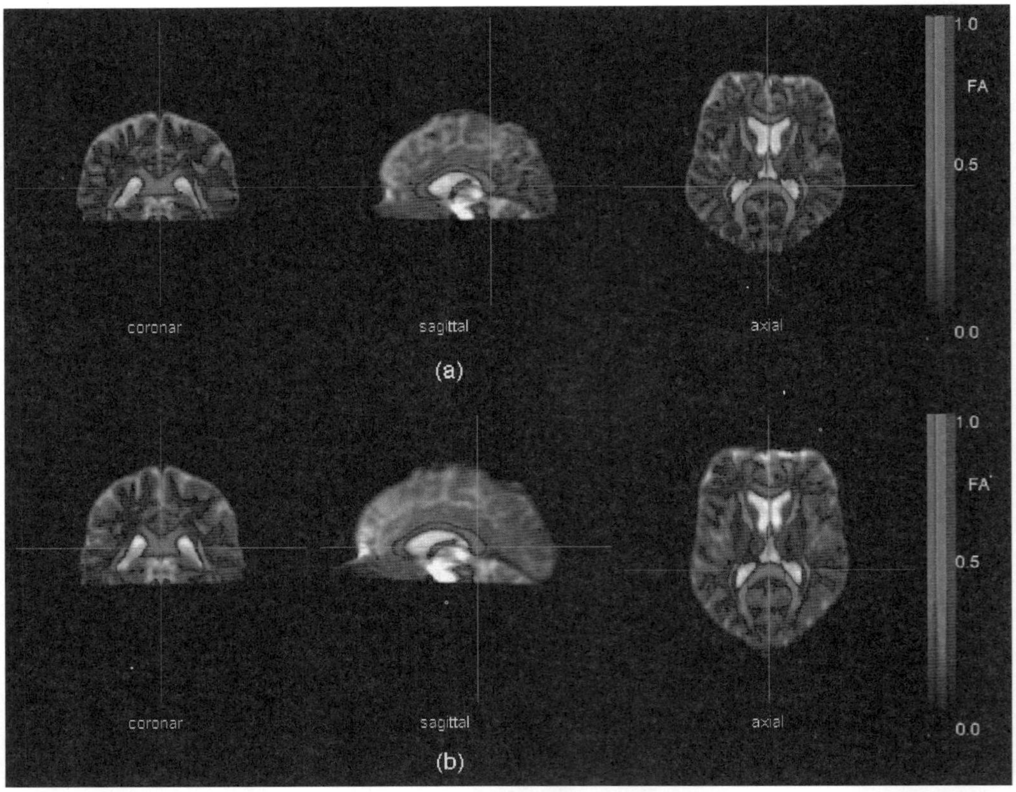

FIGURE 4.13: FA maps of a single subject before (a) and after (b) affine template MNI deformation. Display threshold for FA values is 0.2, background is the b_0 data set.

- For each single subject's data in MNI space, the FA map is calculated separately, and arithmetic averaging of the FA maps is performed afterward (Fig. 4.14a and b). Thus, a statistical test between the FA maps of the groups can be carried out.
- Each DTI data set is normalized and before FA mapping the whole DTI data sets are averaged. FA parameterization of these group-averaged DTI data leads to a group-specific FA map FAM2 (Fig. 4.15).

4.2.5 Examples for Fiber Tracking

From all data sets, an averaged DTI data set can be created by arithmetic averaging. Then, FT can be performed, taking the fine-corrections into account. Fig. 4.16a shows the FT for a healthy subject after MNI normalization (cf. Fig. 4.13). Fig. 4.16b shows FT performed on an averaged DTI data set, composed by arithmetic averaging of several MNI normalized data sets.

The application of the technique of DTI data averaging on the healthy subject group and the subject group with thinned CC enables detection of differences (Fig. 4.14). The thinning of the CC leads to reduced FA values and thus to restricted possibilities in FT (Fig. 4.17).

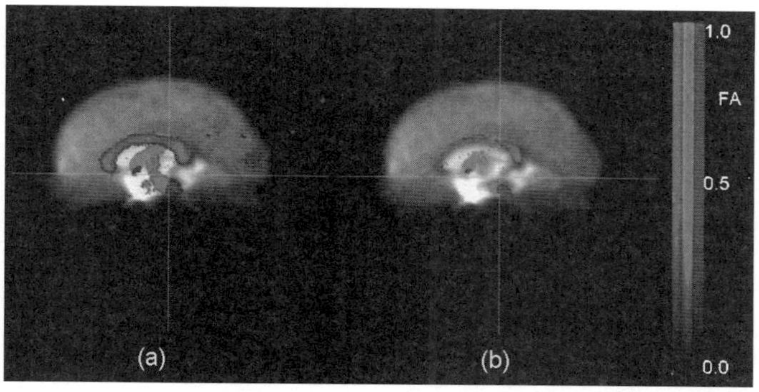

FIGURE 4.14: Arithmetic averaging of the FA maps after MNI normalization (FAM1). Display background is the averaged DTI data set (averaged b_0 volume), display threshold is 0.2, and background is the averaged b_0 data set. (a) Volunteers. (b) Patients with thinned CC.

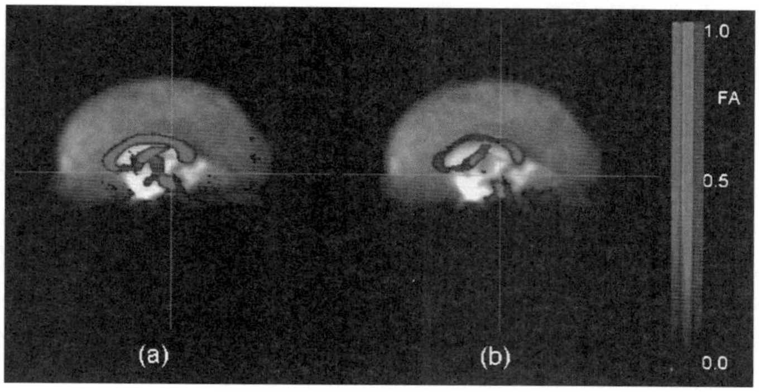

FIGURE 4.15: Arithmetic averaging of the FA maps after MNI normalization (FAM2). Display background is the averaged DTI data set (averaged b_0 volume), display threshold is 0.2, and background is the averaged b_0 data set. (a) Volunteers. (b) Patients with thinned CC.

The fiber tracks can be used as a skeleton mask for a selected statistics. The efforts of tract-based spatial statistics (TBSS) have been described by [62,96,97]. Here, a modified version named tractwise fractional anisotropy statistics (TFAS) is presented.

The resulting voxels from FT are used as skeleton. When FT is performed on averaged DTI data, each voxel that is crossed by a FT is defined as "active" for statistics. The underlying FA voxels contribute to the statistical t-test. The skeleton can be set up in different ways:

- For patients and volunteers skeletons are created independently and the FA voxel selection is performed independently for the volunteer and the patient data sets (skeleton 1).
- As a skeleton a combination of the skeletons from FT for volunteers and from FT for patients is used. In this way, the FA voxel selection is the same for patient data and volunteer data (skeleton 2).

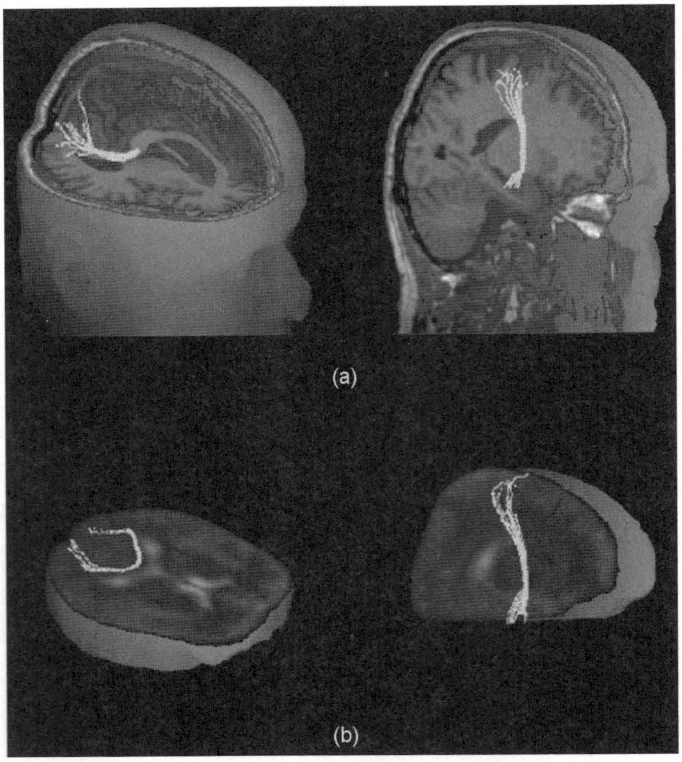

FIGURE 4.16: Fiber tracking results – 3-D views. (a) Healthy subject after MNI deformation displayed on the MP-RAGE as background: starting point in the dorsal corpus callosum (*left*) and starting point in the left internal capsule (*right*). (b) Averaged DTI data set from a group of healthy subjects: starting point in the dorsal corpus callosum (*left*) and starting point in the left internal capsule (*right*).

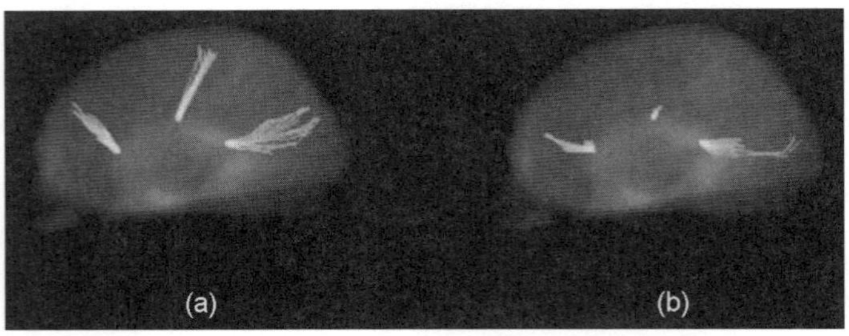

FIGURE 4.17: Examples for FT. Starting points were chosen in the CC. (a) Normal subject groups. (b) Patients with thinned CC. The images are glass-brain plots of the FTs overlaid on glass-brain plots of b_0 data.

Average FA and standard deviation are calculated by

$$F_{\text{avg}} = \frac{1}{N}\sum F_i, \quad \forall\,\text{voxel} \quad i \in \text{skeleton} \tag{4.60}$$

$$F_{\text{std}} = \sqrt{\frac{1}{N-1}\sum (F_{\text{avg}} - F_i)^2}, \quad \forall\,\text{voxel} \quad i \in \text{skeleton} \tag{4.61}$$

where N is the respective skeleton size, which usually is different for different skeletons. Then, TFAS uses F_{avg} and F_{std} for statistical t-test.

For both skeletons significant differences in TFAS analysis can be detected. Averaged FA values for healthy subjects are 0.43 ± 0.19 for skeleton 1 and 0.41 ± 0.19 for skeleton 2, respectively, whereas averaged FA values for patients with thinned CC are reduced to 0.25 ± 0.12 for skeleton 1 and to 0.21 ± 0.10 for skeleton 2, respectively.

4.2.6 Voxel-Based Morphometry

The so-called computational neuroanatomy is a methodology to characterize neuroanatomic configuration of different brains, encompassing voxel-based morphometry (VBM), which compares neuroanatomic differences on a voxel by voxel basis. Other techniques are deformation-based morphometry (DBM), which provides information about global differences in brain shape, and tensor-based morphometry (TBM), which provides information about local shape differences [98–100]. Especially VBM has shown to be a powerful tool for analyzing changes in gray or white matter volumes of the brain since its introduction in 1995 by use of an automated segmentation procedure and standardized parametric statistics in comparison with a normal data base within the Statistical Parametric Mapping (SPM) software (Wellcome Department of Imaging Neuroscience Group, London, UK; http://www.fil.ion.ucl.ac.uk/spm). VBM allows extracting information for each voxel separately from its gray-scale intensity.

For an extensive study of the VBM method with all details, the reader is referred to [101] and references therein. Other works of Ashburner et al. can help to complement the understanding [102,78,103]. A short description of the method is given in the following paragraphs:

The normalization procedure of the individual high-resolution brain data set (voxel size approx 1 mm^3) to a standard stereotactic space is the prerequisite. By nonlinear registration of brain images, a Jacobian matrix field is obtained in which each element is a tensor describing the relative positions of the neighboring elements. This is achieved by registering each of the images to the same template image by minimizing the residual sum of squared differences between them, using a 12 parameter affine transformation [102]. Thus, a Bayesian framework is used taking prior knowledge of the normal variability of brain size into account. In a second step global nonlinear shape differences are modeled by a linear combination of smooth spatial basis functions [78]. It has to be noted that this

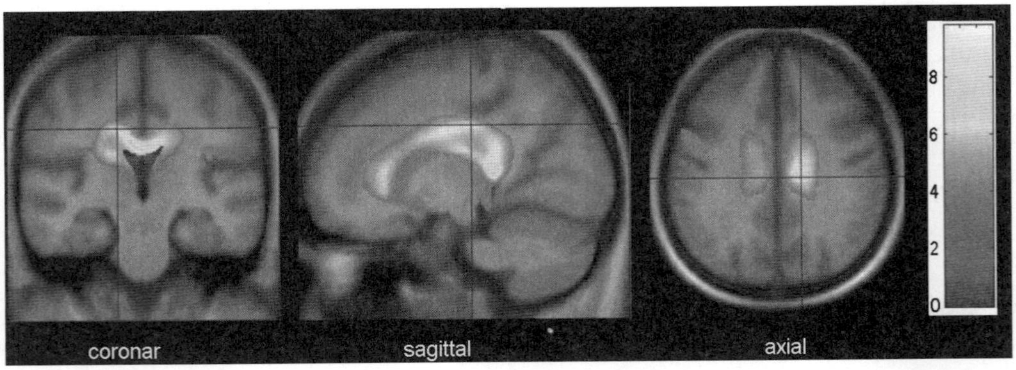

FIGURE 4.18: VBM results for patients with thinned CC versus normal subject groups. Displayed are z-score values on a standardized morphological background.

method of spatial normalization does not attempt to match every cortical feature exactly, but merely corrects for global brain shape differences.

After stereotactic transformation the images are partitioned into gray matter, white matter, cerebrospinal fluid and other background classes. It has to be noted that because the tissue classification is based on voxel intensities, image intensity nonuniformities have to be corrected prior to this segmentation procedure.

Next image processing steps are smoothing by convolving with an isotropic Gaussian kernel, and transformation of the local voxel concentrations (intensities) with the so-called logit function:

$$\text{logit}(I) = \frac{1}{2} \ln \frac{I}{1-I} \qquad (4.62)$$

After all these preprocessing steps, a statistical comparison on a voxel-by-voxel basis can be made. Fig. 4.18 shows an example for comparison of healthy subjects and patients with thinned CC calculated by the SPM software.

4.2.7 Intensity-Based 3-D MRI Analysis

This optimized VBM technique is rather complex. Therefore, in analogy to VBM, intensity-based 3D-MRI analysis (IBA) is used here to demonstrate the complementary results for IBA/VBM and DTI.

The MP-RAGE data sets for each single subject are normalized to the template shown in Fig. 4.11 using the techniques explained in Section 3.2. Arithmetic averaging of the normalized data sets for each group (patients and volunteers) leads to the data displayed in Fig. 4.19a and b.

In Fig. 4.20 Statistical analysis results of the IBA method comparing normal subject groups and patients with thinned CC are shown. Comparison with the VBM method (Fig. 4.18) shows coincidence.

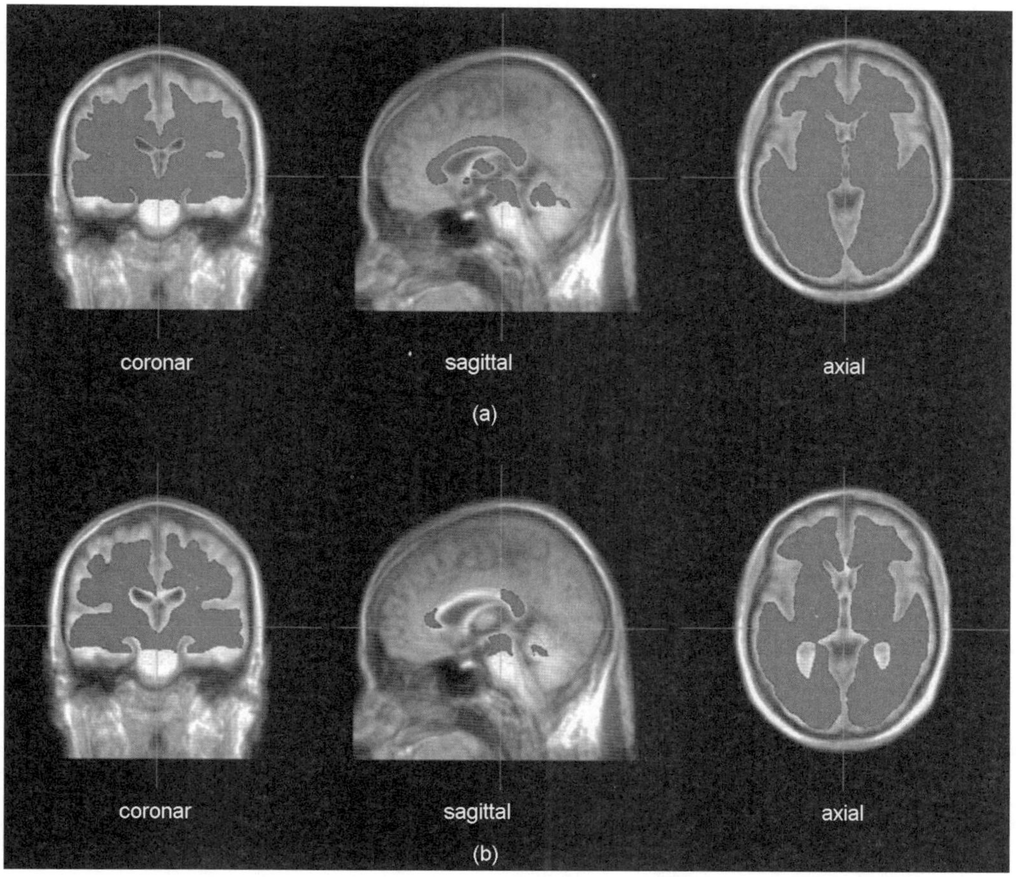

FIGURE 4.19: Intensity-based analysis (IBA) results, display threshold is 20% of maximum intensity, display background is the MP-RAGE template. (a) Averaged intensity image of volunteers. (b) Averaged intensity image of patients with thinned CC.

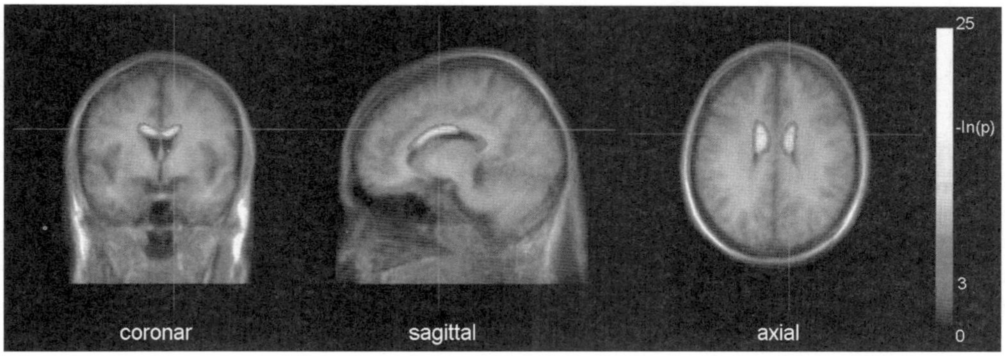

FIGURE 4.20: IBA results for patients with thinned CC versus normal subject groups (t-test analysis in white matter morphology). Display threshold is $p < 0.05$.

4.2.8 Complementary Intramodal Multimodality Results

IBA results and DTI results can provide complementary information displayed on the same MNI normalized background, making comparison feasible. In Fig. 4.21, results from two different MRI techniques are juxtaposed. Fig. 4.21a shows differences in white matter morphology between normal subject groups and patients with thinned CC extracted by IBA whereas Fig. 4.21b displays white matter diffusion anomalies extracted by FA analysis of DTI-weighted data sets.

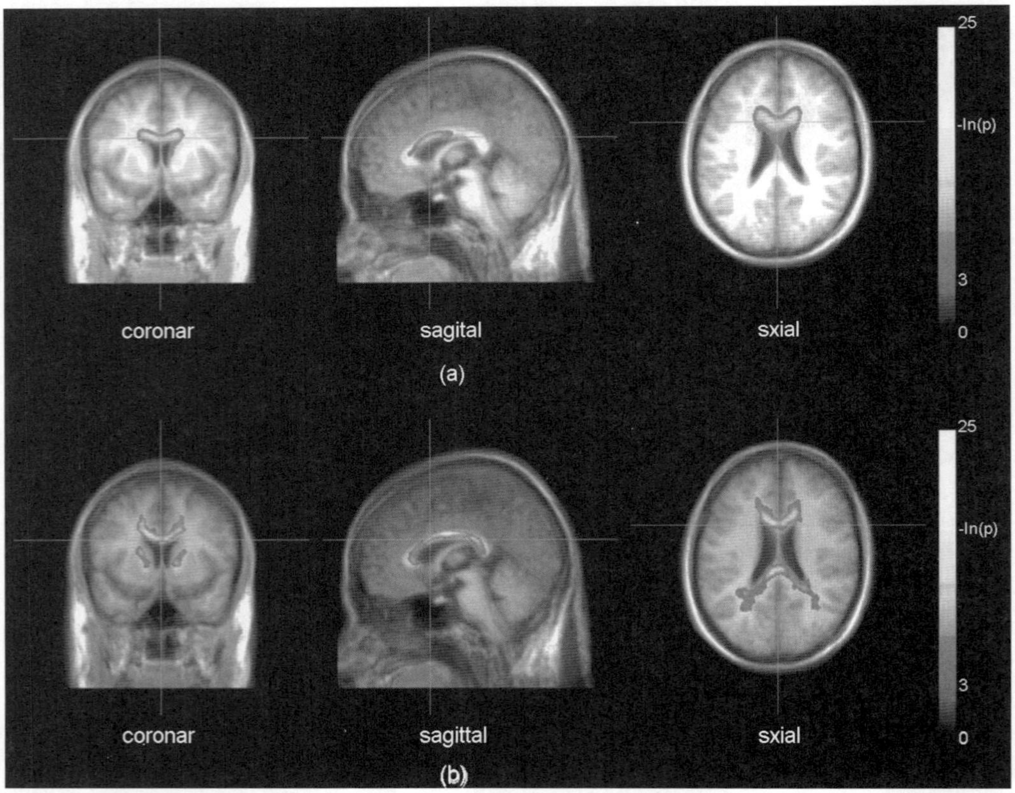

FIGURE 4.21: Complementary intramodal multimodality: (a) Differences in white matter morphology between normal subject groups and patients with thinned CC extracted by IBA. Display threshold difference is 40% of maximum intensity. (b) FA analysis of DTI-weighted data sets for patients with thinned CC versus normal subject groups (t-test analysis of FA maps). Display threshold is $p < 0.05$.

4.3 COMBINATION OF INTERMODAL AND INTRAMODAL MULTIMODALITIES

With the techniques described in Sections 1 and 2, the following comprehensive opportunities are suggested:

- IBA/VBM results might be used as seeds for FT. The starting points of FT might be localized adjacent to the IBA/VBM results and effects on FT might be topic of research. A powerful tool in this respect may be TFAS.
- fMRI functional results are not only complemented by their time development (MEG), but also can be combined to VBM results.
- fMRI active regions can act as seeds for FT. Thus, the FT starting points are localized nearby the fMRI active region. Interconnectivity of brain regions is thereby the major aim of research.

Of course, other neuroimaging techniques can complement the results explained in this chapter, e.g., multimodal imaging of PET, fMRI, and MEG [104], as additional applications of the same basic concepts.

CHAPTER 5

Clinical Aspects of Multimodal Imaging

Beyond general gain of information by multimodal neuroimaging, e.g., on connectivity of the brain [105], the future of multimodal imaging seems to be closely linked to its clinical applications. Two conceptual frameworks are of highest practical relevance: first, investigations with respect to understanding and diagnosis of neurological disorders with changes assessable by structural and/or functional neuroimaging techniques, in particular the large group of neurodegenerative diseases. One example, corpus callosum (CC) thinning due to the neurodegenerative disorder hereditary spastic paraparesis, has already been shown in Chapter 4. The second clinical application is the combination of different noninvasive neuroimaging tools and their cointegration into frameless stereotaxy/ neuronavigational systems after registration as an important element of presurgical diagnostics.

5.1 MULTIMODAL IMAGING IN THE RESEARCH OF NEUROLOGICAL (NEURODEGENERATIVE) DISEASES

Different MRI analysis techniques have been used in order to identify morphological signatures of brain diseases. In particular, the group of the neurodegenerative diseases has been extensively studied. Here, the computerized analysis techniques of structural MRI data (optimally acquired in a volume-rendering mode) are used that can capture the extraordinary morphological variability of the human brain, since these methods such as VBM use mathematical models sensitive to subtle changes in the size, position, shape, and tissue characteristics of brain structures affected by neurodegenerative diseases (cf. [106] for a review). However, beyond this volumetric or morphometric analysis of T_1-weighted MRI, various MRI applications in the sense of intramodal multimodality have proved useful both as diagnostic tools in individual patients and as instruments for the assessment of pathoanatomical and pathophysiological concepts at group level; at its simplest, this means a combination of structural MRI including volumetry and morphometry, DTI, and MRS. Because of this multimodal capability, MR has become the most versatile method for differential diagnosis in dementia with respect to the differential diagnosis and subcategorization and in order to monitor the progression and treatment effects [107]. Also in other important neurodegenerative disorders such as idiopathic Parkinson's disease and other Parkinsonian syndromes (cf. [108] for a review) and in motor neuron disorders (cf. [109] for a review), MRI applications investigating different aspects of the disorder-associated pathoanatomy have massively increased the clinicians' understanding over the recent years. In particular, in Parkinsonian syndromes, but also in other neurodegenerative disorders, the combination of MRI with (multiligand) nuclear medicine techniques is a valuable tool [110]. In multiple sclerosis, i.e., the most important chronic inflammatory disorder of the central nervous system, many different MRI techniques such as conventional MRI [111], DTI [112], fMRI [113], MRS [114], and a technique called magnetization transfer MRI [115] are important elements in diagnostics, differential diagnostics, and in improving our understanding of the lesions and the reparation procedures in the central nervous system. fMRI is still limited in its clinical applicability and has no relevant function in clinical diagnostics to date, but does enable the formulation of neurobiological hypotheses that can be tested clinically, and hence is well suited for testing classic clinical hypotheses about how the brain works so that understanding the mechanisms and sites of pathology might facilitate the development of new therapeutic strategies [116]. Also for MEG/MSI, the clinical use is limited to aspects such as the localization of pathological activity in association with neurological disorders such as certain epilepsy syndromes [117].

5.2 MULTIMODAL PRESURGICAL IMAGING

The potential of MSI for presurgical mapping of functionally important cortex has been extensively demonstrated [118,3,119,120,121,122]. MSI can be used for decision-making and planning of surgical treatment in brain lesions (in particular tumors) [123]. The most frequently used application is MSI with motor evoked fields and somatosensory evoked fields for mapping of the sensorimotor cortex in patients with lesions around the motor cortex [124,125,65]. Also, the determination of primary visual cortex by the investigation of visual evoked fields [126] as well as functional mapping of speech-related brain areas in individuals [4,127] were reported. MEG has a potential to directly detect and localize epilepsy-associated activity, i.e., MSI is used for presurgical evaluation in epilepsy surgery to identify the brain tissue generating epileptic activity [128,129,130,131].

With respect to the multimodality concept in the presurgical evaluation, a combination of functional neuroimaging techniques with basically different underlying neurophysiological correlates such as MEG/MSI and fMRI is most promising. Spatial comparisons between the localizational capability, in particular that of the sensorimotor cortex of MSI and fMRI, have been made by measuring the distances between MEG and fMRI activation sites after coregistration, both in healthy subjects [3] and in patients [4] showing different activation sites for the motor and sensory tasks within a range distance of 10 and 15 mm, probably as a reflection of the correlation between electrophysiologic and hemodynamic responses.

MSI results were obtained by the intraoperative cortical recording of somatosensory evoked potentials (so-called “phase reversal”), the localizations in the pre- and postcentral gyrus were verified in each case [125]. In a study with a similar technical approach using MEG and phase reversal technique in 30 patients with space-occupying lesions in or around the central region, MEG-based functional neuronavigation was also found to be practicable in finding a safe approach to tumors in or adjacent to the central region [132].

Both modalities MSI and fMRI are useful for the estimation of the sensorimotor cortex, and a combination of both methods has been used in order to avoid neurological deficits in certain cases [4]. Consecutively, the value of this integration of these two imaging modalities for spatio-temporal mapping of brain activity was strengthened in several publications [133] and led to the cointegration of both techniques into presurgical protocols [123], especially in patients receiving epilepsy surgery [134,135,136]. As an advanced combination of different modalities, combined data of DTI-based tractography fMRI and MEG have been used for the functional presurgical mapping of language areas [137]. It is useful to perform postprocessing of MEG and fMRI data in one software environment in that context to avoid potential methodologically based problems [79]. For this purpose of integration of fMRI, 3-D MRI, and MSI in one unique coordinate frame, the activation achieved by fMRI analysis can be used as possible sites of current sources for the MEG source reconstruction as an elegant approach for multimodality integration which was suggested by the "open magnetic and electric graphic analysis (OMEGA)" software [83] and which has been described in detail in Chapter 4. Further, the multimodal presurgical concept might be achieved by the combination of MEG/MSI and functional MRS [138]. With respect to the combination of different MRI techniques, quantitative DTI and MRS have been used to determine changes in the pyramidal tract adjacent to gliomas [139].

Regarding presurgical delineation of lesions, in particular epileptogenic lesions prior to epilepsy surgical procedures, voxel-based 3-D MRI analysis tools (e.g., VBM) have been developed towards the goal of automated observer-independent analysis [140,141,142,143]. Furthermore, as a consequent continuation of the multimodality approach in the sense of combined voxel-based mapping [144], a combination of ligand imaging using PET [145] and SPM-based MRI analysis was proposed. It could be demonstrated in a study using the approach of voxel-based MRI analysis and subtraction ictal SPECT coregistered to MRI (SISCOM) that multimodal assessment of malformations of cortical development may contribute to advanced observer-independent preoperative assessment of seizure origin and thus improve presurgical diagnostics in symptomatic epilepsy [146]. Although at the expense of the measurement of absolute quantitative values, the voxel-based comparison of MRI and PET enables the automated objective whole brain analysis, partial volume effects, and mixed tissue sampling in ligand images to be resolved, and disproportionate changes in function and structure to be identified (e.g., [147]). A detailed synopsis of multimodal imaging in presurgical diagnostics is given in [7].

5.3 ALTERNATIVE NEUROIMAGING TECHNIQUES

It has to be kept in mind that additional techniques such as (multichannel) EEG [148,149,150] exist, which is a major presurgical diagnostic element in epilepsy and in the localization prior to epilepsy surgery. However, EEG is rarely used for direct source coregistration with MRI and neuronavigation and was therefore not commented on, although in particular, the spatio-temporal relationship between MSI and EEG in epileptic focus localization has been subject of several studies [151,152], as well as the combination of EEG and fMRI [153] and, recently, the combination of EEG and MRS [154].

Furthermore, the technique of functional transcranial magnetic stimulation might provide an additional tool to add functional information (e.g., in the investigation of the virtual lesion).

5.4 ADVANTAGES AND LIMITATIONS

With respect to the cost effectiveness and the burden of often time-consuming data acquisition procedures in critically ill patients, multimodality imaging does not necessarily mean "the more, the better". Some options such as techniques for the localization of eloquent cortex (fMRI or MSI, respectively) and techniques for the imaging of white matter fiber bundles (DTI) are complementary. For the methods of imaging cortical areas, it has to be held that fMRI can be performed on most regular clinical MRI scanners, whereas MSI, although it seems to be a promising technique, is still limited because of the low distribution of MEG devices which will most probably not increase much over the coming years. Nuclear medicine techniques are valuable tools in the delineation of cerebral neoplasms or other lesions; in particular as multiligand imaging, these nuclear medicine techniques may be complemented by MRS in neurological diseases, especially neurodegenerative disorders.

If different neuroimaging modalities are used, the user has to be aware that the fusion of the results of these techniques is possible after coregistration, but that equivalent localizations in the ideal neuroanatomical sense are very difficult to achieve and sometimes impossible because of different neurophysiological substrates, acquisition methods, and spatial resolution.

5.5 FUTURE

The future of multimodal imaging should be seen in connection to its clinical applicability. Prerequisites for use of multimodal imaging in a clinical setting are listed as follows:

- Measurement devices should be available and measurement time should be comparatively short.
- Optimally, all analyses should be performed in one single software environment.
- Short computation time, i.e., optimal results should be produced within minutes after finishing the last measurement. For example, computation time on a high performance PC is about 1.0 s per volume for fMRI motion correction, 0.8 s for functional analysis and 0.2 s for a dipole fit [83].
- Results should be gained in a data postprocessing as operator-independent as possible, i.e., the software is running automatically with as few observer interactions as possible.

REFERENCES

[1] J. G. Beaumont, *Introduction to Neuropsychology*. New York: The Guilford Press; 1983.

[2] J. S. George, C. J. Aine, J. C. Mosher, D. M. Schmidt, D. M. Ranken, H. A. Schlitt, C. C. Wood, J. D. Lewine, J. A. Sanders, and J. W. Belliveau, "Mapping function in the human brain with magnetoencephalography, anatomical magnetic resonance imaging, and functional magnetic resonance imaging," *Clin. Neurophysiol.*, vol. 12, no. 5, pp. 406–431, 1995, doi:10.1097/00004691-199509010-00002.

[3] C. Stippich, P. Freitag, J. Kassubek, P. Sörös, K. Kamada, H. Kober, K. Scheffler, R. Hopfengärtner, D. Bilecen, E. W. Radü, and J. B. Vieth, "Motor, somatosensory and auditory cortex localization by fMRI and MEG," *NeuroReport*, vol. 9, no. 9, pp. 1953–1957, 1998, doi:10.1097/00001756-199806220-00007.

[4] H. Kober, C. Nimsky, M. Moller, P. Hastreiter, R. Fahlbusch, and O. Ganslandt, "Correlation of sensorimotor activation with functional magnetic resonance imaging and magnetoencephalography in presurgical functional imaging: A spatial analysis," *NeuroImage*, vol. 14, no. 5, pp. 1214–1228, 2001, doi:10.1006/nimg.2001.0909.

[5] N. Fujimaki, T. Hayakawa, M. Nielsen, T. R. Knosche, and S. Miyauchi, "An fMRI-constrained MEG source analysis with procedures for dividing and grouping activation," *NeuroImage*, vol. 17, no. 1, pp. 324–343, 2002, doi:10.1006/nimg.2002.1160.

[6] S. Thees, F. Blankenburg, B. Taskin, G. Curio, and A. Villringer, "Dipole source localization and fMRI of simultaneously recorded data applied to somatosensory categorization," *NeuroImage*, vol. 18, pp. 707–719, 2003, doi:10.1016/S1053-8119(02)00054-X.

[7] J. Kassubek, and F. D. Juengling, "Multimodality functional neuroimaging," in Stippich C, Ed. *Clinical Functional MRI – Presurgical Functional Neuroimaging*. Berlin/Heidelberg/New York: Springer; 2007, pp. 191–210.

[8] E. Fukushima and B. W. Roeder, *Experimental Pulse NMR – A Nuts and Bolts Approach*, 10th ed. Cambridge, MA: Perseus Publishing; 1986.

[9] P. T. Callaghan, *Principles of Nuclear Magnetic Resonance Microscopy*. Oxford: Clarendon Press; 1994.

[10] R. Kimmich, *NMR Tomography, Diffusometry, Relaxometry*. Berlin: Springer; 1997.

[11] S. Stapf, and S. Han, *NMR Imaging in Chemical Engineering*. Weinheim, Germany: Wiley; 2006.

[12] F. Bloch, W. W. Hansen, and M. Packard, "The nuclear induction experiment," *Phys. Rev.*, vol. 70, pp. 474–485, 1946, doi:10.1103/PhysRev.70.474.

[13] A. Haase, J. Frahm, D. Matthaei, W. Hänicke, and K. D. Merbboldt, "FLASH imaging, rapid NMR imaging using low flip angle pulses," *Magn. Reson.*, vol. 67, pp. 258–266, 1986, doi:10.1016/0022-2364(86)90433-6.

[14] A. Mishra, Y. Lu, J. Meng, A. W. Anderson, and Z. Ding Z, "Unified framework for anisotropic interpolation and smoothing of diffusion tensor images," *NeuroImage*, vol. 31, no. 4, pp. 1525–1535, 2006, doi:10.1016/j.neuroimage.2006.02.031.

[15] R. S. Frackowiak, K. J. Friston, C. D. Frith, R. J. Dolan, and J. C. Mazziotta, *Human Brain Function*. New York: Academic Press; 1997.

[16] A. W. Toga, and J. C. Mazziotta, *Brain Mapping – The Methods*. San Diego: Academic Press; 1996.

[17] D. Le Bihan, *Diffusion and Perfusion Magnetic Resonance Imaging – Applications to Functional MRI*. New York: Raven Press; 1995.

[18] S. Ogawa, T. M. Lee, A. R. Kay, and D. W. Tank, "Brain magnetic resonance imaging with contrast dependent on blood oxygenation," *Proc. Natl. Acad. Sci. USA*, vol. 87, no. 24, pp. 9868–9872, 1990, doi:10.1073/pnas.87.24.9868.

[19] S. Ogawa, T. M. Lee, A. S. Nayak, and P. Glynn, "Oxygenation-sensitive contrast in magnetic resonance image of rodent brain at high magnetic fields," *Magn. Reson. Med.*, vol. 14, no. 1, pp. 68–78, 1990, doi:10.1002/mrm.1910140108.

[20] P. C. van Zijl, S. M. Eleff, J. A. Ulatowski, J. M. Oja, A. M. Ulug, R. J. Traystman, and R. A. Kauppinen, "Quantitative assessment of blood flow, blood volume and blood oxygenation effects in functional magnetic resonance imaging," *Nat. Med.* vol. 42, no. 2, pp. 159–167, 1998, doi:10.1038/nm0298-159.

[21] S. Ogawa, T. M. Lee, and B. Barrere, "The sensitivity of magnetic resonance image signals of a rat brain to changes in the cerebral venous blood oxygenation," *Magn. Reson. Med.*, vol. 29, no. 2, pp. 205–210, 1993, doi:10.1002/mrm.1910290208.

[22] K. K. Kwong, "Functional MRI with echo planar imaging," *Mag. Reson. Quarterly*, vol. 11, pp. 1–20, 1995.

[23] R. Turner, "Functional mapping of the human brain with MRI," *Sem. Neurosci.*, vol. 7, pp. 179–194, 1995.

[24] J. V. Hajnal, R. Myers, A. Oatridge, J. E. Schwieso, I. R. Young, and G. M. Bydder, "Artifacts due to stimulus correlated motion in functional imaging of the brain," *Magn. Reson. Med.*, vol. 31, no. 3, pp. 283–291, 1994, doi:10.1002/mrm.1910310307.

[25] R. P. Woods, S. R. Cherry, and J. C. Mazziotta, "Rapid automated algorithm for aligning and reslicing PET images," *Comput. Assist. Tomogr.* vol. 16, no. 4, pp. 620–633, 1992, doi:10.1097/00004728-199207000-00024.

[26] K. J. Friston, J. Ashburner, C. D. Frith, J. B. Poline, J. D. Heather, and R. S. Frackowiak, "Spatial registration and normalization of images," *Hum. Brain Mapp.*, vol. 3, no. 3, pp. 165–189, 1995, doi:10.1002/hbm.460030303.

[27] W. Andrä, H. Nowak, *Magnetism in Medicine – A Handbook*. Wiley-VCH: Berlin; 1998.

[28] P. A. Bandettini, A. Jesmanowicz, E. C. Wong, and J. S. Hyde, "Processing strategies for time-course data sets in functional MRI of the human brain," *Magn. Reson. Med.*, vol. 30, no. 2, pp. 161–173, 1993, doi:10.1002/mrm.1910300204.

[29] K. Kuppusamy, W. Lin, and M. Haacke, "Statistical assessment of crosscorrelation and variance methods and the importance of electrocardiogram gating in functional magnetic resonance imaging," *Magn. Reson. Imaging*, vol. 15, pp. 169–181, 1997, doi:10.1016/S0730-725X(96)00338-4.

[30] P. Filzmoser, R. Baumgartner, and E. Moser, "A hierarchical clustering method for analyzing functional MR images," *Magn. Res. Imaging*, vol. 17, pp. 817–826, 1999, doi:10.1016/S0730-725X(99)00014-4.

[31] X. Ding, J. Tkach, P. Ruggieri, and T. Masaryk, "Analysis of time-course functional MRI data with clustering method without use of reference signal," *Proc. Int. Soc. Magn. Reson.*, pp. 630–635, 1994.

[32] W. H. Press, S. A. Teukolsky, W. T. Vetterling, and B. P. Flannery, *Numerical Recipes in C*. New York: Cambridge University Press; 1998.

[33] K. J. Friston, C. D. Frith, R. S. Frackowiak and R. Turner, "Characterizing dynamic brain responses with fMRI: A multivariate approach," *NeuroImage*, vol. 2, no. 2, pp. 166–172, 1995, doi:10.1006/nimg.1995.1019.

[34] K. J. Friston, C. D. Frith, R. Turner and R. S. Frackowiak, "Characterizing evoked hemodynamics with Fmri," *NeuroImage*, vol. 2, no. 2, pp. 157–165, 1995, doi:10.1006/nimg.1995.1018.

[35] K. J. Friston, A. P. Holmes, J. B. Poline, P. J. Grasby, S. C. Williams, R. S. Frackowiak, and R. Turner, "Analysis of fMRI time-series revisited," *NeuroImage*, vol. 2, no. 1, pp. 45–53, 1995, doi:10.1006/nimg.1995.1007.

[36] K. J. Friston, A. Holmes, J. B. Poline, C. J. Price, and C. D. Frith, "Detecting activations in PET and fMRI: Levels of inference and power," *NeuroImage*, vol. 4, pp. 223–235, 1996, doi:10.1006/nimg.1996.0074.

[37] K. J. Friston, S. Williams, R. Howard, R. S. Frackowiak, and R. Turner, "Movement-related effects in fMRI time-series," *Magn. Reson. Med.*, vol. 35, no. 3, pp. 346–355, 1996.

[38] K. J. Friston, C. D. Frith, P. Fletcher, P. F. Liddle, and R. S. Frackowiak "Functional topography: Multidimensional scaling and functional connectivity in the brain," *Cereb. Cortex*, vol. 6, no. 2, pp. 156–164, 1996, doi:10.1093/cercor/6.2.156.

[39] K. J. Friston, O. Josephs, G. Rees and R. Turner, "Nonlinear event-related responses in Fmri," *Magn. Reson. Med.*, vol. 39, no. 1, pp. 41–52, 1998, doi:10.1002/mrm.1910390109.

[40] D. Le Bihan, F. Breton, D. Lallemand, M. L. Aubin, J. Vignaud, and M. Laval-Jeantet, "Separation of diffusion and perfusionin intravoxel incoherent motion MR imaging," *Radiology*, vol. 168, pp. 497–505, 1988.

[41] E. R. Melhem, S. Mori, G. Mukundan, M. A. Kraut, M. G. Pomper, and P. C. van Zijl, "Diffusion tensor MR imaging of the brain and white matter tractography," *Am. J. Radiol.*, vol. 178, pp. 3–16, 2002.

[42] P. J Basser, J. Mattiello, and D. LeBihan, "Estimation of the effective self-diffusion tensor from the NMR spin echo," *Magn. Reson.*, vol. 103, no. 3, pp. 247–254, 1994, doi:10.1006/jmrb.1994.1037.

[43] P. J. Basser, J. Mattiello, and D. LeBihan, "MR diffusion tensor spectroscopy and imaging," *Biophysics*, vol. 66, no. 1, pp. 259–267, 1994.

[44] A. L. Tievsky, T. Ptak, and J. Farkas, "Investigation of apparent diffusion coefficient and diffusion tensor anisotropy in acute and chronic multiple sclerosis lesions," *Neuroradiology*, vol. 20, pp. 1491–1499, 1999.

[45] K. Yamada, O. Kizu, H. Ito, T. Kubota, W. Akada, M. Goto, A. Takada, J. Konishi, H. Sasajima, K. Mineura, S. Mori, and T. Nishimura, "Tractography for arteriovenous malformations near the sensorimotor cortices," *AJNR Neuroradiol.*, vol. 26, no. 3, pp. 598–602, 2005.

[46] S.-K. Lee, D. I. Kim, J. Kim, D. J. Kim, H. D. Kim, D. S. Kim, and S. Mori, "Diffusion-tensor MR imaging and fiber tractography: a new method of describing aberrant fiber connections in developmental CNS anomalies," *Radiographics*, vol. 25, pp. 53–65, 2005, doi:10.1148/rg.251045085.

[47] C. R. Tench, P. S. Morgan, M. Wilson, and L. D. Blumhardt, "White matter mapping using diffusion tensor MRI," *Magn. Reson. Med.*, vol. 47, pp. 967–972, 2002, doi:10.1002/mrm.10144.

[48] T. L. Chenevert, J. A. Brunberg, and J. G. Pipe, "Anisotropic diffusion in human white matter: demonstration with MR techniques in vivo," *Radiology*, vol. 177, no. 2, pp. 401–405, 1990.

[49] R. Turner, D. Le Bihan, J. Maier, R. Vavrek, L. K. Hedges, and J. Pekar, "Echo-planar imaging of intravoxel incoherent motion," *Radiology*, vol. 177, no. 2, pp. 407–414, 1990.

[50] E. O. Stejskal, and J. F. Tanner, "Spin diffusion measurements: spin echos in the presence of a time-dependent field gradient," *Chem. Phys.*, vol. 42, pp. 288–292, 1965, doi:10.1063/1.1695690.

[51] C. Pierpaoli, P. Jezzard, P. J. Basser, A. Barnett, and G. Di Chiro, "Diffusion tensor MR imaging of the human brain," *Radiology*, vol. 201, pp. 637–648, 1996.

[52] M. Neeman, J. P. Freyer, and L. O. Sillerud, "Pulsed-gradient spin-echo studies in NMR imaging. Effects of the imaging gradients on the determination of diffusion coefficients," *J. Magn. Reson.*, vol. 90, pp. 303–312, 1990, doi:10.1016/0022-2364(90)90136-W.

[53] D. Le Bihan, J.-F. Mangin, C. Poupon, C. A. Clark, S. Pappata, N. Molko, and H. Chabriat, "Diffusion tensor imaging: concepts and applications," *Magn. Reson.*, vol. 13, pp. 534–546, 2001, doi:10.1002/jmri.1076.

[54] P. J. Basser, and D. K. Jones, "Diffusion-tensor MRI: theory, experimental design and data analysis—a technical review," *NMR Biomed.*, vol. 15, no. 7–8, pp. 456–467, 2002, doi:10.1002/nbm.783.

[55] Y. Shen, D. J. Larkman, S. Counsell, I. M. Pu, D. Edwards, and J. V. Hajnal, "Correction of high-order eddy current induced geometric distortion in diffusion-weighted echo-planar images," *Magn. Reson. Med.* vol. 52, no. 5, pp. 1184–1189, 2004, doi:10.1002/mrm.20267.

[56] B. Herrnberger, "Diffusionstensor-Magnetresonanztomographie," *Nervenheilkunde*, vol. 23, pp. 50–59, 2004.

[57] P. J. Basser, S. Pajevic, C. Pierpaoli, J. Duda, and A. Aldroubi, "In vivo fiber tractography using DT-MRI data," *Magn. Reson. Med.*, vol. 44, no. 4, pp. 625–632, 2000, doi:10.1002/1522-2594(200010)44:4<625::AID-MRM17>3.0.CO;2-O.

[58] T. E. Conturo, N. F. Lori, T. S. Cull, E. Akbudak, A. Z. Snyder, J. S. Shimony, R. C. McKinstry, H. Burton, and M. E. Raichle, "Tracking neuronal fiber pathways in the living human brain," *Proc. Natl. Acad. Sci. USA*, vol. 96, no. 18, pp. 10422–10427, 1999, doi:10.1073/pnas.96.18.10422.

[59] S. Mori, B. Cain, V. P. Chacko, and P. C. van Zijl, "Three dimensional tracking of axonal projections in the brain by magnetic resonance imaging," *Ann. Neurol.*, vol. 45, pp. 265–269, 1999, doi:10.1002/1531-8249(199902)45:2<265::AID-ANA21>3.0.CO;2-3.

[60] M. Lazar, D. M. Weinstein, J. S. Tsuruda, K. M. Hasan, K. Arfanakis, M. E. Meyerand, B. Badie, H. A. Rowley, V. Haughton, A. Field, and A. L. Alexander, "White matter tractography using diffusion tensor deflection," *Hum. Brain Mapp.* vol. 18, no. 4, pp. 306–321, 2003, doi:10.1002/hbm.10102.

[61] D. K. Jones, A. Simmons, S. C. R. Williams, and M. A. Horsfield, "Noninvasive assessment of axonal fiber connectivity in the human brain via diffusion tensor MRI," *Magn. Reson. Med.*, vol. 42, pp. 37–41, 1999, doi:10.1002/(SICI)1522-2594(199907)42:1<37::AID-MRM7>3.0.CO;2-O.

[62] T. E. Behrens, H. J. Berg, S. Jbabdi, M. F. Rushworth, and M. W. Woolrich, "Probabilistic diffusion tractography with multiple fibre orientations: What can we gain?" *NeuroImage*, vol. 34, no. 1, pp. 144–155, 2007, doi:10.1016/j.neuroimage.2006.09.018.

[63] M. Hämäläinen, R. Hari, R. J. Ilmoniemi, J. Knuutila, and O. V. Lounasmaa, "Magnetoencephalography – theory, instrumentation, and applications to noninvasive

studies of the working human brain," *Rev. Mod. Phys.*, vol. 65, pp. 413–497, 1993, doi:10.1103/RevModPhys.65.413.

[64] K. J. Friston, K. M. Stephan, J. D. Heather, C. D. Frith, A. A. Ioannides, L. C. Liu, M. D. Rugg, J. Vieth, H. Keber, K. Hunter, and R. S. Frackowiak, "A multivariate analysis of evoked responses in EEG and MEG data," *NeuroImage*, vol. 3, pp. 167–174, 1996, doi:10.1006/nimg.1996.0018.

[65] O. Ganslandt, R. Fahlbusch, C. Nimsky, H. Kober, M. Moller, R. Steinmeier, J. Romstock, and J. Vieth, "Functional neuronavigation with magnetoencephalography: Outcome in 50 patients with lesions around the motor cortex," *Neurosurgery*, vol. 91, pp. 73–79, 1999.

[66] J. Kassubek, C. Stippich, P. Sörös, O. Ganslandt, K. Kamada, R. Hopfengärtner, H. Kober, R. Steinmeierv, and J. Vieth, "A motor field source localization protocol using magnetoencephalography," *Biomedizinische Technik.*, vol. 41, Suppl 1, pp. 334–335, 1996.

[67] M. Hämäläinen, and J. Sarvas, "Realistic conductivity geometry model of the human head for interpretation of neuromagnetic data," *IEEE Trans. Biomed. Eng.*, vol. 36, pp. 165–171, 1989, doi:10.1109/10.16463.

[68] D. Geselowitz, "On the magnetic field generated outside an inhomogeneous volume conductor by internal current sources," *IEEE Trans. Magn.*, vol. 6, pp. 346–347, 1970, doi:10.1109/TMAG.1970.1066765.

[69] S. Tissari, J. Rahola, "Error analysis of Galerkin method to solve the forward problem in MEG using the boundary element method," *Comput. Methods Programs Biomed.*, vol. 72, no. 3, pp. 209–222, 2003, doi:10.1016/S0169-2607(02)00144-X.

[70] A. S. Ferguson, X. Zhang and G. Stroink, "A complete linear discretization for calculating the magnetic field using the boundary element method," *IEEE Trans. Biomed. Eng.*, vol. 41, pp. 455–460, 1994, doi:10.1109/10.293220.

[71] J. de Munck, "A Linear discretization of the volume conductor boundary integral equation using analytically integrated elements," *IEEE Trans. Biomed. Eng.*, vol. 39, pp. 986–990, 1992, doi:10.1109/10.256433.

[72] C. Del Gratta, S. N. Erné, and J. Edrich, "A Linear iterative algorithm for dipole localization," in C. Baumgartner, L. Deecke, G. Stroink, and S. J. Williamson, Eds. *Biomagnetism: Fundamental Research and Clinical Applications*. Amsterdam: Elsevier Science; 1995, pp. 313–316.

[73] Polhemus, *3Spacereg Fasttrak User's Manual*. Colchester, Vermont, 1993.

[74] H. Lindenthal, H.-P. Müller, A. Pasquarelli, and S. N. Erné, "Versatile optical co-registration system for multimodal marker localization," *Biomed. Tech. (Berl)*, vol. 48, no. 2, pp. 272–274, 2004.

[75] J. Talairach, and P. Tournoux, *Coplanar Stereotaxic Atlas of the Human Brain*. New York: Thieme Medical; 1988.

[76] D. L. Collins, P. Neeli, T. M. Peters, and A. C. Evans, "Automatic 3D intersubject registration of MR volumetric data in standardized Talairach space," *Comput. Assist. Tomogr.*, vol. 18, no. 2, pp. 192–205, 1994.

[77] M. Brett, I. S. Johnsrude, and A. M. Owen, "The problem of functional localization in the human brain," *Nat. Rev. Neurosci.*, vol. 3, no. 3, pp. 243–249, 2002, doi:10.1038/nrn756.

[78] J. Ashburner, and K. J. Friston, "Nonlinear spatial normalization using basis functions," *Hum Brain Mapp.*, vol. 7, pp. 254–266, 1999, doi:10.1002/(SICI)1097-0193(1999)7:4<254::AID-HBM4>3.0.CO;2-G.

[79] A. M. Dale, A. K. Liu, B. R. Fischl, R. L. Buckner, J. W. Belliveau, J. D. Lewine, and E. Halgren, "Dynamic statistical parametric mapping: Combining fMRI and MEG for high-resolution imaging of cortical activity," *Neuron*, vol. 26, no. 1, pp. 55–67, 2000, doi:10.1016/S0896-6273(00)81138-1.

[80] O. V. Lounasmaa, M. Hämäläinen, R. Hari and R. Salmelin, "Information processing in the human brain: magnetoencephalographic approach," *Proc. Natl. Acad. Sci. USA*, vol. 93, no. 17, pp. 8809–8815, 1996, doi:10.1073/pnas.93.17.8809.

[81] K. K. Kwong, J. W. Belliveau, D. A. Chesler, I. E. Goldberg, R. M. Weisskoff, B. P. Poncelet, D. N. Kennedy, B. E. Hoppel, M. S. Cohen, R. Turner, et al., "Dynamic magnetic resonance imaging of human brain activity during primary sensory stimulation," *Proc. Natl. Acad. Sci. USA*, 1992, vol. 89, pp. 5675–5679, doi:10.1073/pnas.89.12.5675.

[82] A. Riecker, W. Grodd, U. Klose, J. B. Schulz, K. Gröschel, M. Erb, H. Ackermann, and A. Kastrup, "Relation between regional functional MRI activation and vscular reactivity to carbon dioxide during normal aging," *Cereb. Blood Flow Metab.*, vol. 23, no. 5, pp. 565–573, 2003, doi:10.1097/01.WCB.0000056063.25434.04.

[83] H-P Müller, I. de Cesaris, M. de Melis, L. Marzetti, A. Pasquarelli S. N. Erné, A. C. Ludolph, and J. Kassubek, "Open magnetic and electric graphic analysis: comprehensive magnetoencephalographic and functional magnetic resonance imaging in one single software environment," *IEEE Eng. Med. Biol. Mag.*, vol. 24, no. 3, pp. 109–116, 2005.

[84] S. N. Erné, A. Pasquarelli, H. Kammrath, S. Della Penna, K. Torquati, V. Pizzella, R. Rossi, C. Granata, and M. Russo, "*Argos 55—The New MCG System in Ulm*," in T. Yoshimoto, M. Kotani, S. Kuriki, H. Karibe, N. Nakasato, Eds. *Recent Advances in Biomagnetism. BIOMAG'98*, pp. 27–30, 1998.

[85] A. Pasquarelli, H. Kammrath, U. Tenner, and S. N. Erné, "The New Ulm Magnetically Shielded Room," in T. Yoshimoto, M. Kotani, S. Kuriki, H. Karibe, and N. Nakasato, Eds. *Recent Advances in Biomagnetism: BIOMAG'98*, 1998, pp. 55–58.

[86] H. Lindenthal, H.-P. Müller, A. Pasquarelli, and S. N. Erné, "OptiCoS – a new optical co-registration system for multimodal integration," in *Proc. NFSI* 2001, Innsbruck, Austria, 2001, pp. 127–129.

[87] H.-P. Müller, E. Kraft, A. C. Ludolph, and S. N. Erné, "New analysis methods in fMRI analysis: hierarchical cluster analysis for improved signal-to-noise ratio compared to standard techniques," *IEEE Eng. Med. Biol. Mag.*, vol. 21, no. 5, pp. 134–142, 2002.

[88] J. Kassubek, F. D. Juengling, A. Baumgartner, A. C. Ludolph, and A.-D. Sperfeld, "Different regional brain volume loss in pure and complicated hereditary spastic paraparesis: a voxel-based morphometry study, Amyotrophic Leteral Sclerosis 2007," in press.

[89] S. Dreha-Kulaczewski, P. Dechent, G. Helms, J. Frahm, J. Gärtnerand K. Brockmann, "Cerebral metabolic and structural alterations in hereditary spastic paraplegia with thin corpus callosum assessed by MRS and DTI," *Neuroradiology*, vol. 48, no. 12, pp. 893–898, 2006, doi:10.1007/s00234-006-0148-2.

[90] H-P Müller, A. Unrath, A. C. Ludolph AC, and J. Kassubek, "Preservation of diffusion tensor properties during spatial normalization by use of tensor imaging and fiber tracking on a normal brain database," *Phys. Med. Biol.*, vol. 52, no. 6, pp. N99–N109, 2007, doi:10.1088/0031-9155/52/6/N01.

[91] T. E. Behrens, H. Johansen-Berg, M. W. Woolrich, S. M. Smith, C. A. Wheeler-Kingshott, P. A. Boulby, G. J. Barker, E. L. Sillery, K. Sheehan, O. Ciccarelli, A. J. Thompson, J. M. Brady, and P. M. Matthews, "Noninvasive mapping of connections between human thalamus and cortex using diffusion imaging," *Nat. Neurosci.*, vol. 6, no. 7, pp. 750–757, 2003, doi:10.1038/nn1075.

[92] R. A. Kanaan, S. S. Shergill, G. J. Barker, M. Catani, V. W. Ng, R. Howard, P. K. McGuire, and D. K. Jones, "Tract-specific anisotropy measurements in diffusion tensor imaging," *Psychiatry Res.*, vol. 146, no. 1, pp. 73–82, 2006, doi:10.1016/j.pscychresns.2005.11.002.

[93] D. C. Alexander, C. Pierpaoli, P. J. Basser, and J. C. Gee, "Spatial transformations of diffusion tensor magnetic resonance images," *IEEE Trans. Med. Imaging*, vol. 20, no. 11, pp. 1131–1139, 2001, doi:10.1109/42.963816.

[94] H. J. Park, M. Kubicki, M. E. Shenton, A. Guimond, R. W. McCarley, S. E. Maier, R. Kikinis, F. A. Jolesz, and C. F. Westin CF, "Spatial normalization of diffusion tensor MRI using multiple channels," *NeuroImage*, vol. 20, no. 4, pp. 1995–2009, 2003, doi:10.1016/j.neuroimage.2003.08.008.

[95] H. J. Park, C. F. Westin, M. Kubicki, S. E. Maier, M. Niznikiewicz, A. Baer, M. Frumin, R. Kikinis, F. A. Jolesz, R. W. McCarley, and M. E. Shenton, "White matter hemisphere asymmetries in healthy subjects and in schizophrenia: a diffusion tensor MRI study," *NeuroImage*, vol. 23, no. 1, pp. 213–223, 2004, doi:10.1016/j.neuroimage.2004.04.036.

[96] S. M. Smith, M. Jenkinson, H. Johansen-Berg, D. Rueckert, T. E. Nichols, C. E. Mackay, K. E. Watkins, O. Ciccarelli, M. Z. Cader, P. M. Matthews, and T. E. Behrens, "Tract-based spatial statistics: Voxelwise analysis of multi-subject diffusion data," *NeuroImage*, vol. 31, no. 4, pp. 1487–505, 2006, doi:10.1016/j.neuroimage.2006.02.024.

[97] S. M. Smith, H. Johansen-Berg, M. Jenkinson, D. Rueckert, T. E. Nichols, K. L. Miller, M. D. Robson, D. K. Jones, J. C. Klein, A. J. Bartsch, and T. E. Behrens, "Acquisition and voxelwise analysis of multi-subject diffusion data with tract-based spatial statistics," *Nat. Protoc.*, vol. 2, no. 3, pp. 499–503, 2007, doi:10.1038/nprot.2007.45.

[98] A. Mechelli, C. J. Price, K. J. Friston, and J. Ashburner, "Voxel-based morphometry of the human brain: methods and applications," *Curr. Med. Imaging Rev.* vol. 1, pp. 105–113, 2005, doi:10.2174/1573405054038726.

[99] C. D. Good, J. Ashburner, and R. S. Frackowiak, "Computational neuroanatomy: New perspectives for neuroradiology," *Rev. Neurol.*, vol. 157, pp. 797–806, 2001.

[100] C. D. Good, I. S. Johnsrude, J. Ashburner, R. N. Henson, K. J. Friston, and R. S. Frackowiak, "A voxel-based morphometric study of ageing in 465 normal adult human brains," *NeuroImage*, vol. 14, pp. 21–36, 2001, doi:10.1006/nimg.2001.0786.

[101] J. Ashburner, and K. J. Friston, "Voxel-based morphometry – the methods," *NeuroImage*, vol. 11, pp. 805–821, 2000, doi:10.1006/nimg.2000.0582.

[102] J. Ashburner, and K. Friston, "Multimodal image coregistration and partitioning – a unified framework," *NeuroImage*, vol. 6, no. 3, pp. 209–217, 1997, doi:10.1006/nimg.1997.0290.

[103] J. Ashburner, and K. J. Friston, "Why voxel-based morphometry should be used?" *NeuroImage*, vol. 14, no. 6, pp. 1238–1243, Dec. 2001, doi:10.1006/nimg.2001.0961.

[104] R. Peyron, M. Frot, F. Schneider, L. Garcia-Larrea, P. Mertens, F. G Barral, M. Sindou, B. Laurent, and F. Mauguiere, "Role of operculoinsular cortices in human pain processing: converging evidence from PET, fMRI, dipole modeling, and intracerebral recordings of evoked potentials," *NeuroImage*, vol. 17, no. 3, pp. 1336–1346, 2002, doi:10.1006/nimg.2002.1315.

[105] P. A. Valdes-Sosa, R. Kotter, and K. J. Friston, "Introduction: Multimodal neuroimaging of brain connectivity," *Philos. Trans. R. Soc. Lond. Biol. Sci.*, vol. 360, no. 1457, pp. 865–867, 2005, doi:10.1098/rstb.2005.1655.

[106] J. Ashburner, J. G. Csernansky, C. Davatzikos, N. C. Fox, G. B. Frisoni, and P. M. Thompson, "Computer-assisted imaging to assess brain structure in healthy and diseased brains," *Lancet Neurol.*, vol. 2, pp. 79–88, 2003, doi:10.1016/S1474-4422(03)00304-1.

[107] S. R. Felber "Magnetic resonance in the differential diagnosis of dementia," *Neural. Transm.*, vol. 109, no. 7–8, pp. 1045–1051, 2002, doi:10.1007/s007020200088.

[108] W. L. Au, J. R. Adams, A. Troiano, and A. J. Stoessl, "Neuroimaging in Parkinson's disease," *Neural. Transm. Suppl.*, vol. 70, pp. 241–248, 2006.

[109] S. Kalra, and D. Arnold, "neuroimaging in amyotrophic lateral sclerosis," *Amyotroph. Lateral Scler. Motor Neuron. Disord.*, vol. 4, no. 4, pp. 243–248, 2003, doi:10.1080/14660820310011269.

[110] P. Bartenstein, "PET in neuroscience: dopaminergic, GABA/benzodiazepine, and opiate system." *Nuklearmedizin*, vol. 43, no. 1, pp. 33–42, 2004.

[111] M. Filippi and M. A. Rocca, "Conventional MRI in multiple sclerosis," *NeuroImaging*, vol. 17, Suppl. 1, pp. 3S–9S, 2007.

[112] M. Rovaris, and M. Filippi, "Diffusion tensor MRI in multiple sclerosis," *NeuroImaging*, vol. 17, Suppl 1, pp. 27S–30S, 2007.

[113] N. De Stefano, and M. Filippi, "MR spectroscopy in multiple sclerosis," *J NeuroImaging*, vol. 17, Suppl. 1, pp. 31S–35S, 2007.

[114] M. A. Rocca and M. Filippi, "Functional MRI in multiple sclerosis," *NeuroImaging*, vol. 17, Suppl 1, pp. 36S–41S, 2007.

[115] M. Filippi and F. Agosta, "Magnetization transfer MRI in multiple sclerosis," *NeuroImaging*, vol. 17, Suppl. 1, pp. 22S–26S, 2007.

[116] C. Weiller, A. May, M. Sach, C. Buhmann, and M. Rijntjes, "Role of functional imaging in neurological disorders," *Magn. Reson. Imaging*, vol. 23, no. 6, pp. 840–850, 2006, doi:10.1002/jmri.20591.

[117] K. Kamada, M. Moller, M. Saguer, J. Kassubek, M. Kaltenhauser, H. Kober, M. Uberall, H. Lauffer, D. Wenzel, and J. Vieth, "Localization analysis of neuronal activities in benign rolandic epilepsy using magnetoencephalography," *Neurol. Sci.*, vol. 154, no. 2, pp. 164–172, 1998, doi:10.1016/S0022-510X(97)00226-8.

[118] F. Babiloni, D. Mattia, C. Babiloni, L. Astolfi, S. Salinari, A. Basilisco, P. M. Rossini, M. G. Marciani, and F. Cincotti, "Multimodal integration of EEG, MEG, and fMRI data for the solution of the NeuroImage puzzle," *Magn. Reson. Imaging*, vol. 22, no. 10, pp. 1471–1476, 2004, doi:10.1016/j.mri.2004.10.007.

[119] B. D. Gonsalves, I. Kahn, T. Curran, K. A. Norman, and A. D. Wagner, "Memory strength and repetition suppression: Multimodal imaging of medial temporal cortical contributions to recognition," *Neuron*, vol. 47, no. 5, pp. 751–761, 2005, doi:10.1016/j.neuron.2005.07.013.

[120] C. H. Im, H. K. Jung, and N. Fujimaki, "fMRI-constrained MEG source imaging and consideration of fMRI invisible sources," *Hum. Brain Mapp.*, vol. 26, no. 2, pp. 110–118, 2005, doi:10.1002/hbm.20143.

[121] P. Jannin, O. J. Fleig, E. Seigneuret, C. Grova, X. Morandi, and J. M. Scarabin, "A data fusion environment for multimodal and multi-informational neuronavigation," *Comput. Aided Surg.*, vol. 5, no. 1, pp. 1–10, 2000, doi:10.1002/(SICI)1097-0150(2000)5:1<1::AID-IGS1>3.0.CO;2-4.

[122] P. Jannin, X. Morandi, O. J. Fleig, E. Le Rumeur, P. Toulouse, B. Gibaud, and J. M. Scarabin, "Integration of sulcal and functional information for multimodal neuronavigation," *Neurosurgery*, vol. 96, no. 4, pp. 713–723, 2002.

[123] O. Ganslandt, R. Fahlbusch, H. Kober, J. Gralla, and C. Nimsky, "Use of magnetoencephalography and functional neuronavigation in planning and surgery of brain tumors," *Nervenarzt*, vol. 73, pp. 155–161, 2002.

[124] H. Hund, A. R. Rezai, E. Kronberg, J. Cappell, M. Zonenshayn, U. Ribary, P. J. Kelly, and R. Llinas, "Magnetoencephalographic mapping: Basic of a new functional risk profile in the selection of patients with cortical brain lesions," *Neurosurgery*, vol. 40, pp. 936–942, 1997, doi:10.1097/00006123-199705000-00011.

[125] O. Ganslandt, R. Steinmeier, H. Kober, J. Vieth, J. Kassubek, J. Romstock, C. Strauss and R. Fahlbusch, "Magnetic source imaging combined with image-guided frameless stereotaxy: A new method in surgery around the motor strip," *Neurosurgery*, vol. 41, pp. 621–627, 1997, doi:10.1097/00006123-199709000-00023.

[126] O. Ganslandt, M. Buchfelder, P. Hastreiter, P. Grummich, R. Fahlbusch, and C. Nimsky, "Magnetic source imaging supports clinical decision making in glioma patients." *Clin. Neurol. Neurosurg.*, vol. 107, pp. 20–26, 2004, doi:10.1016/j.clineuro.2004.02.027.

[127] P. Ossenblok, F. S. Leijten FS, J. de Munck, G. J. Huiskamp, F. Barkhof, and P. Boon, "Magnetic source imaging contributes to the presurgical identification of sensorimotor cortex in patients with frontal lobe epilepsy," *Clin. Neurophysiol.*, vol. 114, pp. 221–232, 2003, doi:10.1016/S1388-2457(02)00369-3.

[128] J. S. Ebersole, "Magnetoencephalography/magnetic source imaging in the assessment of patients with epilepsy," *Epilepsia*, vol. 38, Suppl 4, pp. S1–S5, 1997, doi:10.1111/j.1528-1157.1997.tb04533.x.

[129] M. J. Fischer, G. Scheler, and H. Stefan, "Utilization of magnetoencephalography results to obtain favourable outcomes in epilepsy surgery," *Brain*, vol. 128, pp. 153–157, 2005, doi:10.1093/brain/awh333.

[130] H. Stefan, C. Hummel, G. Scheler, A. Genow, K. Druschky, C. Tilz, M. Kaltenhauser, R. Hopfengartner, M. Buchfelder, and J. Romstock, "Magnetic brain source imaging of focal epileptic activity: a synopsis of 455 cases," *Brain*, vol. 126, pp. 2396–2405, 2003, doi:10.1093/brain/awg239.

[131] W. J. Marks Jr., "Utility of MEG in presurgical localization," *Epilepsy Curr.*, vol. 4, pp. 208–209, 2004, doi:10.1111/j.1535-7597.2004.04516.x.

[132] R. Firsching, I. Bondar, H. J. Heinze, H. Hinrichs, T. Hagner, J. Heinrich, and A. Belau, "Practicability of magnetoencephalography-guided neuronavigation," *Neurosurg. Rev.*, vol. 25, pp. 73–78, 2002, doi:10.1007/s101430100161.

[133] A. M. Dale, and E. Halgren, "Spatiotemporal mapping of brain activity by integration of multiple imaging modalities," *Curr. Opin. Neurobiol.*, vol. 11, pp. 202–208, 2001, doi:10.1016/S0959-4388(00)00197-5.

[134] B. A. Assaf, K. M. Karkar, K. D. Laxer, P. A. Garcia, E. J. Austin, N. M. Barbaro, and M. J. Aminoff, "Magnetoencephalography source localization and surgical outcome in temporal lobe epilepsy," *Clin. Neurophysiol.*, vol. 115, pp. 2066–2076, 2004, doi:10.1016/j.clinph.2004.04.020.

[135] F. Duffner, D. Freudenstein, H. Schiffbauer, H. Preissl, R. Siekmann, N. Birbaumer, and E. H. Grote, "Combining MEG and MRI with neuronavigation for treatment of an epileptiform spike focus in the precentral region: a technical case report," *Surg. Neurol.*, vol. 59, pp. 40–45, 2003, doi:10.1016/S0090-3019(02)00972-2.

[136] S. Knake, P. E. Grant, S. M. Stufflebeam, L. L. Wald, H. Shiraishi, F. Rosenow, D. L. Schomer, B. Fischl, A. M. Dale, and E. Halgren, "Aids to telemetry in the presurgical evaluation of epilepsy patients: MRI, MEG and other noninvasive imaging techniques," *Suppl. Clin. Neurophysiol.*, vol. 57, pp. 494–502, 2004.

[137] K. Kamada, T. Todo, Y. Masutani, S. Aoki, K. Ino, A. Morita, and N. Saito, "Visualization of the frontotemporal language fibers by tractography combined with functional magnetic resonance imaging and magnetoencephalography," *Neurosurgery*, vol. 106, no. 1, pp. 90–98, 2007.

[138] C. Ramon, J. Haueisen, T. Richards, and K. Maravilla "Multimodal imaging of somatosensory evoked cortical activity," *Neurol. Clin. Neurophysiol.*, vol. 96, 2004.

[139] A. Stadlbauer, C. Nimsky, S. Gruber, E. Moser, T. Hammen, T. Engelhorn, M. Buchfelder, and O. Ganslandt, "Changes in fiber integrity, diffusivity, and metabolism of the pyramidal tract adjacent to gliomas: a quantitative diffusion tensor fiber tracking and MR spectroscopic imaging study," *AJNR Am. J. Neuroradiol.*, vol. 28, no. 3, pp. 462–469, 2007.

[140] M. P. Richardson, K. J. Friston, S. M. Sisodiya, M. J. Koepp, JAshburner, S. L. Free, D. J. Brooks, and J. S. Duncan, "Cortical grey matter and benzodiazepine receptors in malformations of cortical development. A voxel-based comparison of structural and functional imaging data." *Brain*, vol. 120, pp. 1961–1973, 1997, doi:10.1093/brain/120.11.1961.

[141] J. Kassubek, H. J. Huppertz, J. Spreer, and A. Schulze-Bonhage, "Detection and localization of focal cortical dysplasia by voxel-based 3-D MRI analysis," *Epilepsia*, vol. 43, pp. 596–602, 2002, doi:10.1046/j.1528-1157.2002.41401.x.

[142] S. B. Antel, A. Bernasconi, N. Bernasconi, D. L. Collins, R. E. Kearney, R. Shinghal, and D. L. Arnold, "Computational models of MRI characteristics of focal cortical dysplasia improve lesion detection," *NeuroImage*, vol. 17, pp. 1755–1760, 2002, doi:10.1006/nimg.2002.1312.

[143] H. J. Huppertz, C. Grimm, S. Fauser, J. Kassubek, I. Mader, A. Hochmuth, J. Spreer, and A. Schulze-Bonhage, "Enhanced visualization of blurred gray-white matter junctions in focal cortical dysplasia by voxel-based 3D MRI analysis," *Epilepsy Res.*, vol. 67, pp. 35–50, 2005, doi:10.1016/j.eplepsyres.2005.07.009.

[144] J. Kassubek, F. D. Juengling, and A. Otte, "Improved interpretation of cerebral PET with respect to structural MRI changes using combined voxel-based statistical mapping," *Eur. J. Nucl. Med.*, vol. 27, pp. 880, 2000.

[145] J. H. Cross, "Functional neuroimaging of malformations of cortical development, " *Epileptic Disord*, vol. 5, Suppl 2, pp. S73–S80, 2003.

[146] R. Wiest, J. Kassubek, K. Schindler, T. J. Loher, C. Kiefer, L. Mariani, M. Wissmeyer, G. Schroth, J. Mathis, B. Weder, and F. D. Juengling, "Comparison of voxel-based 3-D MRI analysis and subtraction ictal SPECT coregistered to MRI in focal epilepsy," *Epilepsy Res.* vol. 65, pp. 125–133, 2005, doi:10.1016/j.eplepsyres.2005.05.002.

[147] F. D. Juengling, J. Kassubek, H. J. Huppertz, T. Krause, and T. Els, "Separating functional and structural damage in persistent vegetative state using combined voxel-based analysis of 3-D MRI and FDG-PET," *Neurol. Sci.*, vol. 228, no. 2, pp. 179–184, 2005, doi:10.1016/j.jns.2004.11.052.

[148] J. Daunizeau, C. Grova, G. Marrelec, J. Mattout, S. Jbabdi, M. Pelegrini-Issac, J. M. Lina, and H. Benali, "Symmetrical event-related EEG/fMRI information fusion in a variational Bayesian framework," *NeuroImage*, vol. 36, no. 1, pp. 69–87, 2007, doi:10.1016/j.neuroimage.2007.01.044.

[149] B. Fabio, B. Claudio, C. Filippo, R. Paolo Maria, B. Alessandra, A. Laura, and C. Febo, "Multimodal integration of EEG and functional magnetic resonance recordings," in *Proc Conf IEEE Eng. Med. Biol. Soc.*, vol. 7, pp. 5311–5314, 2004, doi:10.1109/IEMBS.2004.1404483.

[150] P. Ritter and A. Villringer, "Simultaneous EEG-Fmri," *Neurosci. Biobehav. Rev.*, vol. 30, no. 6, pp. 823–838, 2006, doi:10.1016/j.neubiorev.2006.06.008.

[151] F. S. Leijten, G. J. Huiskamp, I. Hilgersom, and A. C. Van Huffelen "High-resolution source imaging in mesiotemporal lobe epilepsy: a comparison between MEG and simultaneous EEG," *Clin. Neurophysiol.*, vol. 20, pp. 227–238, 2003, doi:10.1097/00004691-200307000-00001.

[152] J. Gotman, E. Kobayashi, A. P. Bagshaw, C. G. Benar, and F. Dubeau, "Combining EEG and fMRI: A multimodal tool for epilepsy research," *Magn. Reson. Imaging*, vol. 23, no. 6, pp. 906–920, 2006, doi:10.1002/jmri.20577.

[153] G. Lantz, L. Spinelli, R. G. Menendez, M. Seeck, and C. M. Michel, "Localization of distributed sources and comparison with functional MRI," *Epileptic Disord.* (*Special Issue*), pp. 45–58, 2001.

[154] T. Hammen, M. Schwarz, M. Doelken, F. Kerling, T. Engelhorn, A. Stadlbauer, O. Ganslandt, C. Nimsky, A. Doerfler, and H. Stefan, "1H-NMR spectroscopy indicates severity markers in temporal lobe epilepsy: Correlations between metabolic alterations, seizures, and epileptic discharges in EEG," *Epilepsia*, vol. 48, no. 2, pp. 263–269, 2007, doi:10.1111/j.1528-1167.2006.00856.x.

Author Biography

Dr. rer. nat. Hans-Peter Müller is currently working as researcher at the Department of Neurology, University of Ulm, Germany. He formerly worked at Central Institute for Biomedical Engineering, University of Ulm, Germany and at Physikalisch-Technische Bundesanstalt, Berlin, Germany. He finished his Ph.D. in 1996 in magnetic resonance microscopy. Main areas of interest include software development for magnetic resonance imaging (diffusion tensor imaging as well as functional magnetic resonance imaging) and magnetoencephalography.

Prof. Dr. med. Jan Kassubek is currently working as senior clinician and researcher at the Department of Neurology, University of Ulm, Germany and has the position as the Vice Chairman. He finished his M.D. in 1997. He formerly worked as Medical Doctor/Researcher at the Division of Experimental Neuropsychiatry, Department of Neurology, University of Erlangen, Germany, then at the Center for Magnetic Resonance Research, University of Minnesota Medical School, Minneapolis, MN, USA, and at the Department of Neurology, University of Freiburg, Germany. Main areas of interest are clinical research on neuroimaging, in particular magnetic resonance imaging and magnetoencephalography, with a focus on neurodegenerative diseases.

Estimation of Cortical Connectivity in Humans: Advanced Signal Processing Techniques

Laura Astolfi and Fabio Babiloni
Università degli Studi di Roma La Sapienza Italy

MORGAN & CLAYPOOL PUBLISHERS

ABSTRACT

In the last ten years many different brain imaging devices have conveyed a lot of information about the brain functioning in different experimental conditions. In every case, the biomedical engineers, together with mathematicians, physicists and physicians are called to elaborate the signals related to the brain activity in order to extract meaningful and robust information to correlate with the external behaviour of the subjects. In such attempt, different signal processing tools used in telecommunications and other field of engineering or even social sciences have been adapted and re-used in the neuroscience field. The present book would like to offer a short presentation of several methods for the estimation of the cortical connectivity of the human brain. The methods here presented are relatively simply to implement, robust and can return valuable information about the causality of the activation of the different cortical areas in humans using non invasive electroencephalographic recordings. The knowledge of such signal processing tools will enrich the arsenal of the computational methods that a engineer or a mathematician could apply in the processing of brain signals.

KEYWORDS

Multivariate autoregressive model (MVAR), High resolution EEG, Directed Transfer Function (DTF), Partial Directed Coherence (PDC), Realistic head modeling, cortical imaging

Contents

Foreword

This book describes some advanced mathematical signal processing techniques applied to the estimation of the cortical connectivity in humans from non invasive electroencephalographic recordings. Although it may be thought that mathematics could not be the proper tool for a full comprehension of the brain functions, this is often not the case. In the last 10 years many different brain-imaging devices have conveyed a lot of information about the brain functioning in different experimental conditions. In every case, the biomedical engineers, together with mathematicians, physicists, and physicians are called in to elaborate the signals related to the brain activity in order to extract meaningful and robust information to correlate with the external behavior of the subjects. In such attempts, different signal processing tools used in telecommunications and other fields of engineering or even social sciences have been adapted and reused in the neuroscience field.

The present book offers a short presentation of several methods for the estimation of the cortical connectivity of the human brain. The methods presented here are relatively simple to implement, robust, and can return valuable information about the causality of the activation of the different cortical areas in humans using noninvasive electroencephalographic recordings. For these reasons, they are among the most used tools in the modern neuroscientific research. The knowledge of such signal processing tools will enrich the arsenal of the computational methods that an engineer or a mathematician could apply in the processing of brain signals. In our understanding, a comprehensive presentation of these connectivity estimation methods is lacking in literature and this book could represent a good step in filling the void. With this book we would like to be in agreement with a citation of the famous physicist Ludwig Boltzmann, "There is nothing more practical than a good theory". We think that this is still true a hundred years after him.

Laura Astolfi
Fabio Babiloni

INTRODUCTION

In recent years, great advancements have been made in understanding the mechanisms of the functioning of the human brain. Technological developments such as functional magnetic resonance imaging (fMRI), positron emission tomography (PET), and magnetoencephalography (MEG) have made possible the mapping of the images of cerebral activity from hemodynamic, metabolic or electromagnetic measurements. Among these brain imaging techniques, electroencephalograpy (EEG) is unique in terms of simplicity, accessibility, and temporal resolution, and has been viewed with renewed interest in recent years, thanks to the use of advanced methods of analysis and interpretation of its data. These methods are able to improve the spatial resolution of conventional EEG, making it possible to address the analysis of the brain activity in a noninvasive way using the temporal resolution of brain phenomena (of the order of milliseconds). With high-resolution EEG, it is now possible to obtain cortical activation maps describing the activity of the brain at the cortical level during the execution of a given experimental task.

Simple imaging of regions of the brain activated during particular tasks does not, however, convey the information about how these regions communicate with each other for making the execution of the task possible. The concept of brain connectivity is viewed as central to the understanding of the organized behavior of cortical regions, beyond the simple mapping of their activities [1,2]. Such behavior is thought to be based on the interaction between different cortical sites and differently specialized ones. Cortical connectivity estimation aims to describe these interactions in connectivity patterns, which hold the direction and strength of the information flow between cortical areas. To this purpose, several methods have been developed and applied to data gathered from hemodynamic and electromagnetic techniques [3–7]. Two main definitions of brain connectivity have been proposed during recent years: *functional* and *effective* connectivity [8]. Functional connectivity is defined as the temporal correlation between spatially remote neurophysiologic events. Effective connectivity is defined as the simplest brain circuit which would produce the same temporal relationship between cortical sites as observed experimentally.

As for the functional connectivity, the methods proposed in literature typically involve estimation of some covariance properties between different time series. These properties are measured from different spatial sites during motor and cognitive tasks by EEG and fMRI techniques [4,5,7,9].

Structural equation modeling (SEM) is a technique that has been used recently to assess the connectivity between cortical areas in humans from hemodynamic and metabolic measurements [3,10–12]. The basic idea of SEM considers the covariance structure of the data [10]. The estimation of effective cortical connectivity obtained from fMRI data has, however, a low temporal resolution (of the order of seconds) which is far from the time scale in which the brain normally operates. Hence, it is interesting to know whether the SEM technique can be applied to cortical activity which are obtained by applying linear inverse techniques to high-resolution EEG data [5,13–15].

As important information in the EEG signals are coded in frequency domain rather than time domain (reviewed in [16]), attention was focused on detecting frequency-specific interactions in EEG or MEG signals; for instance, the coherence between the activity of pairs of channels [17–19]. However, coherence analysis does not have a directional nature (i.e., it just examines whether a link exists between two neural structures by describing instances when they are in synchronous activity) and does not provide directly the direction of the information flow. In this respect, multivariate spectral techniques such as directed transfer function (DTF) or partial directed coherence (PDC) were proposed [20,21] for determining the directional influences between any given pair of channels in a multivariate data set. Both DTF and PDC [21,22] rely on the key concept of Granger causality between time series [23], according to which an observed time series $x(n)$ causes another series $y(n)$ if the knowledge of $x(n)$'s past significantly improves the prediction of $y(n)$; this relation between time series is not reciprocal, i.e., $x(n)$ may cause $y(n)$ without $y(n)$ necessarily causing $x(n)$. This lack of reciprocity allows the evaluation of the direction of information flow between structures. These estimators are able to characterize at the same time both the directional and spectral properties of the brain signals, and they require only one multivariate autoregressive (MVAR) model estimated from all the EEG channels. The advantages of MVAR modeling of multichannel EEG signals were stressed recently [24] by demonstrating the advantages of multivariate methods with respect to the pairwise autoregressive approach, in terms of both accuracy and computational cost. In order to fully characterize the techniques presented we test them on simulated EEG data whose connectivity characteristics are known in advance, and then we finally apply such methods to human data obtained from high resolution EEG recordings. As a novelty, the application of all the proposed methodologies (SEM, DTF, and PDC) was performed using the cortical signals estimated from high-resolution EEG recordings which exhibit a higher spatial resolution than conventional cerebral electromagnetic measurements. To correctly estimate the cortical signals we used multicompartment head models (scalp, skull, dura mater, and cortex) constructed from individual MRI, a distributed source model, and a regularized linear inverse source estimates of cortical current density.

Chapters I–III discuss the simulation studies in which different main factors (signal-to-noise ratio, cortical activity duration, frequency band, etc.) are systematically imposed in the generation of test signals, and the errors in the estimated connectivity are evaluated by analysis of variance (ANOVA). In particular, we first explore the behavior of the most advanced estimators of effective and functional connectivity – SEM, DTF, dDTF, and PDC – in a simulation context and under different practical conditions.

For SEM, which involves the definition of an a priori connectivity model, the simulation study is designed to answer the following questions:

1. What is the influence of a variable signal-to-noise ratio (SNR) level on the accuracy of the pattern connectivity estimation obtained by SEM?
2. What is the amount of data necessary to get good accuracy of the estimation of connectivity between cortical areas?
3. How are the SEM performances degraded by an imprecise anatomical model formulation? Is it able to perform a good estimation of connectivity pattern when connections between the cortical areas are not correctly assumed? Which kind of errors should be avoided?

For the three multivariate estimators of functional connectivity – DTF, dDTF, and PDC – the experimental design focused on the following questions:

1. How are the connectivity pattern estimators influenced by different factors affecting the EEG recordings such as the signal-to-noise ratio and the amount of data available?
2. How do the estimators discriminate between the direct or indirect causality patterns?
3. What is the most effective method for estimating a connectivity model under the conditions usually encountered in standard EEG recordings?

These questions are addressed via simulations using predefined connectivity schemes linking several cortical areas. The estimation process retrieves the cortical connections between the areas under different experimental conditions. The connectivity patterns estimated by the four techniques are compared with those imposed on the simulated signals, and different error measures are computed and subjected to statistical multivariate analysis. The statistical analysis showed that during the simulations, SEM, DTF, and PDC estimators are able to estimate the imposed connectivity patterns under reasonable operating conditions. It was possible to conclude that the estimation of cortical connectivity can be performed not only with hemodynamic measurements, but also with EEG signals obtained from advanced computational techniques.

After giving a full description of the properties of these connectivity estimators for high resolution EEG recordings, the results of their application to human data relating to different experimental tasks such as finger tapping, Stroop test, and movement imagination are discussed. After the simulation tests, SEM, DTF, and PDC are applied to different sets of experimental data relating to motor and cognitive tasks (Chapters IV–VI). The motor task examined is a fast repetitive finger tapping, while the cognitive task involved recordings during the Stroop test, often employed in studies of selective attention, and found to be sensitive to prefrontal damage. The data employed are cortical estimates obtained from high-resolution EEG recordings using very advanced techniques which exhibit a higher spatial resolution than conventional cerebral electromagnetic measurements. We also briefly describe the high-resolution EEG techniques, including the use of a large number of scalp electrodes, which are realistic models of the head derived from structural magnetic resonance images (MRIs), and advanced processing methodologies related to solutions of linear inverse problems. The results of the estimation of the effective and functional connectivity from data recorded during finger tapping test, Stroop test and movement imagination test are presented.

One of the possible problems in the approach presented above is the hytpothesis that the gathered EEG data are stationary. However, this property of the data cannot be easily assumed. In fact, under this hypothesis, the methodology proposed for the DTF and PDC techniques is valid. Then the connectivity estimation methods for dealing with nonstationary EEG data are highly desirable. In Chapter VII we propose a methodology for the estimation of cortical connectivity extended to the time–frequency domain, based on the use of adaptive multivariate models. Such an approach allows extention of the connectivity analysis to nonstationary data and monitoring the rapid changes in the connectivity between cortical areas during an experimental task. The performances of the time-varying estimators are tested by means of simulations performed on the basis of a predefined connectivity scheme linking different cortical areas. Cortical connections between the areas are retrieved by the estimation process under different experimental conditions, and the results obtained for the different methods are evaluated by statistical analyses.

Finally, as an example of the results that can be obtained by this technique, the application of the simulation study to real data is proposed in Chapter VIII. For this purpose, we applied the time-varying technique to the cortical activity estimated in particular regions of interest (ROIs) of the cortex, and obtained high-resolution EEG recordings during the execution of a combined foot–lips movement in a group of normal subjects.

The experimental data presented here as practical results of the estimation of cortical connectivity in humans during motor and cognitive tasks were obtained from the Laboratory of High Resolution EEG of the University of Rome "La Sapienza" and from the Laboratory of Neuroelectric Imaging and Brain Computer Interface of the S. Lucia Foundation. In addition, part of the experimental data employed were provided by the Department of Biomedical Engineering, University of Minnesota, Minneapolis, USA, and by the Department of Psychology and Beckman Institute Biomedical Imaging Center, University of Illinois at Urbana-Champaign, Illinois, USA, in the framework of a scientific cooperation with the University of Rome.

REFERENCES

[1] L. Lee, L. M. Harrison, and A. Mechelli, "The functional brain connectivity workshop: Report and commentary," *NeuroImage*, vol. 19, pp. 457–465, 2003.

[2] B. Horwitz, "The elusive concept of brain connectivity," *NeuroImage*, vol. 19, pp. 466–470, 2003, doi:10.1016/S1053-8119(03)00112-5.

[3] C. Buchel and K. J. Friston, "Modulation of connectivity in visual pathways by attention: Cortical interactions evaluated with structural equation modeling and fMRI," *Cereb. Cortex*, vol. 7, no. 8, pp. 768–778, 1997, doi:10.1093/cercor/7.8.768.

[4] A. Urbano, C. Babiloni, P. Onorati, and F. Babiloni, "Dynamic functional coupling of high resolution EEG potentials related to unilateral internally triggered one-digit movements," *Electroencephalogr. Clin. Neurophysiol.*, vol. 106, no. 6, pp. 477–487, 1998, doi:10.1016/S0013-4694(97)00150-8.

[5] A. S. Gevins, B. A. Cutillo, S. L. Bressler, N. H. Morgan, R. M. White, J. Illes, and D. S. Greer, "Event-related covariances during a bimanual visuomotor task. II. Preparation and feedback," *Electroencephalogr. Clin. Neurophysiol.*, vol. 74, pp. 147–160, 1989, doi:10.1016/0168-5597(89)90020-8.

[6] M. Taniguchi, A. Kato, N. Fujita, M. Hirata, H. Tanaka, T. Kihara, H. Ninomiya, N. Hirabuki, H. Nakamura, S. E. Robinson, D. Cheyne, and T. Yoshimine, "Movement-related desynchronization of the cerebral cortex studied with spatially filtered magnetoencephalography," *NeuroImage*, vol. 12, no. 3, pp. 298–306, 2000, doi:10.1006/nimg.2000.0611.

[7] A. Brovelli, M. Ding, A. Ledberg, Y. Chen, R. Nakamura, and S. L. Bressler, "Beta oscillations in a large-scale sensorimotor cortical network: Directional influences revealed by Granger causality," *Proc. Natl. Acad. Sci. U.S.A.*, vol. 101, no. 26, pp. 9849–9854, Jun. 29, 2004, doi:10.1073/pnas.0308538101.

[8] K. J. Friston, "Functional and effective connectivity in neuroimaging: A synthesis," *Human Brain Mapp.*, vol. 2, pp. 56–78, 1994, doi:10.1002/hbm.460020107.

[9] L. Jancke, R. Loose, K. Lutz, K. Specht, and N. J. Shah, "Cortical activations during paced finger-tapping applying visual and auditory pacing stimuli," *Brain Res. Cogn. Brain Res.*, vol. 10, no. 1–2, pp. 51–66, 2000, doi:10.1016/S0926-6410(00)00022-7.

[10] K. A. Bollen, *Structural Equations with Latent Variables*. New York: Wiley; 1989.

[11] A. R. McIntosh and F. Gonzalez-Lima, "Structural equation modeling and its application to network analysis in functional brain imaging," *Hum. Brain Mapp.*, vol. 2, pp. 2–22, 1994, doi:10.1002/hbm.460020104.

[12] R. Schlosser, T. Gesierich, B. Kaufmann, G. Vucurevic, S. Hunsche, J. Gawehn, and P. Stoeter, "Altered effective connectivity during working memory performance in schizophrenia: A study with fMRI and structural equation modeling," *NeuroImage*, vol. 19, no. 3, pp. 751–763, 2003, doi:10.1016/S1053-8119(03)00106-X.

[13] P. L. Nunez, *Neocortical Dynamics and Human EEG Rhythms*. New York: Oxford University Press, 1995.

[14] F. Babiloni, C. Babiloni, L. Locche, F. Cincotti, P. M. Rossini, and F. Carducci, "High-resolution electroencephalogram: Source estimates of Laplacian-transformed somatosensory-evoked potentials using a realistic subject head model constructed from magnetic resonance images," *Med. Biol. Eng. Comput.*, vol. 38, no. 5, pp. 512–519, Sep. 2000, doi:10.1007/BF02345746.

[15] F. Babiloni, C. Babiloni, F. Carducci, G. L. Romani, P. M. Rossini, L. M. Angelone, and F. Cincotti, "Multimodal integration of high-resolution EEG and functional magnetic resonance imaging data: A simulation study," *NeuroImage*, vol. 19, no. 1, pp. 1–15, May 2003.

[16] G. Pfurtscheller and F. H. Lopes da Silva, "Event-related EEG/MEG synchronization and desynchronization: Basic principles," *Clin. Neurophysiol.*, vol. 110, no. 11, pp. 1842–1857, Nov. 1999, doi:10.1016/S1388-2457(99)00141-8.

[17] S. L. Bressler, "Large-scale cortical networks and cognition," *Brain Res. Brain Res. Rev.*, vol. 20, no. 3, pp. 288–304, 1995, doi:10.1016/0165-0173(94)00016-I.

[18] J. Gross, J. Kujala, M. Hämäläinen, L. Timmermann, A. Schnitzler, and R. Salmelin, "Dynamic imaging of coherent sources: Studying neural interactions in the human brain," *Proc. Natl. Acad. Sci. U.S.A.*, vol. 98, no. 2, pp. 694–699, 2001, doi:10.1073/pnas.98.2.694.

[19] J. Gross, L. Timmermann, J. Kujala, R. Salmelin, and A. Schnitzler, "Properties of MEG tomographic maps obtained with spatial filtering," *NeuroImage*, vol. 19, pp. 1329–1336, 2003, doi:10.1016/S1053-8119(03)00101-0.

[20] M. Kaminski and K. Blinowska, "A new method of the description of the information flow in the brain structures," *Biol. Cybern.*, vol. 65, pp. 203–210, 1991, doi:10.1007/BF00198091.

[21] L. A. Baccalà and K. Sameshima, "Partial directed coherence: A new concept in neural structure determination," *Biol. Cybern.*, vol. 84, pp. 463–474, 2001, doi:10.1007/PL00007990.

[22] M. Kaminski, M. Ding, W. A. Truccolo, and S. Bressler, "Evaluating causal relations in neural systems: Granger causality, directed transfer function and statistical assessment of significance," *Biol. Cybern.*, vol. 85, pp. 145–157, 2001, doi:10.1007/s004220000235.

[23] C. W. J. Granger, "Investigating causal relations by econometric models and cross-spectral methods," *Econometrica*, vol. 37, pp. 424–438, 1969, doi:10.2307/1912791.

[24] R. Kus, M. Kaminski, and K. J. Blinowska, "Determination of EEG activity propagation: Pair-wise versus multichannel estimate," *IEEE Trans. Biomed. Eng.*, vol. 51, no. 9, pp. 1501–1510, Sep. 2004, doi:10.1109/TBME.2004.827929.

CHAPTER 1

Estimation of the Effective Connectivity from Stationary Data by Structural Equation Modeling

1.1 STRUCTURAL EQUATION MODELING

In structural equation model (SEM), the parameters are estimated by minimizing the difference between the observed covariances and those implied by a structural or path model [1,2]. In terms of neural systems, a measure of covariance represents the degree to which the activities of two or more regions are related.

The SEM consists of a set of linear structural equations, containing observed variables and parameters defining causal relationships among the variables. Variables in the equation system can be endogenous (i.e., dependent on other variables in the model) or exogenous (independent of the model itself). The SEM specifies the causal relationship among the variables, describes the causal effects, and assigns the explained and the unexplained variances.

Let us consider a set of variables (expressed as deviations from their means) with N observations. In this study, these variables represent the activity estimated in each cortical region of the brain, obtained with the procedures described in the chapters that follow.

The SEM for these variables is the following:

$$\mathbf{y} = \mathbf{B}\mathbf{y} + \mathbf{\Gamma}\mathbf{x} + \boldsymbol{\zeta} \tag{1.1}$$

where

$\mathbf{y}$ is an $m \times 1$ vector of dependent (endogenous) variables;

$\mathbf{x}$ is an $n \times 1$ vector of independent (exogenous) variables;

$\boldsymbol{\zeta}$ is an $m \times 1$ vector of equation errors (random disturbances);

$\mathbf{B}$ is an $m \times m$ matrix of coefficients of the endogenous variables;

$\mathbf{\Gamma}$ is an $m \times n$ matrix of coefficients of the exogenous variables;

$\boldsymbol{\zeta}$ is assumed to be uncorrelated with the exogenous variables;

$\mathbf{B}$ is supposed to have zeros in its diagonal (i.e., an endogenous variable does not influence itself) and to satisfies the assumption that $(\mathbf{I} - \mathbf{B})$ is nonsingular, where $\mathbf{I}$ is the identity matrix.

The covariance matrices of this model are the following:

$\mathbf{\Phi} = E[\mathbf{x}\mathbf{x}^{\mathrm{T}}]$ is an $n \times n$ covariance matrix of the exogenous variables;

$\mathbf{\Psi} = E[\boldsymbol{\zeta}\boldsymbol{\zeta}^{\mathrm{T}}]$ is an $m \times m$ covariance matrix of the errors.

If $\mathbf{z}$ is a vector containing all the $p = m + n$ variables, exogenous and endogenous, in the following order

$$\mathbf{z}^{\mathrm{T}} = [\, x_1 \cdots x_n \quad y_1 \cdots y_m \,] \tag{1.2}$$

the observed covariances can be expressed as

$$\mathbf{\Sigma}_{\mathrm{obs}} = (1/(N-1))\mathbf{z}\mathbf{z}^{\mathrm{T}} \tag{1.3}$$

where $\mathbf{z}$ is the $p \times N$ matrix of the p observed variables for N observations.

The covariance matrix implied by the model can be obtained as follows:

$$\mathbf{\Sigma}_{\mathrm{mod}} = E[\mathbf{z}^{\mathrm{T}}\mathbf{z}] = \begin{bmatrix} E[\mathbf{x}\mathbf{x}^{\mathrm{T}}] & E[\mathbf{x}\mathbf{y}^{\mathrm{T}}] \\ E[\mathbf{y}\mathbf{x}^{\mathrm{T}}] & E[\mathbf{y}\mathbf{y}^{\mathrm{T}}] \end{bmatrix} \tag{1.4}$$

where

$$\begin{aligned} E[\mathbf{y}\mathbf{y}^{\mathrm{T}}] &= E[(\mathbf{I}-\mathbf{B})^{-1}(\mathbf{\Gamma}\mathbf{x}+\zeta)(\mathbf{\Gamma}\mathbf{x}+\zeta)^{\mathrm{T}}((\mathbf{I}-\mathbf{B})^{-1})^{\mathrm{T}}] \\ &= (\mathbf{I}-\mathbf{B})^{-1}(\mathbf{\Gamma}\mathbf{\Phi}\mathbf{\Gamma}+\mathbf{\Psi})((\mathbf{I}-\mathbf{B})^{-1})^{\mathrm{T}} \end{aligned} \tag{1.5}$$

As the errors ζ are not correlated with $\mathbf{x}$

$$E[\mathbf{x}\mathbf{x}^{\mathrm{T}}] = \mathbf{\Phi} \tag{1.6}$$

$$E[\mathbf{x}\mathbf{y}^{\mathrm{T}}] = (\mathbf{I}-\mathbf{B})^{-1}\,\mathbf{\Phi} \tag{1.7}$$

$$E[\mathbf{y}\mathbf{x}^{\mathrm{T}}] = ((\mathbf{I}-\mathbf{B})^{-1}\,\mathbf{\Phi})^{\mathrm{T}} \tag{1.8}$$

and as $\mathbf{\Sigma}_{\mathrm{mod}}$ is symmetric. The resulting covariance matrix, in terms of the model parameters, is given by the following:

$$\mathbf{\Sigma}_{\mathrm{mod}} = \begin{bmatrix} \Phi & (\mathbf{I}-\mathbf{B})^{-1}\mathbf{\Phi} \\ ((\mathbf{I}-\mathbf{B})^{-1}\mathbf{\Phi})^{\mathrm{T}} & (\mathbf{I}-\mathbf{B})^{-1}(\mathbf{\Gamma}\mathbf{\Phi}\mathbf{\Gamma}+\mathbf{\Psi})((\mathbf{I}-\mathbf{B})^{-1})^{\mathrm{T}} \end{bmatrix} \tag{1.9}$$

Without other constraints, the problem of minimizing the differences between the observed covariances and those implied by the model is not determined, because the number of variables (elements of matrices $\mathbf{B}$, $\mathbf{\Gamma}$, $\mathbf{\Psi}$, and $\mathbf{\Phi}$) is greater than the number of equations $(m+n)(m+n+1)/2$. For this reason, the SEM technique is based on *a priori* formulation of a model depending on the anatomic and physiologic constraints. This model implies the existence of just some causal relationships among variables represented by arcs in a "path" diagram; all parameters related to arcs that are not present in the hypothesized model are set to zero. For this reason, all the parameters to be estimated are called free parameters. If t is the number of free parameters, then $t \leq (m+n)(m+n+1)/2$.

These parameters are estimated by minimizing a function of the observed and implied covariance matrices. The most widely used objective function for SEM is the maximum likelihood (ML) function:

$$F_{\mathrm{ML}} = \log|\mathbf{\Sigma}_{\mathrm{mod}}| + \mathrm{tr}(\mathbf{\Sigma}_{\mathrm{obs}}\mathbf{\Sigma}_{\mathrm{mod}}^{-1}) - \log|\mathbf{\Sigma}_{\mathrm{obs}}| - p \qquad (1.10)$$

where $\mathrm{tr}(\cdot)$ is the trace of the matrix. In the context of multivariate normally distributed variables, the minimum of the maximum likelihood function when multiplied by $N-1$, follows a χ^2 distribution with $(p(p+1)/2) - t$ degrees of freedom, where t is the number of parameters to be estimated and p is the total number of observed variables (endogenous + exogenous). The χ^2 statistics test can then be used to infer the statistical significance of the structural equation model obtained. In the present study, the publicly available software LISREL [3] was used for the implementation of the SEM technique.

1.2 SIMULATION STUDY

We adopted an experimental design that analyzes the recovery of the connectivity of an estimated model with respect to an imposed one. This has been built under different levels of the main factors, the signal-to-noise ratio (SNR) and the temporal duration (LENGTH); they have been imposed during the generation of a set of test signals simulating cortical average activations and were obtained by starting from actual cortical data (estimated with the high-resolution EEG procedures available at the High Resolution EEG Laboratory at the University of Rome "La Sapienza").

1.2.1 Signal Generation

Different sets of test signals have been generated in order to fit an imposed connectivity pattern (shown in Figs. 1.1A, 1.2A, and 1.3A) and maintain the imposed levels LENGTH and SNR. In the following, in order to use a more compact notation, the signals have been represented with the $\mathbf{z}$ vector defined in Eq. (1.2), containing both the endogenous and the exogenous variables.

Channel z_1 is a reference source waveform, estimated from a high-resolution EEG (128 electrodes) recording in a healthy subject, during the execution of random self-paced movements of the right finger.

Signals z_2, z_3, and z_4 were obtained by the contribution of signals from all other channels with an amplitude variation and zero mean uncorrelated white noise processes with appropriate variances, as shown in the following:

$$\mathbf{Z}[k] = \mathbf{A} * \mathbf{z}[k] + \mathbf{W}[k] \qquad (1.11)$$

where $\mathbf{z}[k]$ is the 4×1 vector of signals, $\mathbf{W}[k]$ is the 4×1 noise vector and $\mathbf{A}$ is the 4×4 parameters matrix obtained from the $\mathbf{\Gamma}$ and $\mathbf{B}$ matrices as follows:

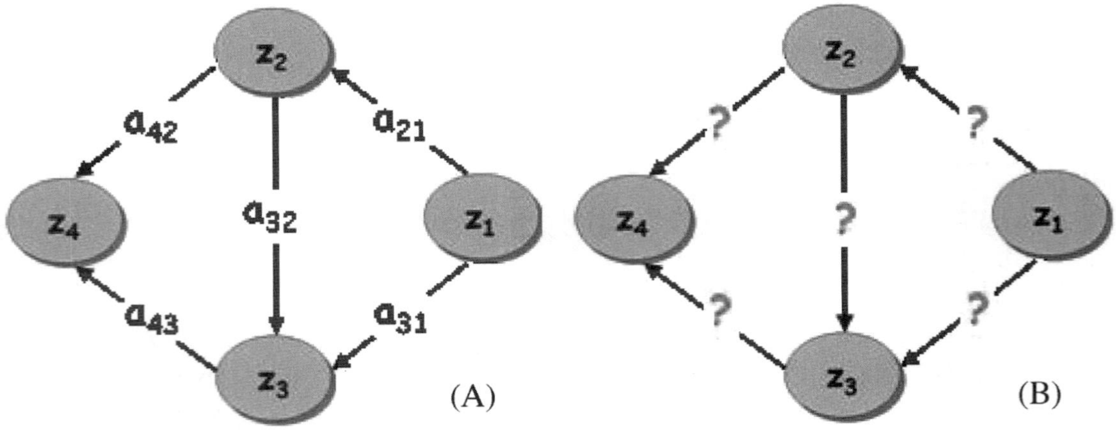

FIGURE 1.1: Correct model. (A) Connectivity patterns imposed in the generation of simulated signals. $z_1, \ldots, z_4$ represent the average activities in the four cortical areas. Values on the arcs represent the strength of the connections ($a_{21} = 1.4$; $a_{31} = 1.1$; $a_{32} = 0.5$; $a_{42} = 0.7$; $a_{43} = 1.2$). (B) Connectivity model used for the parameter estimation. (Published with permission from [6].)

$$\mathbf{A} = \begin{bmatrix} 0 & 0 & 0 & 0 \\ \gamma_1 & \beta_{11} & \beta_{12} & \beta_{13} \\ \gamma_2 & \beta_{21} & \beta_{22} & \beta_{23} \\ \gamma_3 & \beta_{31} & \beta_{32} & \beta_{33} \end{bmatrix} = \begin{bmatrix} a_{11} & \cdots & a_{14} \\ \vdots & \ddots & \vdots \\ & & \ddots \\ a_{41} & \cdots & a_{44} \end{bmatrix} \tag{1.12}$$

where β_{ij} stands for the generic (i, j) element of the $\mathbf{B}$ matrix and γ_i is the ith element of the matrix $\boldsymbol{\Gamma}$.

All procedures of signal generation were repeated under the following conditions:

SNR factor levels = [1, 3, 5, 10, 100];

LENGTH factor levels (in seconds) = [60, 190, 310, 610]. This corresponds, for instance, to [120, 380, 620, 1220] EEG epochs, each of which is 500 ms long.

It is worth noting that the levels chosen for both SNR and LENGTH factors cover the typical range for the cortical activity estimated with high resolution EEG techniques.

1.2.2 Parameter Estimation

The set of simulated signals generated as described above was given as input to the program LISREL for the estimation of SEM parameters. As mentioned in the section 1.1 of this Chapter, SEM needs a model based on a previous information on the anatomic connections on which the estimate is successively performed. For this reason, its performance has been observed in different situations

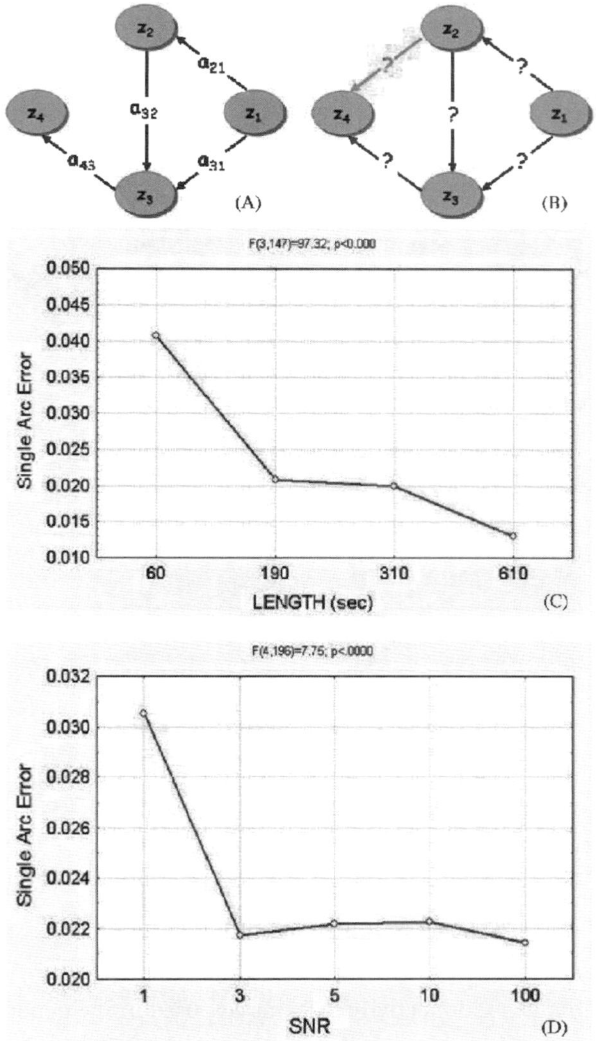

FIGURE 1.2: Arc in excess. (A) Connectivity patterns imposed in the generation of simulated signals. Values on the arcs represent the strength of the connection ($a_{21} = 1.4; a_{31} = 1.1; a_{32} = 0.5; a_{43} = 1.2$). (B) Connectivity model used for the estimation of the parameter. Results of ANOVA performed on the error committed on the arc in excess a_{42} (Single Arc Error). (C) Plot of means with respect to signal LENGTH as a function of time (in seconds). ANOVA shows a high statistical significance of the LENGTH factor ($F = 97.32, p < 0.0001$). Duncan's post hoc test at 5% level of significance shows statistically no significant difference between a signal length of 190 or 310 s (25 or 40 trials, 7.5 s per trial). (D) Plot of means with respect to SNR. A statistical influence of the SNR factor on the error in the evaluation of the presence of arc a_{42} is shown ($F = 7.75, p < 0.0001$). Duncan's post hoc test at 5% level of significance shows no statistically significant differences between levels 3, 5, 10, and 100 of the SNR factor. (Published with permission from [6].)

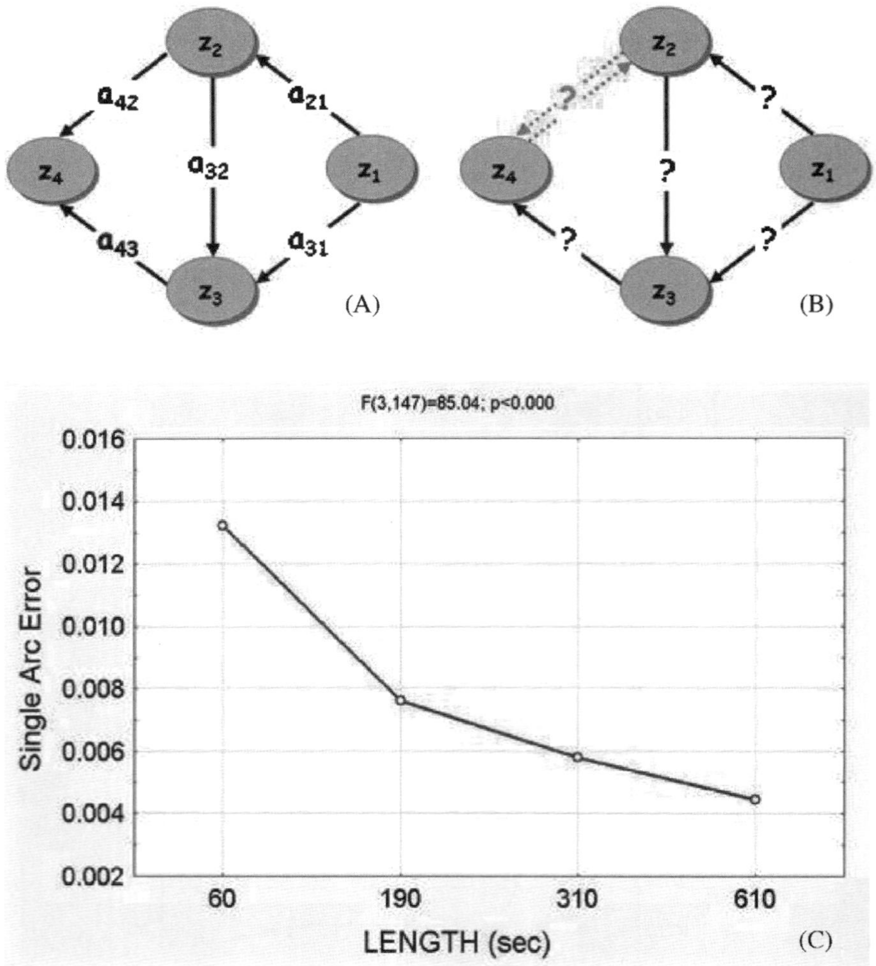

FIGURE 1.3: Arc direction. (A) Connectivity patterns imposed in the generation of simulated signals. Values on the arcs represent the strength of the connections ($a_{21} = 1.4$; $a_{31} = 1.1$; $a_{32} = 0.5$; $a_{42} = 0.7$; $a_{43} = 1.2$). (B) Connectivity model used for the parameter estimation. No assumption has been made on the direction of arc a_{42} (both directions are present in the model). (C) Results of ANOVA performed on the error committed on the wrong direction arc a_{24}, not present in the imposed model (Single Arc Error). Plot of means with respect to signal LENGTH as a function of time (in seconds). ANOVA shows a high statistical significance of the LENGTH factor($F = 85.04, p < 0.0001$). Duncan's post hoc test at 5% level of significance shows statistically significant differences between all levels of LENGTH. (Published with permission from [6].)

where the connections between the four cortical areas are not always correctly assumed. The situations analyzed were as follows:

1. an identical connectivity graph between the generated and the estimated model;
2. a different number of connectivity arcs between the generated and the estimated model; in particular, we analyzed the case of an arc in excess and of an arc missing in the estimated model with respect to the generated one;
3. the same number of connectivity arcs between generated and estimated models, but with an ambiguousness on its orientation.

1.2.3 Evaluation of Performances

In order to evaluate the quality of the estimation performed, the following indexes were computed:

1. the Frobenius norm of the matrix reporting the difference between the values of the estimated (via SEM) and the imposed connections (Relative Error):

$$E_{\text{relative}} = \frac{\sqrt{\sum_{i=1}^{m}\sum_{j=1}^{m}(a_{ij} - \hat{a}_{ij})^2}}{\sqrt{\sum_{i=1}^{m}\sum_{j=1}^{m}(a_{ij})^2}} \tag{1.13}$$

2. the absolute value of the difference between the estimated parameter and the imposed value on a single particular arc (Single Arc Error):

$$E_{\text{single}} = |a_{ij} - \hat{a}_{ij}| \tag{1.14}$$

In order to increase the robustness of the successive statistical analysis, simulations were performed by repeating for 50 runs each connectivity estimation obtained by SEM.

1.2.4 Statistical Analysis

The results obtained were subjected to separate analysis of variance (ANOVA). The main factors of the ANOVA were SNR (with five levels: 1, 3, 5, 10, 100) and LENGTH (with four levels: 60, 190, 310, 610 s). Separate ANOVAs were performed on the error indexes adopted (Relative Error, Single Arc Error). In all the evaluated ANOVAs, the Greenhouse–Gasser corrections for the violation of the spherical hypothesis were used. The post hoc analysis with Duncan's test at $p = 0.05$ statistically significant level was then performed.

1.3 RESULTS OF THE SIMULATIONS

The simulated signals obtained for the different levels of the two SNR and LENGTH factors were analyzed by means of the software LISREL, performing an estimation of the strengths of the connections. Figure 1.1A shows the connection model used in the signal generation and parameter estimation. The arrows represent the existence of a connection directed from the signal z_i toward the signal z_j, while the values on the arcs a_{ij} represent the connection parameters described in Eq. (1.12). The results obtained from 50 repetition of the procedure were subjected to statistical analysis (ANOVA), as reported in detail in the following paragraphs.

1.3.1 Correct Formulation of the Connectivity Model

Figure 1.1 shows the first situation analyzed. A set of signals was generated as described before in order to fit the connectivity pattern shown in Fig. 1.1A. Parameters were estimated on the model shown in Fig. 1.1B, which has exactly the same structure as shown in Fig. 1.1A. We thus test the goodness-of-parameter estimation of model parameters via SEM when no errors are made in the model assumption phase. The appropriate index for this analysis is the Relative Error, as defined in Section 1.2.3, Eq. (1.13). This was computed for each of the 50 runs of the generation–estimation proceedure performed for each level of SNR and LENGTH factors and then subjected to ANOVA, by which was obtained a rather strong statistical significance of both factors for the performance of SEM. In fact, SNR and LENGTH were both highly significant (LENGTH, $F = 288.60$, $p < 0.0001$, SNR, $F = 22.70$, $p < 0.0001$). Figure 1.4A shows the plot of means of Relative Error with respect to the signal length levels, which reveals a decrease of the connectivity estimation error with the increase of the length of the available data. Figure 1.4B shows the plot of means with respect to different SNR levels employed in the simulation. Since the main factors were found to be highly statistically significant, Duncan's post hoc tests at 5% level of significance were then applied. Such tests showed statistically significant differences between all levels of the LENGTH factor, but no statistically significant differences between levels 3, 5, and 10 of the SNR factor.

1.3.2 Hypothesis of a Model with an Arc in Excess or a Missing Arc

Since a perfect formulation of the connectivity model is not always a realistic option, we analyzed several situations in which the connections between the four cortical areas were not correctly assumed in the estimated model.

1.3.3 Arc in Excess

The first situation is described in Fig. 1.2. The SEM parameter estimation was performed on the model shown in Fig. 1.2B, containing an arc which is absent in the imposed pattern (Fig. 1.2A). The aim was to test if the SEM procedure can reject the error made in the model assumption. The appropriate index for this analysis is the Single Arc Error (1.14) on the arc a_{42}, i.e., the one which is not present in the correct model. ANOVA performed on the simulation results showed that both the main factors LENGTH and SNR, have a statistical influence on the ability of SEM to reveal the modeling error. Figure 1.2C and D shows the plot of means with respect to the different levels

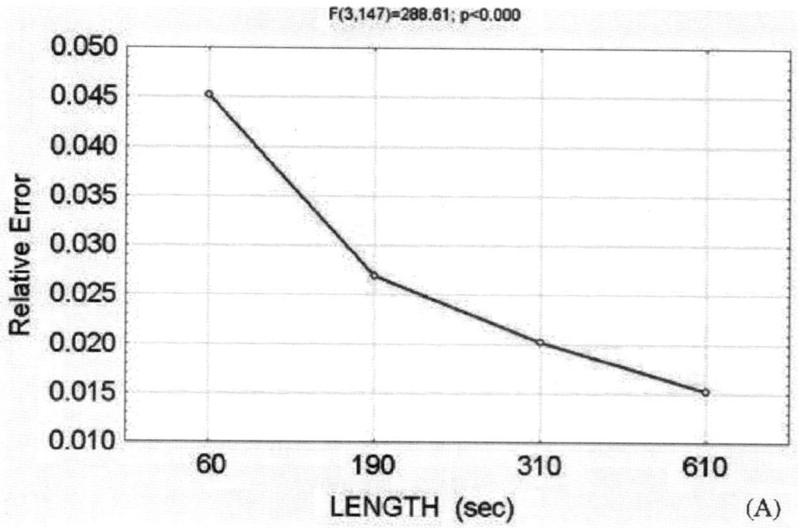

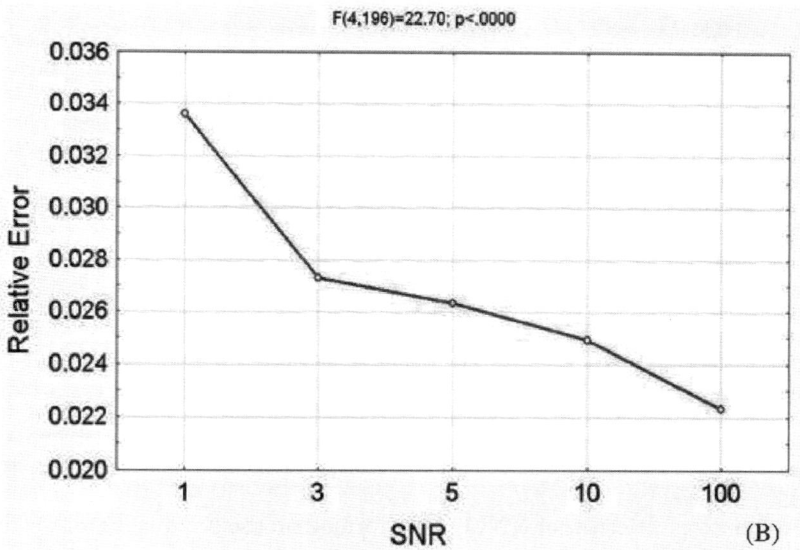

FIGURE 1.4: Correct model. (A) Results of ANOVA performed on the Relative Error. Plot of means with respect to signal LENGTH as a function of time (in seconds). ANOVA shows a high statistical significance for the LENGTH factor ($F = 288.60, p < 0.0001$). Duncan's post hoc test performed at 5% level of significance shows statistically significant differences between all levels. (B) Results of ANOVA performed on the Relative Error. Plot of means with respect to SNR. Here too a high statistical influence of the SNR factor on the error in the estimation is seen ($F = 22.70, p < 0.001$). Duncan's post hoc test performed at 5% level of significance shows no statistically significant differences between levels 3, 5, and, 10 of the SNR factor. (Published with permission from [6].)

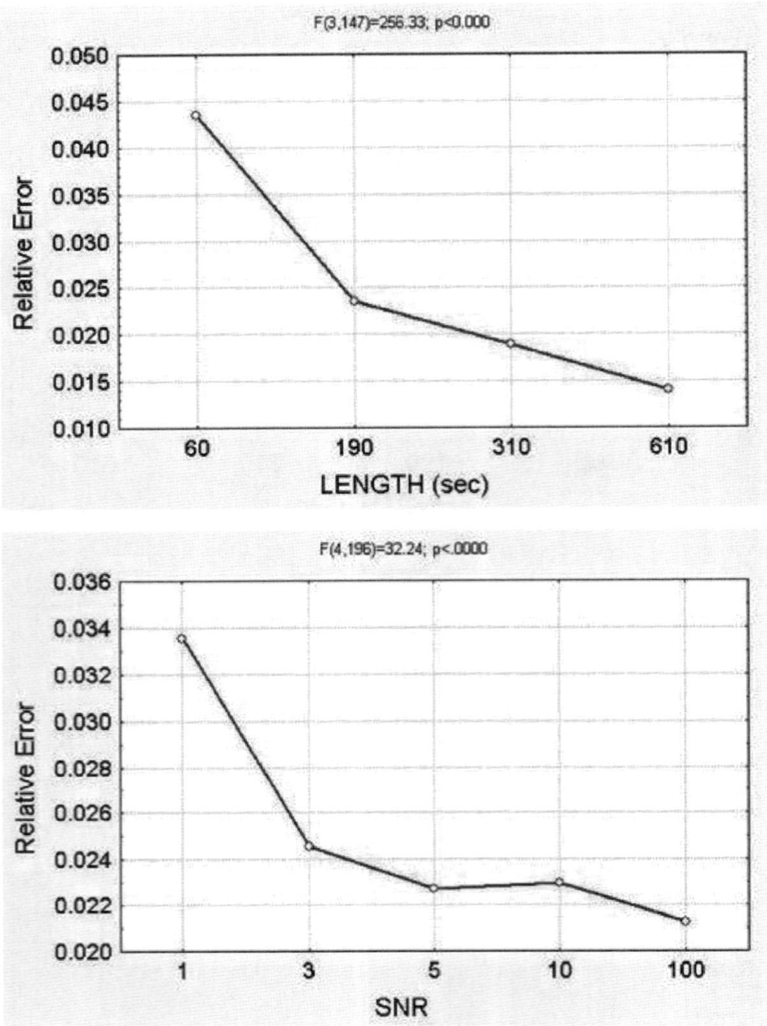

(A)

(B)

FIGURE 1.5: Arc in excess. Results of ANOVA performed on the Relative Error for the same situation of Fig. 1.3. (A) Plot of means with respect to signal LENGTH as a function of time (in seconds). ANOVA shows a high statistical significance for the LENGTH factor ($F = 256.33, p < 0.0001$). Duncan's post hoc test performed at 5% level of significance shows statistically significant differences between all levels. (B) Results of ANOVA performed on the Relative Error. Plot of means with respect to Signal to Noise Ratio. Here too a high statistical influence of the SNR factor on the error in the estimation is seen ($F = 32.24, p < 0.001$). Duncan's post hoc test performed at 5% level of significance shows no statistically significant differences between levels 5 and 10 of the SNR factor. (Published with permission from [6].)

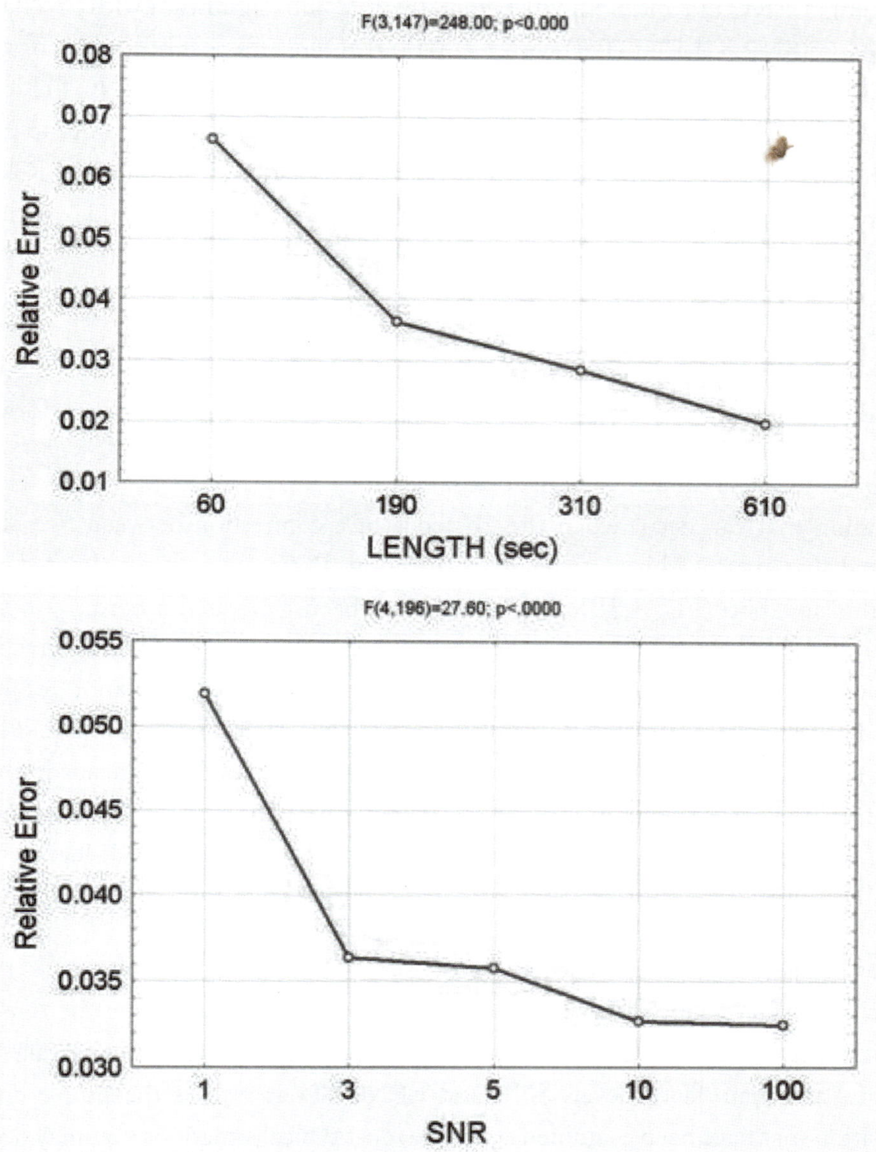

(A)

(B)

FIGURE 1.6: Arc direction. Results of ANOVA performed on the Relative Error for the same situation as in Fig. 1.5. (A) Plot of means with respect to signal LENGTH as a function of time (in seconds). ANOVA shows a high statistical significance for the LENGTH factor ($F = 248.00, p < 0.0001$). Duncan's post hoc test performed at 5% level of significance shows statistically significant differences between all levels. (B) Results of ANOVA performed on the Relative Error. Plot of means with respect to Signal-to-Noise Ratio. Here too a high statistical influence of the SNR factor on the error in the estimation is seen ($F = 27.60, p < 0.001$). Duncan's post hoc test performed at 5% level of significance shows no statistically significant differences between levels 3, 5, and 10 of the SNR factor. (Published with permission from [6].)

of the main factors LENGTH and SNR. As before, they are both significant with $p < 0.001$ as well as their interaction (SNR × LENGTH) with $p < 0.0001$. Duncan's post hoc test performed at 5% level of significance shows no statistically significant difference between the LENGTH levels of 190 or 310 s as well as between levels 3, 5, 10, and 100 of the main SNR factor. In order to evaluate the influence of the exceeding arc in the model on the global parameter estimation, the Relative Error (1.13) was also computed. Figure 1.5A and B shows the plot of mean of this index with respect to the two main factors, with a level of statistical significance lower than 0.001.

1.3.4 Missing Arc

In this case, the χ^2 statistics test indicates no statistical significance of the estimated model. Hence, the corresponding error values were not computed and no statistical analysis was performed.

1.3.5 Ambiguousness on an Arc Direction

This is a situation that can occur when the existence of a connection between two structures is well known, and there is a need to investigate its direction. Parameters were estimated on a model representing this situation (Fig. 1.3B). The signals had been generated according to the pattern of Fig. 1.3A and the Single Arc Error made on the arc representing the wrong direction (a_{24} in this example) was considered. The statistical analysis performed on the simulation results with the ANOVA reported no statistical significance of the main SNR factor, while the LENGTH factor (EEGTRIAL) is still statistically significant (with $p < 0.0001$). The plot of means in the function of the levels of LENGTH is shown in Fig. 1.3C. Figure 1.6A and B shows the plot of means of the Relative Error with respect to the signal LENGTH levels and to different SNR levels employed in the simulation.

1.4 DISCUSSION

The experimental design adopted for the simulation study aimed at analyzing the most common situations in which the proposed application of SEM technique to EEG data may take place. The levels chosen for the main factor levels SNR and LENGTH as well as the simple errors in the model formulation that have been examined cover the most typical situations that may occur in such analysis. The results obtained have shown a significant statistical influence of the factors considered on SEM performances.

On the basis of the simulations performed, we are now able to answer the questions raised in Section 1.1:

1. There is a statistical influence of a variable SNR level imposed on the high-resolution EEG data on the accuracy of the connectivity pattern estimation. In particular, an SNR = 3 seems to be satisfactory in order to obtain good accuracy as there are no significant differences in the performance for higher values.

2. The minimum amount of EEG data necessary to get a usable accuracy of the estimation of connectivity between cortical areas is 190 s of registration (equivalent, for instance, to 380 trials of 500 ms each). However, in this case, an increase in the length of the available EEG data is always related to a decrease in the connectivity estimation error.
3. Different situations, in which the connections between the four cortical areas were not correctly assumed in the estimated model were evaluated in order to analyze their influence on SEM performances. In the first situation, there was a deliberate error in the hypothesized model consisting of the presence of an arc not corresponding to an actual influence between areas. The aim was to test if the SEM procedure can reject the error made in the model assumption and to evaluate the influence of the introduction of such modeling error on the goodness-of-parameter estimation. The analysis of the Single Arc Error on the arc in excess, revealed that an SNR = 3 and an EEG data of 190 s of registration seems to be satisfactory in order to obtain good accuracy. The effect on the global performance of parameter estimation can be inferred by comparing the Relative Error obtained in this situation to the correct one. From Fig. 1.4A and B, compared to Fig. 1.5A and B, it can be seen that the error values remain at the same level in both cases, and the general performance is not decreased by this kind of error. In the second situation analyzed, the voluntary error in the hypothesized model consists of a lack of an arc corresponding to an influence between areas. The performed analysis has not reported any statistical significance, as indicated by the χ^2 to degrees-of-freedom ratio: $\chi^2/df > 1$. This suggests that, in the case of results of this kind, an arc can be added to the putative model in order to decrease the χ^2 to degrees-of-freedom ratio. In the third situation analyzed, the estimated model contained arcs in both directions between two areas, corresponding to a single arc in the model imposed in the signal generation. The Single Arc Error computed on the "wrong direction" arc shows that the error is smaller (less than 1.5% for all factors and levels considered) than in the case of an arc in excess in a single direction in the first situation analyzed. However, it is worth noting that the general performance, as indicated by the Relative Error (Fig. 1.6A and B), is significantly worse in this case than in the case of correct modeling, especially for low values of the LENGTH factor (cf. Fig. 1.4A and B). This means that a simple error such as attributing both directions to a couple of channels causes a significant increase of the error made in the parameter estimation.

In conclusion, the ANOVA results (integrated with the Duncan's post hoc test performed at 5% level of significance indicated a clear influence of different levels of the main factors SNR and LENGTH on the efficacy of the estimation of cortical connectivity via SEM. In particular, it has been noted that at least an SNR = 3 and a LENGTH of the measured cortical data of 190 s are necessary to decrease significantly the errors related to the indexes of the quality adopted.

The simulation study has shown that the ability of SEM to perform a good estimate of connectivity pattern, when connections between the four cortical areas are not correctly assumed, depends on the kind of error made in the model formulation. It seems that the error consisting of a lack of a connection arc is the worst with respect to the parameter estimate, though it can be easily detected by a χ^2 statistical test. Putting in the model an arc not corresponding to an actual influence between areas, on the contrary, does not particularly influence the goodness-of-parameter estimate, and the exceeding arc is attributed a value near to zero. Putting arcs in both directions between two areas, when the influence is directed only from one to the other, causes larger errors in the parameter estimation, though it allows discrimination of the right direction with a precision which does not depend on the signal SNR and which is very high for most levels of the signal LENGTH.

Although the performance seems to be rather good for a correct assumption of the hypothesized model, it decreases when even a simple error is made, depending on the error type. This degradation of the performance seems to indicate the opportunity to use connectivity models that are not too detailed in terms of cortical areas involved as a first step of the network modeling. By using a coarse model of the cortical network to be fitted on the EEG data, there is an increase of the statistical power and a decrease of the possibility of generating an error in a single arc link [4]. In the human study presented in Chapter 2, such an observation was taken into account by selecting a coarse model for the brain areas subserving the task being analyzed. This simplified model does not take complete account of all the possible regions engaged in the task and all the possible connections between them. Elaborate models permitting also cyclical connections between regions can become computationally unstable [5].

In conclusion, the results of the simulation study on SEM performed during this PhD course indicate that the information that a quite accurate estimation of the cortical connectivity patterns can be achieved by using the SEM technique.

REFERENCES

[1] K. Jöreskog and D. Sörbom, LISREL 8.53, software, December 2002. Scientific Software International, Inc. Available: http://www.ssicentral.com, doi:10.1002/hbm.460020107.

[2] B. Horwitz, "The elusive concept of brain connectivity," *Neuroimage*, vol. 19, pp. 466–470, 2003, doi:10.1016/S1053-8119(03)00112-5.

[3] A. R. McIntosh and F. Gonzalez-Lima, "Structural equation modeling and its application to network analysis in functional brain imaging," *Hum. Brain Mapp.*, vol. 2, pp. 2–22, 1994, doi:10.1002/hbm.460020104.

[4] L. Astolfi, et al., "Estimation of the cortical connectivity by high resolution EEG and structural equation modeling: simulations and application to finger tapping data," *IEEE Trans. Biomed. Eng.*, vol. 52, no. 5, pp. 757–768, May 2005, doi:10.1109/TBME.2005.845371.

CHAPTER 2

Estimation of the Functional Connectivity from Stationary Data by Multivariate Autoregressive Methods

2.1 MULTIVARIATE AUTOREGRESSIVE PROCESS

Let Y be a set of cortical waveforms, obtained from several cortical regions of interest (ROI) as described in detail as follows:

$$\mathbf{Y} = [y_1(t), y_2(t), \ldots, y_N(t)]^{\mathrm{T}} \tag{2.1}$$

where t refers to time and N is the number of cortical areas considered.

Suppose that the following multivariate autoregressive (MVAR) process is an adequate description of the data set Y:

$$\sum_{k=0}^{p} \mathbf{\Lambda}(k)\mathbf{Y}(t-k) = \mathbf{E}(t) \text{ with } \mathbf{\Lambda}(0) = \mathbf{I} \tag{2.2}$$

where

- $\mathbf{Y}(t)$ is the data vector in time;
- $\mathbf{E}(t) = [e_1(t), \ldots, e_N]^{\mathrm{T}}$ is a vector of multivariate zero-mean uncorrelated white noise processes;
- $\mathbf{\Lambda}(1), \mathbf{\Lambda}(2), \ldots, \mathbf{\Lambda}(p)$ are the $N \times N$ matrices of model coefficients;
- p is the model order.

In the present study, p was chosen by means of the Akaike information criteria (AIC) for MVAR processes [1] and was used for MVAR model fitting to simulations as well as to experimental signals. It has been noted that, although the sensitivity of MVAR performance depends on the model order, small model order changes do not influence results [2,3].

A modified procedure for the fitting of MVAR on multiple trials was adopted [3–5]. When many realizations of the same stochastic process are available, as in the case of several trials of an event-related potential (ERP) recording, the information from all the trials can be used to increase the reliability and statistical significance of the model parameters. In the present book, the data were in the form of several trials of the same length, both in the simulation and application to real data, as described in detail in the following sections.

After an MVAR model is adequately estimated it becomes the basis for subsequent spectral analysis. To investigate the spectral properties of the examined process, Eq. (2.2) is transformed to the frequency domain

$$\mathbf{\Lambda}(f)\mathbf{Y}(f) = \mathbf{E}(f) \tag{2.3}$$

where

$$\mathbf{\Lambda}(f) = \sum_{k=0}^{p} \mathbf{\Lambda}(k)e^{-j2\pi f\Delta t\, k} \tag{2.4}$$

and Δt is the temporal interval between two samples.

Eq. (2.3) can be rewritten as

$$\mathbf{Y}(f) = \mathbf{\Lambda}^{-1}(f)\mathbf{E}(f) = \mathbf{H}(f)\mathbf{E}(f) \tag{2.5}$$

$\mathbf{H}(f)$ is the transfer matrix of the system, whose element H_{ij} represents the connection between the jth input and the ith output of the system.

2.2 DIRECTED TRANSFER FUNCTION

The directed transfer function (DTF) representing the causal influence of the cortical waveform (estimated in the jth ROI on that estimated in the ith ROI) is defined [6] in terms of elements of the transfer matrix H given by

$$\theta_{ij}^2(f) = |H_{ij}(f)|^2 \tag{2.6}$$

In order to compare the results obtained for cortical waveforms with different power spectra, a normalization can be performed by dividing each estimated DTF by the squared sums of all elements of the relevant row, thus obtaining the so-called normalized DTF [6]

$$\gamma_{ij}^2(f) = \frac{|H_{ij}(f)|^2}{\sum_{m=1}^{N} |H_{im}(f)|^2} \tag{2.7}$$

$\gamma_{ij}(f)$ expresses the ratio of influence of the cortical waveform estimated in the jth ROI on the cortical waveform estimated in the ith ROI with respect to the influence of all the estimated cortical waveforms. Normalized DTF values are in the interval [0, 1], and the normalization condition

$$\sum_{n=1}^{N} \gamma_{in}^2(f) = 1 \tag{2.8}$$

is applied.

From the transfer matrix, we can calculate power spectra $\mathbf{S}(f)$. If we denote by V the variance matrix of the noise $\mathbf{E}(f)$, the power spectrum is defined by

$$S(f) = H(f)\mathbf{V}\mathbf{H}^*(f) \tag{2.9}$$

where the superscript * denotes transposition and complex conjugate.

From $\mathbf{S}(f)$, ordinary coherence can be computed as

$$k_{ij}(f) = \frac{|S_{ij}(f)|^2}{S_{ii}(f)S_{jj}(f)} \tag{2.10}$$

Coherence measures express the degree of synchrony (simultaneous activation) between areas i and j.

2.3 PARTIAL DIRECTED COHERENCE

Partial coherence is another estimator of the relationship between a pair of signals, describing the interaction between areas i and j when the influence due to all $N-2$ time series is discounted. It is defined by the formula

$$|\chi_{ij}(f)|^2 = \frac{|M_{ij}(f)|^2}{M_{ii}(f)M_{jj}(f)} \tag{2.11}$$

where $M_{ij}(f)$ is the minor matrix obtained by removing ith row and jth column from the spectral matrix S.

In 2001, Baccalà proposed the following factorization:

$$\chi_{ij}(f) = \frac{\Lambda_i^*(f)V^{-1}\Lambda_j(f)}{\sqrt{\left(\Lambda_i^*(f)V^{-1}\Lambda_i(f)\right)\left(\Lambda_j^*(f)V^{-1}\Lambda_j(f)\right)}} \tag{2.12}$$

where $\Lambda_n(f)$ is the nth column of the matrix $\mathbf{\Lambda}(f)$. This led to the definition of partial directed coherence (PDC), [7]

$$\pi_{ij}(f) = \frac{\Lambda_{ij}(f)}{\sqrt{\sum_{k=1}^{N}\Lambda_{ki}(f)\Lambda_{kj}^*(f)}} \tag{2.13}$$

The PDC from j to i, $\pi_{ij}(f)$, describes the directional flow of information from the activity in the ROI $s_j(n)$ to the activity in $s_i(n)$, whereupon common effects produced by other ROIs $s_k(n)$ on the latter are subtracted, leaving only a description that is specifically from $s_j(n)$ to $s_i(n)$.

The PDC values are in the interval [0, 1], and the normalization condition

$$\sum_{n=1}^{N}|\pi_{ni}(f)|^2 = 1 \tag{2.14}$$

is verified. According to this condition, $\pi_{ij}(f)$ represents the fraction of the time evolution of ROI j directed to ROI i, compared to all of j's interactions with other ROIs.

For both DTF and PDC high values in a frequency band represent the existence of an influence between any given pair of areas in the data set. However, an important difference is that PDC does not involve the inversion of matrix $\mathbf{\Lambda}$. This leads to several points. In fact, an analysis of the definition of DTF reveals that, due to this matrix inversion, it is a linear combination of both the direct influence from one area to the other and the influence mediated by other areas along various cascade pathways [8]. This becomes immediately clear from an example. Given a three-region model, the nonnormalized DTF from area 1 to area 2 is given by

$$\theta_{21}^2(f) = |H_{21}(f)|^2 = \frac{|\Lambda_{21}(f)\Lambda_{33}(f) - \Lambda_{31}(f)\Lambda_{23}(f)|}{|\mathbf{\Lambda}(f)|^2} \tag{2.15}$$

From this formula it can be noted that even if the direct influence from area 1 to area 2, $\Lambda_{21}(f)$, is zero, $\theta_{21}^2(f)$ may still be different from zero since there is an influence from 1 to 3 ($\Lambda_{31}(f)$) and from 3 to 2 ($\Lambda_{23}(f)$). The link between 1 and 2 will be indicated by DTF as a causal pathway if all the causal influences along the way are nonzero.

The PDC, due to the lack of the matrix inversion, behaves differently. It indicates only the existence of a direct causal influence from area 1 to area 2. If no direct influence exists, PDC_{21} is virtually zero.

2.4 DIRECT DTF

In order to distinguish between direct and cascade flows in DTF, the direct DTF (dDTF) was introduced [9]. It is defined by multiplying the full frequency directed transfer function (ffDTF) given by

$$\eta_{ij}^2(f) = \frac{|H_{ij}(f)|^2}{\sum_f \sum_{m=1}^k |H_{im}(f)|^2} \tag{2.16}$$

by the partial coherence defined in Eq. (2.11). The dDTF from area j to area i is defined as

$$\delta_{ij}(f) = \chi_{ij}(f)\eta_{ij}(f) \tag{2.17}$$

This function describes only the direct relations between channels. The denominator of the ffDTF function (2.16) does not depend on the frequency.

2.5 SIMULATION STUDY

The experimental design involved the following steps:

- *Generation of a set of test signals simulating cortical average activations*: Several sets of simulated data were generated in order to fit a predefined connectivity model and to respect imposed levels of the signal-to-noise ratio (factor SNR) and the length of the data (factor LENGTH).

The data were in the form of multiple trials, and the factor LENGTH indicates the total length of all trials.

- *Estimation of the cortical connectivity pattern obtained under different conditions*: The estimators used were DTF, PDC, and dDTF.
- *Computation of indices of connectivity estimation performance*: These indices were error functions describing the error in the connectivity estimation for the whole pattern and for each single arc. A comparison between the value estimated for the direct and indirect arcs was also performed.
- *Statistical analysis*: Analysis of variance (ANOVA) of the results of the simulations was performed to study the effects of the factors SNR and LENGTH on the recovery of the connectivity pattern resulting from the different methods.

2.5.1 Signal Generation

The connectivity model used in the generation of test signals is shown in Fig. 2.1. It involves five areas linked by both direct and indirect pathways. For example, ROIs 1 and 2 are linked by a *direct* path directed from 1 to 2. ROIs 1 and 5 are not connected by any direct arc but are linked by an *indirect* path from 1 to 2 and from 2 to 5. ROIs 4 and 5 are not linked by either a direct or an indirect arc. As described, these situations are rather different with respect to the estimates obtained by multivariate methods based on MVAR models. In particular, the model shown in Fig. 2.1 has 7 direct arcs, 2 indirect arcs, and 11 "null" arcs (i.e., 11 pairs of ROIs are not linked, either directly or indirectly).

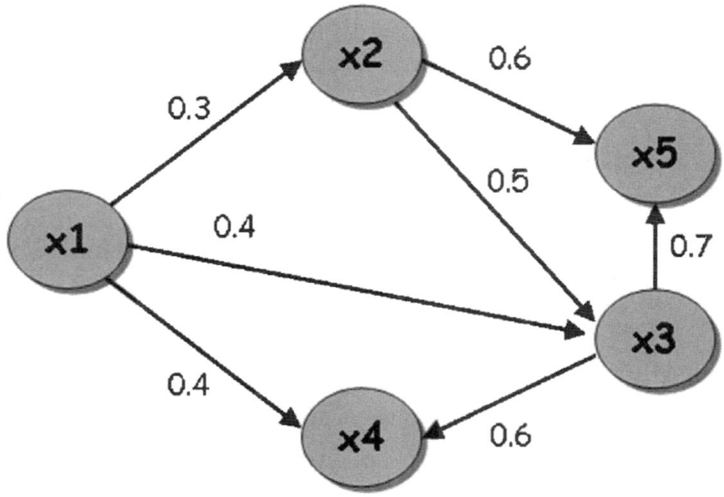

FIGURE 2.1: Connectivity model imposed in the generation of simulated signals. Values on the arrows represent the connections strengths. (Published with permission from [10].)

The simulated signals were obtained starting from a neural mass model of an ROI, fitted to a real cortical estimation of the average activity in an ROI. Signal x_1 was a waveform generated by a model of three neural populations arranged in parallel. Each population simulates neural activity in a specific frequency band: 4–12 Hz, 12–30 Hz, and 30–50 Hz. The model of each population is based on equations proposed by Wendling et al. [11]. The basic idea behind this model is that the oscillations derive from the interactions of pyramidal neurons with three other local neural subsets: excitatory interneurons, slow inhibitory interneurons, and fast inhibitory interneurons. Parameters of the three populations (time constants and synaptic gains) have been set by using an automatic best-fitting procedure to mimic the entire power spectrum density of cortical activity in an ROI.

Subsequent signals $x_2(t)$ to $x_5(t)$ were iteratively obtained according to the imposed connectivity scheme (Fig. 2.1), by adding to signal x_j contributions from the other signals, delayed by intervals τ_{ij}, and amplified by factors a_{ij} and uncorrelated Gaussian white noise. Coefficients of the connection strengths were chosen in a range of realistic values as observed in studies that applied other connectivity estimation techniques, such as structural equation modeling to several memory, motor, and sensory tasks [12,13]. The values used for the connection strengths are given in the legend for Fig. 2.1. The data were generated using different delay schemes on the connectivity pattern imposed. The values used for the delay from the ith ROI to the jth (τ_{ij}) ranged from one sample up to $p-2$, where p is the order of the MVAR model used. These schemes were chosen in order to cover a variety of situations to represent the effect of different delay conditions.

Generation of simulated data was repeated under the following combinations of conditions:

SNR factor levels = [1, 3, 5, 10];

LENGTH factor levels = [2,500, 6,750, 11,250, 15,000, 20,000] data samples, corresponding to a signal length (in seconds) of [10, 27, 45, 60, 80], in the form of several trials of the same length, at a sampling rate of 250 Hz.

The levels chosen for both SNR and LENGTH factors cover a typical range for cortical activity estimated from ERPs with high resolution EEG techniques.

The MVAR model was estimated by means of the Nuttall–Strand method, or multivariate Burg algorithm, which has been demonstrated to provide the most accurate results [14–16].

2.5.2 Evaluation of Performance

A statistical evaluation of the performance of the different estimators required a precise definition of an error function, describing the goodness of the pattern recognition. This was achieved by focusing on the MVAR model structure described in Eq. (2.1) and comparing it with the signal generation scheme. The elements of matrices $\Lambda(k)$ of MVAR model coefficients can be related to the coefficients used in the signal generation and are different from zero only for $k = \tau_{ij}$, where τ_{ij}

is the delay chosen for each pair ij of ROIs and for each direction among them. In particular, for the independent reference source waveform $x_1(t)$, an autoregressive model of the same order of the MVAR was estimated, with coefficients $a_{11}(1), \ldots, a_{11}(p)$ corresponding to the elements $\Lambda_{11}(1), \ldots, \Lambda_{11}(p)$ of the MVAR coefficients matrix. Thus, with the estimation of the MVAR model parameters, we aimed to recover the original coefficients $a_{ij}(k)$ used in signal generation. In this way, reference functions were computed for each of the estimators on the basis of the signal generation parameters. The error function was then computed as the difference between these reference functions and the estimated ones (both averaged in the frequency band of interest).

To evaluate the performance in retrieving connections between areas, we used the Frobenius norm of the matrix reporting the differences between the values of the estimated and the imposed connections (Relative Error)

$$E_{\text{relative}} = \frac{\sqrt{\sum_{i=1}^{m}\sum_{j=1}^{m}\left(\bar{\zeta}_{ij}(f_1, f_2) - \bar{\bar{\zeta}}_{ij}(f_1, f_2)\right)^2}}{\sqrt{\sum_{i=1}^{m}\sum_{j=1}^{m}\left(\bar{\zeta}_{ij}\,(f_1, f_2)\right)^2}} \tag{2.18}$$

where $\bar{\bar{\zeta}}_{ij}(f_1, f_2)$ is the mean value of the estimator function estimated in the frequency band (f_1, f_2), and $\bar{\zeta}_{ij}(f_1, f_2)$ is the mean value of the reference functions obtained from the generation model in the same frequency band. Here, $\bar{\zeta}_{ij}$ can be either DTF or PDC.

Simulations were performed by repeating each generation-estimation procedure 50 times in order to increase the robustness of the successive statistical analysis.

2.5.3 Statistical Analysis

The results obtained were subjected to separate ANOVA study. The first analysis was a three-way ANOVA examining the effect of SNR, LENGTH, and the different methods used to estimate the cortical connectivity (METHODS) on the error for the entire connectivity model estimated. The "within" main factors of the ANOVAs were as follows:

SNR (with four levels: 1, 3, 5, 10),

LENGTH (with six levels: [2,500, 6,750, 11,250, 15,000, 20,000] data samples, corresponding to a signal length (in seconds) of [10, 27, 45, 60, 80], in three trials of the same length, at a sampling rate of 250 Hz) and METHODS (with two levels: DTF and PDC).

The dependent variable was the Relative Error defined in (2.18). The Greenhouse–Geisser correction for the violation of the spherical hypothesis was used. Duncan's post hoc analyses at $p = 0.05$ significance level were then performed.

As explained above, the presence of indirect paths in the network (i.e., a path linking a node to another node on the network not directly but through one or more intermediate nodes) is a critical situation for MVAR-based estimators of causality relations. For this reason, particular attention was paid to the analysis of the estimation error in such indirect relationships. Another ANOVA was performed where the values estimated on the indirect arcs by the three methods, DTF, PDC, and dDTF were compared to the average value of the parameters estimated on the arcs actually present in the generation model. The dependent variable was the absolute level of the connectivity estimates, while the independent factors were METHODS, LENGTH, SNR, and PATHS. The first three main factors had the same levels used before in the other ANOVAs, and the main factor PATHS had three levels, describing the value of the dependent variable for the "indirect" links moving from the area x_1 to the area x_5, that from the area x_2 to the area x_4, and the average value estimated on nonzero arcs in the model. The Greenhouse–Geisser correction was used also in this case.

2.6 RESULTS

Several sets of signals were generated as described in the previous section in order to fit a predefined connectivity pattern involving five cortical areas (shown in Fig. 2.1). The graph depicts the flow of information from area x_1 toward areas x_2–x_5. This connectivity model contains two indirect paths, by which the signal is transmitted to a destination only by indirect relationships with no direct link between the source area and the target one (from area x_2 to area x_4 through x_3, and from area x_1 to area x_5 through x_2 and x_3, by several different paths: $x_1 \rightarrow x_3 \rightarrow x_5$ and $x_1 \rightarrow x_2 \rightarrow x_3 \rightarrow x_5$).

A multivariate autoregressive model of order 10 was fitted to each set of simulated data, which were in the form of about 30 trials of the same length. The procedure of signal generation and connectivity estimation for the different methods was carried out 50 times for each level of the factors SNR and LENGTH for increasing the robustness of the subsequent statistical analysis. The index of performance, i.e., the Relative Error [Eq. (2.18)], and the estimated value on direct/indirect arcs were computed for each generation-estimation procedure and then subjected to two ANOVAs.

In the first ANOVA, the dependent variable was the Relative Error, representing the average error for the entire connectivity pattern estimated. Results revealed a strong influence of the main factors SNR ($F = 205, p < 0.0001$), METHODS ($F = 1190, p < 0.0001$), and LENGTH ($F = 1644, p < 0.0001$), as well as the SNR $\times$ METHOD interaction ($F = 56, p < 0.0001$), on the Relative Error. Figure 2.2 shows the influence of the levels of the main factors LENGTH and SNR on the Relative Error in the entire connectivity graph for each estimator. The errors resulting from DTF, evaluated with respect to the theoretical values that represent the ideal information obtainable by this indicator, were smaller than those resulting from PDC for every level of SNR and LENGTH. This indicates that DTF is more robust with respect to noise and the amount of data available for its estimation. In particular, Fig. 2.2 shows that for each increase in the length of the recordings, there is a constant decrease in estimation error. The influence of the factor SNR is weaker. Post hoc tests revealed that there were no significant differences between levels 3, 5, and 10 of the SNR factor. The bar on each point represents the 95% confidence interval of the mean errors computed across the simulations.

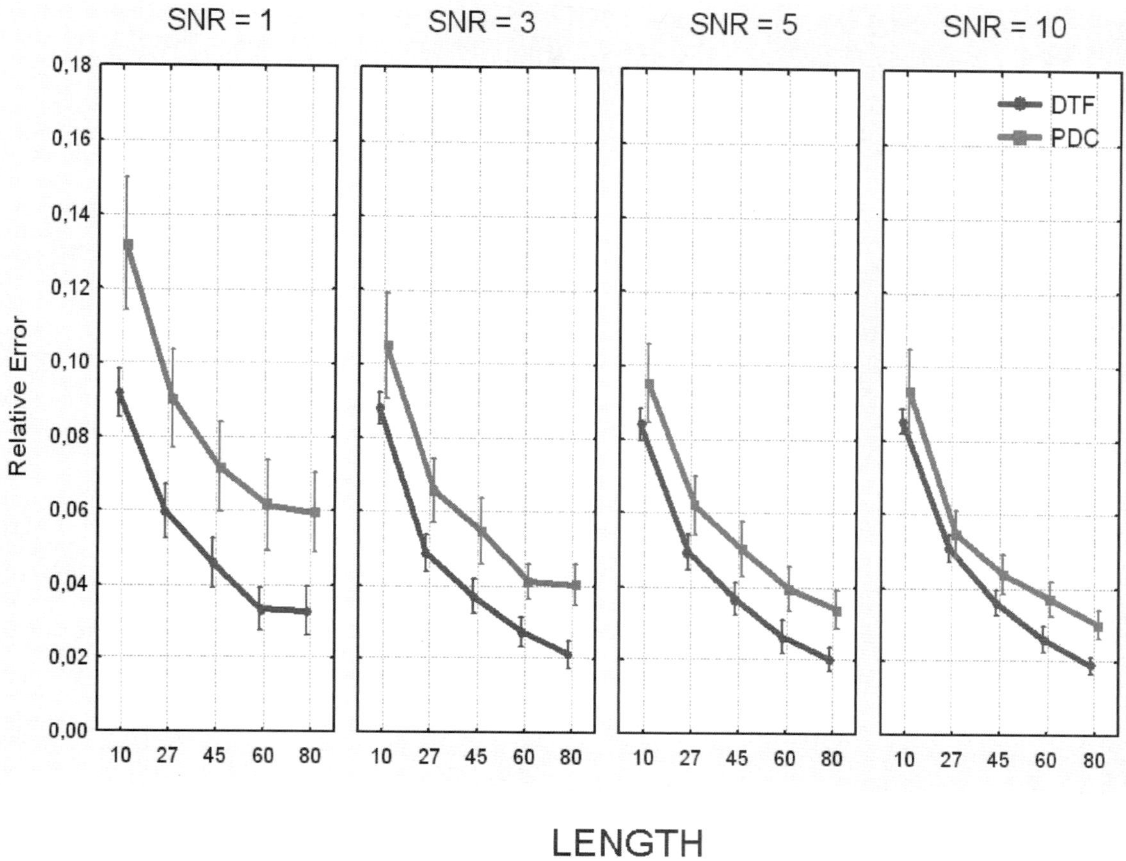

FIGURE 2.2: Results of the ANOVA performed on the relative error made in the estimation of the connectivity flows. The diagram shows the influence of the different levels of the main factors SNR and LENGTH on the estimation of the correct flows in the connection graph employed for the simulation, for the two estimators DTF and PDC. The bar on each point represents the 95% confidence interval of the mean errors computed across the simulations. Duncan's post hoc test (performed at 5%) showed no significant difference between levels 3, 5, and 10 of factor SNR. (Published with permission from [10].)

Particular attention was paid to the ability of the different estimators to distinguish between direct and indirect causality flows. The values estimated for the two indirect pathways in the model were compared to the average value obtained for the direct arcs present in the connectivity model imposed. The values of the connections imposed between cortical areas ranged from 0.3 to 0.7. The two indirect paths analyzed were arc $1 \rightarrow 5$ and arc $2 \rightarrow 4$, indicated by big arrows in Fig. 2.3. As expected, DTF was not always able to attribute a zero value to a nondirect arc. In contrast, PDC and dDTF correctly recognized the indirect path in all cases, estimating values close to zero in these instances (see Fig. 2.4). However, the results obtained by dDTF for the direct arcs were also very

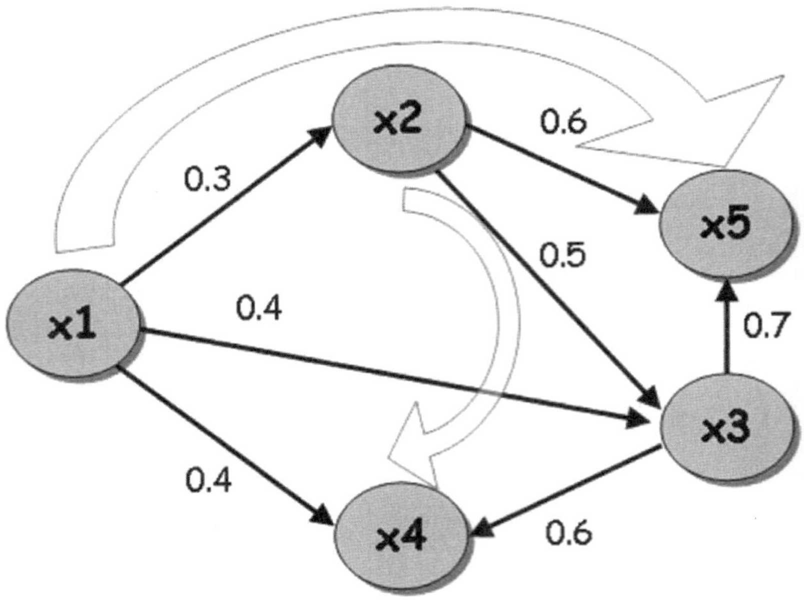

FIGURE 2.3: Connectivity model imposed on the simulated signals. The thick arrows represent the indirect pathways linking the cortical areas. (Published with permission from [10].)

small when compared to the other methods. Post hoc tests revealed no differences between values obtained by dDTF on the average of direct arcs and the values obtained by the same method on indirect arcs. The ANOVA revealed a strong influence of the main factors SNR ($F = 20, p < 0{,}0001$), METHOD ($F = 2980000$, $p < 0{,}0001$), and LENGTH ($F = 493$, $p < 0{,}0001$). Furthermore, all possible interactions of the main factors were significant, with F values not below 3 and p always below 0.001.

2.7 DISCUSSION

The present study examines the performance of three different techniques commonly used to assess information flows between scalp electrodes and local field potentials [7–9] on simulated and real cortical waveforms obtained via the linear inverse problem solution, using the realistic head volume conductor models and high-density EEG recordings. The spatial resolution provided by the techniques presented here has been previously characterized in a series of simulation studies using the present ROI analysis approach [17–20]. The simulation studies involved realistic head models, high-density EEG and MEG setups (with 64 and 128 electrodes as well as 143 magnetic sensors), and standard levels of SNRs (1, 3, 5, 10, 100). The returned errors are lower than 5% for the estimation of the shape (via the cross-correlation measure) and energy (via the Relative Error measurement) of the simulated cortical waveforms. These figures assure that the estimation

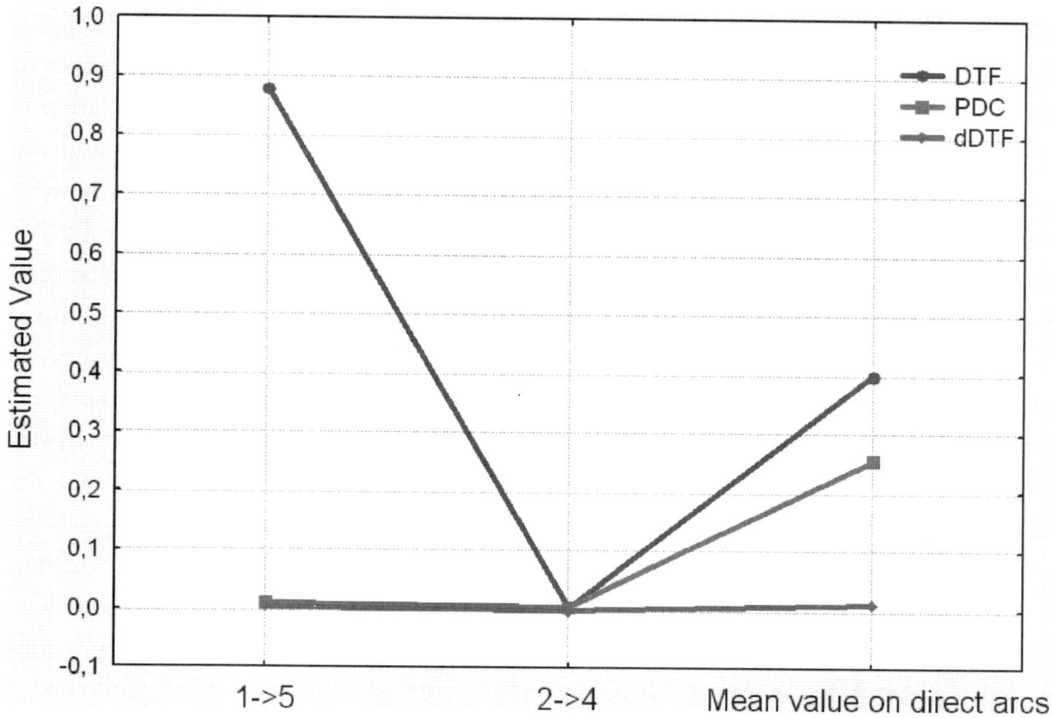

FIGURE 2.4: Average connectivity values estimated on two indirect links ($1 \rightarrow 5$ and $2 \rightarrow 4$) and for all the other existing arcs for the networks by the three methods DTF, PDC, and dDTF during all the simulations. (Published with permission from [10].)

of the cortical current density using high spatial sampling and realistic head models is a reliable process that allows a precise reconstruction of the cortical waveforms in a variety of experimental situations.

All the techniques investigated in the present study are based on the Granger theory and MVAR models. The approach using the DTF, PDC, and dDTF techniques has the advantage of providing connectivity links that can be interpreted in the sense of Granger causality, which includes a concept of directionality. Other techniques have been presented in the literature for the evaluation of functional connectivity of EEG/MEG data. For instance, the technique called dynamic imaging of coherent sources (DICS) [21,22], which uses a spatial filter and a realistic head model, has been recently introduced and employed to assess connectivity between cortical areas from MEG data [21,22]. This technique has the advantage, when compared to the DTF, PDC, and dDTF methods investigated here, of a direct mathematical characterization of its spatial resolution of the point spread function [22]. However, spectral coherence or DICS techniques do not return directly the direction of the flow between cortical areas, though in the latter case, DICS is usually coupled with another technique able to estimate such directional flow, e.g., directionality index [23]. The main difference between DICS and the approach presented here is in the estimation of cortical activity, specifically the fact that the DICS technique is a beamformer [24]. Differences in the performance of beamformers and weighted minimum norm linear inverse techniques depend on the particular experimental setup used. In particular, it has been shown in simulation studies that EEG/MEG beamformers can reconstruct rather precisely the spatial location and the time series of neuronal sources when they exhibit a transient correlation in time [25]. If, however, the correlation between source activities exceeds 30–40% of the total duration of the period over which the beamformer weights are computed, effects of temporal distortion and signal cancellation will be observed [21,26,27]. This suggests that, if the beamformer is obtained using covariance time windows that are long enough with respect to the duration of transient linear interaction between sources, it will return an accurate estimate of spatiotemporal source activity. However, if the covariance windows are sufficiently long, the portion of the stimulus will be small in comparison with the baseline state, and a decrease in the SNR will occur. This would ultimately result in a loss of spatial resolution [21,26].

An interesting issue is related to the possibility of applying the connectivity estimators to multimodal neuroelectric and hemodynamic data (i.e., from EEG/MEG and fMRI measurements). Recent simulation and experimental studies [3,19,28,29] stated that the multimodal integration of EEG, MEG, and fMRI improves the quality of the cortical estimation when compared to any single modality alone. This was obtained by using MEG and fMRI estimation [28,29], EEG and fMRI [3,19,30], or EEG and MEG [18,31]. It is reasonable to expect that, with the improvement of the quality of the cortical estimation given by multimodal integration, the estimation of connectivity could improve as well. However, the effect of multimodal integration in terms of connectivity is not yet addressed in literature, due to the lack of a precise model of electrovascular coupling. It is also worth noting that all the techniques (EEG, MEG, and fMRI) can detect the activity of a particular set of neural sources and are blind to others. For instance, the activity of stellate neurons in the cortex can be detected by fMRI because of their metabolic demand, but not by neuroelectromagnetic measurements to which they are invisible due to the closed field they generate. In contrast, transient (milliseconds) synchronous activity of a small subset of neurons can be detected by EEG and MEG but is invisible to fMRI [32]. Hence, the use of multimodal integration can provide information about cortical activity that moves beyond that offered by a single technique. This is an important point in favor of multimodal integration of EEG, MEG, and fMRI, especially in the perspective of the estimation of functional cortical connectivity.

We performed a series of simulations to evaluate the use of connectivity estimators on test signals generated to simulate the average electrical activity of cerebral cortical regions, as it can be estimated from high resolution EEG recordings gathered under different conditions of noise and length of the recordings. The information on performances, limits of applicability, and range of errors under different levels of the several factors that are of interest in normal EEG recordings were inferred from statistical analysis (ANOVA and Duncan's post hoc tests on the Relative Error and

Single Arc Error). The values used for the strength coefficients in simulations are consistent with the ones estimated in previous studies for a large sample of subjects performing memory, motor, and sensory tasks [12,13].

The simulations provided the following answers to the questions raised in the Introduction chapter:

1. Decreased SNR impairs the accuracy of the connectivity pattern estimation obtained by the DTF, PDC, and dDTF estimators.
2. The length of the EEG recordings has a reliable effect on the accuracy of connectivity pattern estimations. A length corresponding to 27 s of nonconsecutive recordings with an SNR of at least 3 ensures that connectivity patterns can be accurately recovered with an error below 7% for PDC and 5% for DTF.
3. The error variance observed for the DTF estimator is lower than that for PDC or dDTF. However, DTF has the highest bias in the estimate of the connectivity pattern as it includes values for the indirect paths that were not generated in the simulations. It has been noted that there is a higher bias for dDTF than for PDC, as the first estimator often removes from the estimated connectivity pattern some direct paths that were present in the original modeling. In this respect, PDC is characterized by lower bias in the estimation of connectivity patterns under the present conditions of SNR and LENGTH.

In conclusion, results indicated a clear influence of different levels of SNR and LENGTH on the efficacy of the estimation of cortical connectivity in each of the methods. In particular, it has been noted that an SNR equal to or greater than 3 and an overall LENGTH of the estimated cortical data of 6750 data samples (27 s at 250 Hz), even in several short trials, are sufficient to significantly decrease the errors on the indices of quality adopted in this study. These conditions are common in recordings of event-related activity in humans. These recordings are usually characterized by SNR ranging from 3 (movement-related potentials) to 10 (sensory evoked potentials) [33].

The present simulations allowed an evaluation of the level of error expected for different arcs, related to direct or indirect pathways. In addition, comparisons were made between the relative errors obtained on single arcs characterized by only direct connections and those related to multiple paths between the source and the target. The results showed that the error was generally greater when the signal was transmitted to a destination by more than one path. This result (not presented here) is in agreement with the results of the study on indirect paths, according to which such transmission may induce an error in the MVAR estimation.

ANOVA was also performed on the error values obtained for different delay schemes imposed during the signal generation for covering a variety of situations. The ANOVA results indicated that there was no significant influence of the delay on the performances of the methods.

The information obtained from the simulations was used to evaluate the applicability of these methods to actual event-related recordings. The ERP signals, from a Stroop task, showed an SNR between 3 and 5 in the five subjects examined. Therefore, according to the simulation results, a small amount of error in the estimation of cortical connectivity patterns was expected.

All three estimators provided *directional* information (i.e., each of them allowed the establishment of the direction of information flow between two cortical areas) and *directed* information (i.e., they discriminated between direct and indirect connection paths). This information is not available using other techniques to assess the coupling between signals such as standard coherence (which lacks directionality). An evaluation of several methods for the computation of the functional connectivity between EEG/MEG signals was recently performed [34]. It was concluded that, although nonlinear methods such as mutual information, nonlinear correlation, and generalized synchronization [35–37] might be preferred when studying EEG broadband signals that are sensitive to dynamic coupling and nonlinear interactions expressed over many frequencies, the linear measurements (like those presented here) afford a rapid and straightforward characterization of functional connectivity.

REFERENCES

[1] H. Akaike, "A new look at statistical model identification", IEEE Trans Automat Control AC-19, 1974, pp. 716–723, doi:10.1109/TAC.1974.1100705.

[2] P. J. Franaszczuk, K. J. Blinowska, and M. Kowalczyk, "The application of parametric multichannel spectral estimates in the study of electrical brain activity," *Biol. Cybern.*, vol. 51, pp. 239–247, 1985, doi:10.1007/BF00337149.

[3] F. Babiloni, F. Cincotti, C. Babiloni, F. Carducci, A. Basilisco, P. M. Rossini, D. Mattia, L. Astolfi, L. Ding, Y. Ni, K. Cheng, K. Christine, J. Sweeney, and B. He, "Estimation of the cortical functional connectivity with the multimodal integration of high resolution EEG and fMRI data by Directed Transfer Function," *Neuroimage*, vol. 24, no. 1, pp. 118–131, Jan. 1 2005, doi:10.1016/j.neuroimage.2004.09.036.

[4] M. Ding, S. L. Bressler, W. Yang, and H. Liang Ding, "Short-window spectral analysis of cortical event-related potentials by adaptive multivariate autoregressive modeling: data preprocessing, model validation, and variability assessment," *Biol. Cybern.*, vol. 83, pp. 35–45, 2000, doi:10.1007/s004229900137.

[5] L. Astolfi, F. Cincotti, D. Mattia, C. Babiloni, F. Carducci, A. Basilisco, P. M. Rossini, S. Salinari, L. Ding, Y. Ni, B. He, and F. Babiloni, "Assessing cortical functional connectivity by linear inverse estimation and directed transfer function: simulations and application to real data," *Clinical Neurophysiology*, vol. 116 no. 4, pp. 920–932, Apr. 2005b, doi:10.1016/j.clinph.2004.10.012.

[6] M. Kaminski, K. Blinowska, "A new method of the description of the information flow in the brain structures," *Biol. Cybern.*, vol. 65, pp. 203–210, 1991, doi:10.1007/BF00198091.

[7] L. A. Baccalà, and K. Sameshima, "Partial Directed Coherence: a new concept in neural structure determination," *Biol. Cybern.*, vol. 84, pp. 463–474, 2001, doi:10.1007/PL00007990.

[8] M. Kaminski, M. Ding, W. A. Truccolo, and S. Bressler, "Evaluating causal relations in neural systems: Granger causality, directed transfer function and statistical assessment of significance," *Biol. Cybern.*, vol. 85, pp. 145–157, 2001, doi:10.1007/s004220000235.

[9] A. Korzeniewska, M. Manczak, M. Kaminski, K. Blinowska, and S. Kasicki, "Determination of information flow direction between brain structures by a modified Directed Transfer Function method (dDTF)," *Journal of Neuroscience Methods*, vol. 125, pp. 195–207, 2003, doi:10.1016/S0165-0270(03)00052-9.

[10] L. Astolfi, F. Cincotti, D. Mattia, M. G. Marciani, L. A. Baccala, F. De Vico Fallani, S. Salinari, M. Ursino, M. Zavaglia, L. Ding, J. C. Edgar, G. A. Miller, B. He, and F. Babiloni, "A

comparison of different cortical connectivity estimators for high resolution EEG recordings," *Human Brain Mapping*, vol. 28, no. 2, pp. 143–157, Feb 2007, doi:10.1002/hbm.20263.

[11] F. Wendling, F. Bartolomei, J. J. Bellanger, and P. Chauvel, "Epilepsy fast activity can be explained by a model of impaired GABAergic dendritic inhibition," *Eur. J. Neurosci.*, vol. 15, no. 9, pp. 1499–1508, May 2002, doi:10.1046/j.1460-9568.2002.01985.x.

[12] C. Buchel, and K. J. Friston, "Modulation of connectivity in visual pathways by attention: cortical interactions evaluated with structural equation modeling and fMRI," *Cereb. Cortex*, vol. 7, no. 8, pp. 768–778, 1997, doi:10.1093/cercor/7.8.768.

[13] L. Fa-Hsuan, *Spatio temporal brain imaging and modeling*. PhD Thesis, MIT Press, December 2003.

[14] S. L. Marple, *Digital Spectral Analysis with Applications*, Prentice Hall, 1987.

[15] M. S. Kay, *Modern Spectral Estimation*, Prentice Hall, 1988.

[16] A. Schlögl (2003). Comparison of Multivariate Autoregressive Estimators. Available online at: http://www.dpmi.tugraz.ac.at/~schloegl/publications/TR_MVARcomp201.pdf.

[17] F. Babiloni, C. Babiloni, L. Locche, F. Cincotti, P. M. Rossini, and F. Carducci, "High-resolution electroencephalogram: source estimates of Laplacian-transformed somatosensory-evoked potentials using a realistic subject head model constructed from magnetic resonance images," *Med. Biol. Eng. Comput.* vol. 38, no. 5, pp. 512–519, Sep. 2000, doi:10.1007/BF02345746.

[18] F. Babiloni, F. Carducci, F. Cincotti, C. Del Gratta, V. Pizzella, G. L. Romani, P. M. Rossini, F. Tecchio, and C. Babiloni, "Linear inverse source estimate of combined EEG and MEG data related to voluntary movements," *Human Brain Mapping*, vol. 14, no. 3, 2001, doi:10.1002/hbm.1052.

[19] F. Babiloni, C. Babiloni, F. Carducci, G. L. Romani, P. M. Rossini, L. M. Angelone, and F. Cincotti, "Multimodal integration of high-resolution EEG and functional magnetic resonance imaging data: a simulation study," *Neuroimage*, vol. 19, no. 1, pp. 1–15, May 2003, doi:10.1016/S1053-8119(03)00052-1.

[20] F. Babiloni, C. Babiloni, F. Carducci, G. L. Romani, P. M. Rossini, A. Basilisco, S. Salinari, L. Astolfi, and F. Cincotti, "Solving the neuroimaging puzzle: the multimodal integration of neuroelectromagnetic and functional magnetic resonance recordings," *Suppl. Clin. Neurophysiol.*, vol. 57, pp. 450–457, 2004c.

[21] J. Gross, J. Kujala, M. Hämäläinen, L. Timmermann, A. Schnitzler, R. Salmelin, "Dynamic imaging of coherent sources: studying neural interactions in the human brain," *Proc. Natl. Acad. Sci. USA*, vol. 98, no. 2, pp. 694–699, 2001, doi:10.1073/pnas.98.2.694.

[22] J. Gross, L. Timmermann, J. Kujala, R. Salmelin, and A. Schnitzler, "Properties of MEG tomographic maps obtained with spatial filtering," *NeuroImage* vol. 19, pp. 1329–1336, 2003, doi:10.1016/S1053-8119(03)00101-0.

[23] M. G. Rosenblum, and A. S. Pikovsky, "Detecting direction of coupling in interacting oscillators," *Phys. Rev., E Stat. Nonlinear Soft. Matter. Phys.*, vol. 64, p. 045202, 2001, doi:10.1103/PhysRevE.64.045202.

[24] M. X. Huang, J. Shih, R. R. Lee, D. L. Harrington, R. J. Thoma, M. P. Weisend, F. M. Hanlon, K. M. Paulson, T. Li, K. Martin, G. A. Miller, and J. M. Cañive, "Commonalities and differences among vectorized beamformers in electromagnetic source imaging," *Brain Topography*, vol. 16, pp. 139–158, 2003, doi:10.1023/B:BRAT.0000019183.92439.51.

[25] A. Hadjipapas, A. Hillebrand, I. E. Holliday, K. D. Singh, and G. R. Barnes, "Assessing interactions of linear and nonlinear neuronal sources using MEG beamformers: a proof of concept," *Clin. Neurophysiol.*, vol. 116, no. 6, pp. 1300–1313, Jun 2005. Epub 2005 Mar 28, doi:10.1016/j.clinph.2005.01.014.

[26] VanVeen, B. D. vanDrongelen, W. Yuchtman, and M. Suzuki, "A Localization of brain electrical activity via linearly constrained minimum variance spatial filtering," *IEEE Trans. Biomed. Engl.*, vol. 44, pp. 867–880, 1997, doi:10.1109/10.623056.

[27] K. Sekihara, S. Nagarajan, D. Poeppel, and A. Marantz, "Performance of an MEG adaptive-beamformer technique in the presence of correlated neural activities: effects on signal intensity and time-course estimates," *IEEE Trans. Biomed. Eng.*, vol. 49, pp. 1534–1546, 2002, doi:10.1109/TBME.2002.805485.

[28] A. K. Liu, J. W. Belliveau, and A. M. Dale, "Spatiotemporal imaging of human brain activity using functional MRI constrained magnetoencephalography data: Monte Carlo simulations," *Proc. Nat. Acad. Sc.*, vol. 95, no. 15, pp. 8945–8950, 1998, doi:10.1073/pnas.95.15.8945.

[29] A. Liu Dale, B. Fischl, R. Buckner, J. W. Belliveau, J. Lewine, and E. Halgren, "Dynamic statistical parametric mapping: combining fMRI and MEG for high-resolution imaging of cortical activity," *Neuron*, vol. 26, pp. 55–67, 2000, doi:10.1016/S0896-6273(00)81138-1.

[30] F. Babiloni, D. Mattia, C. Babiloni, L. Astolfi, S. Salinari, A. Basilisco, P. M. Rossini, M. G. Marciani, and F. Cincotti, "Multimodal integration of EEG, MEG and fMRI data for the solution of the neuroimage puzzle," *Magnetic Resonance Imaging*, vol. 22, no. 10, pp. 1471–1476, Dec 2004a, doi:10.1016/j.mri.2004.10.007.

[31] F. Babiloni, C. Babiloni, F. Carducci, G. L. Romani, P. M. Rossini, L. M. Angelone, and F. Cincotti, "Multimodal integration of EEG and MEG data: a simulation study with variable signal-to-noise ratio and number of sensors," *Hum. Brain Mapp.*, vol. 22, no. 1, pp. 52–62, May 2004b, doi:10.1002/hbm.20011.

[32] P. L. Nunez, *Neocortical Dynamics and Human EEG Rhythms*, New York: Oxford University Press, 1995.

[33] D. Regan, *Human Brain Electrophysiology. Evoked Potentials and Evoked Magnetic Fields in Science and Medicine*, New York: Elsevier Press, 1989.

[34] O. David, D. Cosmelli, and K. J. Friston, “Evaluation of different measures of functional connectivity using a neural mass model,” *NeuroImage*, vol. 21, pp. 659–673, 2004, doi:10.1016/j.neuroimage.2003.10.006.

[35] M. S. Roulston, “Estimating the errors on measured entropy and mutual information,” *Physica D* vol. 125, pp. 285–294, 1999, doi:10.1016/S0167-2789(98)00269-3.

[36] C. J. Stam, and B. W. van Dijk, “Synchronization likelihood: an unbiased measure of generalized synchronization in multivariate data sets,” *Physica, D*, vol. 163, pp. 236–251, 2002, doi:10.1016/S0167-2789(01)00386-4.

[37] C. J. Stam, M. Breakspear, A. M. van Cappellen van Walsum, and B. W. van Dijk, “Nonlinear synchronization in EEG and whole head MEG recordings of healthy subjects,” *Hum. Brain Mapp.*, vol. 19, no. 2, pp. 63–78, 2003, doi:10.1002/hbm.10106.

CHAPTER 3

Estimation of Cortical Activity by the use of Realistic Head Modeling

3.1 THE PROBLEMS OF CONVENTIONAL ELECTROENCEPHALOGY RECORDINGS

Among the different noninvasive brain imaging techniques, electroencephalography (EEG) and magneto encephalography (MEG) alone directly reflect neuronal firing and exhibit a remarkable temporal resolution (in milliseconds) despite a poor spatial resolution (of the order of fewsquare centimeters). This lack of spatial resolution occurs essentially in the case of EEG because of the spread of brain signals due to the low conductivity of the skull and the rather low signal-to-noise ratio of the data [1]. These potentials originate mainly in the radially oriented cortical pyramidal neurons. The potential distribution arising from these sources is quite wide over the scalp surface because of the different conductivities of cerebrospinal fluid, meninges, skull, and scalp. Furthermore, the distortion of the scalp potential distribution is increased by the ear and eyeholes, which represent shunt paths for intracranial currents [1,2]. As a result, the distribution of the scalp potential shows a low spatial resolution not allowing a reliable localization of the cortical generators of the event-related potentials. Moreover, the variations of electrical reference may enhance or attenuate the spatial components of the potential distribution over the scalp acting as a spatial filter of the cortical generators [2]. For these reasons, the addition of more electrodes is not sufficient per se to improve the spatial information content of an EEG record significantly [1].

3.2 THE HIGH-RESOLUTION ELECTROENCEPHALOGRAPHY

High-resolution EEG technologies have been developed to enhance the spatial information content of EEG activity [1,3]. These technologies consist essentially of high spatial sampling (with 64–128 channels) and surface Laplacian (SL) [4] or spatial deconvolution (SD) estimations [5]. The estimation of the SL of the potential needs the modeling of the scalp surface, while the SD estimation is based on the construction of a multicompartment head volume conductor for simulating cortex, dura mater, skull and scalp surfaces. Most recently, the developed high-resolution EEG enhancement technologies use realistic MRI-constructed subject's head models [5,6]. SL is computed by a spline Laplacian estimator, and SD by a linear inverse estimation method based on boundary-element (BEM) mathematics. The important point of high-resolution EEG technologies is the availability of an accurate model of the head as a volume conductor to be used with advanced computational techniques such as SL or SD. However, appropriate techniques have to be used in order to register the electrode positions on the scalp model. Several authors have shown that it is possible to improve the spatial resolution of EEG by using sophisticated computational algorithms and detailed geometrical models of the head as a volume conductor with the help of the MRI data [1,3,7–10].

3.3 THE SEARCH FOR THE CORTICAL SOURCES

However, the ultimate goal of any EEG or MEG recording is to supply useful information about the brain activity of a subject during a particular task. In order to obtain such information we have to start from these EEG or MEG recordings to arrive at an estimate of cortical activity, by using a body of mathematical techniques known as inverse procedures. Examples of these inverse procedures are the dipole localization, the distributed source and the deblurring or cortical imaging techniques [1,8,11]. Mathematical models for the head as a volume conductor as well as for neural sources are employed by linear and nonlinear minimization procedures to localize putative sources of EEG data. Several studies have indicated the adequacy of the equivalent current dipole as a model for cortical sources [1,2], while the importance of realistic geometry head volume conductor models for the localization of cortical activity was stressed recently [1,8,9,12]. The results of previous intracranial EEG studies support the idea that high-resolution EEG techniques (including head/source models and proper regularization inverse procedures) might model with an acceptable approximation the strengths and the extension of cortical sources of surface EEG data, at least in certain conditions [5,6,10,13].

3.4 HEAD AND CORTICAL MODELS

The estimation of the cortical activity from EEG scalp recordings, based on the solution of the linear inverse problem (showed in the next paragraph) requires the use of electrical models of the structures of the head, interposed between the sensors (electrodes on the scalp) and the electrical sources (the neurons in the cortex). Realistic head models can be reconstructed from T1-weighted MRIs of each experimental subject and used to improve such estimation. In the data analysis performed in this book, scalp, skull, and dura mater compartments were segmented from MRIs and tessellated with about 5000 triangles for each surface (see Fig. 3.1). A source model of the cortex was built using the following procedure: (1) The cortex compartment was segmented from MRIs and tessellated to obtain a fine mesh with about 100,000 triangles; (2) A coarser mesh was obtained by resampling the fine mesh down to about 5000 triangles (this was done by preserving the general features of the neocortical envelope, especially in correspondence of precentral and postcentral gyri and frontal mesial area); (3) An orthogonal unitary equivalent current dipole was placed in each node of the tessellated surface, parallel to the vector sum of the normals to the surrounding triangles. Such a model was used to define the dole source configuration that is the result of the estimation procedure.

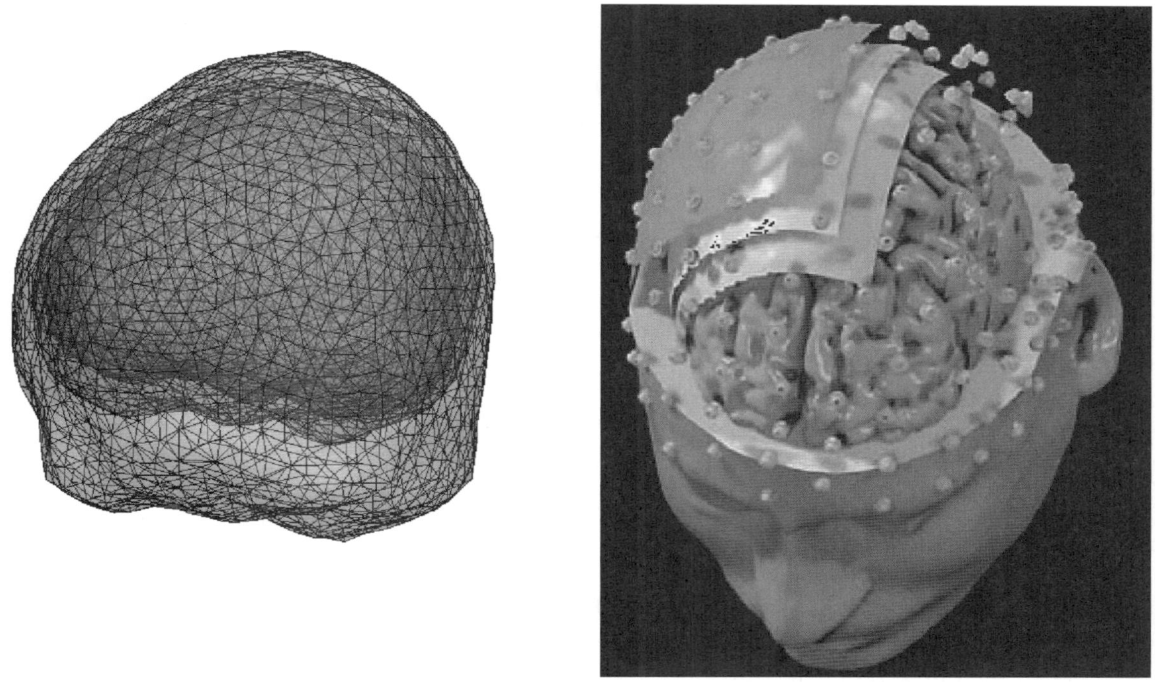

FIGURE 3.1: Realistic models of scalp, skull, dura mater and cortex, tessellated from MRI images of the experimental subject. The electrodes were superimposed to the model by means of a stereophotogrammetric method with a precision <1 mm.

3.4.1 Regions of Interest (ROI)

Thousands of dipoles are used to describe adequately the complex folding of the cortical surface from a mathematical point of view. However, particular regions of the cortex present similar features in terms of the cytoarchitecture (basically, the appearance of the cortex under the light microscope). Then, it could be convenient to use such particular regions in order to estimate the cortical activity, in order to decrease the number of the waveforms to be treated with the connectivity estimators described in the previous chapters. Korbinian Brodmann (1868–1918) was an anatomist who divided the cerebral cortex into numbered subdivisions on the basis of cell arrangements, types, and staining properties (Brodmann classified brain regions based on their cytoarchitecture). In some instances there is a clear link between the microscopic appearance of a region and its function. For example, the "stripe" of the striate cortex delineates the first main cortical area of the visual system (today this area is usually referred to as V1, Brodmann called it area 17). However, it is important to remember that Brodmann's areas (BAs) were identified purely based on visual appearance, which is not necessarily related to function. Figure 3.2 presents the original Brodmann maps. As described in the next pages, several BAs are important in the context of the motor and cognitive tasks.

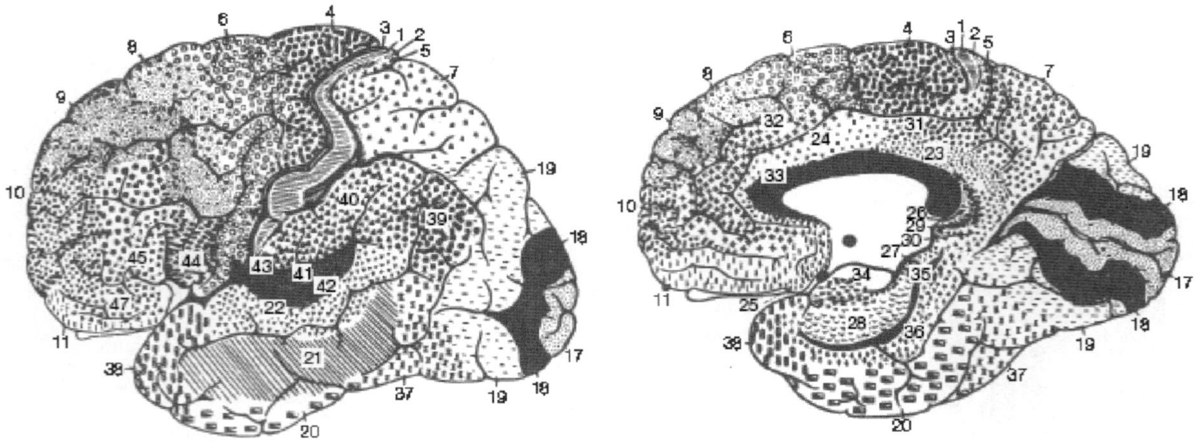

FIGURE 3.2: Brodmann's original cytoarchitectonic maps of the human brain with a ventral view of the brain on the left and a medial view of the cortical areas on the medial wall of the brain on the right.

It is useful to estimate the cortical activity for the cortical patches that have the same cytoarchitectonic properties, i.e., BAs, instead of using data from any single dipole.

The functional connectivity will then be a computing device between the estimated data for the ROIs, depicted along BAs identified on individual cortical model. This strategy uses *a priori* information according to the role of the BAs in the brain functions. The presented technique could also be applied by drawing the ROIs around the cortical estimated peaks of the power spectra activity in the different frequency bands with a post hoc procedure [14]. In this book, the ROIs depicted on the base of the BAs were employed to allow comparison of the functional connectivity patterns elicited by the same experimental behavior across subjects.

The present analysis was based on a cortical source model, namely the source space where the cortical activity generated was identified. Hence, it may be argued that because only cortical sources are modeled, if a deep source is active, then the source reconstruction (and likely the connectivity estimates) could fail. In this context, a widely accepted notion that the main sources for the scalp-recorded EEG signals is derived from the cortex, whereas the thalamus and the basal ganglia can hardly produce appreciable contribution to the scalp EEG should be taken into account [1]. However, even if a subcortical neural source contributed markedly to scalp-recorded EEG, this deep contribution would be distributed over the source space lying on the cortical surface by the employed model. This phenomenon would result in an increase in the low spatial frequency component in the recorded EEG.

3.5 DISTRIBUTED SOURCES ESTIMATE

Accurate estimates of the cortical current density could be obtained by using adequately detailed geometrical reconstruction of the main compartments lying between the cortical generator sources and the EEG or MEG sensors. The estimate of the cortical current density from noninvasive EEG and/or MEG data can be obtained by solving a linear problem. In this problem, the cortical sources to be estimated are related to the noninvasive measurements by means of a transfer matrix (lead field matrix) that mimics the effects of the volume conductor [15,16]. In mathematical terms, the relationship between the modeled sources **x**, the lead field matrix **A**, the EEG/ MEG measurements **b** and the noise **n** can be written as

$$\mathbf{Ax} = \mathbf{b} + \mathbf{n} \tag{3.1}$$

The solution of this linear system provides an estimation of the dipole source configuration **x** that generates the measured EEG potential distribution **b**. The system includes the measurement noise **n**, which is assumed to be normally distributed. **A** is the lead field or the forward transmission matrix, whose *j*th column describes the potential distribution generated on the scalp electrodes by the *j*th unitary dipole. The current density solution vector $\boldsymbol{\xi}$ was obtained as [16]

$$\boldsymbol{\xi} = \arg\min_{x}(||\mathbf{Ax} - \mathbf{b}||_{\mathbf{M}}^{2} + \lambda^{2}||\mathbf{x}||_{\mathbf{N}}^{2}) \tag{3.2}$$

where **M** and **N** are the matrices associated with the metrics with the data and of the source space, respectively, λ is the regularization parameter and $||\mathbf{x}||_{\mathrm{M}}$ represents the **M** norm of the vector **x**. The solution of Eq. (3.2) is given in the inverse operator **G**:

$$\boldsymbol{\xi} = \mathbf{Gb}, \qquad \mathbf{G} = \mathbf{N}^{-1}\mathbf{A}'(\mathbf{A}\mathbf{N}^{-1}\mathbf{A}' + \lambda\mathbf{M}^{-1})^{-1} \tag{3.3}$$

An optimal regularization of this linear system was obtained by the L-curve approach [17,18]. As a metric in the data space we used the identity matrix, whereas as a norm in the source space we use the following metric:

$$(\mathbf{N}^{-1})_{ii} = ||\mathbf{A}_{i}||^{-2} \tag{3.4}$$

where $(\mathbf{N}^{-1})_{ii}$ is the *i*th element of the inverse of the diagonal matrix **N** and all the other matrix elements N_{ij} are set to 0. The L_2 norm of the *i*th column of the lead field matrix **A** is denoted by $||\mathbf{A}_i||$.

3.5.1 Cortical Estimated Waveforms

Using the relations described earlier, an estimate of the signed magnitude of the dipolar moment for each one of the 5000 cortical dipoles was obtained for each time point. As the orientation of the dipole was defined to be perpendicular to the local cortical surface in the head model, the estimation process returned a scalar vector field. To obtain the cortical current waveforms for all the time points, we used a unique "'quasioptimal" regularization λ value for all the analyzed EEG potential distributions. The quasioptimal regularization value was computed as an average of the several λ values obtained by solving the linear inverse problem for a series of EEG potential distributions.

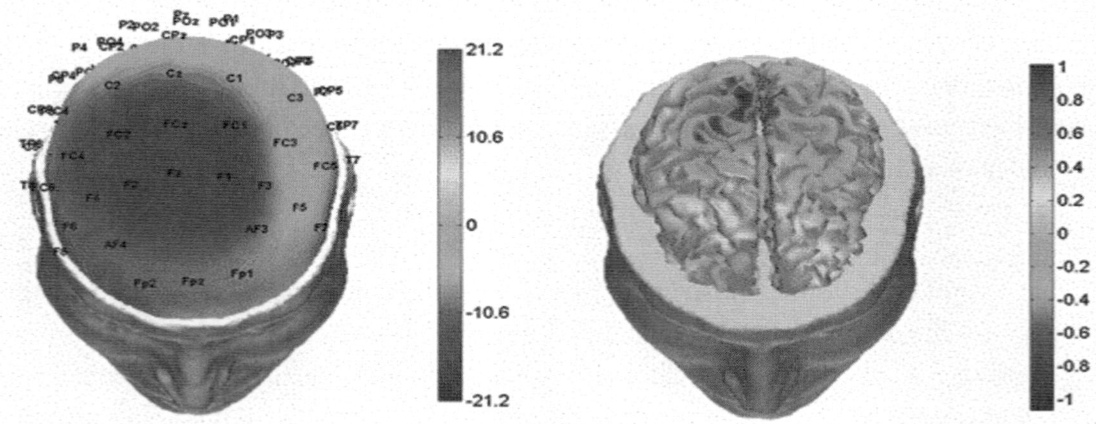

FIGURE 3.3: (*Left*) map of potential recorded on the scalp by means of 58 electrodes. (*Right*) Estimation of the cortical current density obtained from the potentials on the scalp by means of the linear inverse procedure with the use of realistic head models.

These distributions are characterized by an average Global field power (GFP) with respect to the higher and lower GFP values obtained from all the recorded waveforms. An example of the spatial resolution obtained by means of the linear inverse estimation is shown in Fig. 3.3. The instantaneous average of the dipole's signed magnitude belonging to a particular ROI generates the representative time value of the cortical activity in that given ROI. By iterating this procedure on all the time instants of the gathered ERP, the cortical ROI current density waveforms were obtained and they could be taken as representative of the average activity of the ROI, during the task performed by the experimental subjects.

REFERENCES

[1] P.L. Nunez, *Neocortical Dynamics and Human EEG Rhythms*. New York: Oxford University Press; 1995.

[2] P. Nunez, *Electric Fields of the Brain*. New York: Oxford University Press; 1981.

[3] A. Gevins, P. Brickett, B. Costales, J. Le, and B. Reutter, "Beyond topographic mapping: towards functional–anatomical imaging with 124-channel EEG and 3-D MRIs," *Brain Topogr.*, vol. 1, pp. 53–64, 1990, doi:10.1007/BF01128862.

[4] P.L. Nunez, R.B. Silberstein, P.J. Cadiush, J. Wijesinghe, A.F. Westdorp, and R. Srinivasan, "A theoretical and experimental study of high resolution EEG based on surface Laplacians and cortical imaging," *Electroenceph. Clin. Neurophysiol.*, vol. 90, pp. 40–57, 1994, doi:10.1016/0013-4694(94)90112-0.

[5] J. Le, and A. Gevins, "A method to reduce blur distortion from EEG's using a realistic head model," *IEEE Trans. Biomed. Eng.*, vol. 40, pp. 517–528, 1993, doi:10.1109/10.237671.

[6] F. Babiloni, C. Babiloni, F. Carducci, L. Fattorini, C. Anello, P. Onorati, and A. Urbano, "High resolution EEG: a new model-dependent spatial deblurring method using a realistically-shaped MR-constructed subject's head model," *Electroenceph. Clin. Neurophysiol.*, vol. 102, pp. 69–80, 1997, doi:10.1016/S0921-884X(96)96508-X.

[7] F. Babiloni, C. Babiloni, L. Locche, F. Cincotti, P.M. Rossini, and F. Carducci, "High-resolution electroencephalogram: source estimates of Laplacian-transformed somatosensory-evoked potentials using a realistic subject head model constructed from magnetic resonance images," *Med. Biol. Eng. Comput.*, vol. 38, no. 5, pp. 512–519, Sep 2000, doi:10.1007/BF02345746.

[8] A. Gevins, "Dynamic functional topography of cognitive task," *Brain Topography*, vol. 2, pp. 37–56, 1989b, doi:10.1007/BF01128842.

[9] A. Gevins, P. Brickett, B. Reutter, and J. Desmond, "Seeing through the skull: advanced EEGs use MRIs to accurately measure cortical activity from the scalp," *Brain Topogr.*, vol. 4, pp. 125–131, 1991, doi:10.1007/BF01132769.

[10] A. Gevins, J. Le, N. Martin, P. Brickett, and J. Desmond, B. Reutter, "High resolution EEG: 124-channel recording, spatial deblurring and MRI integration methods," *Electroenceph. Clin. Neurophysiol.*, vol. 39, pp. 337–358, 1994, doi:10.1016/0013-4694(94)90050-7.

[11] A.M. Dale, and M. Sereno, "Improved localization of cortical activity by combining EEG and MEG with MRI cortical surface reconstruction:a linear approach," *J. Cogn. Neurosci.*, vol. 5, pp. 162–176, 1993.

[12] A. Gevins, J. Le, H. Leong, L.K. McEvoy, and M.E. Smith, "Deblurring," *J. Clin. Neurophysiol.*, vol. 16, no. 3, pp. 204–213, 1999, doi:10.1097/00004691-199905000-00002.

[13] B. He, Y. Wang, and D. Wu, "Estimating cortical potentials from scalp EEG's in a realistically shaped inhomogeneous head model," *IEEE Trans. Biomed. Eng.*, vol. 46, pp. 1264–1268, 1999, doi:10.1109/10.790505.

[14] J. Gross, J. Kujala, M. Hämäläinen, L. Timmermann, A. Schnitzler, and R. Salmelin, "Dynamic imaging of coherent sources: studying neural interactions in the human brain," *Proc. Natl. Acad. Sci. USA*, vol. 98, no. 2, pp. 694–699, 2001, doi:10.1073/pnas.98.2.694.

[15] R.D. Pascual-Marqui, Reply to comments by Hamalainen, Ilmoniemi and Nunez. In ISBET Newsletter N.6, December 1995. Ed: W. Skrandies., 16–28.

[16] R. Grave de Peralta Menendez, and S.L. Gonzalez Andino, Distributed source models: standard solutions and new developments. In: Uhl, C. (Ed.), Analysis of neurophysiological brain functioning. Springer Verlag, pp.176–201, 1999.

[17] P.C. Hansen, "Analysis of discrete ill-posed problems by means of the L-curve," *SIAM Review*, vol. 34, pp. 561–580, 1992a, doi:10.1137/1034115.

[18] P.C. Hansen, "Numerical tools for the analysis and solution of Fredholm integral equations of the first kind," *Inverse Problems*, vol. 8, pp. 849–872, 1992b, doi:10.1088/0266-5611/8/6/005.

CHAPTER 4

Application: Estimation of Connectivity from Movement-Related Potentials

The estimation of connectivity patterns by using the direct transfer function (DTF) and structural equation model (SEM) on high-resolution electroencephalographic (EEG) recordings has been applied to analyze a simple movement task. In particular, we considered the right hand finger tapping movement, externally paced by a visual stimulus. This task was chosen as it has been very well studied in literature with different brain imaging techniques like EEG or functional magnetic resonance imaging [1–3].

4.1 SUBJECTS AND EXPERIMENTAL DESIGN

Three right-handed healthy subjects (age 23.3 ± 0.58 years, one male and two females) participated in the study after the informed consent was obtained according to University of Illinois at Chicago/Institutional Review Board (UIC/IRB). Subjects were seated comfortably in an armchair with both arms relaxed and resting on pillows and they were requested to perform fast repetitive right finger movements cued by visual stimuli. Ten to fifteen blocks of 2 Hz thumb oppositions were recorded, with each 30 s blocks of finger movement and rest. During motor task, subjects were instructed to avoid eye blinks, swallowing, or any movement other than the required finger movements.

4.2 EEG RECORDINGS

Event-related potential (ERP) data were recorded with 96 electrodes; data were recorded with a left ear reference and submitted to the artifact removal processing. Six hundred ERP trials of 600 ms duration were acquired. The A/D sampling rate was 250 Hz. The surface electromyographic (EMG) activity of the muscle was also recorded. The onset of the EMG response served as zero time. All data were visually inspected, and trials containing artifacts were rejected. Semiautomatic supervised threshold criteria were used for the rejection of trials contaminated by ocular and EMG artifacts, as described in detail elsewhere [4]. After the EEG recording, the electrode positions were digitized using a three-dimensional (3D) localization device with respect to the anatomic landmarks of the head (nasion and two preauricular points). The analysis period for the potentials time locked to the movement execution was set from 300 ms before to 300 ms after the EMG trigger (0 time); the ERP time course was divided in two phases relative to the EMG onset; the first period, labeled as "PRE," marked the 300 ms before the EMG onset and was intended as a generic preparation period; the second labeled as "POST," lasted up to 300 ms after the EMG onset and was intended to signal the arrival of the movements somatosensory feedback. We kept the same PRE and POST nomenclature for the signals estimated at the cortical level.

4.2.1 Selection of Regions of Interest (ROIs)

Several cortical regions of interest (ROIs) were drawn on the computer-based cortical reconstruction of the individual head models (three subjects). The ROIs representing the left and right primary somatosensory (S1) areas included the Brodmann areas (BA) 3, 2, and 1, while the ROIs representing the left and right primary motor (MI) included the BA 4. The ROIs representing the supplementary motor area (SMA) were obtained from the cortical voxels belonging to the BA 6. We further separated the proper and anterior SMA indicated with BA 6P and 6A, respectively. Furthermore, ROIs from the right and the left parietal areas including the BA 5, 7, and the occipital areas (BA 19) were also considered. In the frontal regions, the BA 46, 8, and 9 were also selected. A selection of the gathered ERPs related to the visual paced finger tapping task from one of the subjects is depicted in the left part of Fig. 4.1. The waveforms are relative to the signals gathered from the standard electrode leads of the augmented 10–20 system, represented on the realistic scalp reconstruction of the subject, and represented the average of the artifact-free trials. By means of the linear inverse procedure, the estimation of the current density waveforms in the ROIs of interest was then performed, according to the Eqs. (3.1)–(3.4).

The instantaneous average of the signed magnitude of all the dipoles belonging to a particular ROI was used to estimate the average cortical activity in that ROI, during the entire interval of the experimental task. These waveforms could then be subjected to the MVAR modeling in order to estimate the connectivity pattern between ROIs, by taking into account the time varying increase or decrease of the power spectra in the frequency bands of interest.

The estimated current density waveforms for the same subject are represented in some selected ROIs in the right part of the Fig. 4.1. Relevant cortical activity was observed to be different from baseline in the left ROIs representing parietal (BA 5), premotor (BA 6A), sensorimotor (BA 3, 2, 1, BA 4), and prefrontal (BA 8) cortical areas, whereas similar statistical data were obtained for the ROIs located in the right hemisphere in premotor (BA 6A) and prefrontal (BA 8) cortical areas.

4.2.2 Definition of the *A Priori* Model for the Effective Connectivity

From the results obtained by the simulation study described in the previous chapter, the performances seem to be rather good for a correct assumption of the hypothesized model, but they decrease when even a simple error is made, depending on the error type. This degradation of the performances seems to indicate that detailed cortical areas cannot be used in the first step of the network modeling.

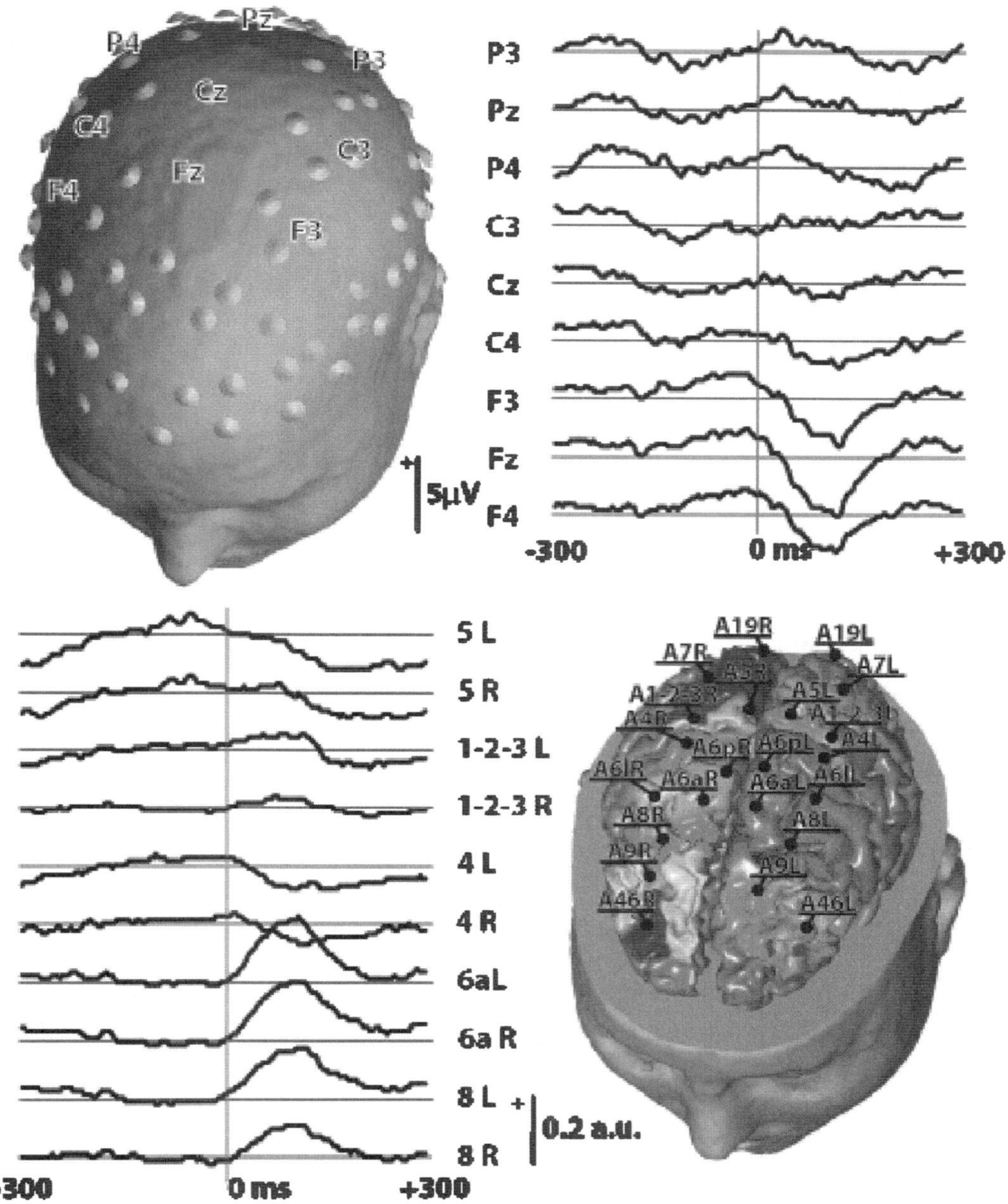

FIGURE 4.1: (*Left*) a selection of the ERPs gathered from the standard electrode leads of the augmented 10–20 system, represented on the realistic geometry scalp reconstruction of the subject (average of the artifact free trials). The onset of the electromyographic (EMG) signal for the start of the movement of the right finger is at the 300 ms from the beginning of the ERP trial. (*Right*) the estimated (via the solution of the linear inverse problem) current density waveforms, represents some selected ROIs on the realistic cortex reconstruction of the same subject. Each ROI is represented in a different grey level. (Published with permission from [5].)

By using a coarse model of the cortical network to be fitted on the EEG data, there is an increase of the statistical power and a decrease of the possibility to generate an error in a single arc link. In the present human study, such observation was taken into account by selecting a coarse model for the brain areas subserving the task being analyzed. This simplified model does not take complete account of all the possible regions engaged in the task, and all the possible connections between them. Elaborate models, permitting also cyclical connections between regions can become computationally unstable [6].

Our model of interactions between cortical areas is based of previous results on similar tasks obtained with different brain imaging methods. It is sufficient to address some key questions regarding the influence of the premotor and motor areas toward the prefrontal cortical areas during the task analyzed. The anatomical model employed is based on the principal cortical areas recognized as active during this simple task in these studies. Cortical areas used in this human study included the prefrontal areas (PF), which includes large BAs 8, 9, and 46; the premotor areas (PM), including the BA 6, the sensorimotor areas (SM) including the BAs 4, 3, 2, and 1, and the parietal areas (P), generated by the union of the BAs 5 and 7. The model employed *a priori* knowledge about the flow of connections between these macroareas, as derived from neuroanatomy and fMRI studies. In particular, information flow was hypothesized to exist from the parietal (P) areas toward the sensorimotor (SM), the premotor (PM) and the prefrontal (PF) ones [1,2,7].

4.2.3 Statistical Evaluation of Connectivity Measurements by SEM and DTF

As described before, the statistical significance of the connectivity pattern estimated with SEM technique was assured by the fact that in the context of the multivariate normally distributed variables, the minimum of the maximum likelihood function F_{ML}, multiplied by $(N-1)$, follows a χ^2 distribution with $(p(p+1)/2)-t$ degrees of freedom, where t is the number of parameters to be estimated and p is the total number of observed variables (endogenous + exogenous). The χ^2 statistic test can then be used to infer statistical significance of the structural equation model obtained.

The situation is different for all areas of the statistical significance of the DTF measurements, as the DTF function has a highly nonlinear relation to the time series data from which they are derived, and the distribution of their estimators is not well established. This makes tests of significance difficult to perform, unless the tests based on this empirical distribution can be performed. A possible solution to this problem was proposed in Kaminski et al. [8]. It consists of the use of a surrogate data technique [9], in which an empirical distribution for random fluctuations of a given estimated quantity is generated by estimating the same quantity from several realizations of surrogate datasets, in which deterministic interdependency between variables were removed. In order to ensure that all features of each dataset are as similar as possible to the original dataset, with the exception of channel coupling, the very same data are used; any time-locked coupling between channels is disrupted by shuffling phases of the original multivariate signal. As the EEG signal had been divided into single trials, each surrogate dataset was built up by scrambling the order of epochs, using different sequences for each channel. The set properties of univariate surrogate signals are not influenced by this shuffling procedure, since only the epoch order is varied. Moreover, since no shuffling was performed between single samples and the temporal correlation, the spectral features, of univariate signals are the same for original and surrogate dataset, thus allowing the estimate of different distributions of DTF fluctuations for each frequency band. One thousand surrogate datasets were generated as described above, and DTF spectra were estimated from each dataset. For each channel pair and for each frequency bin, the 99 percentile was computed and subsequently considered as a significance threshold.

4.2.4 Connectivity Pattern Representation

The connectivity patterns are represented by arrows pointing from one cortical area ("the source") toward another one ("the target"). The grey level and size of the arrow code the strength of the functional connectivity estimated between the source and the target. The bigger and lighter the arrow, the stronger the connection. Only the cortical connections statistically significant at $p < 0.01$ are represented, according to the thresholds obtained as previously described.

The connectivity patterns in the different frequency bands (theta, 4–8 Hz; alpha, 8–12 Hz; beta, 13–30 Hz; gamma, 30–40 Hz), and between the different cortical regions were summarized by using indices representing the total flow from and toward the selected cortical area. The total inflow in a particular cortical region was defined as the sum of the statistically significant connections from all the other cortical regions toward the selected area. A sphere centered on the cortical region, whose radius is linearly related to the magnitude of all the incoming statistically significant links from the other regions, represents the total inflow for each ROI. Inflow information is also coded through a grey level scale. This information depicts each ROI as the target of functional connections from the other ROIs. The same conventions were used to represent the total outflow from a cortical region, generated by the sum of all the statistical significant links.

4.3 RESULTS

The results of the application of the SEM method for the estimation of the connectivity on the event-related potential recordings is depicted in Fig. 4.2, which shows the statistically significant cortical connectivity patterns obtained for the period preceding the movement onset in the subject #1, in the alpha frequency band. Each pattern that connects one cortical area ("the source") to another one ("the target") is represented by arrows. The grey level and sizes of the arrows code the strength level of the functional connectivity observed between ROIs. The labels indicate the names of the ROIs employed. Note that the connectivity pattern during the period preceding the movement in the alpha band involves mainly the parietal left ROI (Pl) coincident with the BAs 5 and 7, functionally connected with the left and right premotor cortical ROIs (PMl and PMr), the left sensorimotor area

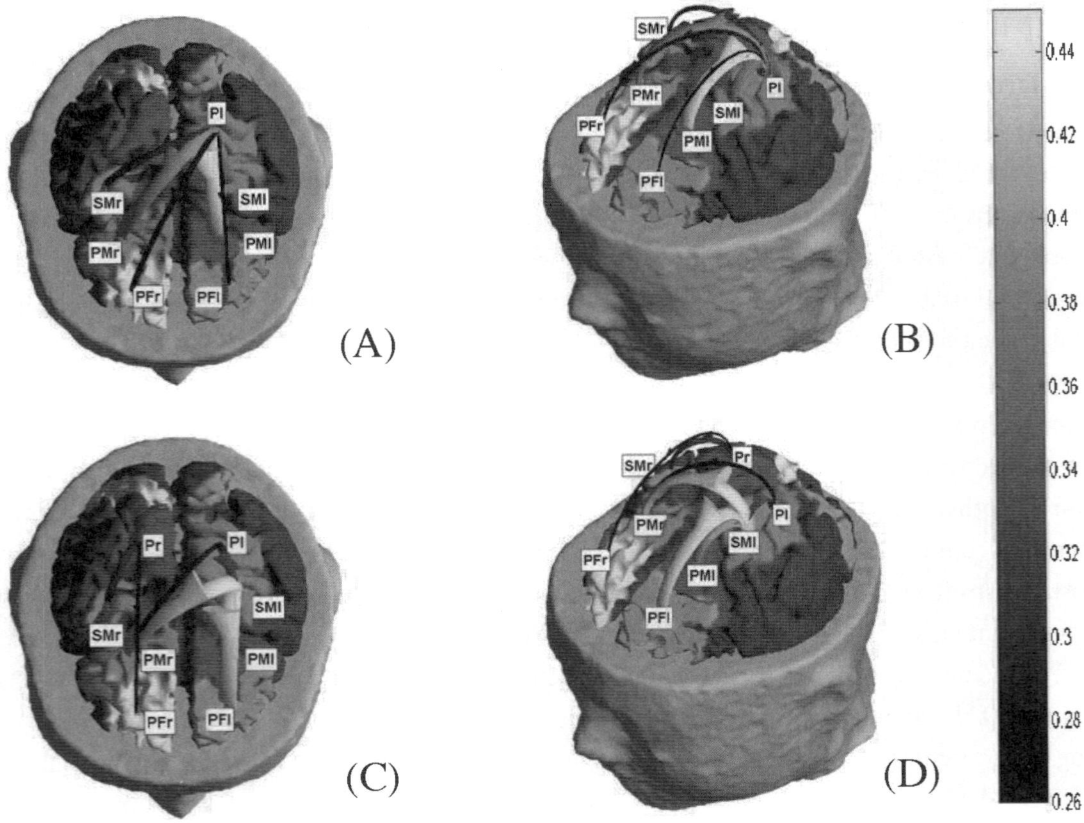

FIGURE 4.2: (A–D) The cortical connectivity pattern obtained for the period preceding and following the movement onset in the subject, in the alpha (8–12 Hz) frequency band. The realistic head model and cortical envelope of the subject analyzed obtained from sequential MRIs is used to display the connectivity pattern. Such a pattern is represented with arrows that move from one cortical area toward another. The grey level and sizes of arrows code the strength level of the functional connectivity observed between ROIs. The labels are relative to the name of the ROIs employed. (A–B) Connectivity patterns obtained from ERP data before the onset of the right finger movement (electromyographic onset; EMG), from above (*left*) and from the left of the head (*right*). (C–D) Connectivity patterns obtained after the EMG onset. (Published with permission from [10].)

(SMl), and both the prefrontal ROIs (PFl and PFr). The stronger functional connections are relative to the link between the left parietal and the premotor areas of both cerebral hemispheres. After the preparation and the beginning of the finger movement POST period changes in the connectivity pattern can be noted. In particular, the origin of the functional connectivity links is positioned in the sensorimotor left cortical areas (SMl). From there, functional links are established with prefrontal left (PFl), and both the premotor areas (PMl. PMr). A functional link emerged in this condition connecting the right parietal area (Pr) with the right sensorimotor area (SMr). The left parietal area

(Pl), much active in the previous condition, was linked instead with the left sensorimotor (SMl) and right premotor (PMr) cortical areas.

Figure 4.2 shows the cortical connectivity patterns obtained for the period preceding the movement onset in the subject #1, in the alpha frequency band. Each pattern is represented by arrows that connect one cortical area to another. The grey level and sizes of arrows code the strength level of the functional connectivity observed between ROIs. The labels indicate the names of the ROIs employed. Note that the connectivity pattern during the period preceding the movement in the alpha band involves mainly the parietal left ROI (Pl) coincident with BAs 5 and 7, which is functionally connected with the left and right premotor cortical ROIs (PMl and PMr), the left sensorimotor area (SMl), and both the prefrontal ROIs (PFl and PFr). The stronger functional connections are relative to the link between the left parietal and the premotor areas of both cerebral hemispheres. After the preparation and the beginning of the finger movement, in the POST period changes in the connectivity pattern can be noted.

In particular, the origin of the functional connectivity links is positioned in the sensorimotor left cortical areas (SMl). From there, functional links are established with prefrontal left (PFl), both the premotor areas (PMl. PMr). A functional link emerged in this condition connecting the right parietal area (Pr) with the right sensorimotor area (SMr). The left parietal area (Pl) much active in the previous condition was linked instead with the left sensorimotor (SMl) and right premotor (PMr) cortical areas.

The connectivity estimations performed by DTF on the movement-related potentials were first analyzed from a statistical point of view via the shuffling procedure described. The order of the MVAR model used for each DTF estimation had to be determined for each subject and in each temporal interval of the cortical waveforms segmentation (PRE and POST interval). The Akaike information criterion (AIC) procedure was used and it returned an optimal order between 6 and 7 for all the subjects, in both PRE and POST intervals. On such cortical waveforms, the DTF computational procedure described in the Methods section was applied. Figure 4.3 shows the cortical connectivity patterns obtained for the period preceding and following the movement onset in the subject #1. Here, we present the results obtained for the connectivity pattern in the alpha band (8–12 Hz), since the ERP data related to the movement preparation and execution are particularly responsive in such frequency interval (for a review, see [12]). The task-related pattern of cortical connectivity was obtained by calculating the DTF between the cortical current density waveforms estimated in each ROI depicted on the realistic cortex model. The connectivity patterns between the ROIs have been represented by arrows pointing from one cortical area toward another one. The grey level and size of the arrow code the strength of the functional connectivity estimated between the source and the target ROI. Labels indicate the ROIs involved in the estimated connectivity pattern. Only the cortical connections statistically significant at $p < 0.01$ are represented, according to the thresholds obtained by the shuffling procedure described earlier. It can be noted that the

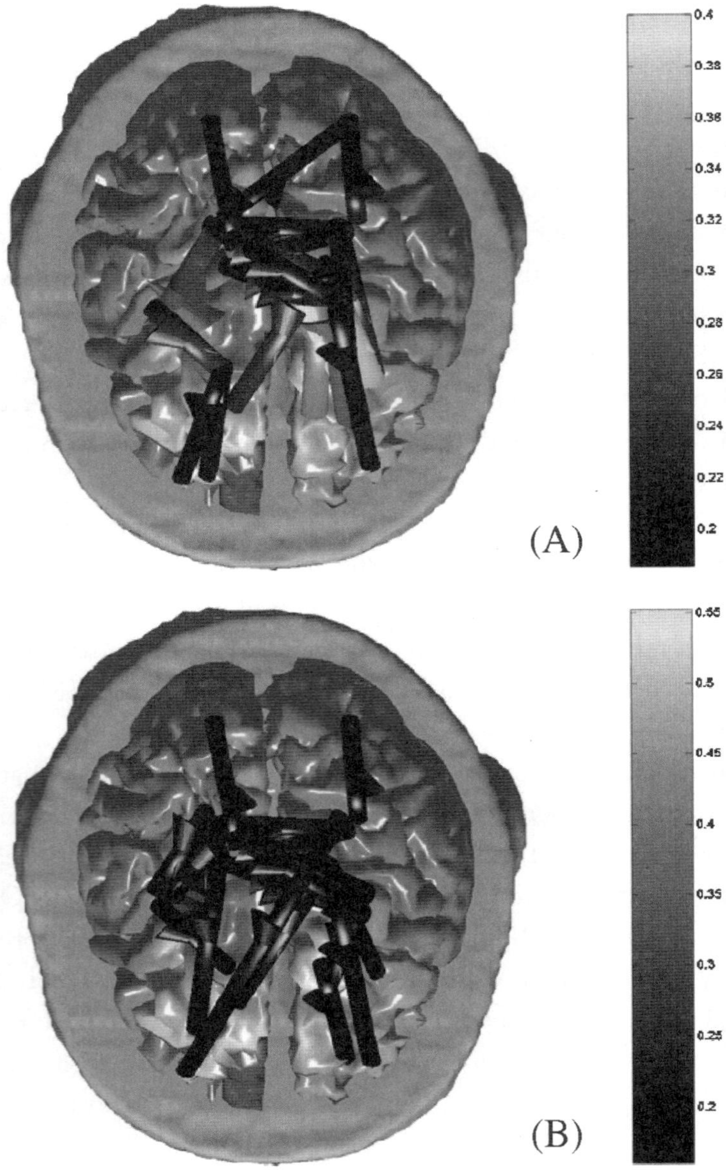

FIGURE 4.3: Cortical connectivity patterns obtained by the DTF method, for the period preceding and following the movement onset, in the alpha (8–12 Hz) frequency band. The patterns are shown on the realistic head model and cortical envelope of the subject analyzed, obtained from sequential MRIs. Functional connections are represented with arrows, moving from a cortical area toward another one. The grey level and sizes of the arrows code the strengths of the connections. (A) Connectivity pattern obtained from ERP data before the onset of the right finger movement (electromyographic onset; EMG). (B) Connectivity patterns obtained after the EMG onset. Same conventions as described earlier. (Published with permission from [11].)

connectivity patterns during the period preceding and following the movement in the alpha band involves bilaterally the parietal and sensorimotor ROIs, which are also functionally connected with the premotor cortical ROIs. A minor involvement of the prefrontal ROIs is also observed. The stronger functional connections are relative to the link between the premotor and prefrontal areas of both cerebral hemispheres. After the preparation and the beginning of the finger movement, in the POST period, slight changes in the connectivity patterns can be noted.

4.4 DISCUSSION

In the case in which the SEM methodology was applied on the recorded high-resolution EEG data the model of interactions between cortical areas is based on previous results on similar tasks obtained with different brain imaging methods. Such a model is sufficient to address some key questions regarding the influence of the premotor and motor areas toward the prefrontal cortical areas during the task analyzed. The finger tapping data analyzed here present a high SNR and a large number of trials, resulting in an extended record of ERP data. Hence, the present simulation results suggest the optimal performance of the SEM method as applied to the human ERP potentials. The connectivity pattern estimated via SEM reveals the potentiality of the employed methodology including the use of high-resolution EEG recordings, the generation of a realistic head model by using sequential MRIs, and the estimation of the cortical activity with the solution of linear inverse problem. With this methodology, it would be possible not only to detect which of the cortical areas are activate during a particular (motor) task but also how these areas are effectively interconnected in subserving that given task. In particular, the influence of the parietal area has been observed toward the premotor cortical areas during the task preparation, consistent with the role that the parietal areas have in the engagement of attentive resources as well as temporization, as assessed by several electrophysiological studies on primate or hemodynamical studies on humans [13] It is interesting to note the cortical areas behaving as the most relevant origin of functional links, occurring when the somatosensory reafferences arrive from the periphery to the cortex. In fact, the left sensorimotor area becomes very active with respect to the left parietal one, which, in turn, is mainly engaged in the time period preceding the finger movement. Connections between the sensorimotor area and the premotor and left prefrontal areas are appropriate to distribute the information related to the movement of the finger to the higher functional regions (prefrontal and premotor).

From a physiological point of view, the results obtained by estimating the connectivity patterns with the DTF are consistent and integrate those already present in literature on simple finger movements, as they have been obtained with neuroelectric measurements. A study employing ERP measurements from scalp electrodes and the assessment of connectivity with the non directional coherence methods has underlined the role of the primary sensorimotor and supplementary motor areas in the processing of the movements [1]. The connectivity patterns depicted in the premotor and prefrontal ROIs here analyzed are in agreement with earlier hypotheses formulated in literature

[14–16]. The aforementioned studies have suggested that the dorsolateral and the ventral premotor cortices are specifically activated by movements guided by sensory information as opposed to movements carried out with no sensory control. The activity noted in the parietal areas (BA 5) in the present study, could be associated with the role that this area has in the somatosensory–motor integration underlying movement performing. In fact, it has been hypothesized that this area could be regarded as a higher order somatosensory zone devoted to the analysis of proprioceptive information from joints for the appropriate motor control [17].

REFERENCES

[1] C. Gerloff, J. Richard, J. Hadley, A. E. Schulman, M. Honda, and M. Hallett, "Functional coupling and regional activation of human cortical motor areas during simple, internally paced and externally paced finger movements," *Brain*, vol. 121, no. Pt 8, pp. 1513–1531, 1998, doi:10.1093/brain/121.8.1513.

[2] A. S. Gevins, B. A. Cutillo, S. L. Bressler, N. H. Morgan, R. M. White, J. Illes, and D.S. Greer, "Event-related covariances during a bimanual visuomotor task. II. Preparation and feedback," *Electroencephalogr. Clin. Neurophysiol.*, vol. 74, pp. 147–160, 1989, doi:10.1016/0168-5597(89)90020-8.

[3] L. Jancke, R. Loose, K. Lutz, K. Specht, and N. J. Shah, "Cortical activations during paced finger-tapping applying visual and auditory pacing stimuli," *Brain Res. Cogn. Brain Res.*, vol. 10, no. 1–2, pp. 51–66, 2000, doi:10.1016/S0926-6410(00)00022-7.

[4] D. V. Moretti, F. Babiloni, F. Carducci, F. Cincotti, E. Remondini, P. M. Rossini, S. Salinari, and C. Babiloni, Computerized processing of EEG-EOG-EMG artifacts for multi-centric studies in EEG oscillations and event-related potentials," *Int. J. Psychophysiol.*, vol. 47, no. 3, 199–216, Mar. 2003, doi:10.1016/S0167-8760(02)00153-8.

[5] L. Astolfi, et al., "Assessing cortical functional connectivity by linear inverse estimation and directed transfer function: simulations and application to real data," *Clinical Neurophysiol.*, vol. 116, no. 4, pp. 920–932, Apr 2005, doi:10.1016/j.clinph.2004.10.012.

[6] A. R. McIntosh and F. Gonzalez-Lima, "Structural equation modeling and its application to network analysis in functional brain imaging," *Hum. Brain Mapp.*, vol. 2, pp. 2–22, 1994, doi:10.1002/hbm.460020104.

[7] A. Urbano, C. Babiloni, P. Onorati, and F. Babiloni, "Dynamic functional coupling of high resolution EEG potentials related to unilateral internally triggered one-digit movements," *Electroencephalogr. Clin. Neurophysiol.*, vol. 106, no. 6, pp. 477–487, 1998, doi:10.1016/S0013-4694(97)00150-8.

[8] M. Kaminski, M. Ding, W.A. Truccolo, and S. Bressler, "Evaluating causal relations in neural systems: Granger causality, directed transfer function and statistical assessment of significance," *Biol. Cybern.*, vol. 85, pp. 145–157, 2001, doi:10.1007/s004220000235.

[9] J. Theiler, S. Eubank, A. Longtin, B. Galdrikian, and J. D. Farmer, "Testing for nonlinearity in time series: the method of surrogate data," *Physica D*, vol. 58, pp. 77–94, 1992, doi:10.1016/0167-2789(92)90102-S.

[10] L. Astolfi, et al., "Estimation of the cortical connectivity by high resolution eeg and structural equation modeling: simulations and application to finger tapping data," *IEEE Trans. Biomed. Eng.*, vol. 52, no. 5, pp. 757–768, May 2005, doi:10.1109/TBME.2005.845371.

[11] L. Astolfi, et al., "Estimation of the effective and functional human cortical connectivity with structural equation modeling and directed transfer function applied on high resolution EEG," *Magn. Reson. Imaging*, vol. 22, no. 10, pp. 1457–1470, Dec. 2004, doi:10.1016/j.mri.2004.10.006.

[12] G. Pfurtscheller and F. H. Lopes da Silva, "Event-related EEG/MEG synchronization and desynchronization: basic principles," *Clin. Neurophysiol.*, vol. 110, no. 11, pp. 1842–1857, Nov 1999, doi:10.1016/S1388-2457(99)00141-8.

[13] J. C. Culham and N. G. Kanwisher, "Neuroimaging of cognitive functions in human parietal cortex," *Curr. Opin. Neurobiol.*, vol. 11, no. 2, pp. 157–163, 2001, doi:10.1016/S0959-4388(00)00191-4.

[14] J. C. Rothwell, P. D. Thompson, B. L. Day, S. Boyd, and C. D. Marsden, "Stimulation of the human motor cortex through the scalp," *Exp. Physiol.*, vol. 76, no. 2, pp. 159–200, 1991.

[15] K. Sekihara and B. Scholz, "Generalized Wiener estimation of three-dimensional current distribution from biomagnetic measurements," *IEEE Trans. Biomed. Eng.*, vol. 43, no. 3, pp. 281–291, 1996, doi:10.1109/10.486285.

[16] J. Classen, C. Gerloff, M. Honda, and M. Hallet, "Integrative visuomotor behavior is associated with interregionally coherent oscillation in the human brain," *J. Neurophysiol.*, vol. 3, pp. 1567–1573, 1998.

[17] G. Rizzolatti, G. Luppino, and M. Matelli, "The organization of the cortical motor system: new concepts," *Electroencephalogr. Clin. Neurophysiol.*, vol. 106, no. 4, pp. 283–296, 1998, doi:10.1016/S0013-4694(98)00022-4.

CHAPTER 5

Application to High-Resolution EEG Recordings in a Cognitive Task (Stroop Test)

5.1 SUBJECTS AND EXPERIMENTAL DESIGN

High-density electroencephalography (EEG) recordings were performed on a group of five normal subjects. We present here the connectivity results related to one representative subject to illustrate the potential of the methods investigated. Subjects were seated in a comfortable chair in a quiet room, which was connected to the adjacent equipment room by intercom. They viewed a screen where the name of a color (e.g., "red") was printed in the same color (e.g., in red ink, congruent condition) or in a different color (e.g., in blue ink, incongruent condition), as shown in Fig. 5.1. Blocks of congruent or incongruent words alternated with blocks of neutral words (not color names). There were 256 trials in 16 blocks (four color congruent, eight neutral, four color incongruent) of 16 trials, with a variable intertrial interval averaging 2000 ms between trial onsets. Within the congruent blocks, half of the words were neutral so as to prevent the development of word reading strategies in the congruent blocks. The trial began with the presentation of a word for 1500 ms, followed by a fixation cross for an average of 500 ms. Each trial consisted of one word presented in one of four ink colors (red, yellow, green, and blue), with each color occurring equally often with each word type (congruent, neutral, incongruent). Subjects were asked to press one of the four buttons that corresponded to the color of the ink in which the word was presented. Data from 0 to 450 ms poststimulus were analyzed.

5.2 EEG RECORDINGS

EEG was recorded using well-established methods. A custom-designed Falk Minow cap located 64 scalp locations for EEG and EOG recordings, with the EEG electrodes spaced equidistantly, with the left mastoid serving as a reference for all other sites. Electrode impedances were below 10 kΩ. Amplifier bandpass was 0.1 to 100 Hz with digitization at 250 Hz, and the EEG data were successively digitally filtered at 50 Hz. Electrode positions were digitized using a Zebris 3D localization device with respect to anatomic landmarks (nasion and two preauricular points). ERP data were visually inspected, and trials containing artifacts were rejected. Semiautomatic, supervised threshold criteria were used for the rejection of trials contaminated by ocular and EMG

FIGURE 5.1: Scheme of the Stroop task. If the meaning of the word is a color written in the color itself, the condition is congruent (*left, word "red" written in red*). If the word means a color and it is written in a different color, the condition is incongruent (*center, word "yellow" written in blue*). If the meaning is not a color, the condition is neutral (*right, word "dog" written in green*). The subject should pronounce the color in which the word is written (irrespective of the meaning of the word) in the shortest time possible.

artifacts, as described in detail elsewhere [1]. After artifact rejection, ERP signals were adjusted in the baseline.

5.3 RESULTS

After obtaining the solution of the linear inverse problem, the estimation of the current density waveforms on each ROI was obtained as described in the preceding paragraphs. Connectivity estimations were performed by DTF, PDC, and dDTF, and the statistical thresholds were evaluated using the shuffling procedure described earlier. The order of the MVAR model used in each estimation was determined by means of the Akaike information criterion (AIC), which returned an optimal order of 13. Details of the electrode montage are shown on the realistic reconstruction of a subject's scalp in Fig. 5.2. The different ROIs selected are shown in different grey tones on the realistic reconstruction of the subject's left cortical hemisphere (regions of the cortex not of interest are shown in gray). By means of the linear inverse procedure, the estimation of the current density waveforms in each ROI of interest was then performed, according to Eqs. (3.3)–(3.6). The DTF, PDC, and dDTF estimators were applied to the cortical waveforms related to the ROIs of interest.

Figure 5.3 shows the cortical connectivity patterns obtained for the congruent stimuli during the period preceding the subject's answer (0–450 ms after the stimulus presentation) in the representative subject. The results are shown for the beta band (12–29 Hz). The DTF (left), PDC (center), and dDTF (right) methods produced similar results. In particular, functional connections between cortical parietofrontal areas were present in all the estimations performed by the DTF, PDC, and dDTF methods. Moreover, connections involving the cingulate cortex are also clearly visible, as well as those involving prefrontal areas, mainly in the right hemisphere. Functional connections in prefrontal and premotor areas tended to be right-sided, whereas the functional activity in the parietal cortices was generally bilateral.

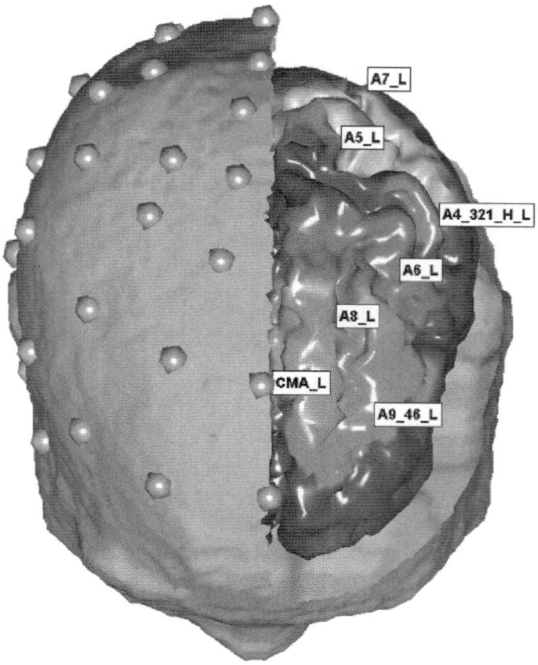

FIGURE 5.2: Over the right hemisphere, the electrode montage (59 electrodes) is shown on the realistic reconstruction of a subject's scalp, obtained from structural MRIs. Over the left hemisphere, the ROIs considered for this study are shown on the realistic reconstruction of the subject's cortex. Each ROI is represented in a different grey level. The ROIs considered are the cingulate motor area (CMA), the Brodmann areas 7 (A7) and 5 (A5), the primary motor area the posterior supplementary motor area (SMAp) and the lateral supplementary motor area (A6_L). (Published with permission from [8].)

Figure 5.4 shows the inflow (first row) and outflow (second row) patterns computed for the same subject in the beta frequency band for the same time period shown in Fig. 5.3. The ROIs that are very active as source or sink (i.e., the source/target of the information flow to/from other ROIs) show results that are generally stable across the different estimators. Values in the left column are related to the inflow and outflow computations obtained with the DTF methods, the central column is related to the flows obtained with the PDC, and the right column to the dDTF. Greater involvement of the right premotor and prefrontal regions is observed across methods.

5.4 DISCUSSION

Although presented here only to demonstrate the capabilities of the estimation procedures with real data, the physiological results shown for a representative subject are consistent with those present in the Stroop literature. The Stroop task is often employed in studies of selective attention and has been found to be sensitive to prefrontal damage. For incongruent stimuli, PET and fMRI studies have shown activation of a network of anterior brain regions. Most studies report

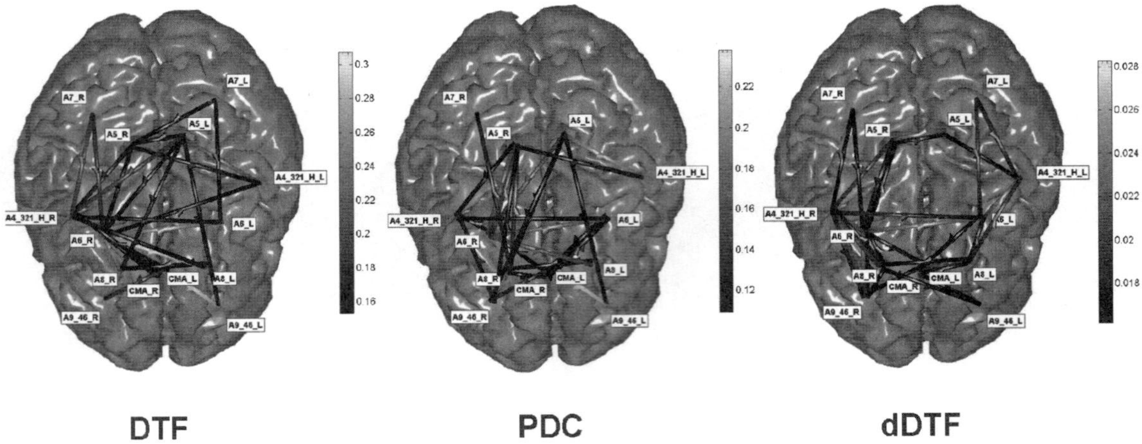

FIGURE 5.3: Cortical connectivity patterns obtained for the period preceding the subject's response during congruent trials in the beta (12–29 Hz) frequency band in a representative subject. The patterns are shown on the realistic head model and cortical envelope of the subject, obtained from sequential MRIs. The brain is seen from above with the left hemisphere represented on the right side. Functional connections are represented by arrows that move from one cortical area toward another. The grey level and sizes of the arrows code the strengths of the connections. The lighter and bigger the arrows are, the stronger the connections. Three connectivity patterns are depicted, estimated in the beta frequency band for the same subject with the DTF (*left*), the PDC (*middle*) and the dDTF (*right*). Only the cortical connections statistically significant at $p < 0.01$ are reported. (Published with permission from [8].)

activation of the anterior cingulate cortex (ACC) and frontal polar cortex, and several authors have hinted at changes in regional cerebral blood flow (rCFB) in posterior cingulate and other posterior regions [2–4].

In the present study, ACC was approximately modeled by the CMA ROI and dorsolateral prefrontal areas by the 9/46 ROI. Connectivity analysis indicated intense, bilateral ACC activity during the task. The number of directed interactions was similar across several frequency bands, but there were differences in the connectivity structure across the congruent–incongruent tasks (not shown here). These results were corroborated by the inflow–outflow analysis that showed agreement between the methodologies used for the derivation of the connectivity patterns. The predominance of outflow from right premotor and prefrontal cortical areas is of interest, and the increased activity in the prefrontal cortical regions is in agreement with previous scalp observations. In an EEG Stroop study, West and Bell [5] reported increased spectral power at medial (F3, F4) and lateral (F7, F8) frontal sites, as well as over parietal regions (P3, P4). They suggested that activation of the parietal cortex resulted from interaction between prefrontal and parietal regions during the suppression of the influence of the irrelevant word meaning. Interaction between parietal and frontal sites has been advocated as an explanation for the activation of posterior areas [3,5,6] Possibly, the minimization

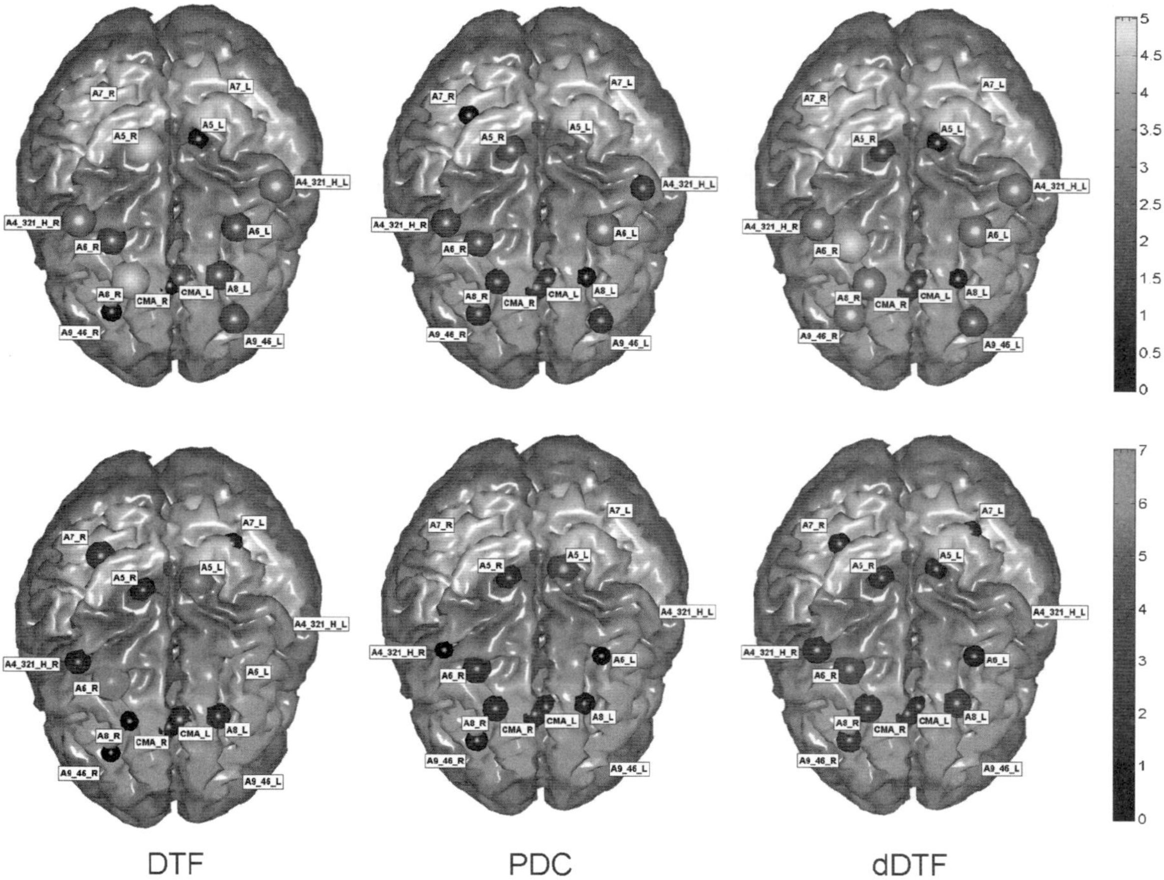

FIGURE 5.4: Figure shows the inflow (*first row*) and the outflow (*second row*) patterns obtained for the beta frequency band, from each ROI during the congruent trials. The brain is seen from above with the left hemisphere represented on the right side. In the first row, different grey hues summarize the behavior of an ROI in terms of reception of information flow from other ROIs, by adding the values of the links arriving on the particular ROI from all the others. The information is coded with the size and the grey level of a sphere, centered on the particular ROI analyzed. The larger the sphere, the higher the value of inflow or outflow for any given ROI. In the second row, the different grey hues code the outflow of information from a single ROI toward all the others. (Published with permission from [8].)

of the influence of irrelevant word information prompts directed interactions from parietal toward frontal sites.

In a previous coherence study [7], increased coherence between parietal and frontal sites was observed late in the trial. This behaviour is not detectable at the scalp level. With the application of advanced high-resolution EEG methodologies, including realistic cortical modeling, solution of the linear inverse problem, and the application to the computed cortical signals of connectivity pattern estimators, it became observable. These data suggest that cognitive control is implemented by medial and lateral prefrontal cortices that bias processes in regions that have been implicated in high-level perceptual and motor processes [8]. It is also striking that in all the frequency bands and for the five subjects analyzed, the differences between the estimated connectivity patterns with the connectivity estimation methods were negligible (results not shown here). This result suggests that, in practical conditions, constant differences of only a few percent observed in the simulation studies between DTF, PDC, and dDTF estimators are not as significant as other factors such as SNR and recording LENGTH.

REFERENCES

[1] D. V. Moretti, F. Babiloni, F. Carducci, F. Cincotti, E. Remondini, P. M. Rossini, S. Salinari, and C. Babiloni, "Computerized processing of EEG-EOG-EMG artifacts for multi-centric studies in EEG oscillations and event-related potentials," *Int. J. Psychophysiol.*, vol. 47, no. 3, pp. 199–216, Mar 2003, doi:10.1016/S0167-8760(02)00153-8.

[2] L. Astolfi, et al., "A Comparison of different cortical connectivity estimators for high resolution EEG recordings," *Hum. Brain Mapp.*, vol. 28, no. 2, pp. 143–157, Feb 2007, doi:10.1002/hbm.20263.

[3] C. S. Carter, M. Mintun, and J. D. Cohen, "Interference and facilitation effects during selective attention: an H215O PET study of Stroop task performance," *NeuroImage*, vol. 2, pp. 264–272, 1995, doi:10.1006/nimg.1995.1034.

[4] C. J. Bench, C. D. Frith, P. M. Grasby, K. J. Friston, E. Paulesu, R. S. J. Frackowiak, and R. J. Dolan, "Investigations of the functional anatomy of attention using the Stroop test," *Neuropsychology*, vol. 31, pp. 907–922, 1993, doi:10.1016/0028-3932(93)90147-R.

[5] M. P. Milham, M. T. Banich, and V. Barad, "Competition for priority in processing increases prefrontal cortex's involvement in top-down control: an event-related fMRI study of the Stroop task," *Cogn. Brain Res.*, vol. 17, pp. 212–222, 2003, doi:10.1016/S0926-6410(03)00108-3.

[6] R. West and M. A. Bell, "Stroop-color word interference and electroencephalogram activation: evidence for age-related decline of the anterior attention system," *Neuropsychology*, vol. 11, pp. 421–427, 1997, doi:10.1037/0894-4105.11.3.421.

[7] B. Schack, A. C. N. Chen, S. Mescha, and H. Witte, "Instantaneous EEG coherence analysis during the Stroop task," *Clin. Neurophysiol.*, vol. 110, pp. 1410–1426, 1999, doi:10.1016/S1388-2457(99)00111-X.

[8] T. Egner and J. Hirsh, "The neural correlates and functional integration of cognitive control in a Stroop task," *NeuroImage*, vol. 24, pp. 539–547, 2005, doi:10.1016/j.neuroimage.2004.09.007.

CHAPTER 6

Application to Data Related to the Intention of Limb Movements in Normal Subjects and in a Spinal Cord Injured Patient

6.1 SUBJECTS AND EXPERIMENTAL DESIGN

Three healthy subjects and one with a spinal cord injury (SCI) participated in the study. Informed consent was obtained from each subject after explanation of the study, which was approved by the local institutional ethics committee. The SCI was of traumatic etiology and was located at the cervical level (C7); the patient had not suffered a head or brain lesion associated with the trauma leading to the injury. The patient was unable to move his upper and lower limbs. For the electroencephalography (EEG) data acquisition, the subjects were comfortably seated on a reclining chair in an electrically shielded, dimly lit room. They were asked to perform a brisk protrusion of their lips (lip pursing) while they were performing (normal subjects) or attempting (SCI patient) a right foot movement. The task was repeated every 6–7 s, in a self-paced manner, and the 100 single trials recorded were used for the estimation of DTF. A 96-channel EEG system (BrainAmp, Brainproducts GmbH, Germany) was used to record electrical potentials by means of an electrode cap accordingly to an extension of the 10–20 international systems. Structural MRIs of the subjects' head were taken with a Siemens 1.5T Vision Magnetom MR system (Germany).

6.2 ESTIMATED CONNECTIVITY PATTERNS

After the solution of the linear inverse procedure, the estimation of the current density waveforms in the employed ROIs were obtained as previously described. Statistical significance of the cortical connections was obtained by comparing the estimated cortical connectivities with the mean values of the distribution of the random connectivity values between the cortical signals after the deterministic interdependency between these signals were removed. Fig. 6.1 shows the cortical connectivity patterns during the period preceding the onset of the lips movement, and hence related to the preparation of the foot and lips movement in all the three normal subjects examined.Here, we present the results obtained for the connectivity pattern in the gamma band. In the inset, the grey level and size of

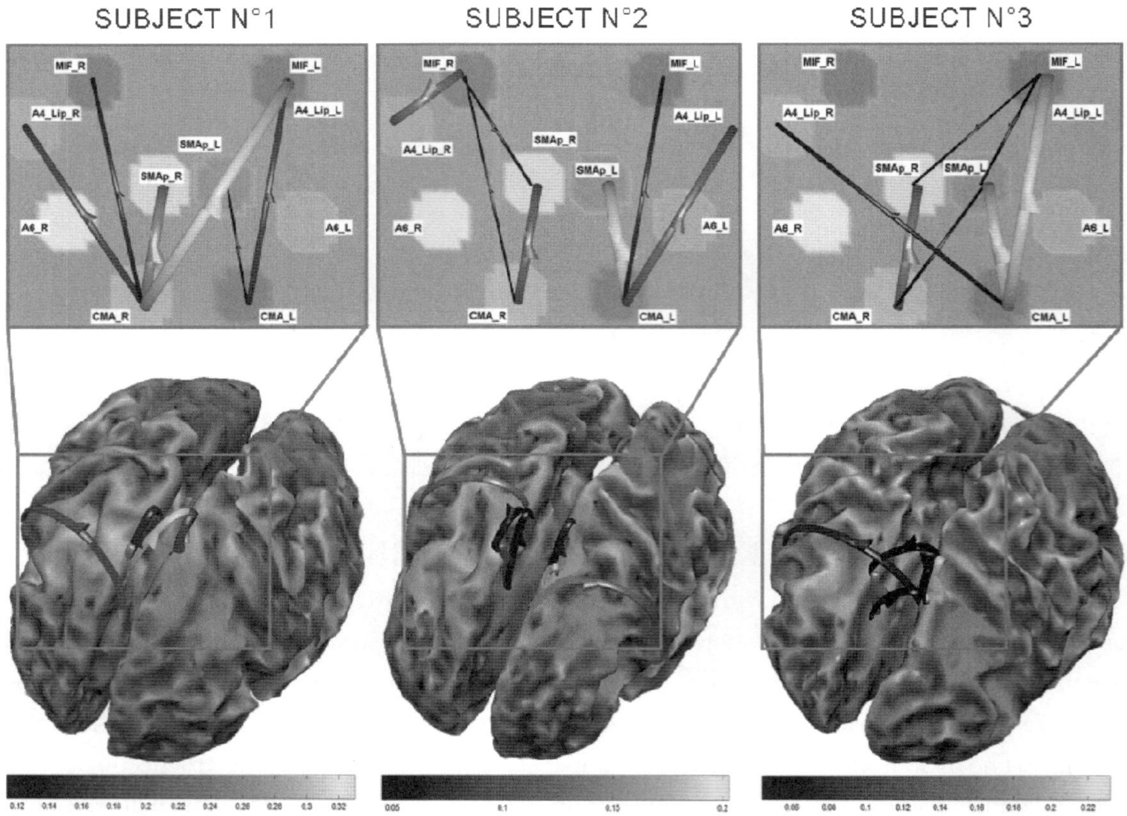

FIGURE 6.1: The cortical connectivity patterns obtained for the period preceding the lips movement in three normal subjects, analyzed in the gamma frequency band. Cortical functional connections are represented with arrows that move from the source cortical area toward the target one. In the inset, the grey level and size of the arrows code the strength level of the connections. (*Bottom*) Connectivity patterns obtained from EEG data represented on the realistic cortical reconstruction of each experimental subject obtained from sequential MRIs, seen from left and above. (*Top*) Details of the connectivity patterns obtained for the central areas. Only the cortical connections statistically significant at $p < 0.01$ are represented. (Published with permission from [1].)

the arrows code the strength level of the connection. In the labels, the names of the ROIs employed are indicated. Only the cortical connections statistically significant at $p < 0.01$ are represented. Note that the connectivity patterns, estimated in the gamma band, presents strong functional connections between the CMA and the premotor areas of both cerebral hemispheres Substantial equivalence of the connectivity patterns estimated for the three normal subjects was found. Hence, the functional directional connections during the preparation of the foot movement are generated in the gamma band from the cingulated areas and spread toward the supplementary motor areas. These patterns have to be compared with those estimated in the SCI patient, during the performance of a similar experimental task.

6.3 CORTICAL CONNECTIVITY PATTERNS IN SPINAL CORD INJURY

The EEG recording and the estimation of the cortical activity and connectivity for the SCI patient during the task was accomplished in the same way already described for the normal subjects. Fig. 6.2 depicts the connectivity pattern in the gamma frequency band before the execution of the lip movement accompanied by the attempt to move the paralyzed limb. It can be noted how this pattern is similar to those generated by the normal subjects during the preparation to the foot movement. Also in this case the connectivity flow is generated from the cingulated areas and spreads to the supplementary motor areas. Similar was the case observed for the connectivity patterns already generated

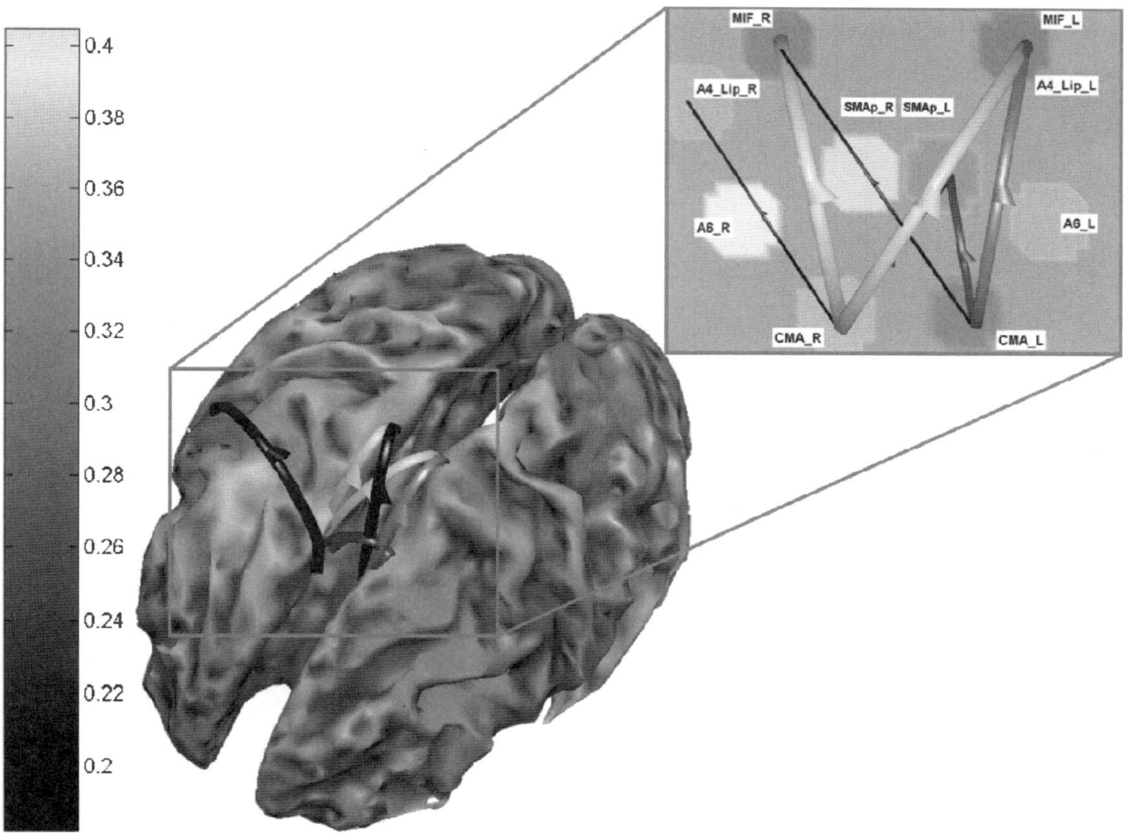

FIGURE 6.2: Cortical connectivity patterns obtained for the period preceding the onset of movement in a SCI patient in the gamma frequency band. Same convention is used as shown in the previous figure. Only the cortical connections statistically significant at $p < 0.01$ are represented. (Published with permission from [1].)

in the other frequency bands examined between the SCI patient and the group of three normal subjects.

6.4 CONCLUSION

The main results provided here highlight the possible existence of a common pattern of cortical connectivity during the execution (normal) or the imagination (SCI patient) of a foot limb movement. The activity noted in the cingulated and supplementary motor areas in the present study is consistent with the role that such cortical areas have in the organization and performance of simple foot movements. This finding, if confirmed in a larger population of normal subjects as well as SCI patients, could open the way for the use of such a feature in a clinical context, for instance, in the brain–computer interface area. It is worth noting that the technology presented here can be applied to retrieve patterns of cortical connectivity during more complex clinically relevant tasks in patients by using noninvasive EEG recordings. Examples include the use of the connectivity pattern study in the analysis of the brain damage caused by a stroke as well as in the analysis of the possible recovery of the brain motor areas during the rehabilitation paths.

REFERENCES

[1] L. Astolfi, et al., "Cortical connectivity patterns during the intention of limb movements in normal and in a spinal cord injured patient estimated with the Directed Transfer Function," *IEEE Eng. Med. Biol. Mag.*, vol. 25, no. 4, pp. 32–38, Jul-Aug 2006.

CHAPTER 7

The Instantaneous Estimation of the Time-Varying Cortical Connectivity by Adaptive Multivariate Estimators

7.1 TIME-VARYING ESTIMATION OF THE CORTICAL CONNECTIVITY

Taken together, the findings of this part of the research indicate that an accurate estimation of the cortical connectivity patterns can be achieved by using realistic models for the head and cortical surfaces, high-resolution EEG recordings, and effective and functional cortical connectivity by using the SEM, DTF, dDTF, and PDC methods, respectively. The simulation findings suggest that under conditions largely met in the ERPs recordings (SNR at least 3 and the EEG recording with 4800 data samples or duration longer than 75 s at 64 Hz), the computation of functional connectivity by all the methods employed can be performed with moderate quantitative errors. The use of high-resolution EEG recordings and the estimation of the cortical waveforms in ROIs through the solution of linear inverse problem allow the evaluation of the functional cortical connectivity patterns related to the task performed. These computational tools (high-resolution EEG, estimation of cortical activity via linear inverse problem, and connectivity methods) can be of interest to assess functional connectivity patterns from noninvasive EEG recordings in humans. However, the classical estimation of these methods requires the stationarity of the signals, and with the estimation of a unique MVAR model on an entire time interval, transient pathways of information transfer remains hidden. This limitation could bias the physiologic interpretation of the results obtained with the connectivity technique employed.

To overcome this limitation, we need to approach a time-varying estimation of cortical connectivity to be performed always within the framework of the Granger causality. This is the aim of this chapter.

As described in Chapter 2, among the multivariate methods, the directed transfer function (DTF) [1,2] and the partial directed coherence [3] are estimators characterizing, at the same time, direction and spectral properties of the interaction between brain signals, and require only one MVAR model to be estimated from all the time series. However, the classical estimation of these methods requires the stationarity of the signals; moreover, with the estimation of a unique MVAR model on an entire time interval, transient pathways of information transfer remains hidden. This limitation could bias the physiologic interpretation of the results obtained with the connectivity technique employed.

To overcome this limitation, different algorithms for the estimation of MVAR with time-dependent coefficients were recently developed. Ding et al. [4] used a short-time windows technique that requires the stationarity of the signal within short-time windows. Moeller et al. [5] proposed an application to MVAR estimation of the extension of the recursive least squares (RLS) algorithm with a forgetting factor. This estimation procedure allows the simultaneous fit of one mean MVAR model to a set of single trials, each one representing a measurement of the same task. In contrast to short-window techniques, the multitrial RLS algorithm does not require the stationarity of the signals, and involves the information of the actual past of the signal, whose influence decreases exponentially with the time distance to the actual samples. The advantages of this estimation technique are an effective computation algorithm and a high adaptation capability. It was demonstrated in [5] that the adaptation capability of the estimation (measured by its adaptation speed and variance) does not depend on the model dimension. Simulations on the efficacy of time-variant Granger causality based on AMVAR computed by RLS algorithm were also provided [6].

In this chapter, we propose the use of the adaptive multivariate approach to define time-varying multivariate estimators based on DTF and PDC, thus making such estimators able to follow rapid changes in the connectivity between cortical areas during an experimental task. Further, the performances of time-varying DTF and PDC were studied by means of simulations performed on the basis of a predefined connectivity scheme linking three cortical areas. Cortical connections between the areas were retrieved by the estimation process under different experimental conditions. The results obtained for the different methods were evaluated by a statistical analysis, with particular attention to the adaptation speed and precision of the pattern retrieved. The simulation study was based on the following questions:

1. What are the performances of the proposed time-varying estimators in retrieving the rapid changes in time of the cortical connectivity pattern?
2. What is the effect of different factors affecting the recordings, such as the signal-to-noise ratio and the amount of trials at disposal?
3. What is the influence of the adaptation constant C on the performances, and what can be a criterion for the choice of its optimum value?
4. Which of the two connectivity estimators applied (i.e., PDC and DTF) is the most effective in reconstructing a connectivity model under the conditions usually met in linear inverse estimations?

Finally, according to the results of the simulation study an application to real data is proposed in order to offer an example of the results that can be obtained by this technique. For this purpose, patterns of cortical connectivity between human brain areas involved in a simple motor task are presented. We applied the time-varying DTF/PDC techniques to the cortical activity estimated in particular regions of interest (ROIs) of the cortex, obtained from high-resolution EEG recordings during the execution of a combined foot–lips movement.

The directed transfer function (DTF) and the partial directed coherence (PDC) estimators have been fully introduced in Chapter 2.

In this chapter, an adaptive formulation of DTF and PDC based on an adaptive MVAR (AMVAR) model is proposed and tested. The time-dependent parameter matrices were estimated by means of RLS algorithm with forgetting factor, as described in [5]. In particular, the RLS algorithm represents a particular variant of the Kalman filter. This recursive estimator for the AMVAR-parameter is characterized by a more universal practicability since it requires less computational effort and it is possible to extend this approach to the presence of multiple realizations of the same process. The extension to multiple trials was introduced by [5]. The fitting procedure of the autoregressive (AR) parameters exploits the RLS technique with the use of a forgetting factor. It is based on the minimization of the sum of exponentially weighted prediction errors of the previous processes. Thereby the weighting depends on an adaptation constant $0 \leq C < 1$ that controls the compromise between adaptation speed and the quality of the estimation. Values close to zero result in a slower adaptation with more stable estimations and vice versa. A mean MVAR model was fitted to a set of trials, each one representing the measurement of the same task. A comprehensive description of the algorithms may be found in [5,6].

7.2 THE SIMULATION STUDY

The simulation was designed in order to test the capability of the two methods to follow rapid changes in the cortical connectivity as well as the precision of the estimation performed under different levels of the factors of interest. Test signals simulating cortical average activations in different regions of the cortex were generated in order to fit an imposed coupling scheme involving three cortical areas (shown in Fig. 7.1A). Simulated signals with different levels of signal-to-noise ratio (SNR) and different number of trials have been systematically generated in order to evaluate the influence of these factors on the estimates produced by the two methods.

Signal x_1 was the average activity of a cortical region of interest obtained by a linear inverse estimation procedure from a real EEG recording. The frequency spectrum of x_1 follows the very typical appearance of the EEG signal with a shape resembling the standard $1/f$ function, f being the EEG frequency with a light peak in the alpha band (8–12 Hz). The other signals were generated as

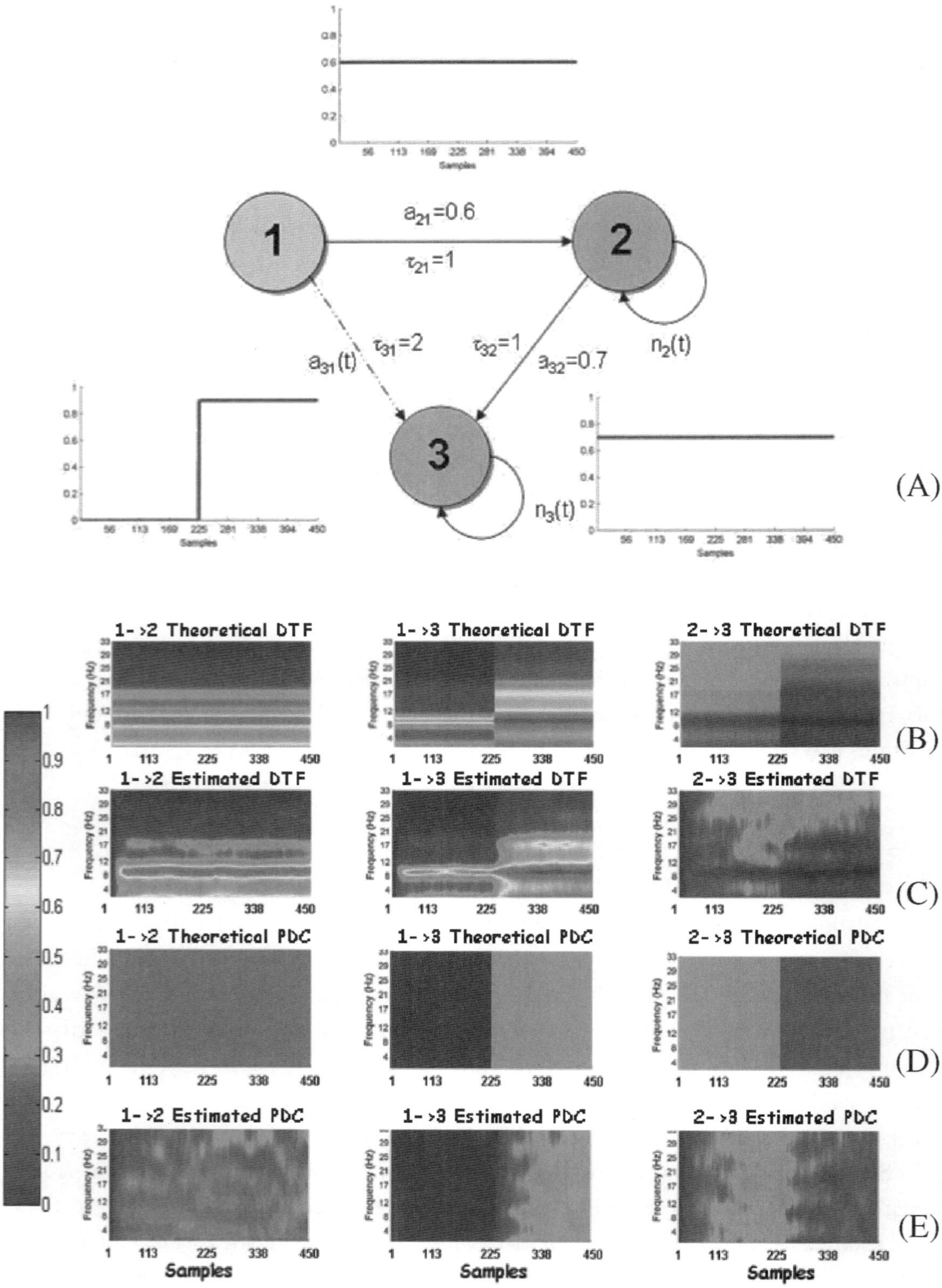

1
2
3
a_{21}=0.6
τ_{21}=1
τ_{31}=2
$a_{31}(t)$
τ_{32}=1
a_{32}=0.7
$n_2(t)$
$n_3(t)$
(A)
1- >2 Theoretical DTF
1- >3 Theoretical DTF
2- >3 Theoretical DTF
(B)
1- >2 Estimated DTF
1- >3 Estimated DTF
2- >3 Estimated DTF
(C)
1- >2 Theoretical PDC
1- >3 Theoretical PDC
2- >3 Theoretical PDC
(D)
1- >2 Estimated PDC
1- >3 Estimated PDC
2- >3 Estimated PDC
(E)
Frequency (Hz)
Samples

follows:

$$x_j(t) = \sum_{i=1}^{N} a_{ji}(t)x_i(t - \tau_{ji}) + n_j(t) \quad \text{for} \quad j = 2, \ldots, N \tag{7.1}$$

where N is the number of ROIs; τ_{ji} is the constant delay in the propagation from the ith to the jth area expressed in samples; $a_{ji}(t)$ is the time-dependent amplitude of the connection between the ith and the jth area; $n_j(t)$ is the residual representing the part of the jth area activation not depending from other areas, here playing the role of noise simulated with an uncorrelated Gaussian white noise. In particular, the SNR levels were obtained as the ratio between the power of the signals and the power of the noise signal. This was obtained by adjusting the amplitude of the white Gaussian noise superimposed to the data, in order to match the desired level of SNR. Delays applied were of one and two samples, corresponding to delays of 4 and 8 ms, respectively, at a sampling rate of 256 Hz.

All procedures of signal generation were repeated under the following conditions:

SNR factor levels = [1, 5, and 10]

TRIALS factor levels = [1, 2, 3, 5, 10, 20, 40, 80] each of 1000 samples, at a sampling frequency of 256 Hz

Adaptation constant C factor levels = [0.01, 0.02, 0.03, 0.04, 0.05, 0.06, 0.07, 0.08, 0.09, 0.10]

The levels chosen for both SNR and TRIALS factors cover the typical range for the cortical activity estimated with high-resolution EEG techniques.

The results obtained by the two estimators were evaluated in four frequency bands, theta (4–7 Hz), alpha (8–12 Hz), beta1 (13–22 Hz), and beta2 (23–30 Hz).

To evaluate both the adaptation speed and the precision of the connectivity patterns estimated by the two methods, two different indices of the performances were defined.

The first index was the time at settling, or t_s, defined as the first instant following the transition after which the error is below the ε% of the transition amplitude (in our case, $\varepsilon = 10$), corresponding to the following condition:

$$|\hat{\sigma}_{ij}(t, b) - \sigma_{ij}(t, b)| \leq \varepsilon \Delta_{ij}^{k}(b)/100, \ \forall t \in k, \quad t > t_s \tag{7.2}$$

FIGURE 7.1: (A) Connectivity pattern imposed between different regions of the cortex during simulation. The values of the connection strengths a_{ij} are time dependent, and are represented by the plots near each connectivity arrow. τ_{ij} represents the constant delay in the propagation from area j to area i, expressed in samples. $\tau_{21} = 1$, $\tau_{31} = 2$ and $\tau_{32} = 1$ (corresponding to 4, 8 and 4 ms, respectively). (B and C) show an example of theoretical and estimated time frequency distribution of the time-varying DTF function for the three significant arcs in the model. (D and E) show the theoretical and the estimated distribution for the time-varying PDC function.

where b represents the generic frequency band, $\sigma_{ij}(t, b)$ and are the theoretical and the estimated average values of the time frequency function (DTF or PDC), in each cortical band b; Δ_{ij}^{k} is the time average computed during the kth transition interval (in the results here presented, there is only a transition in the simulated data, and $k = 1$) and $\boldsymbol{\varepsilon}$ is the percentage of the transition amplitude desired. The values of the DTF or PDC functions analyzed in this study were obtained by averaging the values of DTF or PDC for each frequency belonging to the band considered.

The second index, used to evaluate the precision of the estimation performed, is the average error in each time interval during which the values of connection strengths are kept constant after the transition, and is defined as follows:

$$\overline{E_{ij}^{k}(b)}^{t} = \overline{|\hat{\sigma}_{ij}(t, b) - \sigma_{ij}(t, b)|}^{t} \tag{7.3}$$

Each generation–simulation procedure was repeated 50 times, in order to increase the robustness of the successive statistical analysis.

7.2.1 Statistical Analysis

The results obtained were subjected to separate analysis of variance (ANOVA). The main factors of the ANOVAs were the SNR (three levels: [1, 5, and 10]) the number of TRIALS jointly used for the estimation (eight levels: [1, 2, 3, 5, 10, 20, 40, 80]), the adaptation constant C (10 levels: [0.01, 0.02, 0.03, 0.04, 0.05, 0.06, 0.07, 0.08, 0.09, 0.10], each of 1000 samples, at a sampling frequency of 256 Hz) and the frequency BAND (with four levels: theta = 4–7 Hz; alpha = 8–12 Hz; beta1 = 13–22 Hz; beta2 = 23–30 Hz). Separate ANOVAs for each frequency band were also performed on the error indices defined in the previous paragraph (Relative Error and Time at Settling). In all the evaluated ANOVAs, the correction of Greenhouse–Gasser for the violation of the spherical hypothesis was used. The *post hoc* analysis with Duncan's test at the $p = 0.05$ statistical significance level was then performed.

7.3 RESULTS

Simulated data representing the activations in three cortical areas were generated as described by Eq. (7.1), in order to fit a time-varying connectivity pattern, shown in Fig. 7.1A. The procedure was repeated under all the different conditions of number of trials and SNR. An adaptive MVAR of order $p = 3$ was fitted to each set of simulated data. The model order p was chosen by means of a method proposed in [7]. The multivariate Akaike criterion was applied on each time interval and the maximum order obtained was then adopted for all the recursive estimation.

The estimation of MVAR models with time-varying parameters was repeated with values of the adaptation constant C ranging from 0.01 to 0.1. Time-varying DTF and PDC were then computed, and the ANOVA was performed on the indices of performances described in Eqs. (7.2)

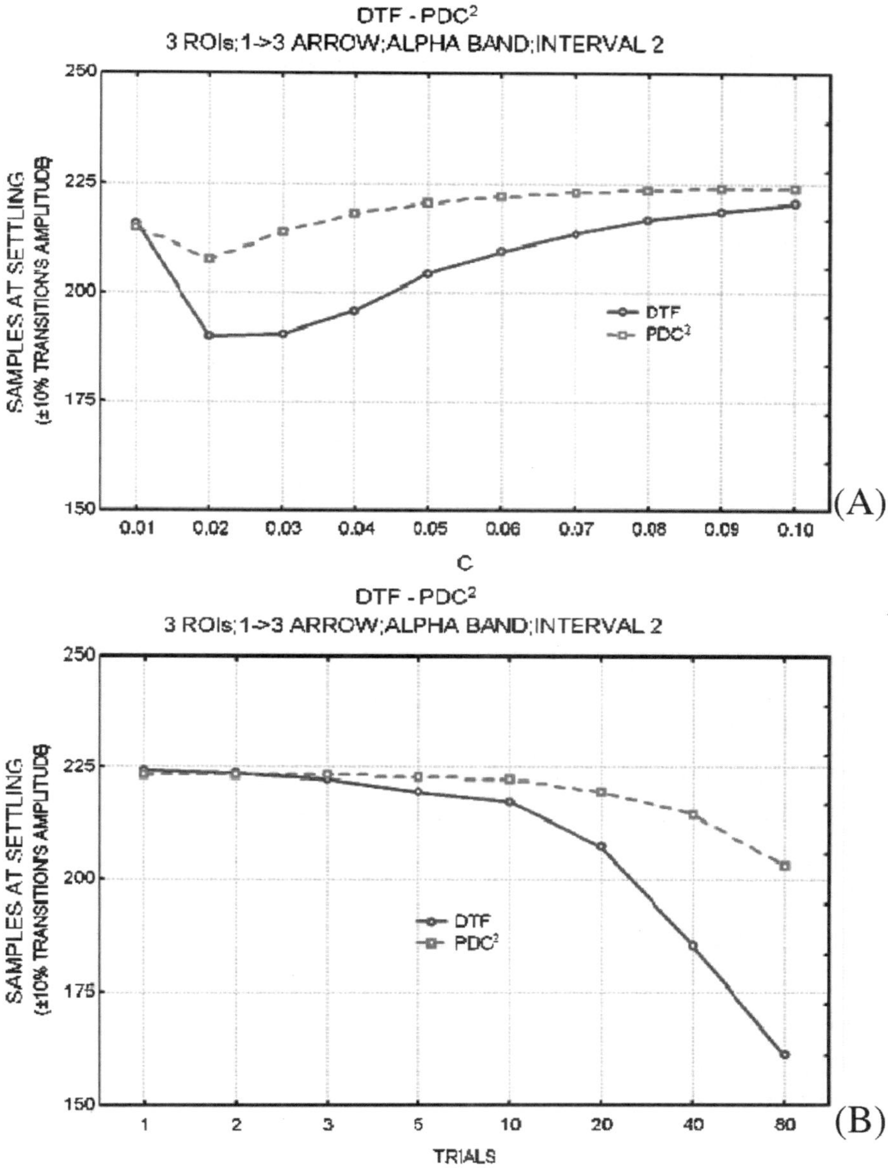

FIGURE 7.2: Results of ANOVA performed on the time at settling (10% of the transition amplitude) for different values of the factors *C* and TRIALS. (A) Plot of means in function of the adaptation factor *C* used in the estimation of the AMVAR. A minimum can be noted for values of 0.02 for DTF and 0.02–0.03 for PDC. (B) Effect of different number of trials used for the recursive estimation. A better adaptation speed is shown for higher number of trials. (C) Two-way interaction between factors *C* and TRIALS. It can be noted that for higher number of trials the minimum shifts to values of *C* comprised between 0.01 and 0.04. Results are reported in the alpha band (8–12 Hz).

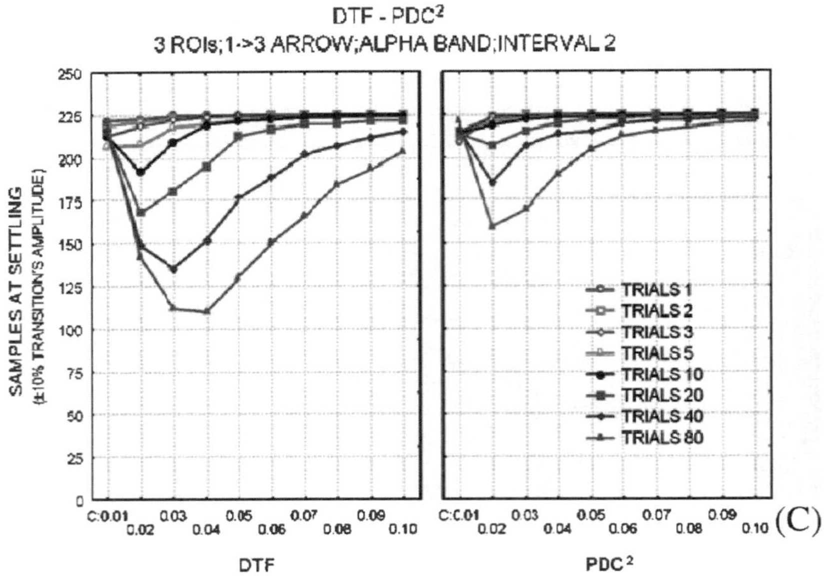

FIGURE 7.2: *(Continued)*

and (7.3), in order to evaluate in a statistically rigorous manner the capability of the method to retrieve the correct connectivity pattern.

Figure 7.1B and C shows an example of theoretical and estimated time frequency distribution of the time-varying DTF function, for the connections imposed to simulated signals ($1 \rightarrow 2$, $1 \rightarrow 3$, and $2 \rightarrow 3$). Fig. 7.1D and E shows the theoretical and the estimated distribution for the time-varying PDC function.

The statistical factors analyzed were the SNR, the number of trials jointly used for the estimation of the AMVAR (TRIALS), and the adaptation factor C. Different ANOVAs addressed the behavior of the indices in the different frequency bands analyzed. However, the results obtained are invariant with respect to the frequency band employed. For this reason, we reported the results in alpha band (8–12 Hz).

The first ANOVA (dependent variable: time at settling) revealed a strong statistical influence of the factors SNR, TRIALS, and C ($p < 0.0001$). In particular, the plots of means for different values of C, TRIALS, and for the two-way interaction between the two factors are shown in Fig. 7.2. It can be noted from Fig. 7.2A that, for both DTF and PDC, there is a minimum in the time at settling for a certain value of C, which is equal to 0.02. This can be explained with a tradeoff between the adaptation speed (which decreases for low values of C) and the variance of the estimation (which decreases with high values of C). Moreover, the adaptation speed of DTF is higher than PDC. From Fig. 7.2B it can be also noted that a high number of trials makes the time at settling decrease.

However, as shown in Fig. 7.2C, it can be seen that an increase in the number of trials makes the minimum of the time at settling shift to higher values of the forgetting factor C. This means that, in practical applications, a higher value of C, and consequently a higher adaptation speed, can be reached if the number of trials at disposal increases.

The second index of performance analyzed was the average error in each time interval during which the values of connection strengths are kept constant after the transition.

ANOVA that was performed revealed a strong influence of the factors SNR, TRIALS, and C in all the frequency bands analyzed. Figure 7.3 shows the effects of different levels of SNR and TRIALS. It can be noted that there is a decrease for higher values of SNR as well as for higher number of trials, for both time-varying DTF and PDC. Moreover, such errors are higher for DTF than for PDC in all conditions. In order to appreciate the comparison with the performances of the classic methods, the results of the application of DTF and PDC based on the estimation of MVAR models are also presented. Figure 7.4 shows the results obtained by the time-varying methods and by the classic MVAR estimation on simulated data with time-dependent connectivity generated according

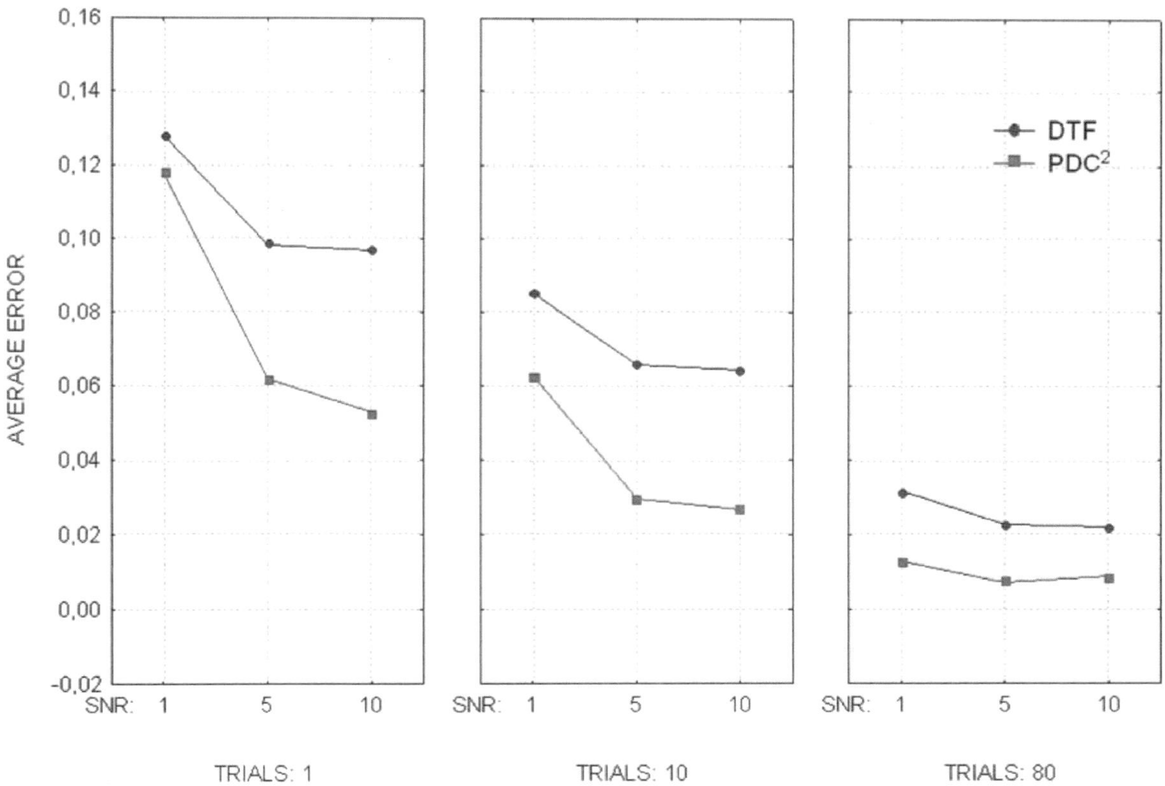

FIGURE 7.3: Results of ANOVA performed on the average error on the arrow $1 \rightarrow 3$ of the model of Fig. 7.1A, in the second time interval (values of connection strengths constant after the transition) on each connectivity arc of the model. $F(4, 196) = 836.44$, $p < 0.00001$.

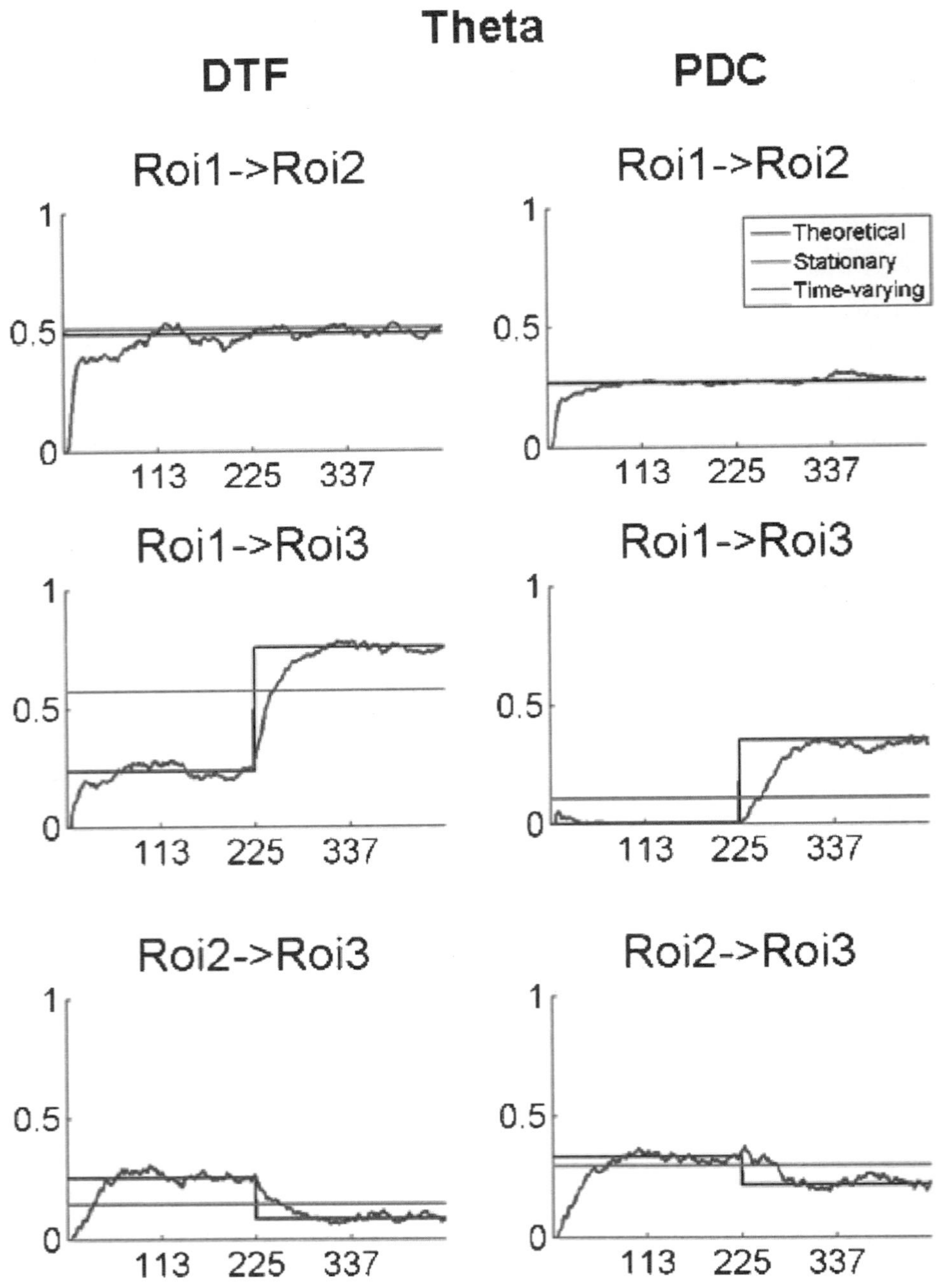

FIGURE 7.4: Comparison of the performances obtained by the time-varying methods and the classic MVAR estimation, on simulated data with time-dependent connectivity. The simulated signals were generated according to the model shown in Fig. 7.1. The time course of the theoretical values of DTF (left column) and PDC (right column) in theta band, for the three connections present in the model (*step line*). The mean value of the results obtained by time-varying DTF and PDC in the same band (*time-varying line*). The mean results obtained by classic DTF and PDC in the same band (*continuous line*).

to the model shown in Fig. 7.1. The time course of the theoretical values of DTF (left column) and PDC (right column), averaged in theta band (4–7 Hz) are shown by the step line. The mean value, in the same band, of the results obtained by time-varying DTF and PDC is shown by the rapid time-varying line. The mean results, in the same band, obtained by classic DTF and PDC are shown by the continuous line. Only the connection imposed in the generation model is shown ($1 \rightarrow 2$, $1 \rightarrow 3$ and $2 \rightarrow 3$).

7.4 DISCUSSION

In this chapter, we propose an evaluation of the performances of two adaptive, multivariate estimators of the cortical connectivity, based on methods usually applied to stationary signals (i.e., PDC and DTF), and to the estimation of an AMVAR with a RLS method. The extension of the RLS algorithm with forgetting factor to MVAR estimation was originally proposed by Moeller et al. [5]. The advantages of this estimation technique are an effective computation algorithm and a high adaptation capability. Simulations on the efficacy of time-variant Granger causality based on AMVAR computed by RLS algorithm were provided previously by [6]. It has been demonstrated [5] that the adaptation capability of the estimation (measured by its adaptation speed and variance) does not depend on the model dimension.

The simulations also address the effect of different factors usually met in the recordings on the performances of the two estimators, and the better choice of the appropriate forgetting factor C under different conditions. The performances of the two estimators were evaluated with particular attention to two aspects: the capability to follow rapid changes in the connectivity patterns (adaptation speed) and the precision in the estimation of connectivity strengths (average error). We performed a series of simulations where test signals were generated to simulate the average electrical activity of cerebral cortical regions, as it can be estimated from high-resolution EEG recordings gathered under different conditions of noise and length of the recordings. Time-varying connectivity was imposed on the simulated signals. The effects of different factors affecting the EEG recordings together with the influence of different choices of the forgetting factor were inferred from statistical analysis (ANOVA and Duncan's *post hoc* tests on the Time at Settling and the Average Error). We also presented the comparison between the performances of the classic and the time-varying formulation of the two estimators, which allows appreciating the improvement obtained with respect to classical methods for non-stationary data.

The results of the simulation study suggested an answer to the questions elicited in the Introduction chapter:

1. The proposed time-varying estimators are effective in retrieving the rapid changes in time of the cortical connectivity pattern. If the operative conditions are properly set, the connectivity pattern can be accurately recovered.

2. The signal-to-noise ratio and the amount of trials at disposal have a statistically significant effect on the performances, both on the side of the adaptation speed and on the side of the accuracy of the estimation. A signal-to-noise ratio of at least 5 is sufficient to obtain a good accuracy, as higher values do not show a significant improvement in the performances. A high number of trials (20 or more) seems to be highly significant, both for improving the accuracy of the estimation and for increasing the adaptation speed.
3. There is an optimum value (between 0.01 and 0.04) for the choice of the adaptation constant C, which varies with the experimental conditions. In particular, a large amount of trials at disposal can increase the value of the optimum choice, thus allowing a higher adaptation speed without loosing the estimation accuracy. In order to obtain a satisfactory adaptation speed, a number of trials not smaller than 20 are required from simulation results.
4. Regarding the comparison between the time-varying DTF and PDC, on the one hand, the DTF method showed better performances of the adaptation speed with lower values of the time at settling. On the other hand, the PDC provided better results of the estimation accuracy with lower values of the average error. The choice between the two methods should take into account the above-mentioned properties, according to the nature of the signals to be examined and of the performances desired, i.e., whether a rapidly changing connectivity is expected or not. It should also be considered that, on the basis of its mathematical formulation, the DTF is unable to discriminate between direct connectivity patterns between areas and indirect links mediated by other areas [3,8–10]. Anyway, for a general purpose, where the interest is in studying the areas that originate the information flow toward the others during a task or a pathological situation, this distinction could not be crucial and DTF could be preferred in order to obtain a higher adaptation speed. Another aspect to be considered is the fact that DTF requires a higher computational effort with respect to PDC, as it involves the inversion of the transfer matrix of the MVAR. So, in order to obtain a faster, more accurate estimation of direct causality links, PDC should be used. This consideration could play a role in the use of these connectivity pattern estimators for on-line applications, such as for instance in the Brain Computer Interface studies.

In conclusion, the statistical results here obtained (by ANOVA integrated with Duncan's *post hoc* tests at $p < 0.05$) indicated an influence of different levels of the main factors SNR, TRIALS, and C on the efficacy of the estimation of time-varying cortical connectivity through time-varying DTF and PDC. In particular, an SNR equal to 5, and at least 20 trials are necessary to decrease significantly the errors on the indices of quality adopted. These conditions are generally obtained in many standard EEG recordings of event-related activity in humans, usually characterized by values of SNR ranging from 3 (movement related potentials) to 10 (sensory evoked potentials) [11].

REFERENCES

[1] M. Kaminski and K. Blinowska, "A new method of the description of the information flow in the brain structures," *Biol. Cybern.*, vol. 65, pp. 203–210, 1991, doi:10.1007/BF00198091.

[2] M. Kaminski, M. Ding, W. A. Truccolo, and S. Bressler, "Evaluating causal relations in neural systems: Granger causality, directed transfer function and statistical assessment of significance," *Biol. Cybern.*, vol. 85, pp. 145–157, 2001, doi:10.1007/s004220000235.

[3] L. A. Baccalà and K. Sameshima, "Partial Directed Coherence: A new concept in neural structure determination," *Biol. Cybern.*, vol. 84, pp. 463–474, 2001, doi:10.1007/PL00007990.

[4] M. Ding, S. L. Bressler, W. Yang, and H. Liang Ding, "Short-window spectral analysis of cortical event-related potentials by adaptive multivariate autoregressive modeling: Data preprocessing, model validation, and variability assessment," *Biol. Cybern.*, vol. 83, pp. 35–45, 2000, doi:10.1007/s004229900137.

[5] E. Möller, B. Schack, M. Arnold, and H. Witte, "Instantaneous multivariate EEG coherence analysis by means of adaptive high-dimensional autoregressive models," *J. Neurosci. Methods*, vol. 105, pp. 143–158, 2001, doi:10.1016/S0165-0270(00)00350-2.

[6] W. Hesse, E. Möller, M. Arnold, and B. Schack, "The use of time-variant EEG Granger causality for inspecting directed interdependencies of neural assemblies," *J. Neurosci. Methods*, vol. 124, pp. 27–44, 2003, doi:10.1016/S0165-0270(02)00366-7.

[7] B. Schack, A. C. N. Chen, S. Mescha, and H. Witte, "Instantaneous EEG coherence analysis during the Stroop task," *Clin. Neurophysiol.*, vol. 110, pp. 1410–1426, 1999, doi:10.1016/S1388-2457(99)00111-X.

[8] M. Winterhalder, B. Schelter, W. Hesse, K. Schwabb, L. Leistritz, D. Klan, R. Bauer, J. Timmer, and H. Witte, "Comparison of linear signal processing techniques to infer directed interactions in multivariate neural systems," *Signal Process.*, vol. 85, no. 11, pp. 2137–2160, Nov 2005, doi:10.1016/j.sigpro.2005.07.011.

[9] R. Kus, M. Kaminski, and K. J. Blinowska, "Determination of EEG activity propagation: Pair-wise versus multichannel estimate," *IEEE Trans. Biomed. Eng.*, vol. 51, no. 9, pp. 1501–1510, Sep 2004, doi:10.1109/TBME.2004.827929.

[10] L. Astolfi, F. Cincotti, D. Mattia, M. Lai, F. de Vico Fallani, S. Salinari, L. A. Baccalà, M. Ursino, M. Zavaglia, and F. Babiloni, "Causality estimates among brain cortical areas by partial directed coherence: Simulations and application to real data," *Int. J. Bioelectromagnetism*, vol. 7, no. 1, 2005.

[11] D. Regan, *Human Brain Electrophysiology. Evoked Potentials and Evoked Magnetic Fields in Science and Medicine*, New York: Elsevier, 1989.

CHAPTER 8

Time-Varying Connectivity from Event-Related Potentials

8.1 EXPERIMENTAL DESIGN AND EEG RECORDINGS

Five right-handed healthy subjects (mean age 24.1 ± 1.5) participated in the study after informed consent was obtained. Subjects were seated comfortably in an armchair with both arms relaxed and resting on pillows. They were asked to perform a brisk protrusion of their lips (lip pursing) while they were performing a right foot movement. The task was repeated every 6–7 s, in a self-paced manner, and 100 single trials were recorded. A 58-channel electroencephalography (EEG) system (BrainAmp, Brainproducts GmbH, Germany) was used to record electrical potentials by means of an electrode cap, to an extension of the 10–20 international systems. Structural MRIs of the subject's head were taken with a Siemens 1.5T Vision Magnetom MR system (Germany). The surface electromyographic (EMG) activity of the muscle was also collected. The onset of the EMG response served as zero time. All data were visually inspected, and trials containing artifacts were rejected. A semiautomatic supervised threshold criteria for the rejection of trials contaminated by ocular and EMG artifacts was used, as described in details elsewhere [1]. After the EEG recording, the electrode positions were digitized using a three dimensional (3D) localization device with respect to the anatomic landmarks of the head (nasion and two preauricular points). The analysis period for the potentials time-locked to the movement execution was set from 1500 ms before to zero time (EMG trigger).

8.2 HEAD AND CORTICAL MODELS

The generation of the head and cortical models as well as the estimation of the cortical activity in the particular region of interest (ROIs) of the obtained model has been adequately treated in the previous chapters and will not be repeated here.

8.2.1 Regions of Interest

The particular cortical regions of interest (ROIs) employed in this study were drawn by an expert neuroradiologist on the computer-based cortical reconstruction of the individual head models of five subjects. Twelve ROIs, thought to be involved in the preparation and execution of simple self-generated movements, were defined on the cortical model: left and right supplementary motor area proper (SMAp); the caudal cingulate motor area (CMAc) from the left and right hemispheres; the primary motor foot (MI-foot) representational area and the primary motor lip (MI-lip) representational area, both from the left and right hemisphere; the superior parietal cortex, SP and the premotor dorsal cortex (PMd), both in the left and right hemispheres.

8.3 RESULTS

The estimation of the current density waveforms in the ROIs employed was performed by means of the linear inverse procedure, for five subjects, according to Eqs. (3.1)–(3.4) of Chapter 3. Relevant cortical activity was different from baseline in SMAp, CMAc, MI-foot, MI-lip, SP, and PMd. On such cortical waveforms, connectivity estimations were performed by the time-varying partial direct coherence/direct transfer function (PDC/DTF) procedure as described in the Methods section. The adaptation constant C was set to 0.02, according to the results of the simulation study. The model order was rather stable for all the subjects investigated; for four of them we selected the order 16 and for the last one the order 17.

Figures 8.1 and 8.2 show the time–frequency distribution of the instantaneous PDC and DTF respectively, obtained for a representative subject from the set of cortical waveforms estimated in the 12 ROIs considered. The time interval is relative to 1500 ms preceding the EMG onset, which served as zero time, while the frequency range is between 0 and 50 Hz (see the inset). The figure should be read from column (areas originating the causal influence) toward the rows (target of the causal influence). The first 50 ms reflect the adaptation epoch of the model. Both methods (PDC and DTF) presented comparable results. High values of connectivity can be noted in particular in the alpha and gamma bands, and involved mainly the PMd areas from the left and right hemispheres, the M1F left and right, the SMAp left and right, and the CMAc left and right.

Figure 8.3 shows the time-varying connectivity patterns in the alpha band, extracted at −500, −250 and 0 ms before the movement onset, for the same representative subject of Figs. 8.1 and 8.2. Results are presented on the realistic reconstruction of the head and cortex of the subject, obtained from sequential MRIs. The different ROIs selected are depicted in different grey levels and described by the labels. The connectivity links are represented by arrows, pointing from one cortical area ("the source") toward another one ("the target"). The grey level and size of the arrows code the interaction strength (see color bar on the right of the figure). From these results, it is possible to note some time-invariant patterns as well as some changes during the period of the movement preparation. In particular, it can be observed that the connections between the PMd_Right area and the M1F_Rigth area, between the PMd_Left to the M1Lips_Left and between the SMA_Left and M1F_Left, reveal a strong increase in their strength during the movement preparation, for both DTF and PDC. On the other side, some connectivity links remain constant during the interval. In particular, strong connections are observed from the PMd_R area to the M1Lips_R area, from the CMA_R to the M1F_R and to the SMA_R, for both DTF and PDC and throughout the period examined.

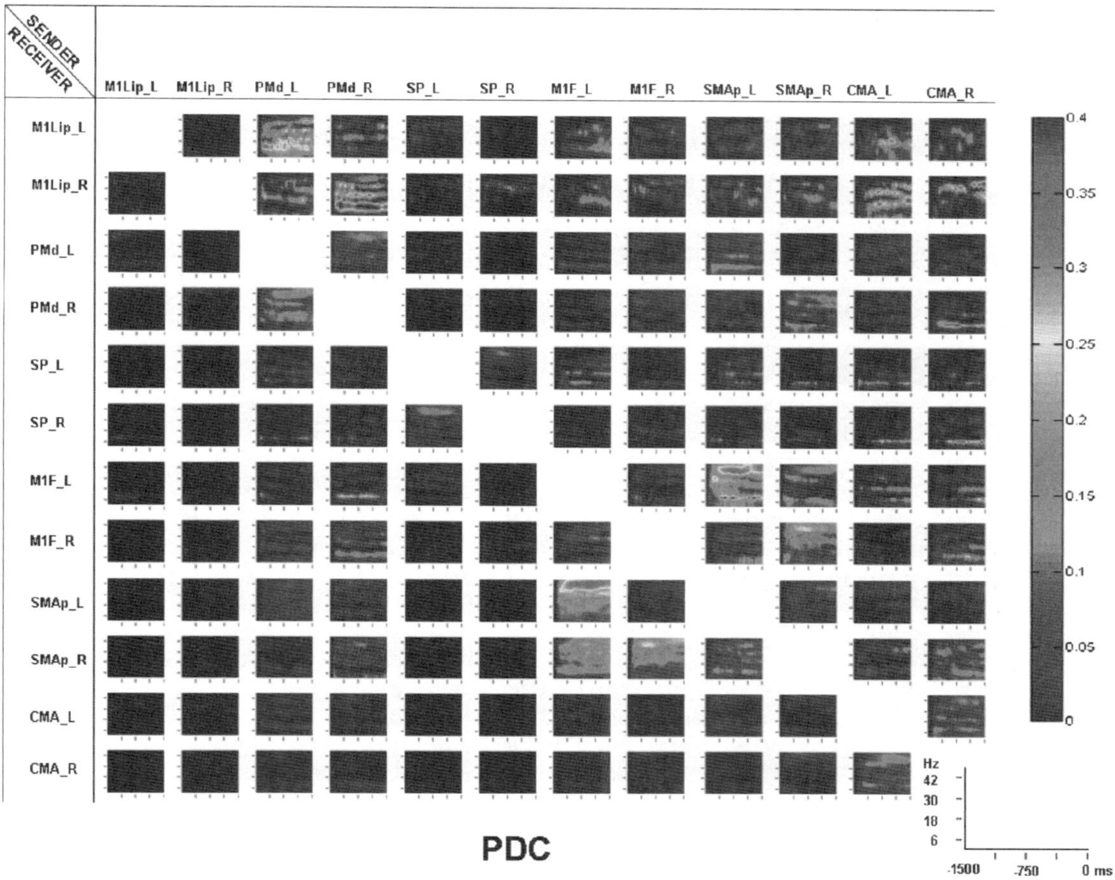

FIGURE 8.1: Time–frequency distribution of the instantaneous PDC obtained for a representative subject, from the set of cortical waveforms estimated in the 12 ROIs considered. On the *x*-axis, the time interval (1500 ms preceding the EMG onset, which served as zero time). On the *y-axis*, the frequencies ranging from 0 to 50 Hz (see the inset at the bottom of the figure). The figure should be read from column (areas originating the causal influence) toward the rows (areas to which the causal influence is directed). The first 50 ms reflect the adaptation epoch of the model. The color bar on the right of the figure represents the intensity of the time-varying connectivity.

Some differences can be detected between the connectivity networks elicited by the two connectivity estimators employed (i.e., DTF and PDC). In particular, it can be observed that DTF showed the existence of a stable connection between the cortical premoter area (PMd) of both hemispheres, as well as a direct link between the right premotor areas (PMd_R) and the left primary motor area for the lips (MILips_L) during all the observed time interval, whereas by using the PDC estimator, these links are not observed at all. In these cases, it could be noted that often the cortical network estimated with the DTF that differs from the PDC networks also admits an alternative path connecting such areas. In particular, the link between the right PMd area and the MILips_L

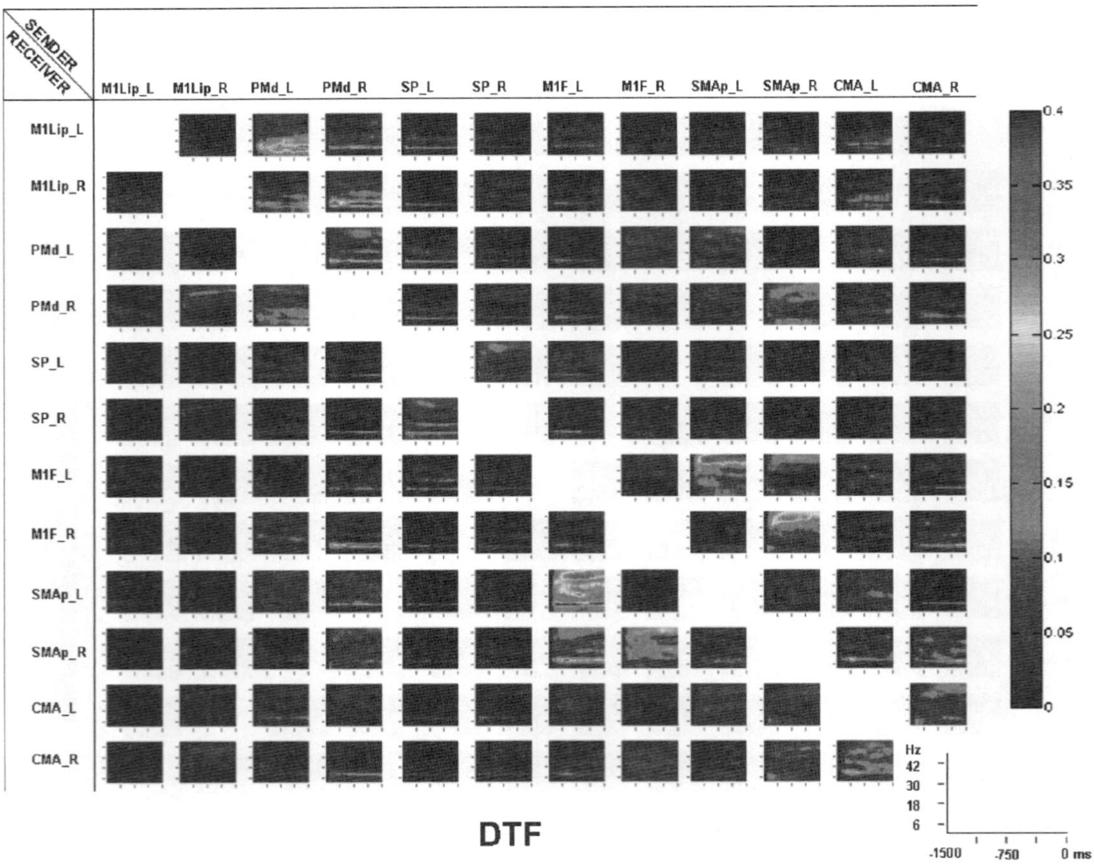

FIGURE 8.2: Time-frequency distribution of the instantaneous DTF obtained for a representative subject, from the set of cortical waveforms estimated in the 12 ROIs considered. Same conventions as in Fig. 8.1.

can also be replaced by a path that starts from PMd_R area, arrives on the SMA_L area and from there toward the PMd_L. In addition, the same path with another step connects the PMd_R with the MILips_L.

Figure 8.4 shows the time course of interaction strength between selected regions of interest from all five subjects analyzed, computed with the PDC estimator. The thin waveforms in different grey levels refer to the connectivity strengths computed for each particular subject of the group. The thick black line is the average time course of connectivity strengths for all the population. Time is expressed in multiseconds from the EMG onset, which served as zero time. The waveforms were normalized to allow comparison between signals with different power spectra, by subtracting their mean values and dividing by their standard deviations. The ROIs presented are the Brodmann area 6 of the left hemisphere (PMd_L), the cingulated motor area left (CMA_L), the primary motor area

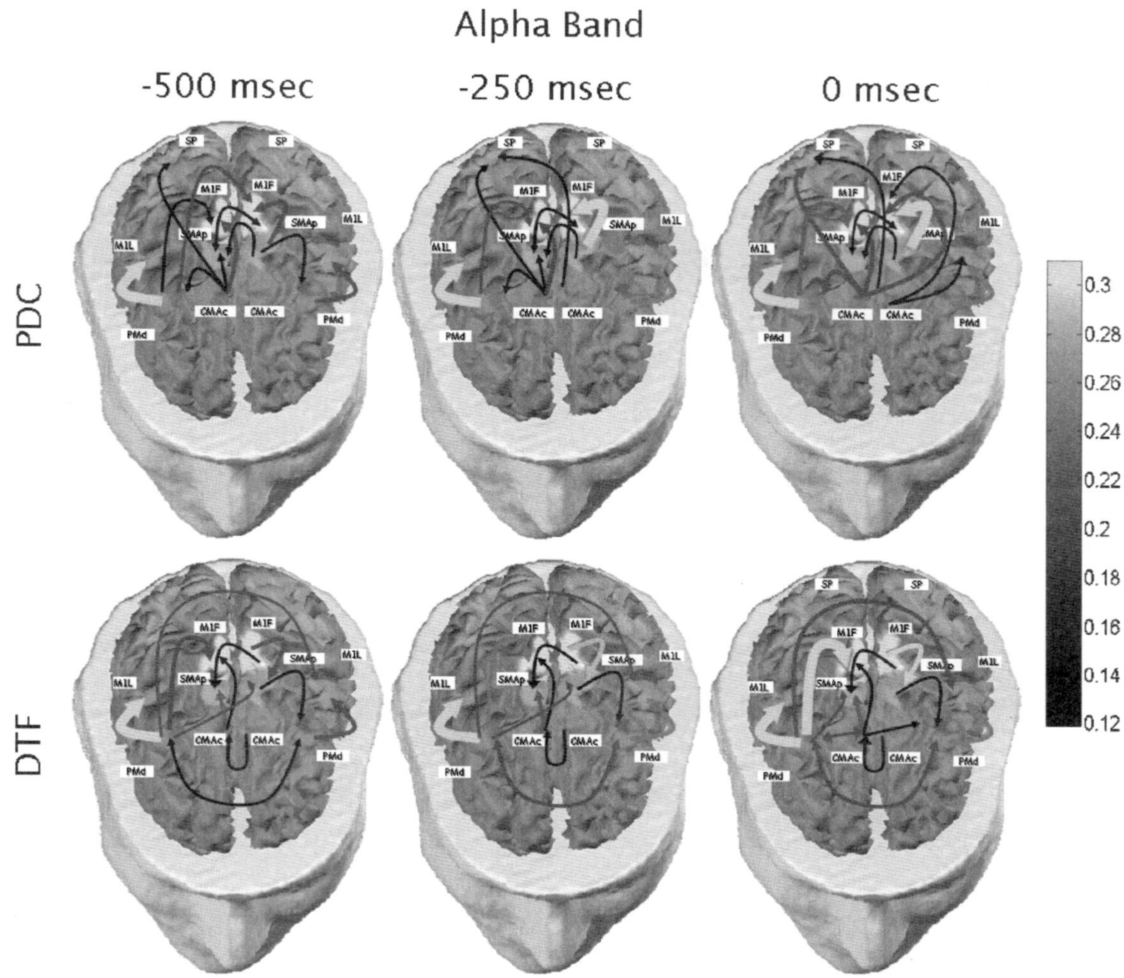

FIGURE 8.3: Time-varying connectivity patterns in the alpha band, extracted at −500, −250 and 0 ms before the movement onset. (*First row*) Results obtained with time-varying DTF. (*Second row*) Results of time-varying PDC. Reconstruction of the head and cortex of the subject, obtained from sequential MRIs. The different ROIs selected are depicted in different grey levels and described by the labels. The grey level and size of the arrows code the interaction strength (see color bar on the right). Similarities can be seen between the results obtained with the two methods as well as an evolution in time of the connectivity during the movement preparation.

for the foot movement, right and left (MIF_R and MIF_L), the primary motor area for the lips, right (A4lip_R). It is possible to recognize a substantial agreement between the subjects across the average connectivity waveform in the framework of the biological variability. In particular, it is of interest that the behavior of the primary motor area of the foot (MIF_R and MIF_L) whose connectivity strengths towards the primary area of the lips (A4Lips) increases starting from 250 ms before the

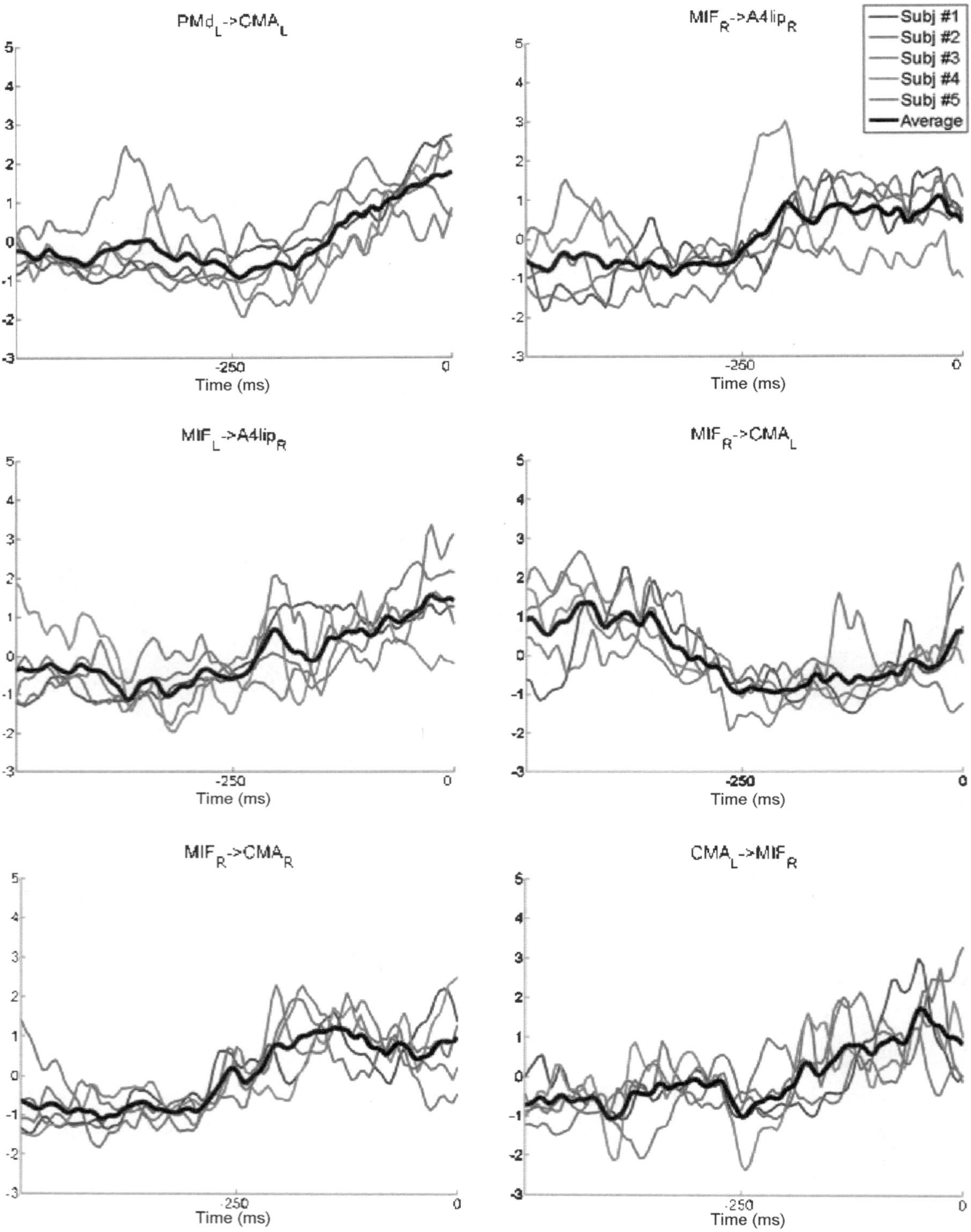
PMd$_L$->CMA$_L$
MIF$_R$->A4lip$_R$
Subj #1
Subj #2
Subj #3
Subj #4
Subj #5
Average
MIF$_L$->A4lip$_R$
MIF$_R$->CMA$_L$
MIF$_R$->CMA$_R$
CMA$_L$->MIF$_R$
5
4
3
2
1
0
-1
-2
-3
-250
0
Time (ms)

EMG onset. A decrease of connectivity strengths is also observed during the period of time from 500 ms to the 250 ms before the EMG onset between the foot motor area and the CMA.

8.4 DISCUSSION

After being tested by the simulation study, the methods analyzed were applied to the cortical activity estimated in some regions of the cortex of the five subjects under study to highlight the possibilities of the methods of analysis. The information obtained by the simulation study was used to evaluate the applicability of this method to actual event-related recordings. The gathered ERP signals related to the combined foot–lips movement showed an SNR between 3 and 5. The trials at disposal for each experimental subject after the artifact rejection were about 100. According to the simulation results, we applied the time-varying PDC/DTF methods on the estimated cortical current density data expecting a limited amount of errors in the estimation of cortical connectivity patterns.

The problem of the linearity of the coupling between the activities in different areas was also addressed. The evaluation of several methods for the computation of the functional connectivity between couple of EEG/MEG signals was recently performed [2]. It has been concluded that although nonlinear methods, such as mutual information, nonlinear correlation and generalized synchronization [3–5], might be preferred when studying EEG broadband signals that are sensitive to dynamic coupling and nonlinear interactions expressed over many frequencies, the linear measurements are still convenient as they afford a rapid and straightforward characterization of functional connectivity. A recent study [6] applied time-varying DTF and PDC to simulated data with nonlinear coupling (generated by a stochastic Roessler system), demonstrating that the PDC and the DTF are sensitive in detecting interactions in nonlinear multivariate systems, and that high model orders (as those employed in this study) are required to describe the nonlinear system sufficiently with a linear MVAR model.

The simple motor task, the preparation to the execution of a foot-lips movement, was chosen, as it is well known and studied in literature. In particular, the time period represented in Figs. 8.2 and 8.3 is not casual. The literature related to the self-paced movement indicate that at the 500 ms there is on average a brisk increase of the potential negativity over the scalp surface in

FIGURE 8.4: Time-courses of connectivity strength between selected ROIs, obtained for the five subjects participating to the study, in the alpha frequency band, by the time varying PDC. Each thin grey waveform refers to the connectivity strengths computed for a particular subject of the analyzed group. The thick black line is the average across subjects of the time-varying connectivity strengths. Time is expressed in milliseconds from the EMG onset (zero time). The waveforms were normalized to allow comparison despite different powers. The ROIs analyzed are the BA 6 of the left hemisphere (PMd_L), the CMA_L, MIF_R and MIF_L, and A4lip_R.

centroparietal regions (last phase of the Bereitschaft potential, BP) and at 250 on average there is a further increase over the central scalp region (negative slope, NS). Furthermore, it has been also shown that at the EMG onset (zero time) the maximum activity is observed on the primary motor area of both hemispheres. Figure 8.4 shows the time-varying PDC connectivity strengths in the population analyzed for a selected couple of ROIs. The average connectivity waveforms (thick black line) showed a consistent pattern of increasing values approaching the EMG onset, suggesting a general increase of connectivity strength between the primary motor areas related to the movement of foot and lips. The observed variability across the population analyzed seems reasonably small, and the same qualitative results were also obtained by computing the time-varying connectivity with the DTF estimator. The physiological results obtained are similar in the five experimental subjects, and seem rather consistent and could integrate those already present in literature on the same movement. The analysis performed with the aid of the time-varying connectivity estimators here derived suggested the existence of two different cortical networks, subserving the preparation and the execution of the joint movement of lip and foot.

In particular, one network is time-invariant during the preparation of the movement involves that PMd, CMA, SMA, M1Lips, and M1F ROIs located on the right hemisphere. It is worth noting that both the connectivity estimators employed (i.e., PDC and DTF) returned the same information. Results obtained suggest that such links connect cortical ROIs that are rather involved in the resource allocation for the behavioral task performed instead of in the monitoring of the actual execution of the joint movement. In this respect, the cortical network depicted by such time-invariant links does not change its activation state during the preparation of the joint movements of lips and foot. In fact, PMd and CMA are ROIs that are activated continuously during the preparation of the movement and generate the sequence of generic activation command towards the primary motor areas responsible for the activation of the specific motorneurons that will trigger the movement for lips and foots.

The second network of cortical areas involved in the preparation and generation of the joint movements is instead dependent on the time-course of the task. In particular, the results suggested the existence of a network that increases in strength during the few hundreds of milliseconds preceding the actual execution of the movement. Such a network involves connections between the PMd_Right area and the M1F_Rigth area, between the PMd_Left to the M1Lips_Left and between the SMA_Left and M1F_Left. Once again, it is worth noting that such increase in connectivity strengths has been detected by both the connectivity estimators applied, i.e., both PDC and DTF. Differences between the cortical networks estimated by the DTF when compared to the PDC are related to the existence of alternative paths between the same cortical areas, which can be modelled by the DTF with also a direct link. This property of the DTF has been established on the basis of its mathematical formulation, as described earlier. However, it must be said that there also exists cortical paths estimated with the PDC that are not estimated with the DTF. For these discrepancies, which are relative to a few connections on the total amount of the estimated cortical networks, possible explanations are relative to the different sensitivities of such algorithms to the amount of noise present on the data. The global conclusion, however, is that the cortical networks estimated by these two methods are rather similar, and allows to conclude positively about the applicability of these estimators on real EEG recordings. It is worth noting that the information on the increasing connectivity strength in the second cortical network would be lost with the application of the standard connectivity estimators, which assumes the stationarity of the analyzed data.

In the last two chapters we presented two multivariate causality estimators to retrieve rapidly changing cortical connectivity. Simulations suggest that the methods were adequate to estimate cortical connectivity under a large range of SNR and TRIALS factors, normally encountered in the standard practice of the high-resolution EEG recordings. An SNR of 5 and a number of trials of at least 20 provide a good accuracy in the estimation. The results pointed out that the DTF estimator can assure a higher adaptation speed than PDC, but PDC ensures a better accuracy in the estimation of connectivity strengths. Possible values for the optimum choice of the adaptation factor *C* have been proposed, according to the different operative conditions. The results of the application of the two methods to real data in a group of subjects revealed time-varying cortical connections during the execution of the task. In particular, two different cortical networks, one constant across the task and the other evolving during the preparation of the joint movement have been detected with the proposed methodology. It is worth noting that the information on the increasing connectivity strength in the second cortical network would be lost with the application of the standard connectivity estimators, which assumes the stationarity of the analyzed data.

In conclusion, this study opens the way to the application to cortical estimations for studying detailed time–frequency patterns describing the evolution of cortical connectivity.

REFERENCES

[1] D. V. Moretti, F. Babiloni, F. Carducci, F. Cincotti, E. Remondini, P. M. Rossini, S. Salinari, and C. Babiloni, "Computerized processing of EEG-EOG-EMG artifacts for multi-centric studies in EEG oscillations and event-related potentials," *Int. J. Psychophysiol.*, vol. 47, no. 3, pp. 199–216, Mar. 2003, doi:10.1016/S0167-8760(02)00153-8.

[2] O. David, D. Cosmelli, and K. J. Friston, "Evaluation of different measures of functional connectivity using a neural mass model," *NeuroImage*, vol. 21, pp. 659–673, 2004, doi:10.1016/j.neuroimage.2003.10.006.

[3] M. S. Roulston, "Estimating the errors on measured entropy and mutual information," *Physica D*, vol. 125, pp. 285–294, 1999, doi:10.1016/S0167-2789(98)00269-3.

[4] C. J. Stam and B. W. van Dijk, "Synchronization likelihood: an unbiased measure of generalized synchronization in multivariate data sets," *Physica D*, vol. 163, pp. 236–251, 2002, doi:10.1016/S0167-2789(01)00386-4.

[5] C. J. Stam, M. Breakspear, A. M. van Cappellen van Walsum, and B. W. van Dijk, "Nonlinear synchronization in EEG and whole head MEG recordings of healthy subjects," *Hum. Brain Mapp.*, vol. 19, no. 2, pp. 63–78, 2003, doi:10.1002/hbm.10106.

[6] M. Winterhalder, B. Schelter, W. Hesse, K. Schwabb, L. Leistritz, D. Klan, R. Bauer, J. Timmer, and H. Witte, "Comparison of linear signal processing techniques to infer directed interactions in multivariate neural systems," *Signal Process.*, vol. 85, no. 11, pp. 2137–2160, November 2005, doi:10.1016/j.sigpro.2005.07.011.

Author Biography

Laura Astolfi is a researcher at the Department of Human Physiology and Pharmacology of the University of Rome "La Sapienza" and at the Neurophysiopathology Unit of the Fondazione Santa Lucia, Rome. Her research interests involve the study of the cortical connectivity, the neuroelectrical inverse problem and the multimodal integration of fMRI and HR EEG data.

She received the degree in Electronic Engineering *cum laude* from the University of Rome "La Sapienza" in 2003 and the PhD in Biomedical Engineering from the University of Bologna in 2007.

Fabio Babiloni was born in Rome in 1961. He got the master degree in Electronic Engineering at the University of Rome "La Sapienza" summa cum laude in 1986, and the PhD in Computational Engineering at the Helsinki University of Technology, Helsinki in the 2000 with a dissertation on the multimodal integration of EEG and fMRI. He is currently Associate Professor of Human Physiology at the Faculty of Medicine of the University of Rome "La Sapienza", Rome, Italy. He is author of more that 85 papers on bioengineering and neurophysiological topics on international peer-reviewed scientific journals, and more than 200 contributions to conferences and books chapter. Currents interest are in the field of multimodal integration of EEG, MEG and fMRI data, cortical connectivity estimation and Brain Computer Interface. Prof. Babiloni is currently Associate Editor of four scientific Journals "*Clinical Neurophysiology*", "*International Journal of Bioelectromagnetism*","*IEEE Trans. On Neural System and Rehabilitation Engineering*", and "*Computational Intelligence and Neuroscience*".

Brain–Machine Interface Engineering

Justin C. Sanchez and José C. Principe
University of Florida

MORGAN & CLAYPOOL PUBLISHERS

ABSTRACT
Neural interfaces are one of the most exciting emerging technologies to impact bioengineering and neuroscience because they enable an alternate communication channel linking directly the nervous system with man-made devices. This book reveals the essential engineering principles and signal processing tools for deriving control commands from bioelectric signals in large ensembles of neurons. The topics featured include analysis techniques for determining neural representation, modeling in motor systems, computing with neural spikes, and hardware implementation of neural interfaces. Beginning with an exploration of the historical developments that have led to the decoding of information from neural interfaces, this book compares the theory and performance of new neural engineering approaches for BMIs.

KEYWORDS
neural interfaces, brain, neural engineering, neuroscience, neural representation, motor systems

Foreword

"Life can only be understood backward, but it must be lived forward."
—Soren Kierkegaard

What has the past decade taught us about mankind's ability to interface with and read information from the brain? Looking back on our experiences, the salient recollection is how ill-prepared the present theories of microelectronic circuit design and signal processing are for building interfaces and interpreting brain's activity. Although there is plenty of room for future improvement, the combination of critical evaluation of current approaches and a vision of nueroengineering are helping us develop an understanding on how to read the intent of motion in brains.

The flow of ideas and discovery conveyed in this book is quite chronological, starting back in 2001 with a multi-university research project lead by Dr. Miguel Nicolelis of Duke University to develop the next-generation BMIs. The series of engineering developments explained in this book were made possible by the collaboration with Miguel, his contagious enthusiasm, vision, and brilliant experimentalism, that have led us in a journey of discovery in new theories for interfacing with the brain. Part of the results presented here also utilize data collected in his laboratory at Duke University.

It was also a journey of innovation shared with colleagues in ECE. Dr. John Harris was instrumental in designing the chips and proposing new devices and principles to improve the performance of current devices. Dr. Karl Gugel helped develop the DSP hardware and firmware to create the new generation of portable systems. We were fortunate to count with the intelligence, dedication, and hard work of many students. Dr. Justin Sanchez came on board to link his biomedical knowledge with signal processing, and his stay at University of Florida has expanded our ability to conduct research here. Dr. Sung-Phil Kim painstakingly developed and evaluated the BMI algorithms. Drs. Deniz Erdogmus and Yadu Rao helped with the theory and their insights. Scott Morrison, Shalom Darmanjian, and Greg Cieslewski developed and programmed the first portable systems for online learning of neural data. Later on, our colleagues Dr. Toshi Nishida and Dr. Rizwan Bashirullah open up the scope of the work with electrodes and wireless systems. Now, a second generation of students is leading the push forward; Yiwen Wang, Aysegul Gunduz, Jack DiGiovanna, Antonio Paiva, and Il Park are advancing the scope of the work with spike train modeling. This current research taking us to yet another unexplored direction, which is perhaps the best indication of the strong foundations of the early collaboration with Duke.

This book is only possible because of the collective effort of all these individuals. To acknowledge appropriately their contributions, each chapter will name the most important players.

Jose C. Principe and Justin C. Sanchez

Contents

CHAPTER 1

Introduction to Neural Interfaces

INTRODUCTION

Neural interfaces are one of the most exciting emerging technologies to impact biomedical research, human health, and rehabilitation. By combining engineering and neurophysiological knowledge with an innovative vision of biointeractive neuroprosthetics, a new generation of medical devices is being developed to functionally link large ensembles of neural structures in the central nervous system (CNS) directly with man-made systems. By communicating directly with the CNS, researchers are acquiring new scientific knowledge of the mechanisms of disease and innovating treatments for disabilities commonly encountered in the neurology clinic. These include spinal cord injury, stroke, Parkinson's disease, deafness, and blindness, to name a few. The seamless integration of brain and body in the healthy human makes us forget that in some diseases, normal brains can be deprived of sensory inputs (vision, audition) or even become locked in by the body (traumatic injury to the spinal cord or by degenerative diseases of the nervous system).

Several approaches have been proposed for dealing sensorimotor injuries of the nervous system, and the technologies fall into one of two categories: biological- and engineering-based solutions. In the biological category,[1] cellular and molecular techniques are being developed to regenerate and repair the damaged or dysfunctioning tissue [1–4]. Although the biological approach may ultimately provide the most natural solution to nervous system injury, the current state of research is at an early stage (i.e., spinal cord regeneration in some cases is limited in spatial scale and often operates over long time scales. Many of the new technologies shown on the right branch of Figure 1.1 have been developed to focus on the facilitation (neurotrophic factors) guidance (scaffolding) of growth of existing neurons or replacement cells (stem cells). This approach primarily works at the cellular or subcellular level of abstraction through modification of biological signaling and growth.

[1]The study of repair and regeneration of the central nervous system is quite broad and includes contributions from molecular/cellular neuroscience, tissue engineering, and materials science. For a comprehensive review of the application of each of these to the repair of the nervous system, see References [1–4].

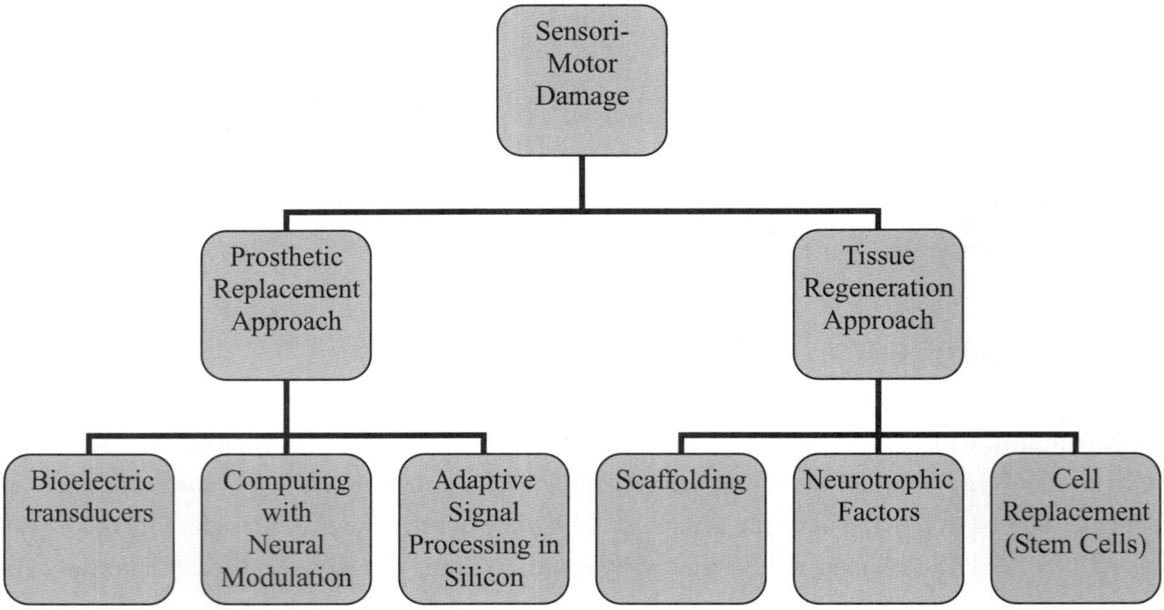

FIGURE 1.1: Roadmap of treatments for neural prosthetics, regeneration, and repair.

In contrast, the engineering approach is based on the artificial neural interface concept to provide therapy options for these individuals. While preserving the same function of the damaged biological systems, the man-made neural interface effectively creates a conduit to transduce the signaling of the nervous systems into quantities that can be used by computing architectures that augment or replace existing neural substrates [5].[2] Neural interface technologies are heavily influenced by electrical and computer engineering principles because they provide the basis and tools for the analysis of time varying quantities such as those produced by the nervous system. On the left branch of Figure 1.1, new processes must be created to mimic the integration, transduction, and interpretation of real biological events that occur internally or through interaction of the organism with the environment. These techniques include the use of adaptive signal processing to compute with neural modulation and the subsequent implementation of that processing in hardware. The overall role of the neural prosthetic are to create one or many of the following: (1) an artificial sensory system (somatosensory, occipital, auditory) by delivering the external world stimuli to the appropriate cortices; (2) an artificial command and control link that allows the motor commands to directly interact with the environment (through robotic or computerized devices); (3) a neural augmentation link within

[2]For a comprehensive review of sensory and motor prosthetics, see Reference [5].

or between neural networks affected by disease. In view of the amazing plasticity of the brain, these man-made devices may, with appropriate sensory feedback, be assimilated and become part of the user's cognitive space, at par with the rest of the body. Because of their man-made qualities, neural interfaces carry special qualities. Potentially, neural interfaces can scale human's natural reaction time, force, and abilities through engineered devices that are much faster and more powerful than biological tissue. In particular, they may enable a higher bandwidth between the human brain and the digital computer, which has conditioned the present research direction and metaphors for the use of computers in our society.

1.1 TYPES OF BRAIN–MACHINE INTERFACES

In the last decade, neural prosthetic technologies have commonly been grouped under the title of brain–computer interfaces (BCIs) or brain–machine interfaces (BMIs), and the names are often used interchangeably.[3] BMIs can be divided into three basic types, depending upon the application: sensory, cognitive, or motor. Sensory BMIs activate sensory systems with artificially generated signals that translate physical quantities. The most common type of sensory BMI, with more than 50,000 implanted devices [6], are cochlear implants that use miniature microphones and signal processing to translate sound wave pressure (physical quantities) in the ear into neuronal firings (neural representation) applied to the auditory nerve's tonotopic organization, allowing deaf people to listen. The same basic concept is being developed for retinal prosthesis, which can deliver to the visual system the appropriate stimulation that indicates the morphology of objects [7]. Cognitive BMIs attempt to reestablish proper neural interactions among neural systems that have damaged internal functioning within the network or do not have the ability to communicate with other networks. Cognitive BMIs have been applied for the treatment of the limbic system that may have long- and short-term memory deficits as in Alzheimer's disease [8]. Motor BMIs, on the other hand, seek to translate brain activity from the central or peripheral nervous system into useful commands to external devices. The idea is to bypass the damaged tissue (as in a spinal cord transection) and deliver the motor control commands either directly control a computer/prosthetic limb or deliver natural motor control of muscles via functional electrical stimulation [9]. Reestablishing the control of lost movement with the peripheral nervous system shares many of the neurotechnology and challenges of BMIs for prosthetic control.

Motor, cognitive, and sensory BMIs for direct neural control have far reaching impact in rehabilitation. If the whole brain or some of its parts are spared from disease or injury and the path of

[3]Historically, BCI has been associated with noninvasive approaches of interfacing with the nervous system as with the electroencephalogram (EEG) (described later). Alternatively, BMI commonly refers to invasive techniques.

transduction from the sensory organs (hair cells in hearing or photoreceptor cells in vision) or to the muscles (spinal cord injury or peripheral neuron damage) is broken, the following neural engineering question is appropriate: "What is the best way to deliver stimuli to the sensory cortex or to extract information about voluntary movement from motor cortex activity?" Obviously, there are two parts to the answer: one relates to the actual design and implementation of the engineering systems that can translate sound or light into electrical stimuli, control motor neurons in the periphery, or directly activate a prosthetic and the other relates to the neurophysiology knowledge to help prescribe the spatiotemporal characteristics of these stimuli to mimic the ones created by the sensory pathway to the brain or, in the case of the motor BMIs, to help model the spatiotemporal interactions of neural activity in the motor cortex. The disciplines of engineering and neuroscience (at least) are involved, and none of them independently will be able to solve the problem. Therefore, this is an imminently multidisciplinary enterprise that requires large research groups to be productive.

1.2 BEYOND STATE-OF-THE-ART TECHNOLOGY

One may think that the engineering side is better prepared to deliver systems in the short term because of the computational speed of modern digital signal processors (DSPs) and technological innovation in sensors, miniaturization, and computers, but this is illusory; interfacing with the central or peripheral nervous systems required beyond state-of-the-art technology and modeling on many fronts. Engineers working in BMI rapidly develop an appreciation for the solutions encapsulated in biological systems because of the demanding specifications of the systems. To relatively assess the relationships between the performance specifications, let us look at a space with three axes (as shown in Figure 1.2) that relates scale, power dissipation, and computation needed to interface brain tissue and microelectromechanical computer technology.

From the figure, we can see that the relationship of the space spanned by the current technology is very different from that of the BMI system, and only a small portion of the spaces overlap (purple). If we compare the scale of our current multiple-input–multiple-output (MIMO) systems, there is a large disparity because the brain is a huge computational system, with, on the order of 10^{12} neurons, about 10^{15} synapses [10]. The density of neurons in the neocortex of a human is about $10^4/\text{mm}^3$ but can vary as a function of the type of animal and the spatial location (i.e., visual, somatosensory, etc.). The conventional electrophysiological techniques (discussed later) present only a coarsely subsampling the neuronal population. The electronics needed to acquire the activity of this small sampling are very sophisticated. First, the amplifiers have to provide high common-mode rejection ratio, high gains with very small power. Second, the system has to transmit very large data rates also with constraints in power. To make the transmission more reliable, the analog signal must be converted to a digital representation, which adds substantially to the total power consumed.

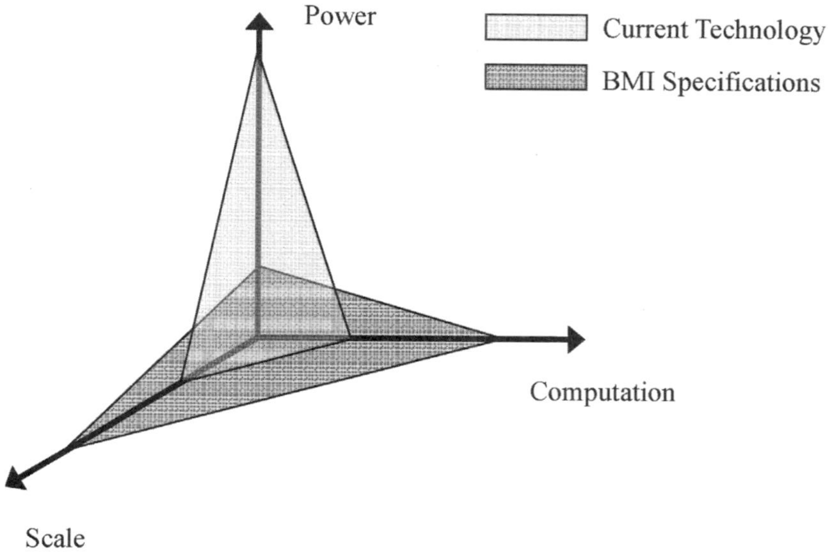

FIGURE 1.2: Comparison of system specifications for current technologies and those needed for BMIs.

Typical numbers for the transmission per channel are around ~144 kbits/channel (12.5 kHz × 12 bits). For a 16-electrode probe, this data rate is around 2.3 Mbits/sec. To compute with the electrophysiological signals at line level, front-end amplifiers (40 dB gain) used with neural probes take about 100 μW of power per channel. If one wants to transmit the data via a wireless implanted device, the total power budget can range between 2.5 and 5 mW. The area of such devices is roughly 0.5–5 mm^2, depending on the number of channels that is absolutely overshadowed by the biological size/power factor [11, 12]. To extend operation times of the BMI over days to weeks using hundreds of neural inputs, the power budgets need to significantly decrease. For the envisioned applications, the systems will have computational demands that exceed by orders of magnitude the capacity of a single powerful workstation. Depending on the scope of the system, it may be necessary to use hundreds or even thousands of processors to provide the needed computational power. Grid-computing infrastructures can potentially deliver these necessary resources on demand. However, the systems will also have stringent real-time requirements, as a result of the need for low latency between brain signaling and sensory feedback.

Improvements using the present approaches are very likely to be evolutionary instead of revolutionary. We have not yet discussed two fundamental and very difficult issues: how to reliably interface with the neural tissue and how to model appropriately the interactions among neurons such that proper signal processing models can be designed and applied. The neural electrode interface is of fundamental importance because invasive BMIs require prolonged contact with the neural tissue and the electrical characteristics of the interface should be stable and reliable to guarantee long life, high signal-to-noise ratios (SNRs), and avoid cell death. By far, the most compelling explanation for signal degradation in chronic neural recordings is the series of events that occurs after implantation, which include inflammatory response, a disruption of the blood–brain barrier, and initiation of astrogliosis and recruitment of microglia and macrophages to the insertion site [13–16]. Each of these problems has been studied extensively, and a variety of approaches have been proposed to improve the electrode tissue interface, which include novel electrode material coatings (conductive polymers, ceramic matrices, timed release of dexamethasone) and the use of microfluidics for engineering the glial response and tissue inflammation [5, 17]. Several groups have demonstrated that it is possible to collect data from neurons for months, but there is a long road still ahead to control tissue responses to the implant.

1.3 COMPUTATIONAL MODELING

The main problem that directly interests the readers of this book is the issue of modeling neural interactions, which will be discussed extensively throughout the text. At this point, we would like to only summarize the state of the art in computational neuroscience and a broad-stroke picture of the signals used in BMIs. There is extensive neuroanatomical knowledge on the excitability properties of neurons as well as the functional mapping of the interconnection among cortical and subcortical networks [18–20]. The issue is that the interaction principles among neurons, which implement the so-called neural code [21], are poorly understood. Moreover, large pieces of a theoretical framework that defines neural representation (i.e., mechanisms of stimulus activated elements), computation (i.e., how the elements optimally use the stimulus), and dynamics (i.e., how the elements change in time) are missing.

From a pragmatic point of view, the fundamental constructs of the brain (or any biological system for this matter) do not come with an instruction manual. The task of finding out how the system is working and what is being computed has been a significant difficulty for computational neuroscientists. Compared with most mechanical or electrical systems, the problem is perhaps even harder because of the self-organizing bottom-up principles embedded in biology versus the top-down approach used in engineering design. The difference in the two perspectives, as shown in Figure 1.3, is the necessary adoption of a big-picture approach by computer scientists and a reductionist approach by many neuroscientists. The top-down approach arises from an engineering perspective where the broad goals are to "design a machine to perform a particular task." The bottom-up perspective is one of the phenomenologist or the taxonomist: "collect data and organize

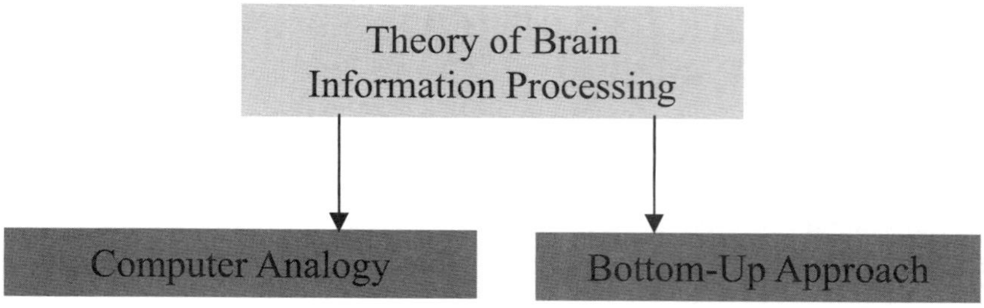

FIGURE 1.3: Bifurcation in the development of computational models of the brain.

it." In either case, the computational approaches used in the development of BMIs contain four pillars: hypothesis of how the brain functions, development of a model, analytically or numerically evaluation of the models, and testing against experimental data. In this framework, modeling plays a big part in the inductive or deductive reasoning used to establish functional associations among neural assemblies. For this reason, the models can take several forms: computational, which focus on which computation is to be performed, in terms of optimality, modularity and others, algorithmic, which look at nature of computations performed, mechanistic, that are based on known anatomy and physiology, descriptive, which summarizes large amounts of experimental data, or interpretive, which explores behavioral and cognitive significance of nervous system function.

The integrative computational neuroscience perspective and the latest technology innovation have been consistently applied to explain brain function, at least since the time of Rene Descartes. Perhaps, the formal computer metaphor of Turing and Ince [22] and Von Neumann [23] is still the one that has more followers today, but others are also relevant, such as the cybernetic modeling of Wiener [24], the universal computer analogy of McCulloch and Pitts [25], the holographic approach of Pribram [26], the connectionist modeling of McClelland and Rumelhart [27], and Rumelhart [28], Rumelhart et al. [29], and Hopfield [30], and the synergetic modeling of Haken [31] and Freeman [32], to name some important examples. This description adapted from Freeman, captures well the difficulty of the modeling task: "Unlike existing computational systems, brains and the sensory cortices embedded within them are highly unstable. They jump from one quasi-stable state to the next at frame rates of 5–10/sec under internal phase transitions. The code of each frame is a 2D spatial pattern of amplitude or phase modulation of an aperiodic (chaotic) carrier wave. The contents of each frame are selected by the input and shaped by synaptic connection weights embodying past learning. These frames are found in the visual, auditory, somatic, and olfactory cortices. Having the same code, they readily combine to form multisensory images (gestalts) that provide a landscape of basins and attractors for learned classes of stimuli."

1.4 GENERATION OF COMMUNICATION AND CONTROL SIGNALS IN THE BRAIN

The generators of the "landscape of basins and attractors," as described by Freeman, are the electrophysiological modulation of neural networks in the living tissue. From a methodological perspective, access to the spatial scale has played an important role in BMI development and includes the use of electrophysiological methods to extract spikes (action potentials) [33], local field potentials (LFPs) [34], electrocorticogram (ECoG) [35], and EEG [36].[4] Figure 1.4 shows the relative relationships of the spatial resolution for the three most common electrophysiological techniques used for BMIs. At the highest resolution, microwire electrode arrays can sense the activity of single neurons simultaneously; however, they are the most invasive to the tissue. ECoG recording is less invasive because the electrodes rest directly on the cortex, and EEG electrodes are the least invasive because they sit on the scalp. Deriving signatures of sensation or intent from electrophysiological recordings carries either the constraints of model selection or the uncertain hypotheses of the signal generation at each level of scale. For instance, microelectrode arrays that sense extracellular action potentials are very specific to neuronal function but are local in space and carry the bottleneck of sensing a representative ensemble of neurons within the technical limitations of electrode spacing and number and the difficulty of bridging the time scale to cognitive events. Because BMI paradigms have developed, the viewpoint from the single neuron has expanded to encompass the time-dependent communication of ensembles of hundreds of neurons [33]. These principles have been directly applied with great success to BMI architectures [37–40] that utilize only a single-unit activity (SUA). Recently, the use of LFPs have also been shown to improve BMI performance when combined with an SUA [34]. Therefore, harnessing both the inputs (dendritic activity) and outputs (action potentials) of neural assemblies seems to be critical for interpreting the intent of the individual.[5] Using a much larger scale, scalp EEG contains a summation of dendritic and axonal currents of millions of neurons over the cortical impedance. The continuous amplitude macroscopic signals modulate at the time scale of behavior but lack specificity and can be much more influenced by noise. The ECoG, when compared with the EEG, has the great advantage of being more spatially specific (smaller electrodes), much less contaminated by noise, and can display better frequency resolution because of its proximity to the neural tissue.

[4]Electrophysiological recordings are one particular subset of brain sensing methodologies. Other techniques such as the magnetoencephalogram, near-infrared spectroscopy, and functional magnetic resonance imaging have been used in BMI studies. However, the aforementioned techniques require quite specialized hardware and provide much different spatial and temporal resolutions compared with the electrophysiological techniques.

[5]In the next chapter, we will present a brief review of the generation of electrical potentials in excitable neural tissue.

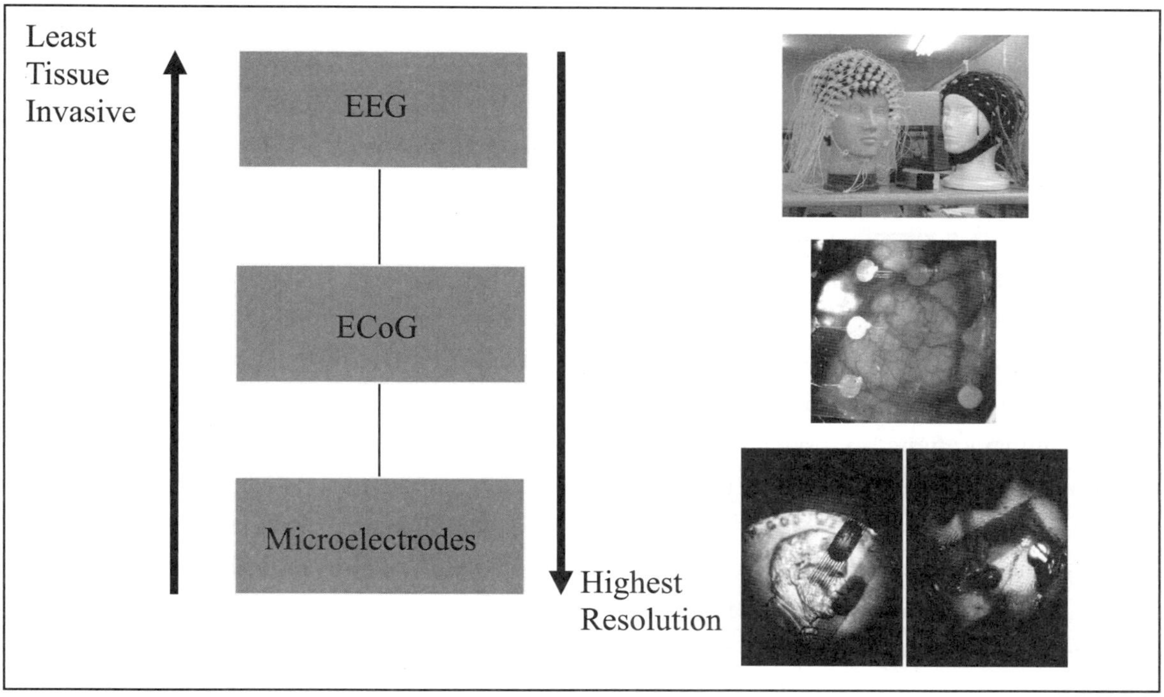

FIGURE 1.4: A comparison of the scales of electrophysiological recording methodologies.

If one analyzes the time course of one channel of data collected from the scalp (EEG), the signal time structure is very complex as shown in Figure 1.5a [41], reminding us of a nonconvergent dynamic system, or outputs of complex systems excited by bursty white noise. One of the characteristics is the rapid fluctuations observed in the signals, which, in the parlance of time series analysis, represents nonstationary phenomena and can depend upon behavioral statistics. Characterizing and extracting features from the EEG with conventional modeling is, therefore, very difficult, and the large number of publications addressing a myriad of methods well exemplifies the difficulty and suggests that the field is still seeking a good methodology to describe brain waves [42–44]. The EEG is called a macroscopic measure of brain activity, and its analysis has been shown to correlate with cognitive state [45] and brain diseases such as epilepsy [46]. The signal is created by a spatial temporal average of the pulse to amplitude conversion created when billions of pulses of current (the spike trains) are translated into voltage by the cortical tissue impedance and recorded by a 1-cm diameter electrode. The EEG frequency band extends from 0.1 to approximately 200 Hz, with a $1/f$ type of spectra (suggesting a scale free origin) and spectral peaks concentrated at a handful of frequencies where the EEG rhythms exist (delta, theta, alpha, sigma, beta, and gamma). Before being collected by the scalp electrode, the signal is attenuated and spatially filtered by the pia, dura, the corticospinal fluid, bone, and scalp. It is also contaminated by many electrical artifacts of biological or nonbiological origins.

Alternatively, if the electrodes are invasively placed at the surface of the cortex as in ECoG, the SNR is much better and the useable bandwidth increases up to at least 1 kHz (as shown in Figure 1.5c) using electrodes of 4 mm in diameter [47]. Theoretical analysis outlined by Freeman [32, 48–52] and Nunez [53–55] has identified the utility of ECoG potentials and attempts to explain how to extract the relevant modulation of neural assemblies. Closely spaced subdural electrodes have been reported to measure the spatially averaged bioelectrical activity of an area much smaller than several square centimeters [56]. The production of potentials is because of the superposition of many aligned and synchronous dipole sources [53]. The coactivation of sources is related to neural "synchrony" and is used to describe the amplitude modulations in extracellular recordings that occur during state changes [36]. The fluctuating cortical potentials have been associated with traveling waves of local dendritic and axonal potentials [57]. The results in Reference [57] indicate that at least two separate sources of signal coherence are produced either through the action of short-length axonal connections or the action of long distance connections. Synchronous activity can occur at different spatial scales and over time lags [53], which requires wide spatial sampling of the cortex and analysis over fine temporal scales.

Being a multiscale system, brain activity can also be collected at an intermediate level, which targets the neural assembly and is named the LFP as shown in Figure 1.5b. LFPs are continuous amplitude signals also in the frequency range of 0.1–200 Hz created by the same principles as the EEG, but they are normally collected with higher impedance microelectrodes with cross-sectional recording diameters of 20–100 μm. Therefore, the spatial averaging operated by the larger EEG/ECoG electrodes is not present. The LFPs differ in structure depending where the microelectrode tip is placed with respect to the dendritic or axonal processes of the neural assembly and therefore better describe the electrical field fluctuations in a more localized volume of tissue. In the frequency domain, LFP tuning has been shown to be manifested also, and differentially, in oscillatory activities in different frequency ranges. Oscillations and spiking in sensorimotor cortex in relation to motor behavior in both humans [36, 58] and nonhuman primates [59–61] in the 15- to 50-Hz range were found to increase in relation to movement preparation and decrease during movement execution [62]. Other researchers have also described the temporal structure in LFPs and spikes where negative deflections in LFPs were proposed to reflect excitatory, spike-causing inputs to neurons in the neighborhood of the electrode [63].

At the microscopic scale, the electrical activity captured by microelectrode arrays placed closer to the neuron cell body displays a pulse (called a spike) whenever the neuron fires an action potential shown in Figure 1.5b. Spikes are approximately 1 msec in duration, 30–200 μV in

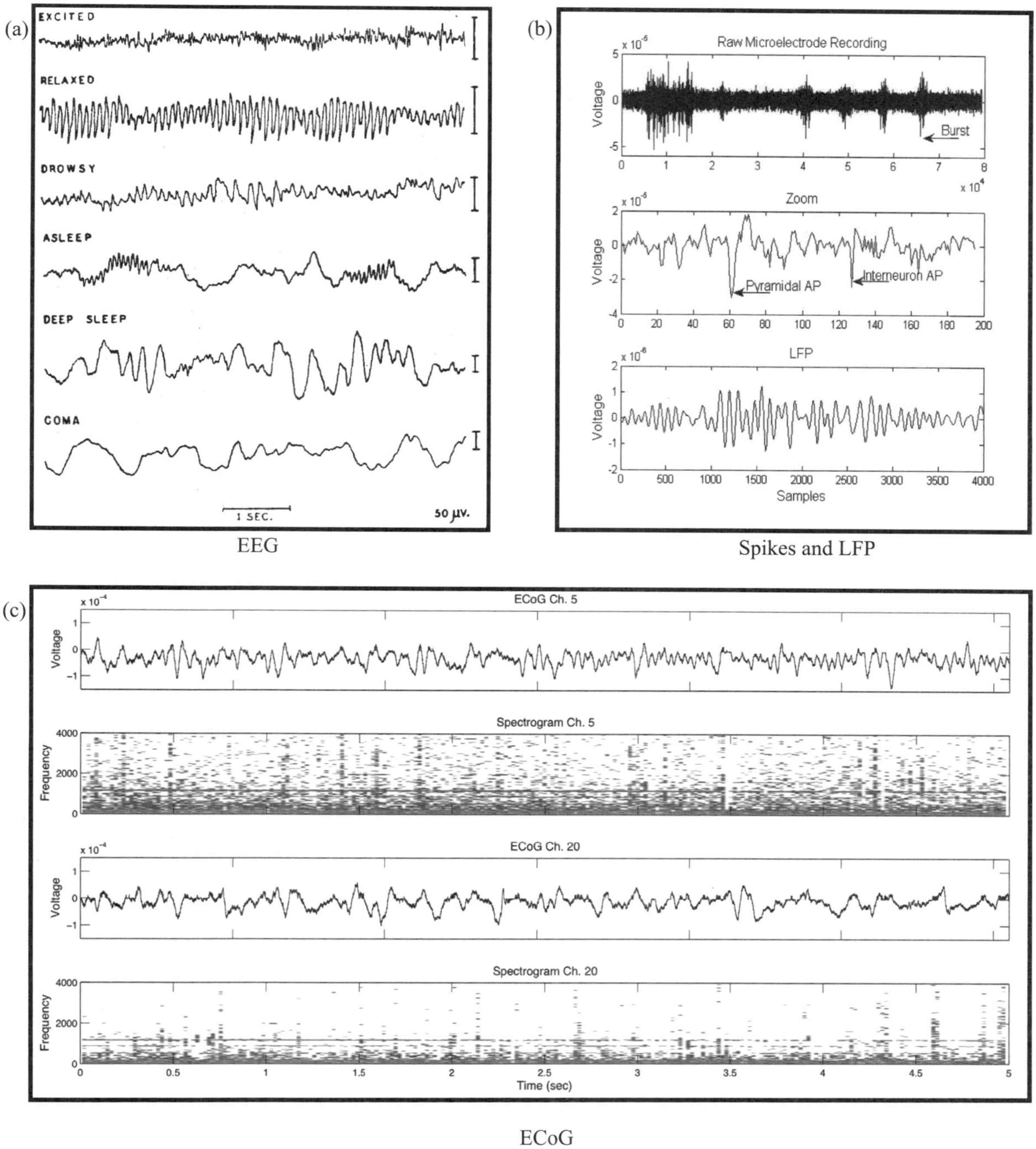

FIGURE 1.5: Examples of electrophysiological recordings. (a) Human EEG (from Jasper and Penfield [41]), (b) spikes and LFPs from a rat (time range: top, 6.5 sec; middle, 16 msec; bottom, 328 msec), (c) ECoG from a human performing a reaching task. The patient implanted with ECoG electrodes was undergoing epilepsy monitoring.

peak-to-peak amplitude on average, and they are biphasic or triphasic in morphology depending upon the relationship of the electrode with respect to the cell body. Because of the electrophysiological morphology of the action potential (1-msec biphasic wave), the signals contain frequency in the range of 300 Hz to 7 kHz; therefore, it is common to high-pass filter the signal collected by the electrode at 300 Hz. The specificity of the spike firing is unparalleled (i.e., one can spike sort [64] the data to attribute it to a specific neuron). SUA can then be modeled as a Poisson-distributed point process [65–68]. The tradeoffs from a signal processing perspective embedded in the LFP and spike representation should be clear: LFPs loosely measure the electrical field characteristics of a region of brain tissue and the time varying activity of one or more neural populations; therefore, they are unspecific about a single neural behavior. On the other hand, the spike trains are very specific about the excitability of a given neuron, but they coarsely subsample the neural population. Unfortunately, the relationships between spike trains and LFPs are not well understood. It is reasonable to model LFP, ECoG, and EEG as nonstationary, perhaps nonlinearly generated random processes, whereas SUAs require a point process modeling.

From this discussion, it should also be apparent that there is insufficient functional knowledge of neural population dynamics in neuroscience to help guide a principled engineering approach to create realistic models of brain function or even of neural population dynamics. Therefore, we are left with the conventional engineering, statistics, and physics tools to quantify physical phenomena. Naturally, the predominance of linear approaches, Gaussian statistics, can be expected, but they should be taken with a grain of salt because they have been developed for a set of constraints likely not operational in brain function. In fact, the golden rule taken from years of working in this area is that justifying with data the assumptions of the modeling approach is critical for progress. The time dimension is of utmost importance for the brain because its role is to anticipate what is going to happen in an unpredictable and complex world. Therefore, assumptions that disregard variability over time (such as the stationarity assumption in random processes) should be avoided. Unfortunately, the tools of nonlinear dynamics are not sufficiently developed to help in this process, but we hope that someday they will come to age.

1.5 MOTOR BMIS

Because of the breadth of the BMI topic, this book will be specifically focusing on motor BMIs and will only describe properties of neurons sampled from the motor system and signal processing models and algorithms for SUA inputs. The model inputs either use point processes per se or a time-to-amplitude conversion that is called *counting* in point processes [67] or *rate coding* (binning) in neurosciences [69] that allows the use of conventional time series analysis as found in signal processing. A brief summary of BMIs for command and control using the EEG will be provided here and in the appendix for completeness.

At the center of motor BMI technology is the ability to infer the user's intent directly from the CNS with sufficient resolution and accuracy to seamlessly control an external robotic device. A conceptual drawing of a motor BMI is shown in Figure 1.6 where neural activity is recorded, conditioned, transmitted (wired or wirelessly), and translated directly into control commands of a prosthetic arm or cursor of a computer. As described earlier, variety of signal processing techniques have been used to analyze the control commands collected from EEG [70], EcOG [35], LFPs [71], and ensembles of single neurons (SUA) [72].

There are two basic types of motor BMIs: command and trajectory control BMIs. Research in command BMIs started in the 1970s with EEG [73] where subjects learned (and practice extensively) to control their regional brain activity in a predetermined fashion so that it could be robustly classified by a pattern recognition algorithm and converted into a set of discrete commands. Often, EEG-based interfaces use evoked potentials that study the impulse response of the brain as a dynamic system [74–76]. Event-related potentials are very important as clinical tools, but they are based on repeatability across stimuli, often abstract imagery, and may not be the most natural interface for real-time motor control. Therefore, the paradigm for communication is mostly based on selection of cursor actions (up/down, left/right) on a computer display. The computer presents to the users a set of possibilities, and they choose one of them through the cursor movement, until

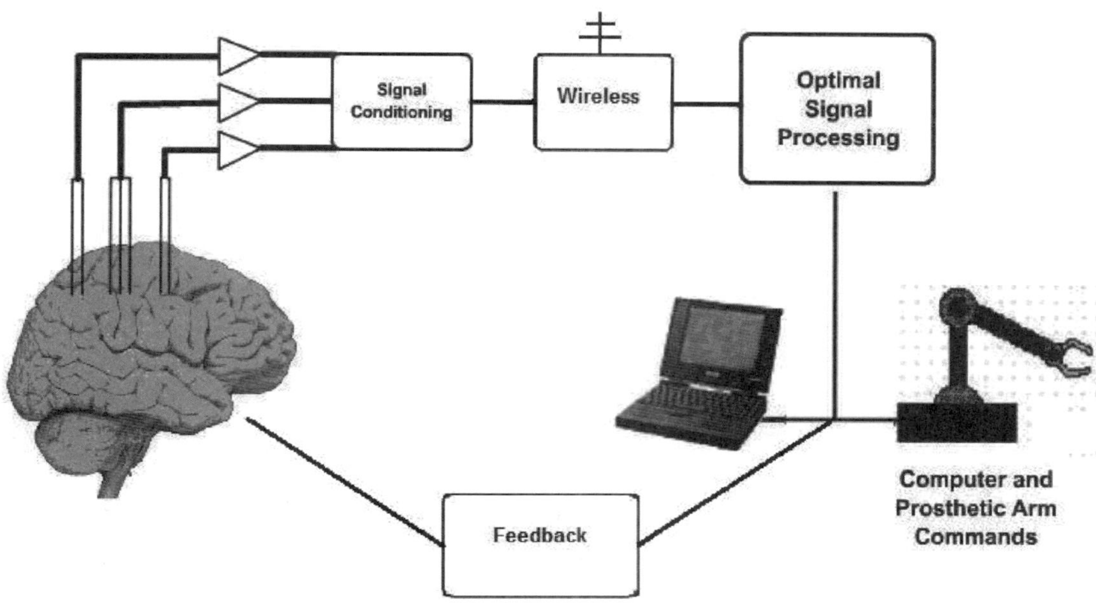

FIGURE 1.6: Motor BMIs derive intent from brain signals to command and control devices such as computers and prosthetics.

an action is completed. BCIs require subject training through biofeedback, and they display a low bandwidth for effective communication (15–25 bits/min) [77], which hinders the speed at which tasks can be accomplished. However, because of their noninvasiveness, BCIs have already been tested with success in paraplegics and locked-in patients. Several groups all over the world have demonstrated working versions of EEG-based interfaces [77], and a system software standard has been proposed. There is a biannual BCI meeting and even a BCI competition to evaluate the progress of algorithms in this area. A special issue of the *IEEE Transactions on Biomedical Engineering* also surveyed the state of the art in its June 2004 issue.

Trajectory control BMIs use brain activity to control directly the path of an external actuator in three-dimensional space, mimicking the role of the motor system when it controls, for instance, the trajectory of a hand movement. The control of trajectories is a common problem encountered by engineers (aircraft stability, gun control, etc.); therefore, feedback controllers in the area of physiological motor control are common. An example block diagram of a simplified motor system is given in Figure 1.7 [78]. The technical control problem here, at least in a first approach, does not lend itself to classify brain signals, but requires the BMI to generate a trajectory directly from brain activity. Here, the motor command generator could be the neural representation of the dynamics of the controlled object, whereas the motor command is the series of signals sent through the motor neurons, brainstem, spinal cord, and muscle fibers. In real-world situations, the controllers are subject to perturbations (loads, noise, moving objects) that can introduce errors in the original commands. To overcome these discrepancies, sensory systems can be used to modify the motor program. The two sensory modalities described here, visual and proprioceptive, operate on very different time scales. Nevertheless, because timing is critical for internal adaptation of this system, it has been difficult so far to extract from the EEG the motor command signatures with sufficient spatiotemporal resolution. Therefore, invasive techniques utilizing directly neuronal firings or LFPs have been utilized. In the last decade, the introduction of new methods for recording and analyz-

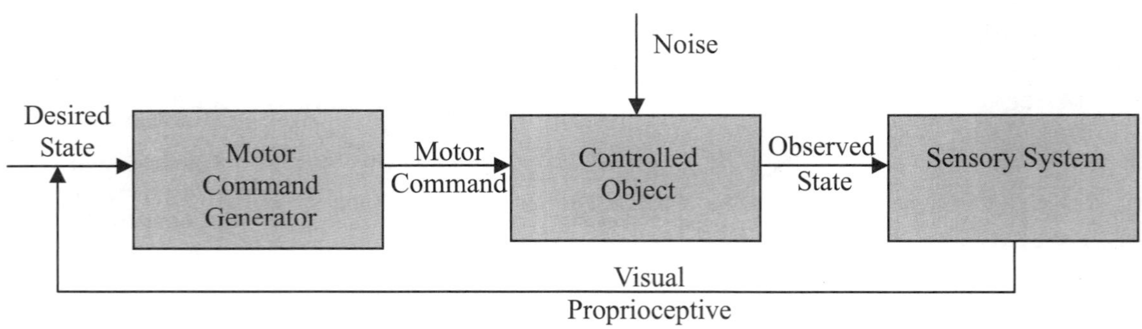

FIGURE 1.7: Elements of a feedback control system.

ing large-scale brain activity and emerging developments in microchip design, signal processing algorithms, sensors, and robotics are coalescing into a new technology—*neurotechnology*—devoted to creating BMIs for controls [33].

From an optimal signal processing viewpoint, BMI modeling is the identification of a MIMO time varying, eventually nonlinear system, which is a challenging task because of several factors: the intrinsic subsampling of the neural activity due to the relative small number of microelectrodes versus the number of neurons in the motor cortex, the unknown aspects of where the information resides (neural coding), the huge dimensionality of the problem, and the need for real-time signal processing algorithms. The problem is further complicated by a need for good generalization in nonstationary environments that depends on model topologies, fitting criteria, and training algorithms. Finally, reconstruction accuracy must be assessed, because it is linked to the choice of linear versus nonlinear and feedforward versus feedback models. In the next chapters, we will dissect the BMI design problem from a signal processing perspective where the choices in the decoding architecture will be evaluated in terms of performance and feasibility. The goal is to review and critique the past and state of the art in BMI design for guiding future development.

REFERENCES

1. Bähr, M., *Brain repair*. Advances in Experimental Medicine and Biology, 2006. **557**: pp. xii, 252.
2. Ingoglia, N.A., and M. Murray, *Axonal regeneration in the central nervous system*. Neurological Disease and Therapy, 2001. **51**: pp. xvii, 711.
3. Marwah, J., H. Teitelbaum, and K.N. Prasad, Neural Transplantation, CNS Neuronal Injury, and Regeneration: Recent Advances. 1994, Boca Raton, FL: CRC Press.
4. Seil, F.J., *Neural regeneration*. Progress in Brain Research, 1994. **103**: pp. xvi, 413.
5. Chapin, J.K., and K.A. Moxon, eds. Neural Prostheses for Restoration of Sensory and Motor Function. Methods and New Frontiers in Neuroscience. 2001, Boca Raton, FL: CRC Press.
6. Coclear, C.A., *http://www.cochlearamericas.com/About/about_index.asp*. 2004.
7. Humayun, M.S., et al., *Visual perception in a blind subject with a chronic microelectronic retinal prosthesis*. Vision Research, 2003. **43**(24): pp. 2573–2581.
8. Berger, T.W., et al., *Brain-implantable biomimetic electronics as the next era in neural prosthetics*. Proceedings of the IEEE, 2001. **89**(7): pp. 993–1012.
9. Ohnishi, K., R.F. Weir, and T.A. Kuiken, *Neural machine interfaces for controlling multifunctional powered upper-limb prostheses*. Expert Review of Medical Devices, 2007. **4**(1): pp. 43–53.
10. Abeles, M., Corticonics: Neural Circuits of the Cerebral Cortex. 1991, New York: Cambridge University Press.

11. Akin, T., et al., *A modular micromachined high-density connector system for biomedical applications.* IEEE Transactions on Biomedical Engineering, 1999. **46**(4): pp. 471–480.
12. Wise, K.D., et al., *Wireless implantable microsystems: High-density electronic interfaces to the nervous system.* Proceedings of the IEEE, 2004. **92**(1): pp. 76–97.
13. Spataro, L., et al., *Dexamethasone treatment reduces astroglia responses to inserted neuroprosthetic devices in rat neocortex.* Experimental Neurology, 2005. **194**(2): p. 289.
14. Szarowski, D.H., et al., *Brain responses to micro-machined silicon devices.* Brain Research, 2003. **983**(1–2): p. 23.
15. Kam, L., et al., *Correlation of astroglial cell function on micro-patterned surfaces with specific geometric parameters.* Biomaterials, 1999. **20**(23–24): p. 2343.
16. Turner, J.N., et al., *Cerebral Astrocyte Response to Micromachined Silicon Implants.* Experimental Neurology, 1999. **156**(1): p. 33.
17. Moxon, K.A., et al., *Ceramic-Based Multisite Electrode Arrays for Chronic Single-Neuron Recording.* IEEE Transactions on Biomedical Engineering, 2004. **51**(4): pp. 647–656.
18. Llinás, R.R., I of the Vortex: From Neurons to Self. 2000, Cambridge, MA: MIT Press.
19. Koch, C., and J.L. Davis, Large-Scale Neuronal Theories of the Brain (Computational Neuroscience). 1995, Cambridge, MA: MIT Press.
20. Buzsáki, G., Temporal Coding in the Brain (Research and Perspectives in Neurosciences). 1994, Berlin: Springer-Verlag.
21. Rieke, F., Spikes: Exploring the Neural Code. 1996, Cambridge: MIT Press.
22. Turing, A.M., and D. Ince, Mechanical Intelligence. 1992, New York: North-Holland.
23. Von Neumann, J., The Computer and the Brain. 1959, New Haven, CT: Yale University Press.
24. Wiener, N., Cybernetics; or, Control and Communication in the Animal and the Machine. 2nd ed. 1961, New York: MIT Press.
25. McCulloch, W.S., and W.H. Pitts, *A Logical Calculus for Ideas Imminent in Nervous Activity.* Bulletin of Mathematical Biophysics, 1943. **5**: pp. 115–133.
26. Pribram, K.H., Brain and Perception: Holonomy and Structure in Figural Processing. 1991, Hillsdale, NJ: Lawrence Erlbaum Associates.
27. McClelland, J.L., and D.E. Rumelhart, Explorations in Parallel Distributed Processing: A Handbook of Models, Programs, and Exercises. 1988, Cambridge, MA: MIT Press.
28. Rumelhart, D.E., Introduction to Human Information Processing. 1977, New York: Wiley.
29. Rumelhart, D.E., J.L. McClelland, and University of California San Diego PDP Research Group, Parallel Distributed Processing: Explorations in the Microstructure of Cognition (Computational Models of Cognition and Perception). 2nd vol. 1986, Cambridge, MA: MIT Press.

30. Hopfield, J.J., *Olfactory computation and object perception.* Proceedings of the National Academy of Sciences of the United States of America, 1991. **88**: pp. 6462–6466.
31. Haken, H., *Synergetic Computers and Cognition: A Top-Down Approach to Neural Nets.* 2nd English ed. Springer Series in Synergetics. 2004, Berlin: Springer: pp. ix, 245.
32. Freeman, W.J., Mass Action in the Nervous System: Examination of the Neurophysiological Basis of Adaptive Behavior Through EEG. 1975, New York: Academic Press.
33. Nicolelis, M.A.L., Methods for Neural Ensemble Recordings. 1999, Boca Raton, FL: CRC Press.
34. Rickert, J., et al., *Encoding of movement direction in different frequency ranges of motor cortical local field potentials.* Journal of Neuroscience, 2005. **25**(39): pp. 8815–8824.
35. Leuthardt, E.C., et al., *A brain–computer interface using electrocorticographic signals in humans.* Journal of Neural Engineering, 2004. **1**: pp. 63–71.
36. Pfurtscheller, G., and F.H.L. da Silva, *Event-related EEG/MEG synchronization and desynchronization: Basic principles.* Clinical Neurophysiology, 1999. **110**(11): pp. 1842–1857.
37. Sanchez, J.C., et al., *Ascertaining the importance of neurons to develop better brain machine interfaces.* IEEE Transactions on Biomedical Engineering, 2003. **61**(6): pp. 943–953.
38. Serruya, M.D., et al., Brain–machine interface: Instant neural control of a movement signal. Nature, 2002. 416: pp. 141–142.
39. Taylor, D.M., S.I.H. Tillery, and A.B. Schwartz, *Direct cortical control of 3D neuroprosthetic devices.* Science, 2002. **296**(5574): pp. 1829–1832.
40. Wessberg, J., et al., *Real-time prediction of hand trajectory by ensembles of cortical neurons in primates.* Nature, 2000. **408**(6810): pp. 361–365.
41. Jasper, H., and W. Penfield, Epilepsy and the Functional Anatomy of the Human Brain. 1954, Boston: Little, Brown & Co.
42. Babiloni, F., et al., *High resolution EEG: A new model-dependent spatial deblurring method using a realistically-shaped MR-constructed subject's head model.* Electroencephalography and Clinical Neurophysiology, 1997. **102**(2): pp. 69–80.
43. Hill, N.J., et al., *Classifying EEG and ECoG signals without subject training for fast BCI implementation: Comparison of nonparalyzed and completely paralyzed subjects.* IEEE Transactions on Neural Systems and Rehabilitation Engineering, 2006. **14**(2): pp. 183–186.
44. Ginter, J., et al., *Propagation of EEG activity in the beta and gamma band during movement imagery in humans.* Methods of Information in Medicine, 2005. **44**(1): pp. 106–113.
45. Kupfermann, *Localization of higher cognitive and affective functions: The association cortices,* in: Principles of Neural Science, E.R. Kandel, J.H. Schwartz, and J.T.M, Eds. 1991, Norwalk, CT: Appleton & Lange: pp. 823–838.
46. Nakasatp, N., et al., *Comparisons of MEG, EEG, and ECoG source localization in neocortical partial epilepsy in humans.* Electroencephalography and Clinical Neurophysiology, 1994. **91**(3): pp. 171–178.

47. Sanchez, J.C., et al., *Extraction and localization of mesoscopic motor control signals for human ECoG neuroprosthetics.* Journal of Neuroscience Methods (in press).
48. Freeman, W.J., *Mesoscopic neurodynamics: From neuron to brain.* Journal of Physiology—Paris, 2000. **94**(5–6): pp. 303–322.
49. Freeman, W.J., *Origin, structure, and role of background EEG activity. Part 1. Analytic Phase.* Clinical Neurophysiology, 2004. **115**: pp. 2077–2088.
50. Freeman, W.J., *Origin, structure, and role of background EEG activity. Part 2. Analytic amplitude.* Clinical Neurophysiology, 2004. **115**: pp. 2089–2107.
51. Freeman, W.J., *Origin, structure, and role of background EEG activity. Part 3. Neural frame classification.* Clinical Neurophysiology, 2005. **116**(5): pp. 1118–1129.
52. Freeman, W.J., *Origin, structure, and role of background EEG activity. Part 4. Neural frame simulation.* Clinical Neurophysiology (in press).
53. Nunez, P.L., Electric Fields of the Brain: The Neurophysics of EEG. 1981, New York: Oxford University Press.
54. Nunez, P.L., *Generation of human EEG by a combination of long and short range neocortical interactions.* Brain Topography, 1989. **1**: pp. 199–215.
55. Nunez, P.L., Neocortical Dynamics and Human EEG Rhythms. 1995, New York: Oxford University Press.
56. Pfurtscheller, G., et al., *Spatiotemporal patterns of beta desynchronization and gamma synchronization in corticographic data during self-paced movement.* Clinical Neurophysiology, 2003. **114**(7): p. 1226.
57. Thatcher, R.W., P.J. Krause, and M. Hrybyk, *Cortico-cortical associations and EEG coherence: a two-compartmental model.* Electroencephalography and Clinical Neurophysiology, 1986. **64**: pp. 123–143.
58. Salenius, S., and R. Hari, *Synchronous cortical oscillatory activity during motor action.* Current Opinion in Neurobiology, 2003. **13**(6): pp. 678–684.
59. MacKay, W.A., and A.J. Mendonca, *Field potential oscillatory bursts in parietal cortex before and during reach.* Brain Research, 1995. **704**(2): pp. 167–174.
60. Sanes, J.N., and J.P. Donoghue, *Oscillations in local-field potentials of the primate motor cortex during voluntary movement.* Proceedings of the National Academy of Sciences of the United States of America, 1993. **90**(10): pp. 4470–4474.
61. Rougeul, A., et al., *Fast somato-parietal rhythms during combined focal attention and immobility in baboon and squirrel-monkey.* Electroencephalography and Clinical Neurophysiology, 1979. **46**(3): pp. 310–319.
62. Donoghue, J.P., et al., *Neural discharge and local field potential oscillations in primate motor cortex during voluntary movements.* Journal of Neurophysiology, 1998. **79**(1): pp. 159–173.

63. Arieli, A., et al., *Coherent spatiotemporal patterns of ongoing activity revealed by real-time optical imaging coupled with single-unit recording in the cat visual cortex.* Journal of Neurophysiology, 1995. **73**: pp. 2072–2093.
64. Lewicki, M.S., *A review of methods for spike sorting: The detection and classification of neural action potentials.* Network: Computation in Neural Systems, 1998. **9**(4): pp. R53–78.
65. Eden, U.T., et al., *Dynamic analysis of neural encoding by point process adaptive filtering.* Neural Computation, 2004. **16**: pp. 971–998.
66. Brown, E.N., et al., *An analysis of neural receptive field plasticity by point process adaptive filtering.* Proceedings of the National Academy of Sciences of the United States of America, 2001. **98**(12): pp. 12 261–12 266.
67. Perkel, D.H., G.L. Gerstein, and G.P. Moore, *Neuronal spike trains and stochastic point processes: I. The single spike train.* Biophysical Journal, 1967. 7(4): pp. 391–418.
68. Perkel, D.H., G.L. Gerstein, and G.P. Moore, *Neuronal spike trains and stochastic point processes: II. Simultaneous spike trains.* Biophysical Journal, 1967. **7**(4): pp. 419–440.
69. Nawrot, M., A. Aertsen, and S. Rotter, *Single-trial estimation of neuronal firing rates—From single neuron spike trains to population activity.* Journal of Neuroscience Methods, 1999. **94**: pp. 81–92.
70. Pfurtscheller, G., and A. Aranibar, *Evaluation of event-related desynchronization (ERD) preceding and following voluntary self-paced movement.* Electroencephalography and Clinical Neurophysiology, 1979. **46**: pp. 138–146.
71. Mehring, C., et al., *Inference of hand movements from local field potentials in monkey motor cortex.* Nature Neuroscience, 2003. **6**(12): pp. 1253–1254.
72. Fetz, E.E., *Are movement parameters recognizably coded in the activity of single neurons.* Behavioral and Brain Sciences, 1992. **15**(4): pp. 679–690.
73. Vidal, J.J., *Towards direct brain–computer communication.* Annual Review of Biophysics and Bioengineering, 1973. **2**: pp. 157–180.
74. Makeig, S., *Gamma-band event-related brain dynamics—Historic perspective.* International Journal of Psychophysiology, 1993. **14**(2): pp. 136–136.
75. Lutzenberger, W., et al., *Dimensional analysis of the human EEG and intelligence.* Neuroscience Letters, 1992. **143**(1–2): pp. 10–14.
76. Johnson, R.N., et al., *Evoked-potentials as indicators of brain dynamics—results from an interactive computer system.* Annals of Neurology, 1977. **1**(5): pp. 500–501.
77. Wolpaw, J.R., et al., *Brain–computer interfaces for communication and control.* Clinical Neurophysiology, 2002. **113**: pp. 767–791.
78. Trappenberg, T.P., Fundamentals of Computational Neuroscience. 2002, New York: Oxford University Press.

CHAPTER 2

Foundations of Neuronal Representations

Additional Contributors: Sung-Phil Kim, Yiwen Wang

The development of BMIs relies upon the ability to decipher neuronal representations expressed by individual neuronal elements operating in distributed ensembles, and communicating using convergent and divergent principles.[1] Understanding the properties of information transmission and determining how this information translates into commands of the motor system as a whole is one of the cornerstones of BMI development. The intricate computational neurobiology of motor BMIs can be elucidated by first specifying the general principles of neural representation, transmission, and signal processing in the system, which are summarized in the following questions:

- What are the functional computational elements?
- What are the relevant mechanisms of information transmission in the system?
- How is activity aggregated between single neurons and other neurons in the system?
- How can one sufficiently model the interaction and function of biologically complex neurons and neural assemblies?

2.1 CYTOARCHITECTURE

Most of what is known about the role of neurons as the computational elements of the nervous system has been derived by measuring the cause and effect of experimental situations on neuronal firing patterns. From these initial functional observations, the neurons belonging to the neocortex that generate movement programs were spatially divided into four lobes (frontal, temporal, parietal, and occipital) and have later been associated with motor imagery [1], visual/tactile manipulation

[1]Groups of neurons have a variety of means of organizing their inputs. *Convergence* refers to neural input from a diverse set of sources, whereas *divergence* refers to a neuron's ability to exert influence on multiple target sites.

of objects [2], spatial coordinate transformations [3], and motor hemispheric specializations [4–6]. Historically though, the theoretical support for these functional divisions in the context of neuronal organization for computational motor control is largely incomplete. Namely, the challenge is to determine what is the neuronal substrate of mental functions [7] and how the system contributes to the principal organization of elements that are supporting mental processes such as perception and motor control. Early on, the likes of Santiago Ramón y Cajal and German anatomist Korminian Brodmann believed that the diversity in cytoarchitecture was sufficient to sustain the complexities in mental abilities. Using this rationale, Cajal studied the intricate morphology of single neuron using the Golgi method [8][2] and suggested that neurons communicate with each other via specialized synaptic junctions [9].[3] This hypothesis became the basis of the neuron doctrine, which states that the individual unit of the nervous system is the single neuron. Broadmann later also classified 52 cortical areas based upon detailed analysis of cytoarchitecture through histological analysis. However, with the advent of new electrophysiological experimental paradigms, it became clear that the structural differences did not reflect the diversity in functional specialization. In other words, the anatomically distinguishable layers of the cortex did not seem to map to all of the information-processing capabilities of neuronal networks for the following reasons. First, the columnar structure of neural assemblies is similar throughout the cortex [10]. The comparison of local processing in each of the cortices revealed similar functions: signal integration and broadcasting.

2.2 CONNECTIONISM

The apparent differences of function seem to be consequences of diverse larger scale interconnectivity and communication, which are implicitly defined in the structure. Hence, structure defines function. From a global perspective of connectivity, the main pathways among cortical and subcortical structures are primarily specified by genetics, but the fine details of the connections are influenced by the interactions of the brain with the body and environment. The fundamental architecture of the pathways consists of multiple recurrent connections of neurons within a nested hierarchy of parallel circuits. If one were to attempt to simulate the connectivity of a real neural system, to link each of the neurons in the brain to each other would require an exceeding large number of connections with impractical volume[4] and energetic requirements. The problem then becomes one of studying the minimum number of connections needed for neural networks. This problem has faced neuroscientists for many years [11], and the general solution was derived by mathematicians Erdos

[2] Cassic staining techniques respond to different cells or parts of cells: Nissl, color cell body; Golgi, nerve cells and fibers; Weigert, myelin, to name a few.

[3] The term *synapse* was coined by Sherrington in 1897.

[4] Axons occupy more volume in the brain than the combined contributions of cell bodies and dendrites.

and Renyi in their study of graph theory [12]. In the biological system, each neocortex neuron communicates with 5000–50,000 other neurons mostly in its proximity, but distant connections are also present, creating a random graph structure that can be modeled as a small world [13]. The density of neurons in the human neocortex is so large (10,000–100,000/mm^3 [14]) that despite of the huge divergent properties, each neuron only communicates with 1% of its neighbors within its dendritic radius. Moreover, as the number of neurons in the network increases, the minimum number needed to connect them decreases.

From a functional perspective, the principal pyramidal cells (~85%) that make up these connections are primarily excitatory, and the neurons predominantly communicate by mutual excitation. However, in the absence of inhibition, the activity of the population could increase in this one direction to an unstable state. In general, excitatory activity begets excitatory activity in a positive direction. Therefore, excitatory population dynamics must be balanced by inhibitory neurons. Inhibitory activity is inherently more sophisticated than excitatory because it has several modes of operation. First, in direct negative feedback of a neuron, stability is enhanced by inhibition. Second, if the inhibitory neuron operates in a feedforward manner on another neuron, then it acts a filtering mechanism by dampening the effects of excitation. Finally, if the inhibitory neuron acts as a segregator (i.e., lateral inhibition) among neural assemblies, it can serve to function as a gating mechanism between the groups of neurons. The intricate relationships between the excitatory principal cells, inhibitory neurons, and the inherent refractory periods (i.e., when a neuron fires, it can not fire again immediately after even if fully excited) are at the root of the admirable stability of the spontaneous cortical activity.

2.3 NEURAL SIGNALING AND ELECTRIC FIELDS OF THE BRAIN

The mechanism of neural connectionism and communication involves neuronal signaling across a synapse (i.e., between its dendritic input and the axonal output) is mediated through a series of changes in cell membrane potential as shown in Figure 2.1. Molecular and voltage-gated signaling occurs at the synapses, which are chemical–electrical connectors between the axon of presynaptic neurons and one of many dendritic inputs of a given neuron. The axon of a firing presynaptic neuron transfers the transient current and subsequent action potential (spike) through a series of changes in membrane potential to its many synaptic endings. At the synapse, ions are released to the attached dendrite, locally increasing the electrical potential at the dendrite's distal portion. This voltage is slowly integrated with contributions from other dendritic branches and propagated until the cell body. When the electrical potential at the base of the dendritic tree crosses a threshold that is controlled by the cell body, the neuron fires its own action potential that is transferred through its axonal output to other postsynaptic neurons. Therefore, the closer the membrane is to the threshold,

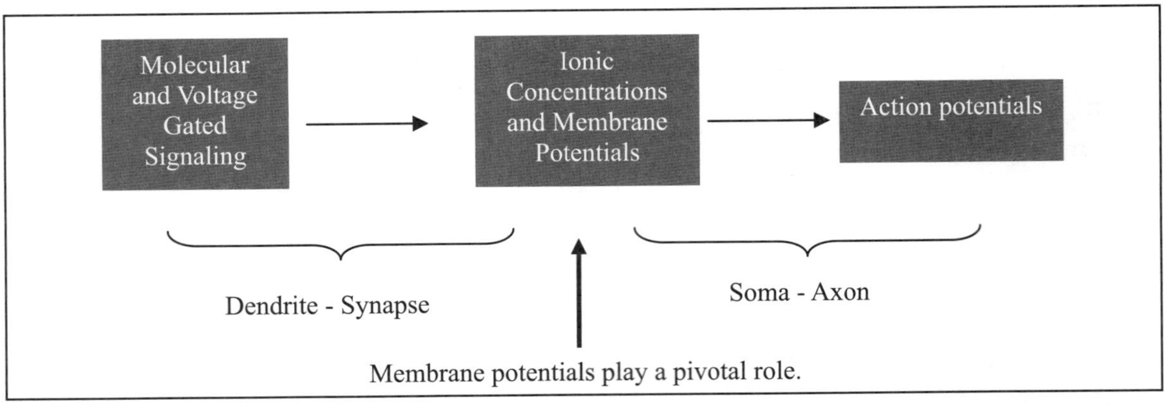

FIGURE 2.1: Relationship among neuron structure, electrochemistry, and generation of cell signaling.

the more likely is the neuron to fire. Most cortical neurons spend 99.9% of their lifetimes with their membrane potentials just below threshold.

Through this signaling procedure, there are two types of electric fields produced by neurons: waves of dendritic synaptic current that establish weak potential fields as the dendritic synaptic current flows across the fixed extracellular cortical resistance and trains of action potentials on the axons that can be treated as point processes.[5] The variation of the membrane potential initiated from a presynaptic spike can be described as a postsynaptic potential as shown in Figure 2.2. Excitatory synapses increase the membrane potential in excitatory postsynaptic potentials, which occur after the diffusion process of a neurotransmitter opens the channels of *N*-methyl-D-aspartate or non–*N*-methyl-D-aspartate receptors.[6] Inhibitory postsynaptic potentials lower the rise of the membrane potential. The electrical potentials superimpose as the sum of the individual potentials as they approach the threshold. Pulse densities are monotonically dependent on wave current densities. The range is bounded by threshold at the low end and refractoriness at the high end, with a near-linear small-signal range. This is similar to the sigmoidal curve in artificial neural networks, with a major and significant difference: the nonlinearity is asymmetric with the maximal slope skewed to the right translating the neuron's excitatory nature. As shown in chapter 1, the LFP in upper layers of neocortex (II-IV), ECoG, and EEG measure the electrical field produced by the dendritic currents (also called the post synaptic potentials) of large groups of neurons. Conventionally, microelectrode

[5] The review of the generation of action potentials is a brief introduction to the minimum components of the process; for a more detailed description, see Reference 15.

[6] *N*-methyl-D-aspartate or glutamatergic receptors are excitatory, which are typically slower than non–*N*-methyl-D-aspartate synapses.

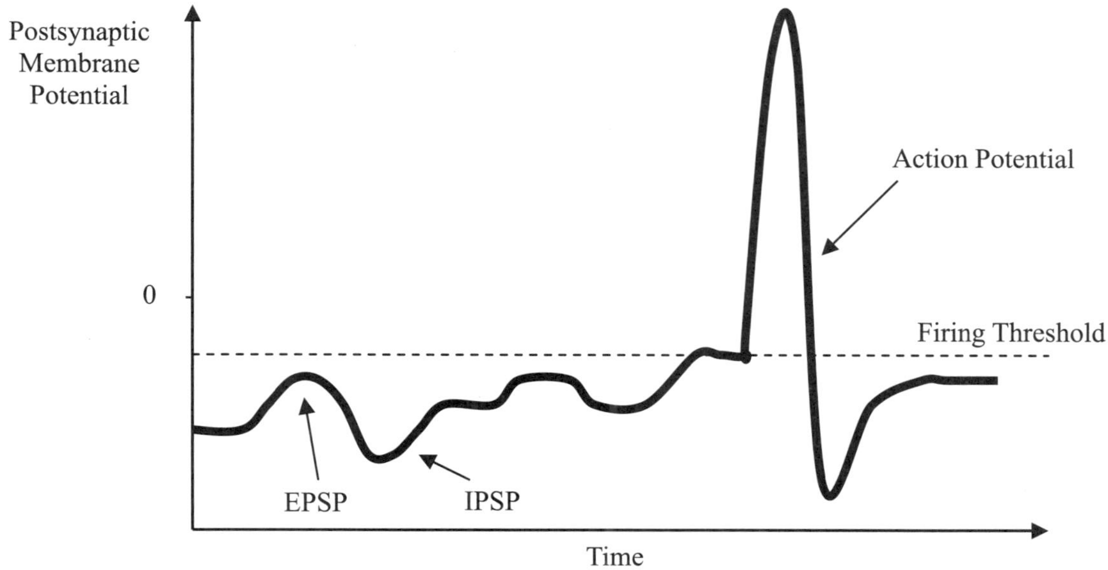

FIGURE 2.2: The time course of postsynaptic potentials. EPSP, excitatory postsynaptic potential; IPSP, inhibitory postsynaptic potential.

arrays are placed in layer V of the neocortex, close to the pyramidal/stellate cell bodies to measure their action potentials. The generation of an action potential, or spike, is considered to be the basic unit of communication among neurons. Measurement of the action potential resulting from ionic current exchanges across the membranes of neurons has been sought after for many years through the use of microelectrode recording technology [16].

An example of a microelectrode recording from a single neuron collected extracellularly near the cell body is shown in Figure 2.3a. Typical cellular potentials have magnitudes ranging from tens of microvolts to tens of millivolts, and time durations of a millisecond. The action potentials in Figure 2.3a are of the same amplitude and have the same shape, which indicates they are from the same neuron. The amplitude and shape of the measured action potential is affected by the electrode–neuron distance and orientation [17]. The ability to automatically discriminate the activity of individual neurons in a given electrode recording has been a formidable computational challenge with respect to the experimental conditional and subjective observations traditionally used to sort action potentials from each other [18]. Accurate spike detection and sorting is a critical step for high-performance BMIs because its role is to identify from the background noise the neuromodulation related to movement intent and execution. To date, accurate and automated spike detection and sorting remains an ongoing research topic [18–21]. One of the complexities of analyzing neuronal activity is the fact that neurons are active in assemblies surrounding the electrode. In Figure 2.3b,

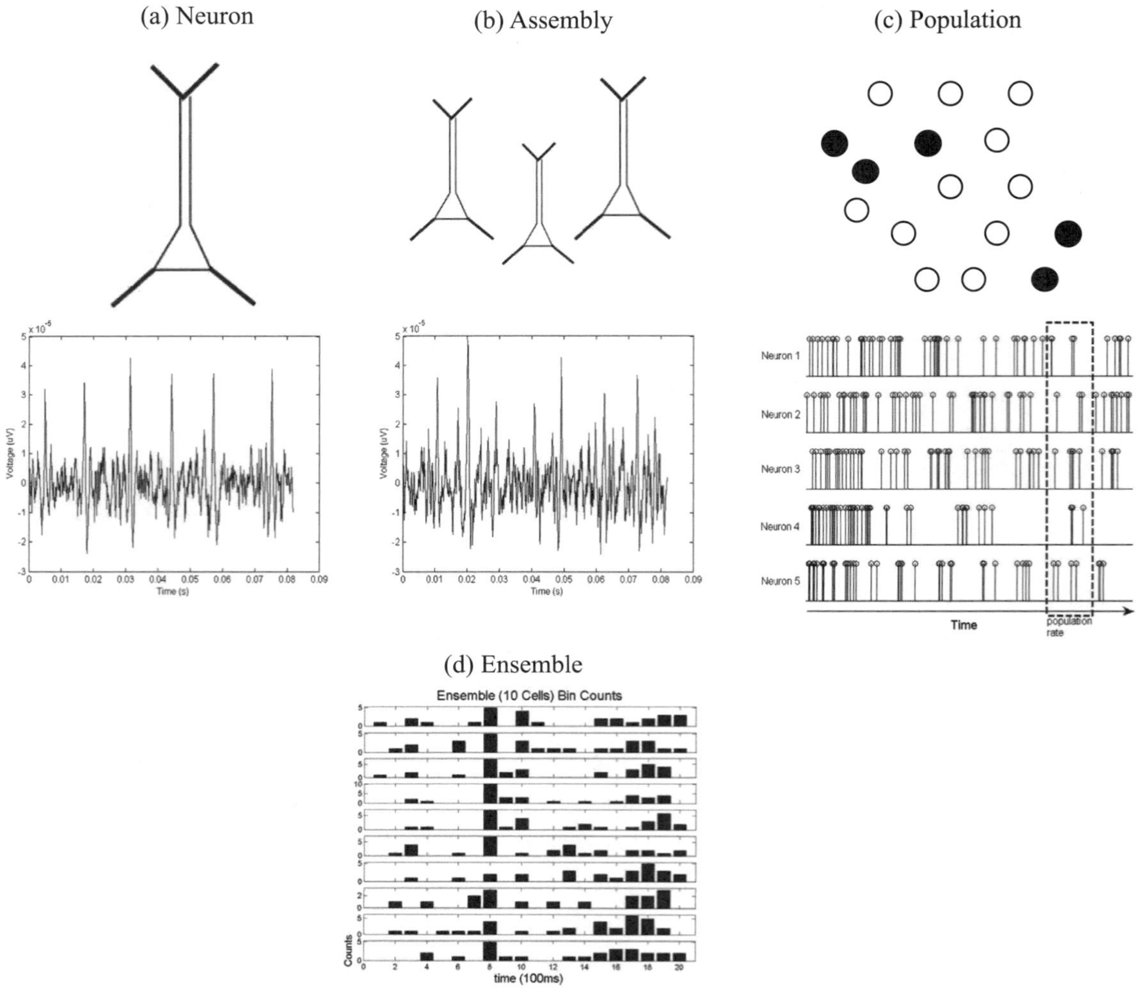

FIGURE 2.3: Levels of abstraction of neuronal representation. (a) Recordings from a single neuron. (b) A single electrode recording from multiple neurons. (c) Spike raster plots from a population of neurons indicating the time of action potentials. (d) Firing rates from an ensemble of neurons.

the action potentials of several neurons (an assembly) appear aggregated in the time series captured by one microelectrode. Here, multiple action potentials can be observed with different heights and widths. The concept of cell assembly was proposed by Hebb [52] in his book *The Organization of Behavior*, where the assembly served as a diffuse group of cells acting in a closed system that delivers facilitation to other system. In multielectrode recordings, there are many (100 are typical) channels of neural data that can translate into hundreds of neurons probed. One of the difficult aspects of interpreting this population activity is to determine information flow in networks of neurons that could be serially connected or excited through diverging or converging chains. Sherrington [9] introduced the concept of neural ensemble as a population of neurons involved in a particular computation in a distributed manner (each neuron shares a part of the processing). Because cortical neurons have large overlaps both in the extracellular potential generated by axonal projections and within dendritic trees, BMI implementations will need to interpret the many-to-one or one-to-many mappings. Moreover, the neural firing is not a deterministic process due to the tiny electrical fluctuations in the membrane produced by the other neurons, and the fact that the neuron normally operates just below threshold, waiting to fire as will be described below.

This microscopic scale of analysis can provide insight to the communication (both rate and time codes) of ensembles of local distant neurons. The variability in extracellular voltage potentials that produce variations in the neuronal recordings are attributed to cell morphology and the resistivity of the surrounding tissue. For cellular environments, the maximum measured potential has been shown to decrease in a spherical volume with a radius of 2 mm as measured from the soma of the pyramidal cell [22–24]. Experimentally and in simulation, the amplitude of the extracellular field drops off sharply at distances of 50–100 μm from the cell body [22]. Experimental access to the raw extracellular potentials enables analysis of the synaptic dendritic activity as LFPs (0.5–200 Hz) as well as spikes (300–6 kHz) originating from the cell body and axon. Depending upon the application, postprocessing of extracellular potentials is enhanced by low- or band-pass filtering. Ultimately, in single-unit neural analysis, it is the action potential that is of great interest to neural ensemble neurophysiologists. Analysis of action potentials provides a consistent waveform signature that can be tracked, which is rooted in the fundamental elements of the nervous system. Therefore, it is critical to detect peaks in the potential above the noise floor from local and distant neurons that must be assessed directly from the raw potentials. Electrophysiological parameters of the action potential (amplitude and width) are then used to identify whether a recording electrode contains any spurious signals (e.g., electrical noise, movement artifact). In a preprocessing procedure called spike sorting, the peak-to-peak amplitude, waveform shape, and interspike interval can be evaluated to ensure that the recorded neurons have a characteristic and distinct waveform shape when compared with other neuronal waveforms in the same and other recording electrodes. The goal of spike sorting is to reliably and confidently select action potentials from single neurons and extract the firing times. Hence, a variety of sorting techniques have been proposed [18], with the most common approaches being template matching and principal component analysis (PCA)–based analysis of waveform variance. Extensions of the basic techniques have been proposed using automated, manual, and multielectrode implementations [20, 25–28]. Despite the great advances in the spike sorting technique, there is often much variability in the analysis due to the experimenter [21].

2.4 SPIKING MODELS FOR THE NEURON

There are two basic types of models for the fluctuations in excitability of the neuron. The so-called conductance models that describe the channel dynamics of the ionic exchanges across the neuron membrane and the threshold-fire models[7] that try to capture the most representative behavior of neurons. One of the first conductance models was developed by Hodgkin and Huxley [29], who described the dynamics of sodium, potassium, and leak channels in the cell membrane of the giant squid axon. Building upon the contributions of Hodgkin and Huxley, there are currently many other more sophisticated compartmental neural models and many excellent software programs to simulate neural excitability [30, 31] and the dynamics of synapses [32]. One of the difficulties of this modeling approach is that it is very computational demanding because they are physics-based modeling approaches.[8] Therefore, they are mostly used to advance the study of the neuron but hardly ever used to study neural assembly activity and its relation to behavior. A notable exception is the Blue Brain Project being conducted at the Brain Science Institute at École Polytechnique Fédérale de Lausanne in Switzerland by Markram et al. who teamed with IBM to build a detailed simulation of a cortical column [33, 34].

Although the Hodgkin–Huxley model provides a detailed mathematic model on the origins of the action potential, we would like to note here that neural spikes are very stereotypical. Therefore, in the context of motor BMIs, the shape is not the information-carrying variable. Moreover, because of the computational infeasibility of modeling large networks of Hodgkin–Huxley neurons, a better modeling path may be to implicitly include the ion channel dynamics and just focus on spikes. Because organism interactions with the world are constantly changing in real time, the question now becomes one of how can discontinuous signals (spikes) be used to represent continuous motor control commands.

The threshold-fire model shown in Figure 2.4 captures the basic behavior of the neuron excitability that we are describing here (i.e., the slow integration of potential at the dendritic tree followed by a spike once the voltage crosses a threshold). In equation (2.1) $u(t)$ is the membrane potential, and $I(t)$ is the input current. In this formulation, tau serves as a membrane constant determined by the average conductances of the sodium and leakage channels. The appeal of this modeling approach is that it is computationally simple, so many neurons can be simulated in a desktop environment to simulate neural assembly behavior. The threshold-fire neuron also naturally incorporates a number of physiological parameters, including membrane capacitance, membrane

[7]The threshold-fire models are also known as the integrate and fire models.

[8]The physics being referred here relates to the set of differential equations that Hodgkin and Huxley derived to model each of the channel dynamics. It is the integration of these equations that adds to the computational complexity. The simulation of even 100 Hodgkin–Huxley neurons for long periods can be a formidable task.

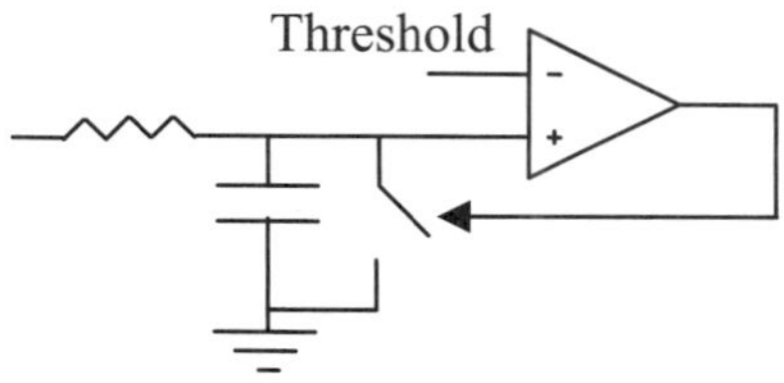

FIGURE 2.4: Threshold and fire or integrate and fire model.

(passive leak) resistance, and absolute refractory period. It is a good approximation over the normal operating range of most neurons, and it introduces the important nonlinearity (threshold) of real neurons (i.e., the neural spike). The threshold-fire neuron also has a few limitations, which include the lack of spatial extent (they are "point" neurons) and the electrochemical properties of dendrites are neglected. Membrane ionic currents tend to be nonlinear functions of both membrane potential and time. The linear summation of dendritic inputs is also inconsistent with the highly irregular firing statistics found in most cortical neurons.

$$\tau \frac{\partial u(t)}{\partial t} = -u(t) + RI(t) \tag{2.1}$$

This figure represents what has been called a generalized linear model, which is much more complex because of the feedback. Although the threshold-fire provides a computational approach to simulating the occurrences of spiking neurons, in practice in the electrophysiology laboratory, neuroscientists have dealt with the complexities of the raw extracellular potentials by developing a discrete representation for neuronal activity. The nature of the "all-or-nothing" firing of the action potential event has lead to the common treatment of action potentials as point processes where the continuous voltage waveform is converted into a series of timestamps indicating the instance when the spike occurred. Using the timestamps, a series of pulses or spikes (zeros or ones) can be used to visualize the activity of each neuron; this time-series (Figure 2.3c) is referred to as a *spike train*. Here, a group of electrodes are capable of capturing a subsampling of the local population (white dots) of neuronal activity indicated by the black dots.

2.5 STOCHASTIC MODELING

Single-neuron recording of this type and the analysis of spike trains reveal that most often neural responses are enormously irregular under constant stimulus. The spike trains of neural populations have several properties including sparsity and nonstationarity. This is a very different result from the threshold-fire model activity discussed previously. How do we quantify and describe the neural response variability due to the unreliable release of neurotransmitters, stochasticity[9] in channel gating, and fluctuations in the membrane potential activity? If the irregularity arises from stochastic forces, then the irregular interspike interval may reflect a random process. If we assume that the generation of each spike depends only on an underlying driving signal of instantaneous firing rate, then it follows that the generation of each spike is independent of all the other spikes; hence, we refer to this as the *independent spike hypothesis*. If the independent spike hypothesis is true, then the spike train would be completely described by a particular kind of random process called a Poisson process. Knowledge of this probability density function (PDF) would uniquely define all the statistical measures (mean, variance, etc.) of the spike train. Although the statistical properties of neural recordings can vary depending on the sample area, the animal, and the behavior paradigm, in general, spike trains modeled as Poisson distribution perform reasonably is well-constrained conditions [35]. The interspike interval from an example neuron of an animal performing a BMI task is presented in Figure 2.5a for comparison. The Poisson model has the great appeal of being defined by a single constant λ, the firing rate, assumed known

$$P_{Poisson}(x;\lambda) = \sum_{i=1}^{x} \lambda^{i} \frac{e^{-\lambda}}{i!} \tag{2.2}$$

Certain features of neuronal firing, however, violate the independent spike hypothesis. Following the generation of an action potential, there is an interval of time known as the absolute refractory period during which the neuron cannot fire another spike. For a longer interval known as the relative refractory period, the likelihood of a spike being fired is much reduced. Bursting is another non-Poisson feature of neuronal spiking.

Perhaps more realistically, the firing rate λ may change over time, yielding a nonhomogeneous Poisson process [36]. One of the simplest neural models that still captures a great deal of the biological detail is the linear–nonlinear–Poisson (LNP) model shown in Figure 2.5b. This model consists of a single linear filter,[10] followed by an instantaneous nonlinear function that accounts for response nonlinearities such as rectification and saturation, followed by a Poisson generative model. The firing rate of the Poisson process is totally determined by the parameters of the linear filter and the nonlinearity. Recently, this model has been used by Simoncelli et al. [37] to estimate better the receptive field, which characterizes the computation being performed by the neuron when it responds to an input stimulus. Later, we are going to delve into more detail on this example to show

[9] A stochastic process is one whose behavior is nondeterministic. Specifically, the next state of the system is not fully determined by the previous state.

[10] The linear filter refers to a linear receptive field. This will be described in the next section.

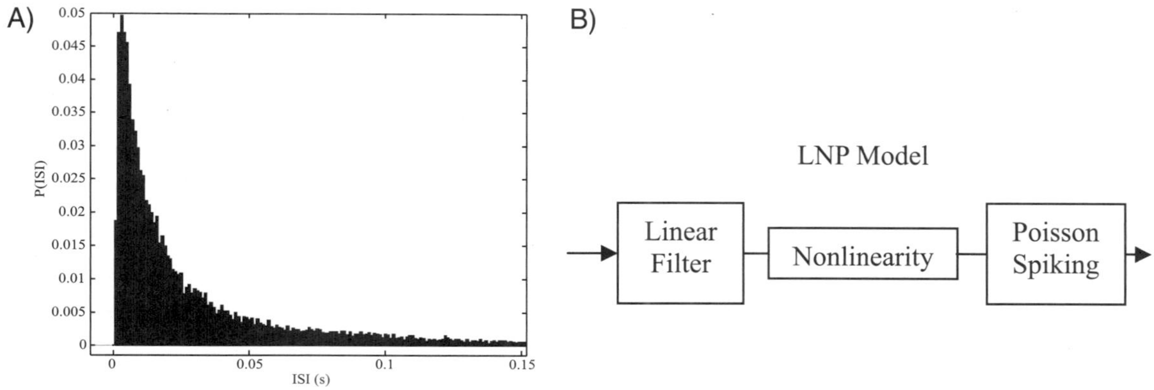

FIGURE 2.5: (a) Poisson distribution of interspike interval for an animal performing a BMI task. (b) LNP model [38].

how modern signal processing can help neuroscience data analysis methods. Indeed, it is our opinion that advanced statistical methods can enormously decrease the amount of data being collected in the wet laboratory simply by utilizing optimal statistical procedures, which are more efficient than heuristic approaches.

2.6 NEURAL CODING AND DECODING

In the previous sections, we described how information is represented in connectionist networks. For BMI motor control applications, we will now discuss how representations in populations of neurons can encode and decode stimuli. The first step is to define what are the basic constructs of neural coding? From a practical point of view, a neural code must be able to support a sufficiently rich computational repertoire. It must encode a sufficient range of values with sufficient accuracy and discriminability. Biological limitations require that the coding scheme must run under a sufficient set of operations implemented with neural mechanics. From the study of cytoarchitecture and connectionism, the representation of motor commands and sensory stimuli may be distributed or localized. The landscape of coding depends on how many components are representing the information, and three classes arise:

- Local representation: One node is active when a stimulus is presented (cardinal or grandmother cells [39]). The number of stimuli encoded increases linearly with the number of cells.
- Fully distributed representation: The stimulus is encoded by the combination of all the neurons.
- Sparsely distributed representation: Only a fraction of the neurons encode the stimuli.

Within each of these representational schemes, several operations are necessary to support a realistically complex computation of sensory information and internal representations, including the ability to combine, match, map, inverse map, compete, attenuate, and scale, to name a few. In general though, all aspects of our behavior are ultimately governed by the responses of neurons, which operate by changing their electrochemical makeup (polarizing and depolarizing). To measure how strong the relationship is between a stimulus and neural response, one can measure the correlated electrical activity between the stimulus and response. Some have framed the functional neural representation problem from a variety of perspectives, which are summarized in Table 2.1. The table is basically broken into two parts (coding and decoding), which depend upon what quantity is to be determined by the experimenter. In neural coding, a stimulus (visual and proprioceptive) is given to the system, and the neural response is to be determined. The decoding problem is the opposite, where a neural response (i.e., motor intent) is provided by the subject, and the experimenter has to determine the physical representation of the neural modulation. There are many experimental and computational approaches to elucidating the neural code that are not mutually exclusive!

The protocol that one uses to deduce the relationships in the input or output neuronal data generally falls into two classes: exploratory analysis and inference of electrophysiological recordings. Exploratory analysis is highly descriptive where graphs and numerical summaries are used to describe the overall pattern in the data. For motor BMIs, some examples of the goals of the exploratory analysis could include characterizing the spiking and LFP correlation for different behavioral conditions or explaining the spiking variation for certain behaviors in a data set. Exploratory analysis often involves characterization of a statistical distribution and can include computations such as spike rates, evoked LFP, spike-triggered averages, LFP spectral analysis, and spike-field coherence. Exploratory analysis often seeks to find patterns in nature, which can be inherently subjective. In contrast, inference is strongly hypothesis driven, where the goal is to produce neural and behavioral data in a specific way on a small scale and then draw more general conclusions about motor control for BMIs. The analysis often involves finding functional relationships in neural recordings with external quantities. Tuning curves (described later) have been used to explain the preferred kinematic spike rate for a neuron

TABLE 2.1: Intersection of experimental variables for neural coding and decoding

	Stimulus	Neural Response
Coding	Given	To be determined
Decoding	To be determined	Given

or relate LFP spectral power in a particular frequency to behavior. Regression techniques are also used to relate observed neural activity to continuous behavioral information (i.e., hand trajectory). Ultimately, hypothesis testing is required to validate inference-based approaches where significance is used to distinguish between behavioral conditions and neural activity.

2.7 METHODS OF KINEMATIC AND DYNAMIC REPRESENTATION

The firing of neurons in the primary motor (M1), premotor (PM), parietal (P), somatosensory (S) cortices, basal ganglia (BG), and cerebellum (CB) contains the representation of many motor and nonmotor signals. Through single-cell and neural population electrophysiological experiments, neuronal firing has been related to joint angles [40], muscle activation [41–43], hand trajectory (path) [44–46], velocity [47, 48], duration [49, 50], as shown in Figure 2.6. From a computational perspective, the challenge has been one of understanding how information flows in different sensory and motor networks, making contributions to each of the parameters in a time-varying manner. Several hypotheses have been proposed about how kinematic and dynamical variables are represented in the nervous system. Up to this point, we have discussed that information processing in the brain revolves around modulation (variation) in neuronal firing. This variation could be locked to some external stimulus or even the firing of another neuron. The three leading hypotheses of neuron information processing include rate coding, correlation coding, and temporal coding. A detailed discussion of the analysis approaches for each hypothesis will be given next.

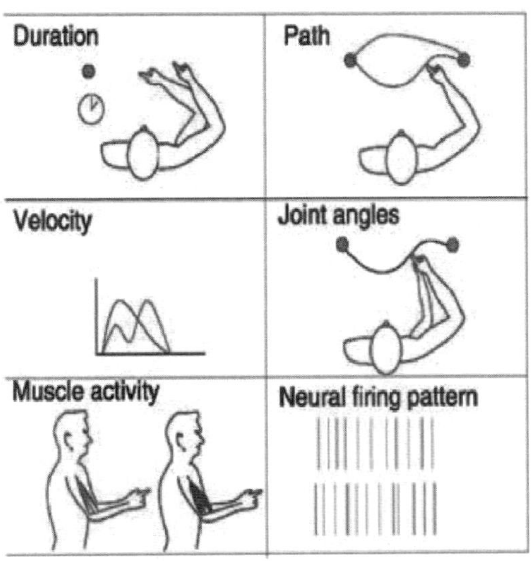

FIGURE 2.6: Parameters of cell signaling in the CNS [51].

2.7.1 Rate Coding

To understand the neuronal firing representations, neuroscientists have correlated the firing rate (2.3) with a wide variety of behaviors. The most celebrated leader of the rate coding hypothesis is English physiologist Edgar Adrian, who determined that neurons tended to modulate their firing rate in response to a stimulus (stretch receptor in the frog leg)[11] [39]. Most microelectrode-based BMIs have also utilized rate coding to map motor parameters. To represent the neural activity, most of experimental designs utilize the local mean firing rate of neurons as the input. The mean firing rate has been estimated by "binning" neural spikes with sliding rectangular time windows of length ranging from 50 up to 100 msec. This method greatly reduces the number of zeros in the spike train point process and also provides a time to amplitude conversion of the firing events. The spike rate response of cells in muscles, visual cortex, and motor cortex have lead to the development of receptive fields and tuning curves that describe the reliability of firing of a neuron as a function of the variation in an experimental condition (i.e., muscle stretch, visual orientation, movement velocity). The search for more complete mappings of receptive fields and tuning curves to the external environment has lead to the need for massively parallel neuronal recordings. These techniques yield the activity of neural ensembles or large groups of neurons from spatially disparate parts of the brain. The known concepts of the computational structure and basis of neuronal representations have been the driving force for deriving models for BMIs [14, 52–57], which will be described in the next chapter.

$$s(t) = \textit{number of spikes in } \Delta t \tag{2.3}$$

2.7.2 Effect of Resolution on Rate Coding

Because the local firing rate may represent the local frequency of a neural spike train, the features can be extracted based on local frequency. One of the methods for the representation of local frequency information is multiresolution analysis [58], usually realized by the wavelet transform. Through multiresolution analysis, it is possible to represent the time-frequency characteristics of a signal. Basically, we can obtain as many local frequency components as we want at a given time instance. Hence, multiresolution analysis of neural spikes may provide richer information about neuronal behavior compared with to the binning using a fixed-width time window as is traditionally performed in the literature.

If we consider multiresolution spike train analysis, it is straightforward to see that the binning process is nothing but a discrete wavelet transform (DWT) with a Haar wavelet [59].

[11]Firing rate increased with increasing load on the leg.

However, because the original DWT is basically a noncausal process, a wavelet transform featuring causality should be considered. For this purpose, we develop here the à trous wavelet transform [60] to implement a causal DWT. With this procedure, multiresolution spike train analysis can be regarded as binning spike trains with multiscale windows.

2.7.3 Multiresolution Analysis of Neural Spike Trains

To facilitate the computation, we apply the DWT using dyadic Haar wavelets, which when integrated with the à trous wavelet transform, yields very effective DSP implementations. The Haar wavelet is the simplest form of wavelet and was introduced in the earliest development of wavelet transform [59]. Here, we only introduce the functional form of the Haar wavelets, whereas the details can be found in Reference [59]. Let us first define the Haar scaling function as

$$\phi(x) = \begin{cases} 1 & \text{if} \quad x \in [0,1) \\ 0 & \text{otherwise.} \end{cases} \tag{2.4}$$

Let V_j be the set of functions of the form

$$\sum_k a_k \phi(2^j x - k), \tag{2.5}$$

where a_k is a real number and k belongs to the integer set. a_k is nonzero for only a finite set of k. V_j is the set of all piecewise constant functions whose supports are finite, where discontinuities between these functions belong to a set,

$$\left\{ \cdots, -\frac{2}{2^j}, -\frac{1}{2^j}, 0, \frac{1}{2^j}, \frac{2}{2^j}, \cdots \right\}. \tag{2.6}$$

Note that $V_0 \subset V_1 \subset V_2 \subset \cdots$. The Haar wavelet function ψ is defined by

$$\psi(x) = \phi(2x) - \phi(2x - 1). \tag{2.7}$$

If we define W_j as the set of functions of the form

$$W_j = \sum_k a_k \psi(2^j x - k), \tag{2.8}$$

then it follows that

$$V_j = W_{j-1} \oplus W_{j-1} \oplus \cdots \oplus W_0 \oplus V_0 , \tag{2.9}$$

where $\oplus$ denotes the union of two orthogonal sets.

The DWT, using a dyadic scaling, is often used because of its practical effectiveness. The output of the DWT forms a triangle to represent all resolution scales because of decimation (holding one sample out of every two) and has the advantage of reducing the computational complexity and storage. However, it is not possible to obtain representation with different scales at every instance with the decimated output. This problem can be overcome by a nondecimated DWT [61], which requires more computations and storage. The nondecimated DWT can be formed in two ways: (1) The successive resolutions are obtained by the convolution between a given signal and an incremental dilated wavelet function. (2) The successive resolutions are formed by smoothing with an incremental dilated scaling function and taking the difference between successive smoothed data.

The à trous wavelet transform follows the latter procedure to produce a multiresolution representation of the data. In this transform, successive convolutions with a discrete filter h is performed as

$$v_{j+1}(k) = \sum_{l=-\infty}^{\infty} h(l) v_j(k + 2^j l), \tag{2.10}$$

where $v_0(k) = x(k)$ is the original discrete time series, and the difference between successive smoothed outputs is computed as

$$w_j(k) = v_{j-1}(k) - v_j(k), \tag{2.11}$$

where w_j represents the wavelet coefficients. It is clear that the original time series $x(k)$ can be decomposed as

$$x(k) = v_S(k) + \sum_{j=1}^{S} w_j(k), \tag{2.12}$$

where S is the number of scales. The computational complexity of this algorithm is $O(N)$ for data length N.

Note that the à trous wavelet transform as originally defined by Sheppard does not necessarily account for a causal decomposition (i.e., where the future data are not incorporated in the present computation of the wavelet transform). To apply the à trous wavelet transform with such a requirement, the Haar à trous wavelet transform (HatWT) was introduced [62]. The impulse response h is now a filter with coefficients (1/2, 1/2). For a given discrete time series, $x(k) = v_0(k)$, the first resolution is obtained by convolution $v_0(k)$ with h, such that

$$v_1(k) = \frac{1}{2}(v_0(k) + v_0(k-1)), \tag{2.13}$$

and the wavelet coefficients are obtained by

$$w_1(k) = v_0(k) - v_1(k). \tag{2.14}$$

For the jth resolution,

$$v_j(k) = \frac{1}{2}(v_{j-1}(k) + v_{j-1}(k - 2^{j-1}) \tag{2.15}$$

$$w_j(k) = v_{j-1}(k) - v_j(k). \tag{2.16}$$

Hence, the computation in this wavelet transform at time k involves only information from the previous time step $k - 1$.

The HatWT provides a set of features from the time series data, such as the wavelet coefficients $[w_1(k), \ldots, w_{S-1}(k)]$ and the last convolution output $[v_{S-1}(k)]$. However, if we seek to associate the HatWT with the binning process a spike trains, the set $[v_0(k), \ldots, v_{S-1}(k)]$ can be translated into the bin count data with multiple bin widths. To yield the multiscale bin count data using (2.13), we only have to multiply $v_j(k)$ by 2^j, such that

$$u_j(k) = 2^j v_j(k), \text{ for } j = 0, \ldots, S-1. \tag{2.17}$$

Hence, the convolution output in the HatWT provides a feature set related with binning with different window widths. In the design of decoding models for BMIs, we may utilize the scaled convolution outputs $[u_j(k)]$ for $j = 0, \ldots, S - 1$, or equivalently, the bin count data with different window widths as input features.

To apply the multiresolution analysis to the BMI data, we must choose a suitable set of scales. Although it is not straightforward to determine a set of scales for the HatWT for spike trains, the firing properties of neuronal data collected from the particular BMI experimental paradigm can be used to help guide the determination of scales. For example, the smallest scale may be chosen to be larger than 1 msec because of the refractory period of neuronal firing. Also, the largest scale may be chosen

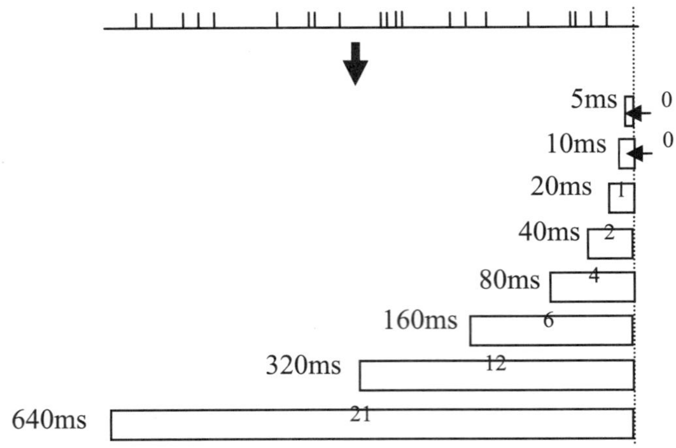

FIGURE 2.7: An illustration of the scaled convolution output from the HatWT; $u_j(k)$ for a given spike train at a time instance k_0. The number in each box denotes the value of $u_j(k_0)$ for $j = 0, \ldots, 7$.

to not exceed 1 sec because it has been reported that only the past neuronal activity up to 1 sec is correlated with the current behavior [44]. In the following case examples, we select eight scales of binning starting at 5 up to 640 msec, with the dyadic scaling of 5, 10, 20, 40, 80, 160, 320, and 640 msec.

With the selected scales, the HatWT is performed on an ensemble of BMI neuronal spike trains in the following way. We first generate the basic bin count data with a 5-msec nonoverlapping window for every neuronal channel. Next, the HatWT is applied to the 5-msec bin count data at each neuronal channel, yielding the convolution output $v_j(k)$ for $j = 0, \ldots, 7$ following (2.15). Each series $v_j(k)$ is then multiplied by 2^j to generate $u_j(k)$ [in (2.17)]. An illustrative example of the generated $u_j(k)$ at specific time instance k_0 is presented in Figure 2.7.

Note that the sampling rate in $u_j(k)$ is 200 Hz for any j. In terms of the binning process, $u_j(k)$ can be interpreted as the bin count data for a given spike train with a $5 \cdot 2^j$-msec time window that slides over time by step of 5 msec. Therefore, $u_j(k)$ with a larger j will add spikes from a longer time window and will overlap more with the previous bin $u_j(k - 1)$, which then yields a smoother temporal patterns for larger j.

The top panel in Figure 2.8 demonstrates an example of $u_j(k)$ of a specific neuron for 5-second period. $u_j(k)$ for each j is normalized to have the maximum value of 1. Darker pixels denote larger values. The set of $u_j(k)$ are temporally aligned with the associated hand trajectories plotted in the bottom panel. To view the correlation of $u_j(k)$ with the movement for each j, the discrete time series $u_j(k)$ is separately plotted on top of the hand trajectory (the x coordinate) in Figure 2.9 [both $u_j(k)$ and the hand trajectory are scaled to be in the similar dynamic range for visualization purposes].

This figure is very telling about the difficulties and opportunities of data modeling for BMIs. It demonstrates that the correlation of $u_j(k)$ with the hand trajectory increases with larger j; in fact, only the 320- and 640-msec scales have reasonable correlations with the movements. Therefore,

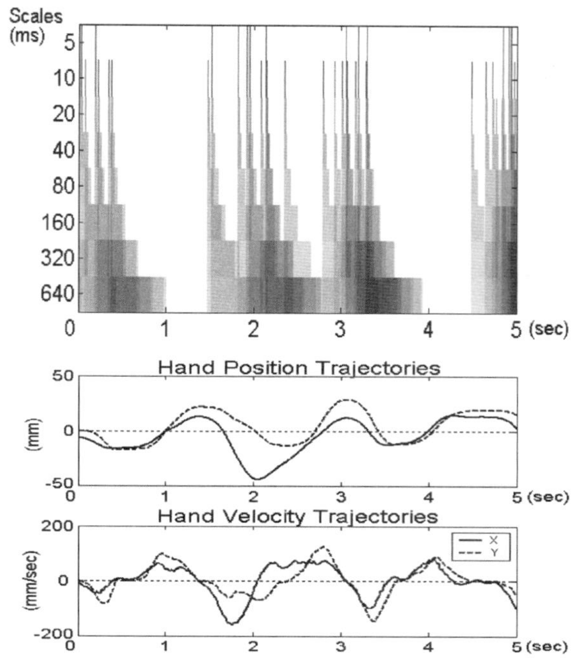

FIGURE 2.8: An example of the series of $u_j(k)$ along with the corresponding hand trajectories: (top) Each row represents the scale j for $j = 0, \ldots, 7$ nd each column represents the time indices with 5-msec interval during a 5-second duration. (bottom) The trajectories of hand position and velocity at x (solid), and y (dotted) coordinates.

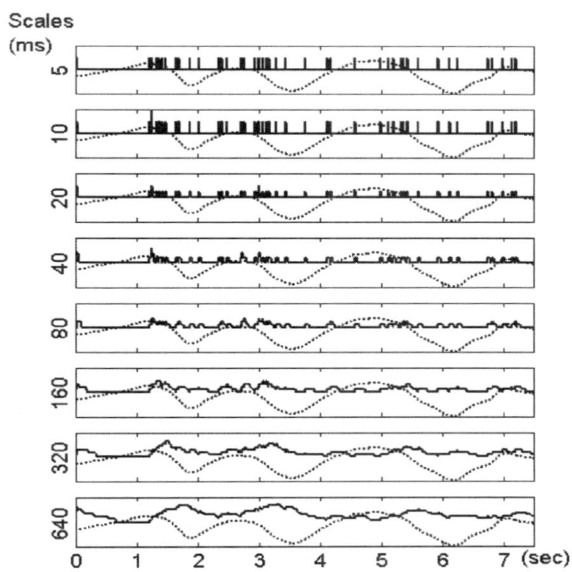

FIGURE 2.9: The demonstration of the relation between the neuronal firing activity representation at each scale (solid lines) and the hand position trajectory at x-coordinate (dotted lines).

binning at 100 msec is a reasonable value for motor BMIs when signal processing models based on correlation are used. We can expect that there are neurons that modulate their firing rates concurrently with the movement, and therefore, linear or nonlinear input output models should provide acceptable performance. But this picture also shows the difficulties between the tremendous gap in resolution between the spike trains and the movement, that are difficult to bridge with pure data driven operations (i.e., without involving some form of physiological modeling). Therefore, it opens up many other interesting possibilities to improve the present state of the art in BMI modeling.

2.7.4 Using Firing Rates to Compute Tuning

Classically, a single motor neuron's physiological response to a behavioral task has been described using directional tuning curves. As originally derived from a standard center-out task by Georgopoulos et al. [63], the tuning curve relates the mean of movement-related cell activity to movement direction. The preferred direction of a neuron, measured in degrees, is the direction that yields the maximal firing response over many trials. Tuning curves convey the expected value of a PDF indicating the average firing a cell will exhibit given a particular movement direction [63] as shown in Figure 2.10. Cells that have cosine-shaped tuning curves participate in neural computation by forming a set of basis functions[12] that represent the output. It has been shown in the literature that a variety of hand trajectories can be reconstructed by simply weighting and summing the vectors indicating preferred directions of cells in an ensemble of neurons [56, 64–66]. Weighing the preferred directions with the neural activities in the population gives a resultant direction vector called the "population vector," which has been shown to be correlated with the actual movement direction [56], and the modeling technique will be described later. To use this measurement, the kinematic data histograms must first be computed by using the desired behaviors from which the corresponding position, velocity, and acceleration vectors (magnitude and direction) between successive points in the trajectory is computed. The quantity that is commonly used in BMI experiments is the hand movement direction measured as an angle between 0° and 360° [67]. Because cellular tuning can produce properties where the average of angle 0° and 360° is not 180° (this results from the wrap-around effect of the measurements), the mean of each cell's hand direction tuning is computed using circular statistics as in (2.1)[13] [68],

$$\text{circular mean} = \arg\left(\sum_N r_N e^{i\theta_N}\right) \tag{2.1}$$

where r_N is the cell's average firing rate for angle θ_N, where N is from 1 to 360.

[12] A term used in vector spaces where a group of functions can be linearly combined to encode any function in a subspace.

[13] In computing the circular mean, we used the four quadrant inverse tangent.

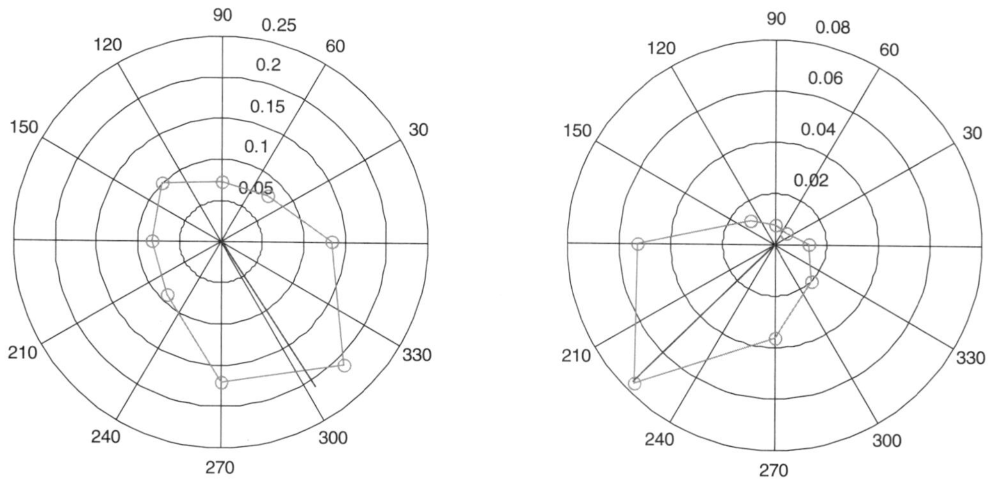

FIGURE 2.10: Examples of neuron tuning. Left plot is the tuning plot with neuronal tuning at 315° and depth of 1. The right plot is for a neuron with tuning at 225° and a depth of 0.93.

To assess the degree of tuning, a metric called the "tuning depth" is often used. This quantity is defined as the difference between the maximum and minimum values in the cellular tuning. For an impartial comparison of the cellular tuning depth, the tuning curves are usually normalized by the standard deviation of the firing rate. This measurement relates how "peaky" the tuning curve is for each cell and is an indicator of how well modulated the cell's firing rate is to the kinematic parameter of interest. The normalized tuning depth for 185 neurons computed from three kinematic vectors, hand positions, hand velocities, and hand accelerations, is presented in Figure 2.11. The cortices where the microelectrodes were implanted are also shown in the figure. From this analysis, one can clearly see that most tuned neurons are in the primary motor cortex for all of the kinematic vectors used to calculate the tuning depth.

In one-dimensional kinematic spaces, the vector space is restricted to vector magnitude. Typical histograms are ramp functions that saturate because neurons tend to increase their activity to the maximum firing rate, so the range is still finite [69]. Despite these differences, the tuning depth of each cell can still be used as a measure of information content provided by all neurons.[14] Because computing the tuning depth from finite data is noisy, the curves are also usually smoothened by

[14]We have observed that the tuning of a cell can vary as a function of the delay between the generation of the neural activity and the physical movement of the hand. Because we are interested in tuning depth, we are looking for cells that have the smallest variance in their tuning function. After computing tuning variance across all cells for delays up to 1 sec, we found that the sharpest tuning occurs at the 0th delay or the instantaneous time alignment.

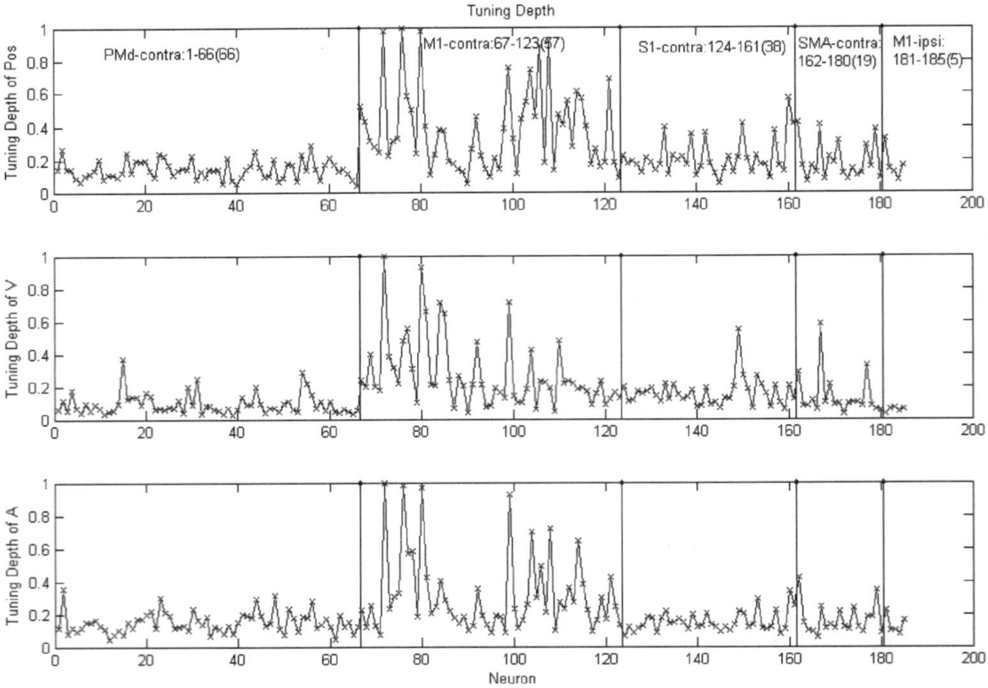

FIGURE 2.11: Tuning depth for all the neurons computed from three kinematics respectively. The upper plot is tuning depth computed from positions. The middle plot is for velocities. And the bottom plot is for accelerations.

counting the number of times that each cell fires in a window. One can expect that for BMI decoding, highly tuned cells will be more relevant. But tuning depth is not the only parameter of interest for a BMI because the input must also supply sufficient information to reconstruct all of the kinematics in the desired trajectories, that is, the set of deeply tuned neural vectors must span the space of the desired kinematic variable.

2.7.5 Information Theoretic Tuning Metrics

The tuning depth is a very coarse metric for the evaluation of the neuron receptive properties. The mutual information between neural spikes and direction angles can alternatively embody the idea of an information theoretic metric to evaluate if a neuron is tuned to a certain kinematic variable such as the angle of the velocity vector.

$$I(\text{spike};\ \text{angle}) = \sum_{\text{angle}} p(\text{angle}) \sum_{\text{spike}=0,1} p(\text{spike}|\text{angle}) \log_2 \left(\frac{p(\text{spike}|\text{angle})}{p(\text{spike})}\right) \tag{2.2}$$

where $p(\text{angle})$ is the probabilistic density of all the direction angles, which can be easily estimated by the Parzen window [70]. The probability of a spike, $p(\text{spike})$, can be simply obtained as the percentage of the spikes occurring during a large segment of the spike train such that the estimate is reliable, and the value is not very small. $p(\text{spike} \mid \text{angle})$ is the conditional probability density of the spike given the direction angle.

For each neuron, the histogram of the spike-triggered angle can be drawn and normalized to approximate $p(\text{spike} = 1, \text{angle})$. In other words, only when there is a spike, the direction angle is counted for the histogram during the corresponding direction angle bin. Then $p(\text{spike} = 1 \mid \text{angle})$ is approximated by dividing the histogram of (spike = 1, angle) by the histogram of (angle), which corresponds actually to the Bayesian formulation.

$$p(\text{spike}=1 \mid \text{angle}) = \frac{p(\text{spike} = 1,\text{angle})}{p(\text{angle})} \tag{2.3}$$

and $p(\text{spike} = 0 \mid \text{angle}) = 1 - p(\text{spike} = 1 \mid \text{angle})$.

In this way, the mutual information between the kinematic direction angle and the neural spike train can be estimated for each neuron. Figure 2.12 shows the mutual information tuning calculated from three different kinematic vectors for 185 neurons. The M1 (primary motor cortex) tuning clearly shows that the velocity (the middle plot) conveys more tuning information than position or acceleration, which has been suggested in the literature, but it is impossible to evaluate in Figure 2.11 because the conventional method normalizes to the same value all the responses. Therefore, because of the self-normalized definition, the information theoretical tuning metric found that particular neurons are more tuned to the position, whereas other neurons are more tuned to the velocity. Because the mutual information is a nonlinear measure, experience has shown that the information theoretic tuning depth should be plotted in logarithmic scale for BMIs.

2.7.6 Timing Codes

Finding appropriate methodologies to analyze in time the spatial neuronal organization during motor control, auditory/visual processing, and cognition is a challenging question. Methods for identifying the fine timing relationships among co-activated neurons have focused on statistical relationships among observed single spike trains [71–76] (see Reference [77] for a review). For pairs of neuronal recordings, several techniques have been proposed such as cross-intensity functions [78] or the method of moments [79], joint peristimulus time histograms [80], and methods based on maximum likelihood (ML) estimation [81].

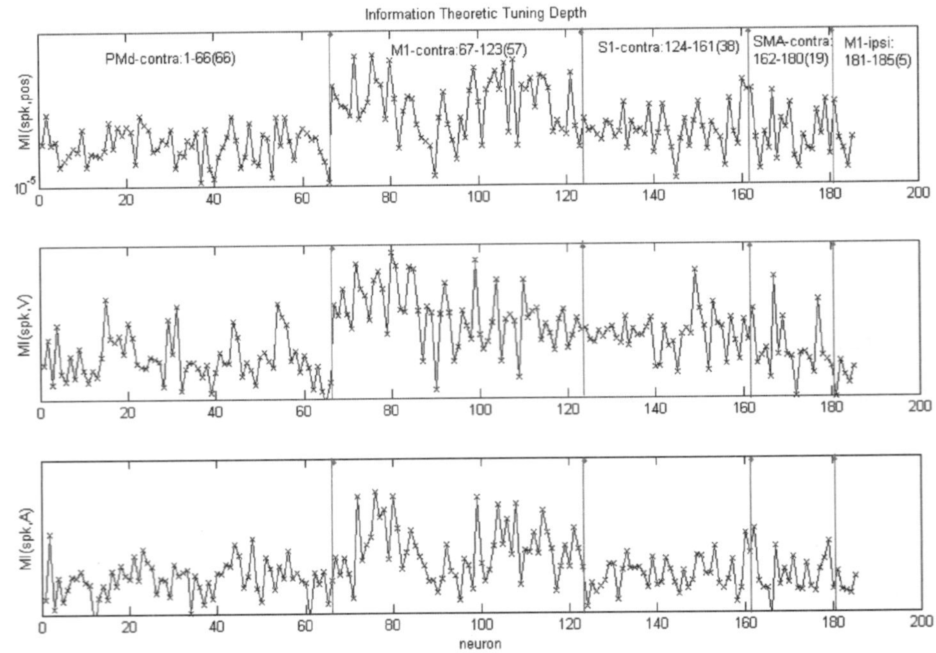

FIGURE 2.12: Information theoretic tuning depth in logarithmic scale for 185 neurons computed from three kinematic variables respectively. The upper plot is tuning depth computed from position, the middle plot is for velocity, and the bottom plot is for acceleration.

However, in many applications such as the design of neural interfaces, the timing of large ensembles of cells (hundreds of cells) must be studied in a principled way. Unfortunately, many of the aforementioned techniques become extremely time-consuming (as all pairs of electrode combinations must be considered) and alternative pattern recognition [74] or clustering [82] methods to quantify multichannel prespecified spike timing relations (patterns) have been proposed.

Most statistical analysis of neuronal spike timing assumes Poisson spike generation where the neural response only depends upon the short-term past of the input stimulus due to the independence assumptions embedded in the Poisson model. To estimate the response of a given neuron to these stimuli, one needs simply to quantify the likelihood of a spike firing in each region of the stimulus space, which can be estimated by the ratio of multidimensional histograms of spike triggered stimuli. To accurately estimate the histograms, large amounts of data are required to isolate the analysis to regions of the space where the neuron shows a large response (many counts to evaluate the histograms) or to tailor the stimuli, parameterized by only a few parameters to produce large neural responses. To overcome data issues, one can time lock the averaging to the spike, a procedure known as spike-triggered averaging. Once the spike-triggered average is obtained, the neural response can be inferred with a nonlinear input output neural model.

In conjunction with these techniques, a second class of nonlinear models for BMIs was derived from system theory using the concept of Volterra series expansions in Hilbert spaces. Volterra series have been proposed to model neuronal receptive fields using the reverse correlation or white noise approach [83], and more recently they have been applied to cognitive BMIs to model the CA3 region of the hippocampus [84]. A Volterra series is similar to a Taylor series expansion in functional spaces, where the terms are convolution integrals of products of impulse responses called kernels. The Volterra models are universal approximators, but they require the estimation of many parameters. Therefore, when applied in practice, one of the design difficulties is how to minimize the parameters, either by limiting the expansion to the second or third term or by using polynomial approximations, such as Laguerre polynomials [85] or, preferably, Kautz functions, to decrease the number of free parameters for the impulse responses.

However, because the neural response tends to be nonlinear, and higher order terms of the Volterra expansion have many parameters, lots of experimental data are needed to estimate the parameters well. The LNP (as shown in Figure 2.5) can reduce the collection of large data sets by using a more efficient estimation of the linear filter term, followed by the instantaneous nonlinearity. The assumptions for the method to work are mild: basically, one assumes that the stimuli components are uncorrelated and they are Gaussian distributed with the same variance (i.e., that the D-dimensional distribution is spherically symmetric).

In the LNP model, the linear filter collapses the multidimensional input space to a line. Spike-triggered averaging provides an estimate of this axis, under the assumption that the raw stimulus distribution is spherically symmetric [86]. Once the linear filter has been estimated, one may compute its response and then examine the relationship between the histograms of the raw and spike-triggered ensembles within this one-dimensional space to estimate the nonlinearity that now can be done with histograms without requiring large data sets. The model can be generalized to take into consideration several directions at the same time by using the covariance. Such approaches have been proposed by Moran and Schwartz introduced an exponential velocity and direction tuned motor cortical model [67]. Eden et al. used a Gaussian tuning function for the hippocampal pyramidal neurons [87]. These nonlinear mathematical models can encounter difficulties when dealing with the real data because the neurons can have very different tuning properties and probably change over time. The accuracy of the tuning function estimation can directly affect the prior knowledge of the Bayesian approach and, therefore, the results of the kinematic estimation. Simoncelli et al. built upon the approach of Marmarelis and proposed a statistical method to model the neural responses with stochastic stimuli [85]. Through parametric model identification, the nonlinear property between the neural spikes and the stimuli was directly estimated from data, which is more reliable. For motor BMIs, one is seeking sequential state estimation of the point process algorithm to infer the kinematic vectors from the neural spike train, which is the opposite of sensory neurons. However, one can regard the proper kinematic vector as the outcome of the motor cortex neurons. The tuning function between the kinematic vector and neural spike train is exactly the system model between the state and observation in our algorithm. Therefore, these methods can provide a very practical way to acquire the prior knowledge (the tuning function) in sequential estimation for adaptive point process filtering algorithms.

2.8 MODELING AND ASSUMPTIONS

2.8.1 Correlation Codes

The coincidence structure among spike trains can be used to suggest the functional connectivity among neurons (not necessarily the anatomic connectivity) in a minimal equivalent circuit [16]. Moreover, correlation and synchrony among neurons can greatly influence postsynaptic neurons, help define cell assemblies, and play a role in information transmission. Often, BMI studies seek to quantify the conditions of full synchrony and total independence among the recorded neurons. As a control, the cross correlation is often used to determine the similarities and differences in spatio-temporal organization in the data using the zero-lag cross-correlation over time, averaged through all pairwise combinations of neuron types. The cross-correlation function [88] is probably the most widely used technique to measure similarity between spike trains. With access to the activity of large neuronal ensembles one can begin to simultaneously describe the spatial and temporal varying nature of the data by computing local neuronal correlations. In this scheme, information contained in cell assemblies can be observed in cortical areas that mutually excite each other [52]. To extract useful information in the local bursting activity of cells in the ensemble, local correlations among the cells is analyzed. Traditionally, the correlogram is a basic tool to analyze the temporal structure of signals. However, applying the correlogram to spike trains is nontrivial because they are point processes, thus the signals information is not in the amplitude but only on the time instances when spikes occur. A well-known algorithm for estimating the correlogram from point processes involves histogram construction with time interval bins [88]. The binning process is effectively transforming the uncertainty in time to amplitude variability. However, this time quantization introduces binning error and leads to coarse time resolution. Furthermore, the correlogram does not take advantage of the higher temporal resolution of the spike times provided by current recording methods. However, one of the disadvantages of continuous time estimation is its computational cost. The joint peristimulus time histogram [89] requires trial averaging and, therefore, is only applicable to experimental paradigms where trial repetition is performed, and it assumes stationarity between trials. Similar to the cross-correlation, the partial directed coherence (PDC) [90–93] and the methods by Hurtado et al. [94] and Samonds and Bonds [95] suffer from the problem of windowing and large variations of the covariance matrix of the stochastic noise, which might lead to wrong conclusions about the underlying interdependence structure of the data. Typically, these methods are applied to time binned (or continuous) signals; hence, they constrain the time resolution of synchrony. Moreover, the extension of the analysis of the methods for more than two spike trains is not a trivial problem. Finally, the algorithms are difficult to apply in online applications or are computationally complex for large ensembles of neurons. The gravity transform [96] successfully tackles some of the above problems, but it lacks a statistical basis. Most importantly, none of the mentioned measures is appropriate to analyze transient synchronization because a measure in the time or spike domain is needed. As one can conclude from the discussion, still more work needs to be done to create a useful similarity metric for spike trains.

The cross-correlation function in (2.4) can be used to quantify synchronized firing among cells in the ensemble. In this case, small sliding (overlapping) windows of data are defined by $\mathbf{A}(t)$, which is a matrix-containing L-delayed versions of firing patterns from 185 neurons. At each time tick t in the simulation, the cross-correlation between neurons i and j for all delays l is computed. Because we are interested in picking only the most strongly synchronized bursting neurons in the local window, we simply average over the delays in (2.5). We define $\bar{C}_{:j}$ as a vector representing how well correlated the activity of neuron j is with the rest of the neuronal ensemble. Next, a single 185 × 1 vector at each time tick t is obtained in (2. 6) by summing cellular assembly cross-correlations only within each sampled cortical area, M1 (primary motor), PMd (premotor dorsal), SMA (supplementary motor associative), and S1 (somatosensory).

$$C(t)_{ijl} = E\left[\mathbf{A}(t)_i \mathbf{A}(t-l)_j\right] \tag{2.4}$$

$$\bar{C}(t)_{ij} = \frac{1}{L}\sum\nolimits_{l=1}^{L}\left|C(t)_{ijl}\right| \tag{2.5}$$

$$\tilde{C}(t)_j = \sum_{i=\text{cortex}} \bar{C}(t)_{ij} \tag{2.6}$$

The neural correlation measure in (2.6) was then used as a real-time marker of neural activity corresponding to segments of movements of a hand trajectory. In Figure 2.13, highly correlated neuronal activity is shown to be time-varying in synchrony with changes in movement direction as indicated by the vertical dashed line. Figure 2.13 shows that the correlations contained in the data are highly variable even for similar movements.

We can immediately see that different neurons present different level of involvement in the task as measured by correlation. Some participate heavily (hot colors), whereas others basically do not correlate with the hand motion (blue). But even for the ones that participate heavily, their role across time is not constant as depicted by the vertical structure of the raster. This is stating that the time structure of neural firings is nonstationary and heavily dependent upon the task.

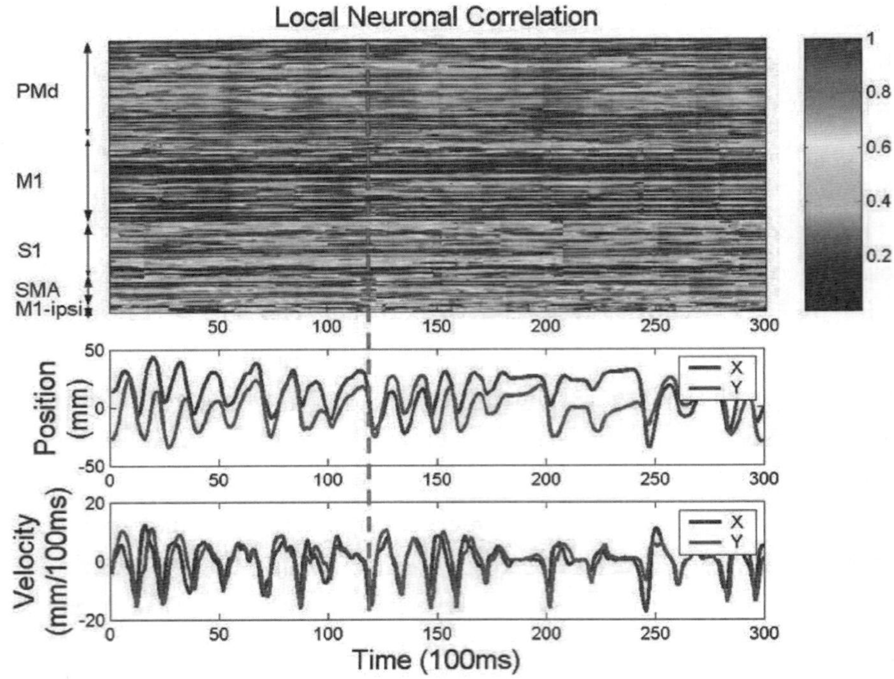

FIGURE 2.13: Local neuronal correlations time-synchronized with hand position and velocity.

2.9 IMPLICATIONS FOR BMI SIGNAL PROCESSING

One of the most important steps in implementing BMI optimal signal processing techniques for any application is data analysis and the understanding of neural representations. Here, the reader should take note that optimality in the signal processing technique is predicated on the matching between the statistics of the data and the a priori assumptions inherent in any signal processing technique [97]. In the case of BMIs, the statistical properties of the neural recordings and the analysis of neural ensemble data are not fully understood. Hence, this lack of information means that the neural-motor translation is not guaranteed to be the best possible, even if optimal signal processing is utilized (because the criterion for optimality may not match the data properties). Despite this reality, through the development of new neuronal data analysis techniques we can improve the match between neural recordings and BMI design [16, 35]. For this reason, it is important for the reader to be familiar with the characteristics of neural recordings that would be encountered.

Some interesting issues to consider include the following:

- How much information is encoded in the neural population response? Noise and the finite number of neurons sampled can influence how one interprets the representation.

- What is the most efficient decoding strategy (i.e., the one that has the minimum decoding error)?
- What are the relationships between resolution and synchronization in the neural representation? We have shown a variety of techniques that can alter the interpretation of the neural representation and depend upon time resolution incorporated in the neural recordings and the synchronization techniques with behavioral or neural events.
- What is the biological plausibility of the encoding or decoding strategy that is assumed? The constraints used in biological systems are very different than those used in computational environments. One must identify and discuss these limitations within the constructs of the BMI computational goals.

In this chapter, we have observed that the neuronal and ensemble responses are highly nonstationary (i.e., local correlations over time) and sparse (spike trains). Moreover, firing rate analysis and its correlation with behavior highly depend on the appropriate choice of the rate window. The important message of this analysis is that standard optimal signal processing techniques (linear filters, neural networks, etc.) were not designed for data that is nonstationary and discrete valued. Ideally, we would like our optimal signal processing techniques to capture the changes observed in Figure 2.13. However, the reader should be aware that in the environment of BMI datasets, applying any of the "out-of-the-box" signal processing techniques means that the neural to motor mapping may not be optimal. More importantly, any performance evaluations and model interpretations drawn by the experimenter can be directly linked and biased by the mismatch between data and model type.

REFERENCES

1. Crammond, D.J., *Motor imagery: never in your wildest dream.* Trends in Neurosciences, 1997. **20**(2): pp. 54–57.
2. Kupfermann, *Localization of Higher Cognitive and Affective Functions: The Association Cortices*, in Principles of Neural Science, E.R. Kandel, J.H. Schwartz, and T.M. Jessel, eds. 1991, Norwalk, CT: Appleton & Lange. pp. 823–838.
3. Andersen, R.A., et al., *Multimodal representation of space in the posterior parietal cortex and its use in planning movements.* Annual Review of Neurosciences 1997. **20**: pp. 303–330.
4. Chen, R., L.G. Cohen, and M. Hallett, *Role of the ipsilateral motor cortex in voluntary movement.* Can. J. Neurol. Sci., 1997. **24**: pp. 284–291.
5. Cisek, P., D.J. Crammond, and J.F. Kalaska, *Neural activity in primary motor and dorsal premotor cortex In reaching tasks with the contralateral versus ipsilateral arm.* Journal of Neurophysiology, 2003. **89**(2): pp. 922–942.

6. Tanji, J., K. Okano, and K.C. Sato, *Neuronal activity in cortical motor areas related to ipsilateral, contralateral, and bilateral digit movements of the monkey.* Journal of Neurophysiology, 1988. **60**(1): pp. 325–343.
7. Edelman, G.M., V.B. Mountcastle, and Neurosciences Research Program. The Mindful Brain: Cortical Organization and the Group-Selective Theory of Higher Brain Function. 1978, Cambridge, MA: MIT Press.
8. Polak, J.M., and S. Van Noorden, Introduction to Immunocytochemistry. 3rd ed. 2003, Oxford: BIOS Scientific Publishers.
9. Sherrington, C.S., The Integrative Action of the Nervous System. Classics in Psychology. 1973, New York: Arno Press.
10. Mountcastle, V., *The columnar organization of the neocortex.* Brain, 1997. **120**: pp. 702–722.
11. Buzsáki, G., Rhythms of the Brain. 2006, New York: Oxford University Press.
12. Erdos, P. and R. A., *On the evolution of random graphs.* Publication of the Mathematical Institute of the Hungarian Academy of Science, 1960. **5**: pp. 17–61.
13. Watts, D. and S. Strogatz, *Collective dynamics of "smal-world" networks.* Nature, 1998. **393**: pp. 440–442.
14. Abeles, M., Corticonics: Neural Circuits of the Cerebral Cortex. 1991, New York: Cambridge University Press.
15. Kandel, E.R., J.H. Schwartz, and T.M. Jessell, eds. Principles of Neural Science. 4th ed. 2000, New York: McGraw-Hill.
16. Nicolelis, M.A.L., Methods for Neural Ensemble Recordings. 1999, Boca Raton, FL: CRC Press.
17. Holt, G.R., and C. Koch, *Electrical Interactions via the Extracellular Potential Near Cell Bodies.* Journal of Computational Neuroscience, 1999. **6**: pp. 169–184.
18. Lewicki, M.S., *A review of methods for spike sorting: The detection and classification of neural action potentials.* Network: Computation in Neural Systems, 1998. **9**(4): pp. R53–78.
19. Fee, M.S., P.P. Mitra, and D. Kleinfeld, *Variability of extracellular spike waveforms of cortical neurons.* Journal of Neurophysiology, 1996. **76**(6): pp. 3823–3833.
20. Fee, M.S., P.P. Mitra, and D. Kleinfeld, *Automatic sorting of multiple-unit neuronal signals in the presence of anisotropic and non-Gaussian variability.* Journal of Neuroscience Methods, 1996. **69**: pp. 175–188.
21. Wood, F., et al., *On the variability of manual spike sorting.* IEEE Transactions on Biomedical Engineering, 2004. **51**(6): pp. 912–918.
22. Koch, C. and G.R. Holt, *Electrical interactions via the extracellular potential near cell bodies.* Journal of Computational Neuroscience, 1999. **6**: pp. 169–184.
23. Somogyvari, Z., et al., *Model-based source localization of extracellular action potentials.* Journal of Neuroscience Methods, 2005. **147**: pp. 126–137.

24. Varona, P., et al., *Macroscopic and subcellular factors shaping population spikes.* Journal of Neurophysiology, 2000. **83**(4): pp. 2192–2208.
25. Bierer, S.M., and D.J. Anderson, *Multi-channel spike detection and sorting using an array processing technique.* Neurocomputing, 1999(26–27): pp. 946–956.
26. Takahashi, S., Y. Anzai, and Y. Sakurai, *A new approach to spike sorting for multi-neuronal activities recorded with a tetrode—How ICA can be practical.* 2003. **46**(3): p. 265.
27. Rogers, C.L., et al. *A Pulse-Based Feature Extractor for Spike Sorting Neural Signals,* in 3rd International IEEE EMBS Conference on Neural Engineering. 2007. Kohala Coast, HI.
28. Rao, S., et al. *Spike Sorting Using Non Parametric Clustering Via Cauchy Schwartz PDF Divergence,* in ICASSP. 2006.
29. Hodgkin, A. and A. Huxley, *A quantitative description of membrane current and its application to conduction and excitation in nerve.* Journal of Physiology, 1952. **117**: pp. 500–544.
30. Koch, C., Biophysics of computation; information processing in single neurons. 1999, New York: Oxford University Press.
31. Koch, C. and I. Segev, eds. Methods in Neural Modelling. 1998, New York: MIT Press.
32. Wilson, H.R., Spikes Decision and Actions: Dynamical Foundations of Neuroscience. 1999, New York: Oxford University Press.
33. Grillner, S., et al., *Microcircuits in action—From CPGs to neocortex.* Trends in Neurosciences, 2005. **28**(10): pp. 525–533.
34. *Blue Brain Project.* 2007 (available from: http://bluebrain.epfl.ch/).
35. Rieke, F., Spikes: Exploring the Neural Code. 1996, Cambridge, MA: MIT Press.
36. Brown, E.N., et al., *The time-rescaling theorem and its application to neural spike train data analysis.* Neural Computation, 2002. **14**(2): pp. 325–346.
37. Simoncelli, E.P., et al., *Characterization of neural responses with stochastic stimuli.* 3rd ed. The New Cognitive Neuroscience. 2004, Cambridge, MA: MIT Press.
38. Pillow, J.W., et al., *Prediction and decoding of retinal ganglion cell responses with a probabilistic spiking model.* Journal of Neuroscience, 2005. **25**(47): pp. 11003–11013.
39. Adrian, E., Nobel Lectures, Physiology or Medicine 1922–1944. 1965, Amsterdam: Elsevier.
40. Evarts, E.V., *Representation of movements and muscles by pyramidal tract neurons of the precentral motor cortex,* in Neurophysiological Basis of Normal and Abnormal Motor Activities, M.D. Yahr and D.P. Purpura, eds. 1967, New York: Raven Press. pp. 215–253.
41. Todorov, E., *Direct cortical control of muscle activation in voluntary arm movements: A model.* Nature Neuroscience, 2000. **3**(4): pp. 391–398.
42. Fetz, E.E., and D.V. Finocchio, *Correlations between activity of motor cortex cells and arm muscles during operantly conditioned response patterns.* Experimental Brain Research, 1975. **23**(3): pp. 217–240.

43. Sanes, J.N., S. Suner, and J.P. Donoghue, *Dynamic organization of primary motor cortex output to target muscles in adult rats. I. Long-term patterns of reorganization following motor or mixed peripheral nerve lesions.* Experimental Brain Research, 1990. **79**: pp. 479–491.
44. Wessberg, J., et al., *Real-time prediction of hand trajectory by ensembles of cortical neurons in primates.* Nature, 2000. **408**(6810): pp. 361–365.
45. Sanchez, J.C., et al. *Learning the contributions of the motor, premotor, and posterior parietal cortices for hand trajectory reconstruction in a brain machine interface*, in IEEE EMBS Neural Engineering Conference. 2003. Capri, Italy.
46. Lin, S., J. Si, and A.B. Schwartz, *Self-organization of firing activities in monkey's motor cortex: Trajectory computation from spike signals.* Neural Computation, 1997. **9**: pp. 607–621.
47. Carmena, J.M., et al., *Learning to control a brain–machine interface for reaching and grasping by primates.* PLoS Biology, 2003. **1**: pp. 1–16.
48. Sanchez, J.C., et al. *Simultaneous prediction of five kinematic variables for a brain-machine interface using a single recurrent neural network*, in International Conference of Engineering in Medicine and Biology Society. 2004.
49. Scott, S.H., *Role of motor cortex in coordinating multi-joint movements: Is it time for a new paradigm?* Canadian Journal of Physiology and Pharmacology, 2000. **78**: pp. 923–933.
50. Kalaska, J.F., et al., *Cortical control of reaching movements.* Current Opinion in Neurobiology, 1997. 7: pp. 849–859.
51. Shadmehr, R. and S.P. Wise, The Computational Neurobiology of Reaching and Pointing: A Foundation for Motor Learning. 2005, Cambridge, MA: MIT Press.
52. Hebb, D.O., The Organization of Behavior: A Neuropsychological Theory. 1949, New York: Wiley.
53. Lilly, J.C., *Correlations between neurophysiological activity in the cortex and short-term behavior in monkey*, in Biological and Biochemical Bases of Behavior, H.F. Harlow and C.N. Woolsey, eds. 1958, Madison, WI: University Wisconsin Press. pp. 83–100.
54. Strumwasser, F., *Long-Term recording from single neurons in brain of unrestrained mammals.* Science, 1958. **127**(3296): pp. 469–470.
55. Grossberg, S., *Studies of mind and brain: Neural principles of learning, perception, development, cognition, and motor control.* Boston Studies in the Philosophy of Science. Vol. 70. 1982, Dordrecht, The Netherlands: Boston. pp. 223–225.
56. Georgopoulos, A.P., A.B. Schwartz, and R.E. Kettner, *Neuronal population coding of movement direction.* Science, 1986. **233**(4771): pp. 1416–1419.
57. Buzsaki, G., *Large-scale recording of neuronal ensembles.* Nature Neuroscience, 2004. **75**(5): pp. 446–451.

58. Murtagh, F., J.L. Starck, and O. Renaud, *On neuro-wavelet modeling.* Decision Support System Journal, 2004. **37**: pp. 475–484.
59. Daubechies, I. *Ten lectures on wavelets,* in Society for Industrial and Applied Mathematics. 1992. Philadelphia, PA: SIAM.
60. Shensa, M.J., *Discrete wavelet transforms: Wedding the à trous and Mallat algorithm.* IEEE Transactions on Signal Processing, 1992. **40**: pp. 2464–2482.
61. Aussem, A., J. Campbell, and F. Murtagh, *Wavelet-based feature extraction and decomposition strategies for financial forecasting.* Journal of Computational Intelligence in Finance, 1998. **6**: pp. 5–12.
62. Zheng, G., et al., *Multiscale transforms for filtering financial data streams.* Journal of Computational Intelligence in Finance, 1999. 7: pp. 18–35.
63. Georgopoulos, A., et al., *On the relations between the direction of two-dimensional arm movements and cell discharge in primate motor cortex.* Journal of Neuroscience, 1982. **2**: pp. 1527–1537.
64. Georgopoulos, A.P., et al., *Mental rotation of the neuronal population vector.* Science, 1989. **243**(4888): pp. 234–236.
65. Schwartz, A.B., D.M. Taylor, and S.I.H. Tillery, *Extraction algorithms for cortical control of arm prosthetics.* Current Opinion in Neurobiology, 2001. **11**(6): pp. 701–708.
66. Taylor, D.M., S.I.H. Tillery, and A.B. Schwartz, *Direct cortical control of 3D neuroprosthetic devices.* Science, 2002. **296**(5574): pp. 1829–1832.
67. Moran, D.W., and A.B. Schwartz, *Motor cortical representation of speed and direction during reaching.* Journal of Neurophysiology, 1999. **82**(5): pp. 2676–2692.
68. Jammalamadaka, S.R., and A. SenGupta, Topics in Circular Statistics. 1999, River Edge, NJ: World Scientific Publishing Company.
69. Fu, Q.G., et al., *Temporal encoding of movement kinematics in the discharge of primate primary motor and premotor neurons.* Journal of Neurophysiology, 1995. **73**(2): pp. 836–854.
70. Parzen, E., *On the estimation of a probability density function and the mode.* Annals on Mathematical Statistics, 1962. **33**(2): pp. 1065–1076.
71. Perkel, D.H., G.L. Gerstein, and G.P. Moore, *Neuronal spike trains and stochastic point processes. II. Simultaneous spike trains.* Biophysical Journal, 1967. **7**(4): pp. 419–40.
72. Gerstein, G.L., and D.H. Perkel, *Simultaneously recorded trains of action potentials: Analysis and functional interpretation.* Science, 1969. **164**(881): pp. 828–30.
73. Gerstein, G.L., D.H. Perkel, and K.N. Subramanian, *Identification of functionally related neural assemblies.* Brain Research, 1978. **140**(1): pp. 43–62.
74. Abeles, M., and G.L. Gerstein, *Detecting spatiotemporal firing patterns among simultaneously recorded single neurons.* Journal of Neurophysiology, 1988. **60**(3): pp. 909–24.

75. Palm, G., A.M. Aertsen, and G.L. Gerstein, *On the significance of correlations among neuronal spike trains.* Biological Cybernetics, 1988. **59**(1): pp. 1–11.
76. Grun, S., M. Diesmann, and A. Aertsen, *Unitary events in multiple single-neuron spiking activity: I. Detection and significance.* Neural Computation, 2002. **14**(1): pp. 43–80.
77. Gerstein, G.L., and K.L. Kirkland, *Neural assemblies: Technical issues, analysis, and modeling.* Neural Networks, 2001. **14**(6–7): pp. 589–98.
78. Cox, D.R., and P.A.W. Lewis, *Multivariate point processes.* Proceedings of the Sixth Berkeley Symposium on Probability and Mathematical Statistics, 1972. **3**: pp. 401–448.
79. Brillinger, D.R., *The identification of point process systems.* Annals of Probability, 1975. **3**: pp. 909–929.
80. Gerstein, G.L., and D.H. Perkel, *Mutual temporal relationships among neuronal spike trains. Statistical techniques for display and analysis.* Biophysical Journal, 1972. **12**(5): pp. 453–473.
81. Borisyuk, G.N., et al., *A new statistical method for identifying interconnections between neuronal network elements.* Biological Cybernetics, 1985. **52**(5): pp. 301–306.
82. Gerstein, G.L., and A.M. Aertsen, *Representation of cooperative firing activity among simultaneously recorded neurons.* Journal of Neurophysiology, 1985. **54**(6): pp. 1513–1528.
83. Marmarelis, P.Z., and V.Z. Marmarelis, Analysis of Physiological Systems: The White Noise Approach. 1978, New York, Plenum Press.
84. Song, D., V.Z. Marmarelis, and T.W. Berger. *Parametric and non-parametric models of short-term plasticity*, in Second Joint EMBS/BMES Conference. 2002. Houston, TX.
85. Marmarelis, V.Z., *Identification of nonlinear biological systems using Laguerre expansions of kernels.* Annals of Biomedical Engineering, 1993. **21**: pp. 573–589.
86. Chichilnisky, E.J., *A simple white noise analysis of neuronal light responses.* Network: Computation in Neural Systems, 2001. **12**: pp. 199–213.
87. Eden, U.T., et al., *Dynamic Analysis of Neural Encoding by Point Process Adaptive Filtering.* Neural Computation, 2004. **16**: pp. 971–998.
88. Dayan, P., and L. Abbott, Theoretical Neuroscience: Computational and Mathematical Modeling of Neural Systems. 2001, Cambridge, MA: MIT Press.
89. Aertsen, A., et al., *Dynamics of neuronal firing correlation: Modulation of "effective connectivity".* Journal of Neurophysiology, 1989. **61**(5): pp. 900–917.
90. Baccala, L.A., and K. Sameshima, *Partial directed coherence: A new concept in neural structure determination.* Biological Cybernetics, 2001. **84**(6): pp. 463–474.
91. Baccala, L.A., and K. Sameshima, *Overcoming the limitations of correlation analysis for many simultaneously processed neural structures.* Brain Research, 2001. **130**: pp. 33–47.
92. Sameshima, K., and L.A. Baccala, *Using partial directed coherence to describe neuronal ensemble interactions.* Journal of Neuroscience Methods, 1999. **94**: pp. 93–103.

93. Baccala, L.A., and K. Sameshima, *Directed coherence: A tool for exploring functional interactions among brain structures*, in Methods for Neural Ensemble Recordings, M. Nicolelis, ed. 1999, Boca Raton, FL: CRC Press. pp. 179–192.
94. Hurtado, J.M., L.L. Rubchinsky, and K.A. Sigvardt, *Statistical method for detection of phase-locking episodes in neural oscillations.* Journal of Neurophysiology, 2004. **91**(4): pp. 1883–1898.
95. Samonds, J.M., and A.B. Bonds, *Real-time visualization of neural synchrony for identifying coordinated cell assemblies.* Journal of Neuroscience Methods, 2004. **139**(1): pp. 51–60.
96. Gerstein, G.L., and A. Aertsen, *Representation of cooperative firing activity among simultaneously recorded neurons.* Journal of Neurophysiology, 1985. **54**(6): pp. 1513–1528.
97. Haykin, S., Adaptive filter theory. 3rd ed. 1996, Upper Saddle River, NJ: Prentice-Hall International.

• • • •

CHAPTER 3

Input–Output BMI Models

Additional Contributors: Sung-Phil Kim, Yadunandana Rao, and Deniz Erdogmus

The BMI experimental paradigm lends itself to statistical signal processing methodologies used to derive optimal models from data. Indeed, in the BMI setting, the researcher has available synchronously both the input to the BMI (neuronal activity - spike trains) as well as the desired response (kinematics of movement). The problem can be then framed in terms of "decoding," by which spike occurrences of individual neurons are translated into intended movement. In terms of modeling, the aim is to find a functional relationship between neural activity and the kinematic variables of position, velocity, acceleration, and force. Here, we will describe how to translate the BMI decoding problem into a system identification framework, where a parametric linear or nonlinear system can be trained directly from the collected data to achieve outputs close to the kinematics as in Figure 3.1. Model building has been extensively studied in control theory and signal processing so there are a wealth of methods that can be utilized [1].

Because the data are collected by multielectrode arrays, and each electrode can potentially sense several neurons, spike sorting is commonly utilized to identify individual neurons. Provided that the spike features have been accurately extracted, the neuronal spike firings become the decoding model input. Because the desired responses are continuous amplitude signals with unknown structure, we will treat them as random processes. As for the neural data, this chapter will deal with spike rate coding in short windows of 100 msec (called binned data), and a stochastic model for the data will also be assumed. This problem can therefore be framed as system identification with stochastic processes, which is a very well developed area in optimal signal processing and controls [2]. However, its application to BMI is not without problems for the following reasons:

1. There is a huge difference in the time scale of spike trains (msec) and that of behavior (sec) that the model has to bridge. Binning the spikes firing of each neuron over an appropriate time window (50 ~ 100 msec) has been widely used to smooth the spike trains and provide a time scale closer to behavior. It also implements a simple method for time to amplitude conversion of the neural modulation.

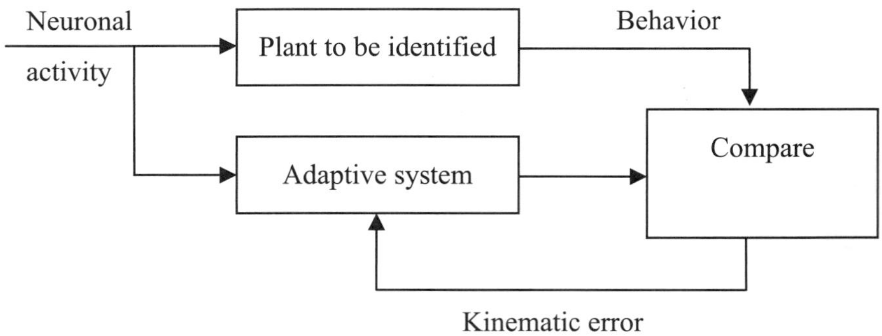

FIGURE 3.1: System identification framework for BMIs.

2. The system is MIMO, with a very large input (hundred of neurons), and the data are limited; therefore, there is a need for proper regularization of the solution.
3. Although the input dimensionality is high, it still coarsely samples in unknown ways the large pool of neurons involved in the motor task (missing data).
4. Not all the channels are relevant for the task, and because of the spike sorting and other difficulties, the data are noisy.
5. Because of the way that neurons are involved in behavior, the data are nonstationary.

With these remarks in mind, one has to choose the class of functions and model topologies that best match the data while being sufficiently powerful to create a mapping from neuronal activity to a variety of behaviors. There are several candidate models available, and based on the amount of neurophysiological information that is utilized about the system, an appropriate modeling approach shall be chosen. Three types of I/O models based on the amount of prior knowledge exist in the literature [3]:

- "White box": The physical system principles of operation are perfectly known, except for the parameters that are learned from the data.
- "Gray box": Some physical insight is available but model is not totally specified, and other parameters need to be determined from the data.
- "Black box": Little or no physical insight is available, so the chosen model is picked on other grounds (robustness, easy implementation, etc).

The choice of white, gray, or black box is dependent upon our ability to access and measure signals at various levels of the motor system, the type of questions one is interested to answer, as well as the computational cost of implementing the model in current computing hardware.

The first modeling approach, the "white box," would require the highest level of physiological detail. Starting with behavior and tracing back, the system comprises muscles, peripheral nerves, the spinal cord, and ultimately the brain. This would be a prohibitively difficult task with the present level of knowledge because of the complexity, interconnectivity, and dimensionality of the involved neural structures. Model implementation would require the parameterization of a complete motor system [4] that includes the cortex, cerebellum, basal ganglia, thalamus, corticospinal tracts, and motor units. Because all of the details of each component/subcomponent of the described motor system remain largely unknown and are the subject of study by many neurophysiological research groups around the world, it is not presently feasible to implement white-box modeling for BMIs. Even if it was possible to parameterize the system to some high level of detail, the task of implementing the system in our state-of-the-art computers and DSPs would be an extremely demanding task. The appeal of this approach when feasible is that a physical model of the overall system could be built, and many physiological questions could be formulated and answered.

The "gray box" model requires a reduced level of physical insight. In the "gray box" approach, one could take a particularly important feature of the motor nervous system or of one of its components, incorporate this knowledge into the model, and then use data to determine the rest of the unknown parameters. One of the most common examples in the motor BMI literature is Georgopoulos' population vector algorithm (PVA) [5]. Using observations that cortical neuronal firing rates were dependent on the direction of arm movement, a model was formulated to incorporate the weighted sum of the neuronal firing rates. The weights of the model are then determined from the neural and behavioral recordings.

The last model under consideration is the "black box." In this case, it is assumed that no physical insight is available to accomplish the modeling task. This is perhaps the oldest type of modeling where only assumptions about the data are made, and the questions to be answered are limited to external performance (i.e., how well the system is able to predict future samples, etc.). Foundations of this type of time series modeling were laid by Wiener for applications in antiaircraft gun control during World War II [22]. Although military gun control applications may not seem to have a natural connection to BMIs, Wiener and many other researchers provided the tools for building models that correlate unspecified time series inputs (in our case, neuronal firing rates) and desired targets (hand/arm movements). We will treat black box modeling in this chapter.

The three input–output modeling abstractions have gained a large support from the scientific community and are also a well-established methodology in control theory for system identification and time series analysis [2]. Here, we will concentrate on the last two types that have been applied by engineers for many years to a wide variety of applications and have proven that the methods produce viable phenomenological descriptions when properly applied [23, 24]. One of the advantages of the techniques is that they quickly find, with relatively simple algorithms, optimal mappings (in the sense of minimum error power) between different time series using a nonparametric approach (i.e., without requiring a specific model and only mild assumptions for the time series generation). These advantages have to be counter weighted by the abstract (nonstructural) level of the modeling and the many difficulties of the method, such as determining what is a reasonable fit, a model order, and a topology to appropriately represent the relationships among the input and desired response time series.

3.1 MULTIVARIATE LINEAR MODELS

The first black box model that we will discuss is the linear model, which has been the workhorse for time series analysis, system identification, and controls [2]. Perhaps, the first account of a filter-based BMI [25] used a linear model trained by least squares with one second (i.e., the current and NINE past bins) of previous spike data per neuron as a memory buffer to predict a lever press in a rat model. It turns out that least squares with past inputs can be shown equivalent to the Wiener solution for a finite impulse response filter. Therefore, the first BMI was trained with a Wiener filter. The bulk of the results reported in the literature use this very simple but powerful linear solution [8, 10, 26, 27]. Since for our BMI application the filter takes a slightly different form because of the multidimensional input and output, it will be presented more carefully than the others. Let us assume that an M-dimensional multiple time series is generated by a stationary stable vector autoregressive (VAR) model

$$x(n)=\boldsymbol{b}+\mathbf{W}_1 x(n-1)+\ldots+\mathbf{W}_L x(n-L)+u(n) \tag{3.1}$$

where $\boldsymbol{b}$ is a vector of intercept terms $\boldsymbol{b} = [b_1,\ldots, b_M]^T$ and the matrices $\mathbf{W}_i$ are coefficient matrices of size $M \times M$ and $u(n)$ is white noise with nonsingular covariance matrix $\mathbf{S}$. We further assume that we observe the time series during a period of T samples. The goal of the modeling is to determine the model coefficients from the data. Multivariate least square estimation can be applied to this problem without any further assumption. Let us develop a vector notation to solve the optimization problem.

$$\begin{aligned}
&X=[x_1,\ldots x_T] \quad MxT\\
&A=[b,W_1,\ldots,W_L] \quad Mx(ML+1)\\
&Z_n=[1,x_n,\ldots,x_{n-L+1}]^T \quad (ML+1)x1\\
&Z=[Z_0,\ldots,Z_{T-1}] \quad (ML+1)xT\\
&U=[u_1,\ldots,u_T] \quad MxT\\
&\chi=\text{vec}[X] \quad MTx1\\
&\alpha=\text{vec}[A] \quad (M^2L+M)x1\\
&\nu=\text{vec}[U] \quad MTx1
\end{aligned} \tag{3.2}$$

where vec is a vector stacking operator that creates a vector from a matrix, one column at a time. Note that the ***Z*** vector is composed of the previous values of the input that are in the filter taps. Using this notation, the VAR (L) model can be written as

$$X = AZ + U \tag{3.3}$$

Thus, the multivariate least square estimation chooses the estimator that minimizes

$$J(\alpha) = v^{\mathrm{T}}\Sigma^{-1}v = tr\{(X - AZ)^{\mathrm{T}}\Sigma^{-1}(X - AZ)\} \tag{3.4}$$

To find the minimum of this function, we have to take the derivative with respect to the unknown parameters α

$$\frac{\partial J(\alpha)}{\partial \alpha} = 2(\mathbf{ZZ}^{\mathrm{T}} \otimes \Sigma^{-1})\alpha - 2(\mathbf{Z} \otimes \Sigma^{-1})\chi \tag{3.5}$$

where ⊗ is the Kronecker product of matrices [28] (which creates a matrix whose elements are given by the elements of the first matrix multiplied by the second matrix). The normal equations are obtained equating the derivative to zero, and the optimal solution becomes

$$\hat{\alpha} = ((ZZ^{\mathrm{T}})^{-1}Z \otimes \mathbf{I})\chi \tag{3.6}$$

where **I** is the identity matrix of size M. Equation (3.6), written back into matrix form, reads

$$A = X\mathbf{Z}^{\mathrm{T}}(\mathbf{ZZ}^{\mathrm{T}})^{-1}, \tag{3.7}$$

which has the same form as the conventional least square solution of the univariate time series case. Therefore, we can conclude that the multivariate LS estimation is equivalent to the univariate least square solution of each one of the M equations in (3.7), which means that the conventional adaptive filtering algorithms [2] that use least squares can be used by simply renaming the quantities involved. One can establish the asymptotic normality of the vector LS estimator provided the data matrix $\mathbf{ZZ}^{\mathrm{T}}$ is nonsingular, as well as the Gaussianity of the errors in the limit of infinite long data windows provided the noise is the standard vector white noise (i.e., mean zero, nonsingular covariance, independent, and with finite kurtosis). It is therefore possible to use the conventional t-statistics to check confidence intervals for model coefficients (the number of degrees of freedom can be estimated by T-ML-1). Furthermore, it can be shown that the vector LS estimator coincides with the maximum likelihood estimator for Gaussian distributed stochastic processes [29].

Unfortunately, the small sample behavior of the multivariate least square estimator is a lot harder to establish, and normally, Monte Carlo methods are used to establish empiric coefficient variances, and the decision about the adequacy of the fitting is left to the experimenter. In the practical situation of BMIs, more important issues require our attention. In fact, the spike train time series are not even stationary, and the mappings to the desired response are probably not linear. The approaches presented thus far assume that the spike train statistics do not change over time. This is an unrealistic assumption that can be counteracted using sample by sample adaptation [the famous least mean square (LMS) algorithm] to track the optimal solution through time as will be shown in 3.1.2 or by partitioning the solution space into a set of local linear models that switch automatically among themselves [30]. Moreover, the MIMO models have many parameters, and so generalization is a major concern. Therefore, one must proceed with great caution.

3.1.1 Linear Modeling for BMIs and the Wiener Filter

Because the VAR decomposes into a parallel set of independent linear models (one for each output), we will present here this more familiar notation. Moreover, because this solution coincides with the Wiener–Hopf solution applied to discrete time data and finite impulse responsive filters (FIRs), it will be called here the digital Wiener filter [2]. Consider a set of spike counts from M neurons, and a hand position vector $\mathbf{d}\in\Re^C$ (C is the output dimension, C = 2 or 3). The spike count of each neuron is embedded by an L-tap discrete time-delay line. Then, the input vector for a linear model at a given time instance n is composed as $\mathbf{x}(n) = [x_1(n), x_1(n-1) \ldots x_1(n-L+1), x_2(n) \ldots x_M(n-L+1)]^T$, $\mathbf{x}\in\Re^{L\cdot M}$, where $x_i(n-j)$ denotes the spike count of neuron i at a time instance $n-j$. A linear model estimating hand position at time instance n from the embedded spike counts can be described as

$$y^c = \sum_{i=0}^{L-1}\sum_{j=1}^{M} x_i(n-j)w_{ij}^c + b^c \tag{3.8}$$

where y^c is the c coordinate of the estimated hand position by the model, w_{ij}^c is a weight on the connection from $x_i(n-j)$ to y^c, and b^c is a bias for the c coordinate. The bias can be removed from the model when we normalize $\mathbf{x}$ and $\mathbf{d}$ such that $E[\mathbf{x}] = \mathbf{0}, \mathbf{0}\in\Re^{L\cdot M}$, and $E[\mathbf{d}] = \mathbf{0}, \mathbf{0}\in\Re^C$, where $E[\cdot]$ denotes the mean operator. Note that this model can be regarded as a combination of three separate linear models estimating each coordinate of hand position from identical inputs. In a matrix form, we can rewrite (3.8) as

$$\mathbf{y} = \mathbf{W}^T\mathbf{x} \tag{3.9}$$

where $\mathbf{y}$ is a C-dimensional output vector, and $\mathbf{W}$ is a weight matrix of dimension $(L\cdot M+1)\times C$. Each column of $\mathbf{W}$ consists of $[w_{10}^c, w_{11}^c, w_{12}^c \ldots, w_{1L-1}^c, w_{20}^c, w_{21}^c \ldots, w_{M0}^c, \ldots, w_{ML-1}^c]^T$.

Figure 3.2 shows the topology of the MIMO linear model for the BMI application, which will be kept basically unchanged for any linear model through this book. The most significant dif-

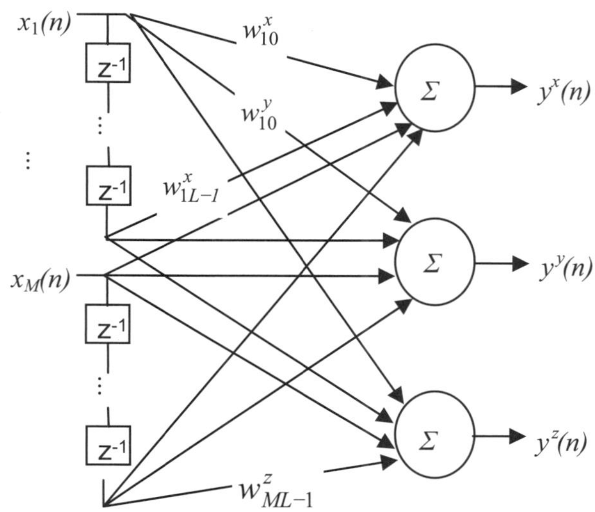

FIGURE 3.2: The topology of the linear filter designed for BMIs in the case of the 3D reaching task. $x_i(n)$ are the bin counts input from the ith neuron (total M neurons) at time instance n, and z^{-1} denotes a discrete time delay operator. $y^c(n)$ is the hand position in the c coordinate. w^c_{ij} is a weight on $x_i(n-j)$ for $y_c(n)$, and L is the number of taps.

ferences will be in the number of parameters, and in the way, the parameters w_{ij} of the model are computed from the data.

For the MIMO case, the weight matrix in the Wiener filter system is estimated by

$$\mathbf{W}_{\text{Wiener}} = \mathbf{R}^{-1}\mathbf{P} \tag{3.10}$$

R is the correlation matrix of neural spike inputs with the dimension of $(L{\cdot}M)\times(L{\cdot}M)$,

$$\mathbf{R} = \begin{bmatrix} \mathbf{r}_{11} & \mathbf{r}_{12} & \cdots & \mathbf{r}_{1M} \\ \mathbf{r}_{21} & \mathbf{r}_{22} & \cdots & \mathbf{r}_{2M} \\ \vdots & \vdots & \ddots & \vdots \\ \mathbf{r}_{M1} & \mathbf{r}_{M2} & \cdots & \mathbf{r}_{MM} \end{bmatrix}, \tag{3.11}$$

where r_{ij} is the $L\times L$ cross-correlation matrix between neurons i and j $(i \neq j)$, and r_{ii} is the $L\times L$ autocorrelation matrix of neuron i. **P** is the $(L{\cdot}M)\times C$ cross-correlation matrix between the neuronal bin count and hand position as

$$\mathbf{P} = \begin{bmatrix} \mathbf{p}_{11} & \cdots & \mathbf{p}_{1C} \\ \mathbf{p}_{21} & \cdots & \mathbf{p}_{2C} \\ \vdots & \ddots & \vdots \\ \mathbf{p}_{M1} & \cdots & \mathbf{p}_{MC} \end{bmatrix}, \tag{3.12}$$

where $\mathbf{p}_{ic}$ is the cross-correlation vector between neuron i and the c coordinate of hand position. The estimated weights $\mathbf{W}_{\text{Wiener}}$ are optimal based on the assumption that the error is drawn from white Gaussian distribution, and the data are stationary. The predictor $\mathbf{W}^{\mathrm{T}}_{\text{Wiener}}\mathbf{x}$ minimizes the mean square error (MSE) cost function,

$$J = E[\|\mathbf{e}\|^2], \mathbf{e} = \mathbf{d} - \mathbf{y}, \tag{3.13}$$

Each sub-block matrix r_{ij} can be further decomposed as

$$\mathbf{r}_{ij} = \begin{bmatrix} r_{ij}(0) & r_{ij}(1) & \cdots & r_{ij}(L-1) \\ r_{ij}(-1) & r_{ij}(0) & \cdots & r_{ij}(L-2) \\ \vdots & & \ddots & \vdots \\ r_{ij}(1-L) & r(2-L) & \cdots & r(0) \end{bmatrix}, \tag{3.14}$$

where $r_{ij}(\tau)$ represents the correlation between neurons i and j with time lag τ. These correlations, which are the second-order moments of discrete time random processes $x_i(m)$ and $x_j(k)$, are the functions of the time difference $(m - k)$ based on the assumption of wide sense stationarity (m and k denote discrete time instances for each process). Assuming that the random process $x_i(k)$ is ergodic for all i, we can utilize the time average operator to estimate the correlation function. In this case, the estimate of correlation between two neurons, $r_{ij}(m - k)$, can be obtained by

$$r_{ij}(m-k) = E[x_i(m)x_j(k)] \approx \frac{1}{N-1}\sum_{n=1}^{N} x_i(n-m)x_j(n-k), \tag{3.15}$$

The cross-correlation vector $\mathbf{p}_{ic}$ can be decomposed and estimated in the same way.

$r_{ij}(m - k)$ is estimated using (3.15) from the neuronal bin count data with $x_i(m)$ and $x_j(k)$ being the bin count of neurons i and j, respectively. From (3.15), it can be seen that $r_{ij}(m - k)$ is equal to $r_{ji}(k - m)$. Because these two correlation estimates are positioned at the opposite side of the diagonal entries of **R**, the equality leads to a symmetric **R**. The symmetric matrix **R**, then, can be inverted effectively by using the Cholesky factorization [28]. This factorization reduces the computational complexity for the inverse of **R** from $O(N^3)$ using Gaussian elimination to $O(N^2)$ where N is the number of parameters.

In the application of this technique for BMIs, we must consider the nonstationary characteristics of the input, which can be investigated through observation of the temporal change of the

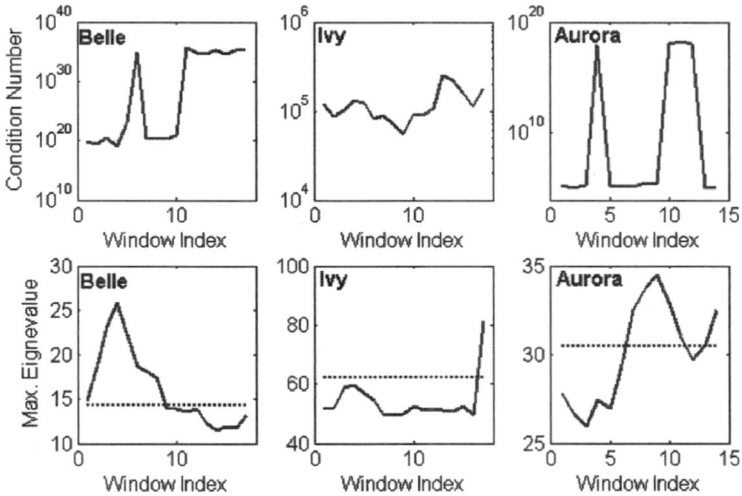

FIGURE 3.3: Illustrations of condition number and nonstationary properties of the input autocorrelation matrix. The dotted lines in the bottom panel indicate the reference maximum eigenvalue that is computed over all the data samples.

input autocorrelation matrix shown in Figure 3.3. The autocorrelation matrix of the multidimensional input data (here 104 neurons) is estimated based on the assumption of ergodicity.[1] In order to monitor the temporal change, the autocorrelation matrix is estimated for a sliding time window (4000-sample length), which slides by 1000 samples (100 second). For each estimated autocorrelation matrix, the condition number and the maximum eigenvalue are computed as approximations of the properties of the matrix. The experimental results of these quantities for three BMI data sets (three different animals) are presented in Figure 3.3. As we can see, the temporal variance of the input autocorrelation matrix is quiet large. Moreover, the inverse of the autocorrelation is poorly conditioned. Notice that in the formulation presented here, **R** must be a nonsingular matrix to obtain the optimal solution from (3.10). However, if the condition number of **R** is very large, the optimal weight matrix $\mathbf{W}_{\text{Wiener}}$ may be inadequately determined due to round off errors in the arithmetic. This usually happens when the number of samples is too small or the input variables are linearly dependent on each other (as may occur with neural recordings in a small volume of tissue). In such a case, we can reduce the condition number by adding an identity matrix multiplied by some constant to **R** before inversion. This procedure is called ridge regression in statistics [31], and the solution obtained

[1]A property of some stationary random processes in which the statistical moments obtained by statistical operators equal the moments obtained with time operators.

by this procedure turns out to minimize a cost function that linearly combines the one in (3.13) and a regularization term. This will be addressed in more detail later under regularized solutions.

Provided that the weights of the Wiener filter are adequately trained, one can begin to interpret their meaning as shown in the Hinton diagram in Figure 3.4 for a BMI model trained for a hand-reaching task. Each column of $\mathbf{W}_{\text{Wiener}}$ (i.e., the weight vector of the Wiener filter for each coordinate) is rearranged in a matrix form to show spatiotemporal structure of weight vectors. In this matrix form, the neuron indices are aligned in the x axis and the time lags are in the y axis. Note that the first row of the matrix corresponds to the zero lag (for the instantaneous neuronal bin counts), followed by the successive rows corresponding to the increasing lags (up to nine). In the Hinton diagram, white pixels denote positive signs, whereas black ones do negative signs. Also, the size of pixels indicates the magnitude of a weight. From the Hinton diagram, we can probe the contribution of individual neurons to the output of the Wiener filter. For this purpose, the weights represented in the Hinton diagrams are yielded from the input in which each neuronal bin count time series $x_j(n)$ is normalized to have unit variance. We can see the sign of the correlation between a particular neuronal input and the output. For instance, the weights for neurons indexed by 5, 7, 21, 23, and 71 exhibit relatively large positive values for reaching indicating that those neuronal activities are positively correlated with the output. On the other hand, the weights for neurons 26, 45, 74, and 85 exhibit large negative values indicating the negative correlation between neuronal inputs and the output. There are also some neurons for which the weights have positive and negative values

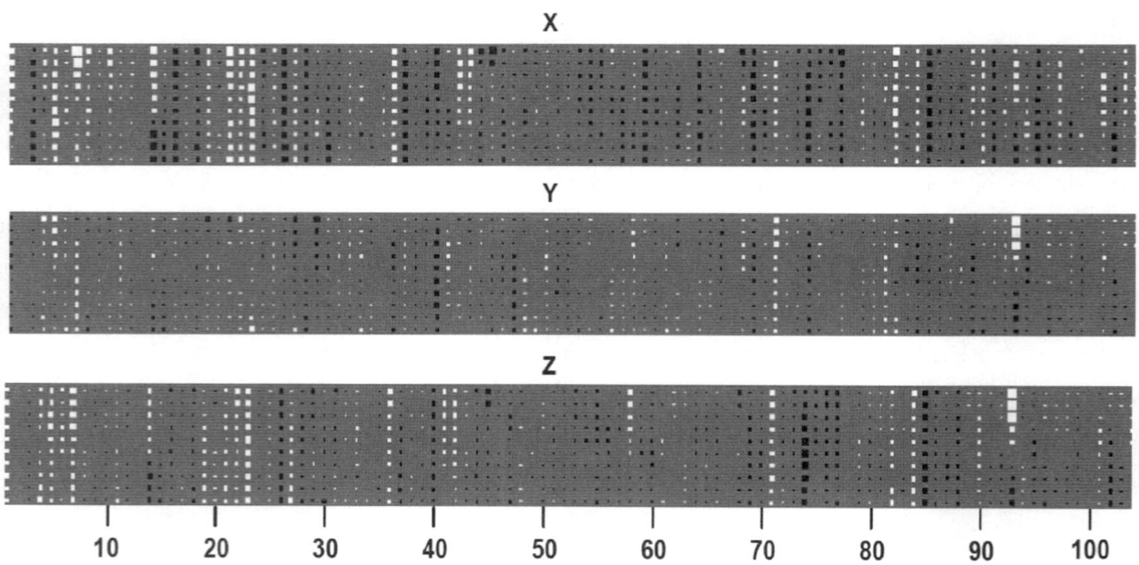

FIGURE 3.4: Hinton diagram of the weights of the Wiener filter for hand reaching.

(e.g., 14 and 93). It is possible from these diagrams to examine the significant time lags for each neuron in terms of the contribution to the filter output. For instance, in the case of neuron 7 or 93, the recent bin counts seem to be more correlated with the current output. However, for neuron 23 or 74, the delayed bin counts seem to be more correlated with the current output. The corresponding reconstruction trajectories for this particular weight matrix are presented in Figure 3.5. For the two reaches presented, the Wiener filter does a fair job at reconstructing the general shape of the trajectory but has a difficult time reaching the peaks and reproducing the details of the movement.

3.1.2 Iterative Algorithms for Least Squares: The Normalized LMS

As is well-known [32], there are iterative or adaptive algorithms that approximate the Wiener solution sample by sample. The most widely known family of algorithms is based on gradient descent, where the weights are corrected at each sample proportionally to the negative direction of the gradient.

$$w(n+1) = w(n) - \eta \nabla J(n) \tag{3.16}$$

where η is called the stepsize. The most famous is without a doubt the LMS algorithm, which approximates locally the gradient of the cost function yielding

$$w(n+1) = w(n) + \eta e(n) x(n) \tag{3.17}$$

One of the amazing things about this procedure is that the weight update is local to the weights and requires only two multiplications per weight. It is therefore very easy to program and cuts the complexity of the calculations to $O(N)$. It is therefore very appropriate for DSP hardware implementations.

The price paid by this simplifying alternative is the need to set an extra parameter, the stepsize that controls both the speed of adaptation and the misadjustement, which is defined as the normalized excess MSE (i.e., an added penalty to the minimum MSE obtained with the optimal LS solution). The stepsize is upper bounded by the inverse of the largest eigenvalue of the input autocorrelation matrix for convergence. Practically, the stepsize can be estimated as 10% the inverse of the trace of the input autocorrelation matrix [33]. The steepest descent algorithms with a single stepsize possess an intrinsic compromise between speed of adaptation and misadjustment captured by the relations:

$$M = \frac{\eta}{2} Tr(R) \tag{3.18}$$

$$\tau = \frac{1}{2\eta\lambda_{\min}} \tag{3.19}$$

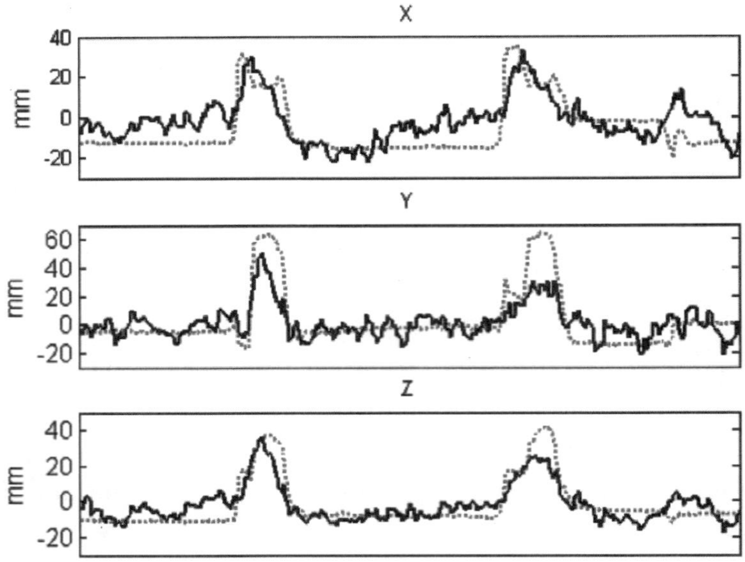

FIGURE 3.5: Wiener filter testing. The actual hand trajectory (dotted red line) and the estimated hand trajectory (solid black line) in the x, y, and z coordinates for a 3D hand-reaching task during a sample part of the test data set.

where τ is the settling time (four times the time constant), λ_{min} is the smallest of the eigenvalues of the input correlation matrix, and M is the misadjustment. Therefore, to speed up the convergence (i.e., less samples to reach the neighborhood of the optimal weight vector), the misadjustment increases proportionally (i.e., the algorithm rattles around the optimal weight value with a larger radius).

Practically, the LMS algorithm has other difficulties, in particular, when the input signal has large dynamical range and the stepsize is chosen for fast convergence. Indeed, in any adaptive algorithm, the weight vector contains all the information extracted from previous data. If the algorithm diverges for some reason, all the past data information is effectively lost. Therefore, avoiding divergence is critical. Since Because the theory for setting the stepsize is based on temporal or statistical averages, an algorithm with a stepsize that converges on average, can transiently diverge. This will happen in times of large errors (difficult to identify) or when the input signal has abrupt large values. To avoid this transient behavior, the normalize stepsize algorithm was developed [2].

The normalized stepsize algorithm is the solution of an optimization problem where the norm of the weight update is minimized subject to the constraint of zero error, that is,

$$J = \|w(n+1) - w(n)\|^2 + \lambda(d(n) - w(n+1)x(n)) \tag{3.20}$$

where λ is the Lagrange multiplier and $d(n)$ is the desired response. Solving this optimization problem with Lagrange multipliers, the weight update becomes

$$w(n+1) = w(n) + \frac{1}{\|x(n)\|^2} x(n)e(n) \tag{3.21}$$

Practically, in our BMI linear model, (3.21) becomes

$$\mathbf{w}^c_{\text{NLMS}}(n+1) = \mathbf{w}^c_{\text{NLMS}}(n) + \frac{\eta}{\gamma + \|\mathbf{x}(n)\|^2} e^c(n)\mathbf{x}(n) \tag{3.22}$$

where η satisfies $0<\eta<2$, and γ is a small positive constant. $e^c(n)$ is an error sample for the c coordinate, and $x(n)$ is an input vector. If we let $\eta(n) \equiv \eta/(\gamma+\|\mathbf{x}(n)\|^2)$, then the NLMS algorithm can be viewed as the LMS algorithm with a time-varying learning rate such that,

$$\mathbf{w}^c_{\text{NLMS}}(n+1) = \mathbf{w}^c_{\text{NLMS}}(n) + \eta(n)e^c(n)\mathbf{x}(n) \tag{3.23}$$

Although the weights in NLMS converge in the mean to the Wiener filter solution for stationary data, they may differ for nonstationary data. Therefore, this algorithm may, in fact, have better tracking performance when compared with the Wiener filter applied to segments of the data. Therefore, one should compare its performance with the Wiener filter for BMIs. For the same data set presented in Section 3.1.1, a 10-tap linear filter was trained with LMS (stepsize = 0.0001) using a data set of 10000 pts. To investigate the effect of an online update, the absolute value of the update for each weight was saved at each time t. As shown in Figure 3.6, histogram bins are placed in the range of 0 to 0.01, with centers placed every 0.001. Each coordinate direction for the 3D reaching task has 1040 weights associated with it. Each color in the histogram represents one of the 1040 weights. For example, dark blue represents weight 1, and it was updated 10 000 times, most of which had a value of zero. Any occurrences of an update greater than 0.01 is placed in the 0.01 bin. Only 15% of the total number of weight updates had a value other than zero. For this particular stepsize, this result indicates that one may be able to reduce the number of calculations because many of the weights are not contributing significantly to the mapping. The experiment was also repeated with the normalized LMS algorithm (time-varying stepsize), and the same trend can be observed.

Once the model was trained it was also tested for the trajectories presented in Section 3.1.1, as shown in Figure 3.6. For this BMI application, the performance between the Wiener filter and the filter trained using LMS were essentially the same. The performance was quantified by using the correlation coefficient (CC) between the model output and the true trajectory. Here, the Wiener filter had a CC of 0.76 ± 0.19) while the LMS generated a CC of 0.75 ± 0.20.

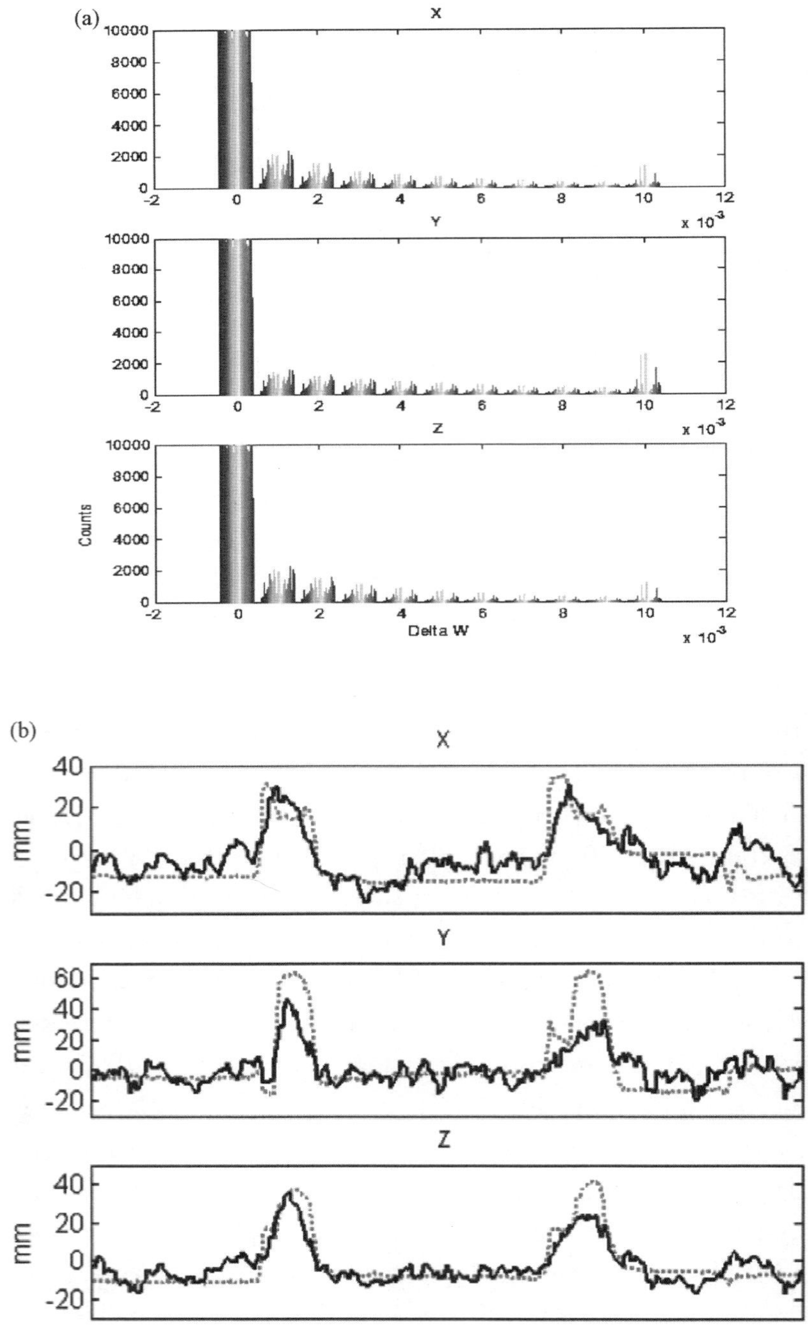

FIGURE 3.6: (a) Histogram of weight updates for the LMS algorithm. LMS training of a linear model. (b) The actual hand trajectory (dotted red line) and the estimated hand trajectory (solid black line) in the x, y, and z coordinate for a 3D hand-reaching task during a sample part of the test data set. (c) Learning curves for LMS training as a function of the stepsize.

(c)

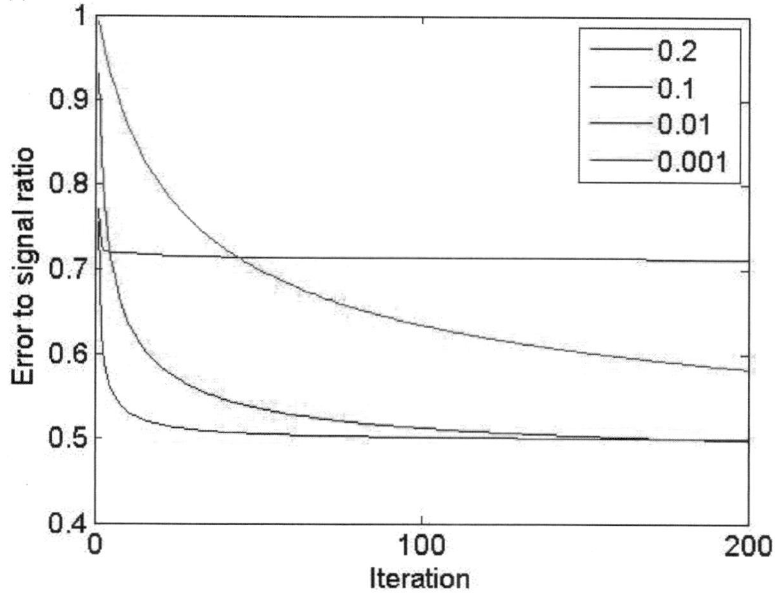

Figure continued

3.2 NONLINEAR MODELS

3.2.1 Time Delay Neural Networks

The primary assumption in the Wiener filter topology is that the input output map is linear, which may not be the case for the neural representation of movement control. To move beyond this assumption for biological systems, dynamic neural networks (a time delay neural network or recurrent neural networks [11, 12, 19, 25, 34, 35]) can be utilized to enhance the simple linear solution at the expense of more sophisticated topologies and training algorithms. In time series analysis, static nonlinear architectures such as the multilayer perceptron (MLP) or the radial basis function network cannot be utilized. Indeed, the hallmark of time series is that there are dependencies hidden in the time structure of the signal, and this holds true for the dynamics of neuronal modulation. Therefore, instantaneous mappers will not capture these dependencies over the lags. Spatiotemporal nonlinear mappings of neuronal firing patterns to hand position can be constructed using an input based time-delay neural networks (TDNNs) [23]. The input based TDNN is an MLP preceded by a delay line that feeds each processing element (PE) of the hidden MLP layer with a vector of present and past input samples. Normally, the nonlinearity is of the sigmoid type (either a logistic

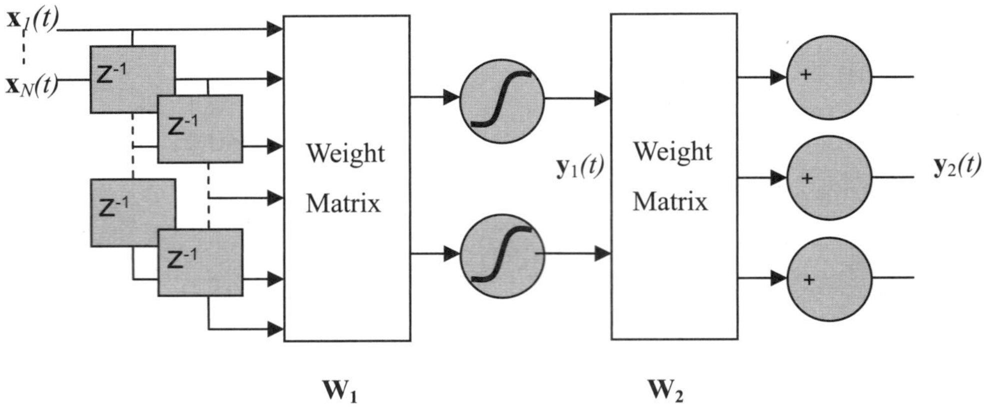

FIGURE 3.7: TDNN topology.

function or a hyperbolic tangent [23]). A full description of the TDNN can be found in References [23, 33]. One of the appeals of the input based TDNN is that it is a dynamic neural network but still utilizes the conventional backpropagation training of the MLP [33].

The TDNN topology (Figure 3.7) is more powerful than the linear FIR filter. In fact, looking at the hidden layer PE just before the nonlinearity, we can recognize an adaptive FIR filter. Therefore, the TDNN is a nonlinear combination of adaptive FIR filters, and has been shown to be a universal approximator in functional spaces [36], provided there is enough memory depth and sufficient number of hidden PEs. Alternatively, each of the hidden PEs outputs can be thought of as an adaptive basis obtained nonlinearly from the high dimensional input data, which defines the projection space for the desired response. Then, the best (orthogonal) projection of the desired hand movements can be obtained on this projection space. Notice that the TDNN can be considered as a "double" adaptive system, where both the axes of the projection space and the optimal projections are adapted. This conceptual picture is important to both help configure the TDNN topology and the training algorithmic parameters.

The output of the first hidden layer of the network can be described with the relation $\mathbf{y}_1(t) = f(\mathbf{W}_1\mathbf{x}(t))$ where $f(.)$ is the hyperbolic tangent nonlinearity ($\tanh(\beta\mathbf{x})$).[2] The input vector $\mathbf{x}$ includes L most recent spike counts from N input neurons. In this model the delayed versions of the firing counts, $\mathbf{x}(t - L)$, are the bases that construct the output of the hidden layer. The number of delays in the topology should be set that there is significant coupling between the neuronal input and desired signal. The output layer of the network produces the hand trajectory $\mathbf{y}_2(t)$ using a linear combination of the hidden states and is given by $\mathbf{y}_2(t) = \mathbf{W}_2\mathbf{y}_1(t)$. The weights ($\mathbf{W}_1$, $\mathbf{W}_2$) of this network can

[2]The logistic function is another common nonlinearity used in neural networks.

be trained using static backpropagation[3] with MSE as the learning criterion. In the TDNN static backpropagation algorithm, the output of the first hidden layer is given by (3.24).

$$y_i^l(n) = f(\mathrm{net}_i^l(n)) \tag{3.24}$$

where

$$\mathrm{net}_i^l(n) = \sum_{j=1}^{l} w_{ij}^l y_j^{l-1}(n) \tag{3.25}$$

Here, $y_0^l = x_j$ or the neuronal inputs.

The output layer delta rule is given in (3.26). Note the derivative of output is equal to one for a linear output as used here.

$$\delta_i^L(n) = e_i(n) f'(\mathrm{net}_i^L(n)) \tag{3.26}$$

$$\delta_i^l(n) = f'(\mathrm{net}_i^l(n)) \sum_k \delta_k^{l+1}(n) w_{ki}^{l+1}(n) \ \textit{(hidden layer delta rule)} \tag{3.27}$$

The overall weight update for the TDNN is given in (3.28)

$$w_{ij}(n+1) = w_{ij}(n) + \eta f'(\mathrm{net}_i(n)) (\sum_k e_k(n) f'(\mathrm{net}_k(n)) w_{ki}(n)) y_j(n) \tag{3.28}$$

Although the nonlinear nature of the TDNN may seem an attractive choice for BMIs, putting memory at the input of this topology presents a difficulty in training and model generalization because the high-dimensional input implies a huge number of extra parameters. For example, if a neural ensemble contains 100 neurons with 10 delays of memory, and the TDNN topology contains 5 hidden PEs, 5000 free parameters are introduced in the input layer alone. Large data sets and slow learning rates are required to avoid overfitting [24]. Untrained weights can also add variance to the testing performance thus decreasing accuracy.

As in any neural network topology, the issues of optimal network size, learning rates, stopping criterion need to be addressed for good performance. There are well-established procedures to control each one of these issues [24]. The size of the input layer delay line can be established from the knowledge of the signal structure. In BMIs, it is well established that one second of neural data preceding the motor event should be used [8]; however, the optimal embedding and the optimal delay should still be established using the tools from dynamic modeling. The rule of thumb to build the architecture is to always start simple (i.e., the smallest number of hidden PEs, which as we saw defines the size of the projection space). As far as training is concerned, one has to realize that both the bases and

[3] Backpropagation is a simple application of the chain rule that propagates the gradients through the topology.

the projection onto the projection space are being adapted at the same time. Another rule of thumb is to adapt the bases at a slower rate than the projection, which means that the learning rate of the first layer weights should be smaller than of the second layer weights (even after recognizing that the errors are attenuated by passing through the hidden PE nonlinearity) and to train the network with early stopping criterion (i.e., set aside 10% of the data for a cross-validation (CV) set, and stop training when the error in the CV increases. The error in the CV set should always be comparable to the error in the training set, otherwise, overtraining occurred and generalization will be compromised.

Here, we present the training of a TDNN for the BMI data described in Section 3.1.1. The networks implemented consisted of one hidden layer with a hyperbolic tangent nonlinearity ($\tanh(\beta x)$). Because the dynamic range of the input data was large (0–21), the slope of the nonlinearity had to be carefully chosen. A beta value of 0.5 provided a sufficiently steep slope, which with slow learning (over many epochs) nonlinearities would saturate to produce flat regions in the trajectory. The slope was also shallow enough to prevent instantaneous jumps in firing rate from saturating the nonlinearity and decreasing learning. The output layer of the network consisted of a linear PE because the desired signal was hand position in space. The network described was trained using static backpropagation. Weights were updated in online mode. Because the data have a large variability, the weights needed to be updated at each time instance; batch updates tend to average out the updates making learning difficult for this problem.

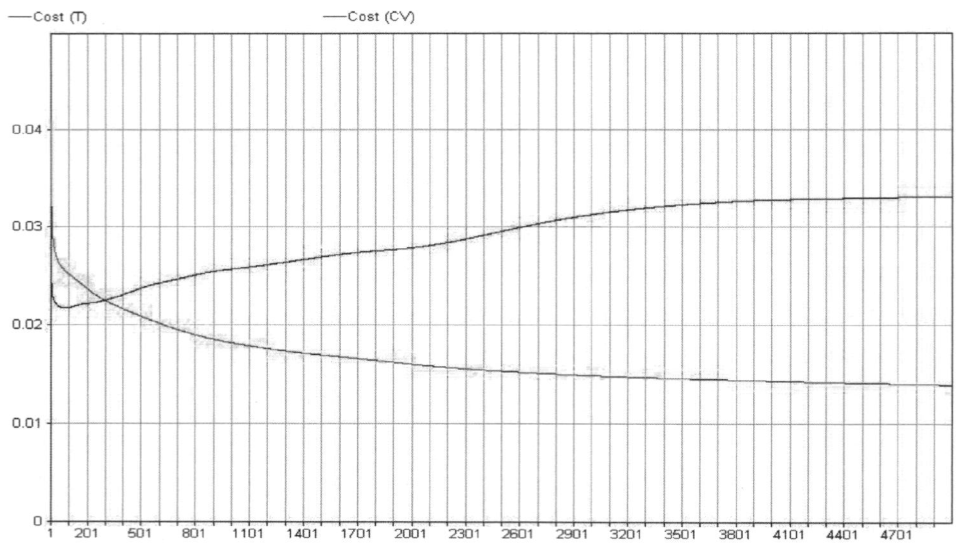

FIGURE 3.8: Learning curve for a BMI TDNN. Here, the y axis relates the MSE to the number of training epochs. Included in the figure is the CV curve. Theoretical analysis indicates that training should be stopped at the minima of the CV curve.

The size of the topology was chosen by experimentation: 3, 5, 15, and 30 PEs were chosen for the hidden layer. Because we seek a model that accurately maps all novel neuronal firing patterns to hand position, we must search for a network topology that minimizes the output bias and variance. The operator must choose learning rates that slowly adjust the weights so that maximum information is captured and the network does not simply track the trajectory. Overfitting the network should also be avoided. The training should be stopped at the point of maximum generalization, which is based on the minimum error produced by a validation set of data [33]. A plot of the CV over epochs is presented in Figure 3.8. For neuronal data, the early stopping method indicates that training should be stopped at approximately 50 epochs of data. This method did not produce good models. Research has verified CVs poor performance in some problems [24]. Other methods will be presented later to quantify maximum generalization for neuronal modeling.

The testing outputs for the TDNN BMI are presented in Figure 3.9. The TDNN performs better than the linear filter in peak accuracy (CC = 0.77 ± 0.17) but suffers in smoothness. The large number of free parameters in the TDNN is not only difficult to train but produce outputs with large variances. The biggest difficulty with BMI models of this type is the enormous set of free parameters produced by the large neural input vector (104 dimensions) and the multiplication by the number of hidden PEs. Training the first layer of the TDNN with backpropagation is a difficult task because of the attenuation of the error by the hidden layer PEs. Decreased input data dimensionality will limit the number of free parameters in the adaptive system architecture. Simulations express the need for preprocessing which prune neurons with the most information in the sparse data set. This problem calls for an adaptive system with the nonlinear power of a TDNN but with a

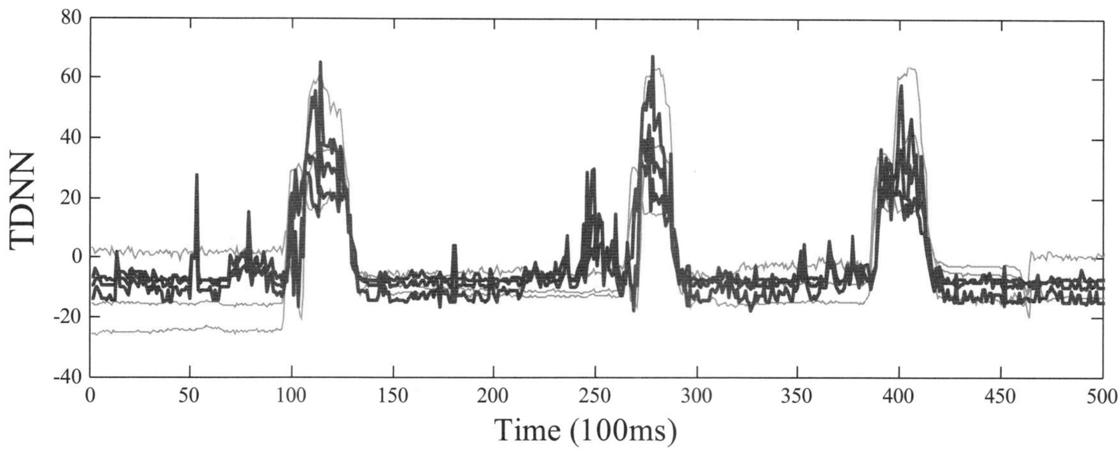

FIGURE 3.9: Testing performance for a TDNN for a reaching task. Here, the red curves are the desired *x*, *y*, and *z* coordinates of the hand trajectory, whereas the blue curves are the model outputs.

fewer number of free parameters at the input layer. Finally, a test for generalization is crucial to the successful production of useful models.

3.2.2 Recursive MLPs

Here, we continue evaluation of nonlinear dynamical models for BMIs by introducing the recursive MLP (RMLP). First, we would like to discuss in detail the list below indicating why the RMLP is an appropriate choice for BMI applications.

- RMLP topology has the desired biological plausibility needed for BMI design
- The use of nonlinearity and dynamics gives the RMLP a powerful approximating capability

Although the RMLP may first appear to be an "off-the-shelf" black box model, its dynamic hidden layer architecture can be compared with a mechanistic model of motor control proposed by Todorov [18, 37], revealing that it has the desired biological plausibility needed for BMI design. The formulation for the RMLP is similar to a general (possibly nonlinear) state space model implementation that corresponds to the representation interpretation of Todorov's model for neural control. The RMLP architecture in Figure 3.10 consists of an input layer with N neuronal input channels, a fully connected hidden layer of nonlinear PEs, (in this case tanh), and an output layer of linear PEs, one for each output. The RMLP has been successfully utilized for difficult control applications [38], although here the more traditional backpropagation-through-time (BPTT) training [39] was utilized.

Each hidden layer PE is connected to every other hidden PE using a unit time delay. In the input layer equation of (3.29), the state produced at the output of the first hidden layer is a nonlinear function of a weighted combination (including a bias) of the current input and the previous

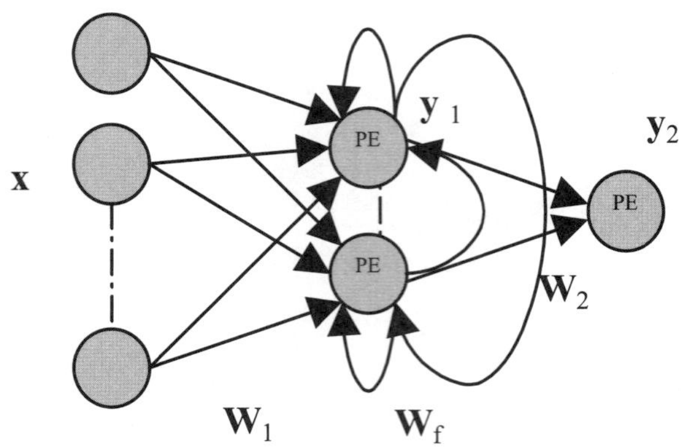

FIGURE 3.10: Fully connected, state recurrent neural network.

state. The feedback of the state allows for continuous representations on multiple timescales and effectively implements a short-term memory mechanism of the state values. Here, $f(.)$ is a sigmoid nonlinearity (in this case, tanh), and the weight matrices $\mathbf{W}_1$, $\mathbf{W}_2$, and $\mathbf{W}_f$, as well as the bias vectors $\mathbf{b}_1$ and $\mathbf{b}_2$ are again trained using synchronized neural activity and hand position data. Again, as in the TDNN, each of the hidden PEs outputs can be thought of as a nonlinear adaptive basis partially created from the input space and utilized to project the high-dimensional input data. These projections are then linearly combined to form the outputs of the RMLP that will predict the desired hand movements as shown in (3.30). It should be now apparent the elegance of feedback in decreasing the number of free parameters when compared with the input-based TDNN. Indeed, the state feedback in the RMLP allows for an input layer weights defined by the number of inputs while requiring only $K^2\mathbf{W}_f$ weights, where K is the number of hidden PEs for a total of $N + 2K^2$ total weights per output. One of the disadvantages of the RMLP when compared with the TDNN is that the RMLP cannot be trained with backpropagation and requires either BPTT or real-time recurrent learning [33]. Either of these algorithms is much more time-consuming than standard backpropagation, and on top of this, the dynamics of learning are harder to control with the stepsize, requiring extra care and more experience for good results.

$$\mathbf{y}_1(t) = f(\mathbf{W}_1\mathbf{x}(t) + \mathbf{W}_f\mathbf{y}_1(t-1) + \mathbf{b}_1), \tag{3.29}$$

$$\mathbf{y}_2(t) = \mathbf{W}_2\mathbf{y}_1(t) + \mathbf{b}_2, \tag{3.30}$$

The performance of RMLP input–output models in BMI experiments is dependent upon the choice of the network topology, learning rules, and initial conditions. Therefore, a heuristic exploration of these settings is necessary to evaluate the performance.

The first consideration in any neural network implementation is the choice of the number of PEs. Because the RMLP topology studied here always consisted of only a single hidden layer of tanh PEs, the design question becomes one of how many PEs are required to solve the neural to motor mapping. The first approach to this problem was a brute-force scan of the performance across a range of PEs, as shown in Table 3.1. It can be seen for a reaching task that the topology with five PEs produced the highest testing correlation coefficients. Although the brute-force approach can

TABLE 3.1: RMLP performance as a function of the number of hidden PEs

	1 PE	5 PE	10 PE	20 PE
Average CC testing	0.7099	0.7510	0.7316	0.6310

immediately direct the choice of topology, it does not explain why the number is appropriate for a given problem. We later came to find out that the number of hidden PEs should be chosen to span the space of the desired trajectory [34]. In the context of BMIs, this knowledge can help to avoid the computational and time costs of the brute-force approach.

The RMLP presented here was trained with BPTT [33, 39] using the NeuroSolutions software package [40]. Training was stopped using the method of CV (batch size of 1000 points) (Figure 3.11) to maximize the generalization of the network [41]. The BPTT training procedure involves unfolding the recurrent network into an equivalent feedforward topology over a fixed interval, or trajectory. For our BMI applications, we are interested in learning the dynamics of the trajectories of hand movement; therefore, for each task, the trajectory was chosen to match a complete movement. For the reaching task, this length was, on average, 30 samples. In Table 3.2, this selection of trajectory was compared again with a brute-force scanning of the testing performance as a function of the trajectory length (samples per exemplar). Included in the BPTT algorithm is the option to update the weights in online (every trajectory), semibatch, or batch mode. Choosing to update the network weights in semibatch mode can protect against noisy stochastic gradients that can cause the network to become unstable by averaging the gradients over several trajectories. Table 3.2 also provides the average testing correlation coefficient as a function of the frequency of updates. We can see that a range of good choices [15–30 samples/exemplar (s/e) and 5–15 exemplar/update (e/u)] exist for the trajectory length and update rule.

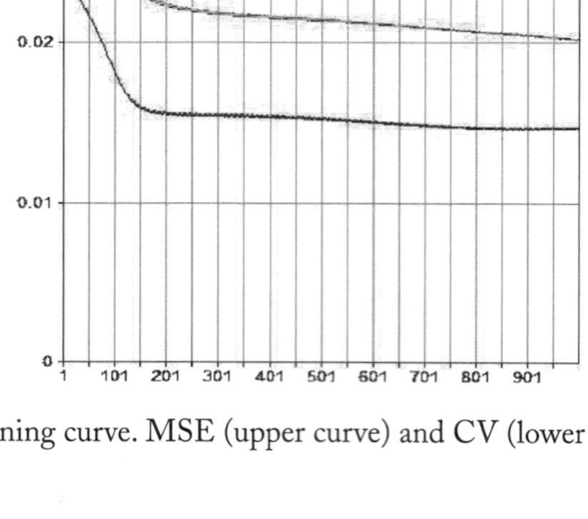

FIGURE 3.11: RMLP learning curve. MSE (upper curve) and CV (lower curve).

TABLE 3.2: Average testing correlation coefficients as a function of trajectory length

	5 S/E	15 S/E	30 S/E	60 S/E
1 e/u	0.6514	0.7387	0.7302	0.7135
5 e/u	0.6458	0.7006	0.7263	0.7091
15 e/u	0.6544	0.7389	0.6680	0.6855
30 e/u	0.6807	0.7482	0.7116	0.6781
60 e/u	0.6457	0.6383	0.6772	0.6624

To test the effect of the initial condition on model performance, 100 Monte Carlo simulations with different initial conditions were conducted with 20 010 consecutive bins (2001 sec) of neuronal data to improve the chances of obtaining the global optimum. In Figure 3.12, the training MSE is presented for all simulations, and it can be seen that all initial conditions reach the approximately the same solution. The greatest effect of the initial condition can be seen in the time it takes for each model to converge. In Figure 3.13, the average and standard deviations of the curves in Figure 3.12

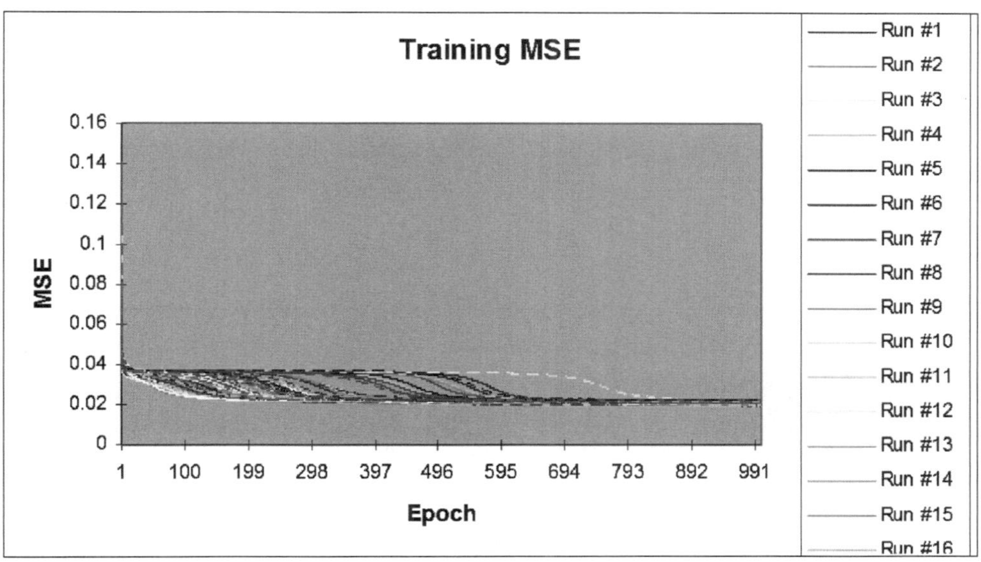

FIGURE 3.12: Training MSE curves for 100 Monte Carlo simulations.

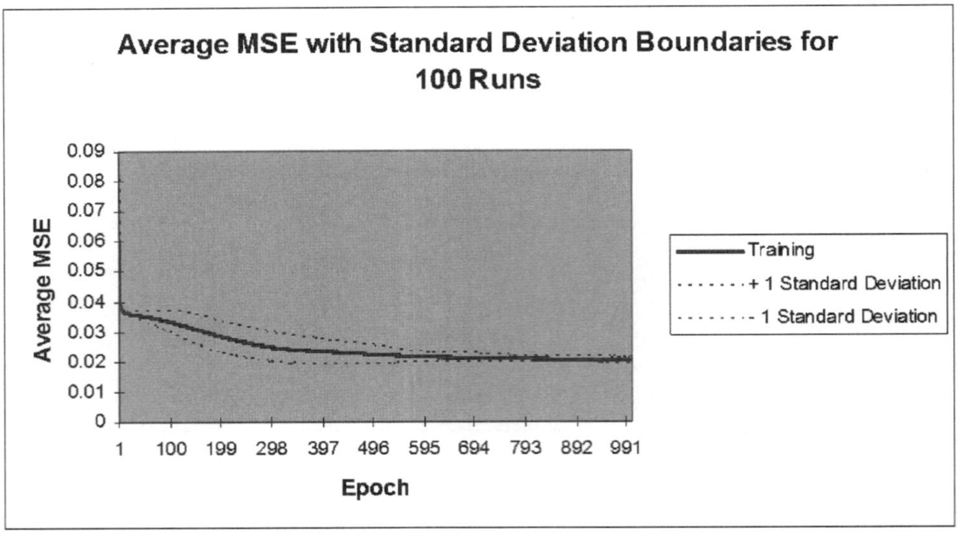

FIGURE 3.13: Standard deviation in training MSE for 100 Monte Carlo simulations.

are presented. This figure is characterized by a large standard deviation in the initial epochs resulting from the initial condition and a small standard deviation in the final epochs signifying that the models are converging to the same solution. These relationships are quantified in Table 3.3 where the average final MSE had a value of 0.0203 ± 0.0009. Again, a small training standard deviation indicates the networks repeatedly achieved the same level of performance. Of all the Monte Carlo simulations, the network with the smallest error achieved an MSE of 0.0186 (Table 3.4).

We can see in the testing outputs for the RMLP shown in Figure 3.14 that the performance of a parsimonious nonlinear dynamical network can improve the testing performance for trajectory reconstruction in a hand-reaching BMI task. The trajectories are clearly smoother and distinctly reach each of the three peaks in the movement (CC = 0.84 ± 0.15).

TABLE 3.3: Training performance for 100 Monte Carlo simulations

ALL RUNS	TRAINING MINIMUM	TRAINING STANDARD DEVIATION
Average of minimum MSEs	0.020328853	0.000923483
Average of final MSEs	0.020340851	0.000920456

TABLE 3.4: Best performing network

BEST NETWORK	TRAINING
Run	85
Epoch	999
Minimum MSE	0.018601749
Final MSE	0.018616276

3.2.2.1 Echo-State Networks. The improvement in performance of the RMLP is gained at the expense of a significant leap in the computational costs of training. The problem is accentuated by the fact that these models may have to be retrained occasionally to adjust to the changing statistics and environment. This can severely restrict practical, real-world application of BMIs. We show here a different architecture that has performance similar to that of an RMLP, but the training has linear complexity as opposed to the (BPTT) RMLP training algorithm [39].

Echo state networks (ESNs) were first proposed by Jaeger [42] and are appealing because of their computational abilities and simplified learning mechanisms. ESNs exhibit some similarities to the "liquid state machines" proposed by Maas et al. [43], which possess universal approximation

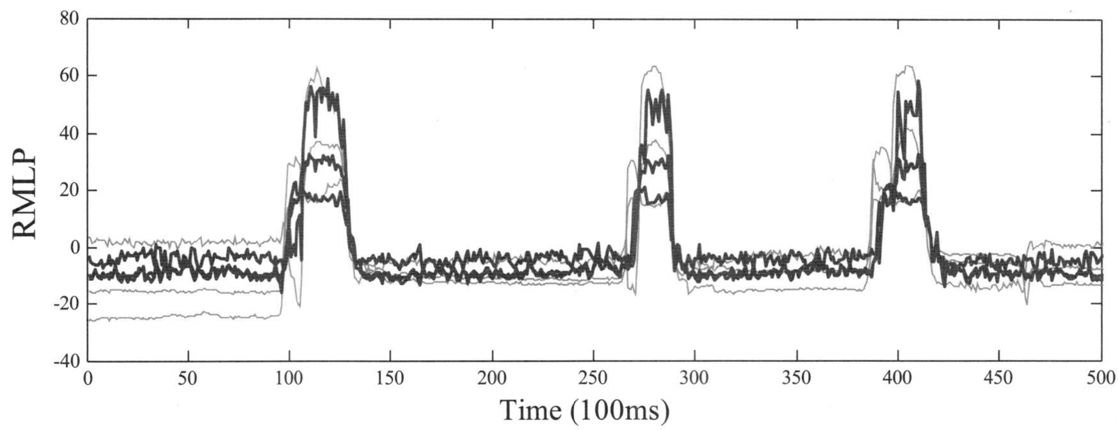

FIGURE 3.14: Testing performance for a RMLP for a reaching task. Here, the red curves are the desired x, y, and z coordinates of the hand trajectory, whereas the blue curves are the model outputs.

capabilities in myopic functional spaces. The fundamental idea of an ESN is to use a "large-reservoir" recurrent neural network that can produce diversified representations of an input signal, which can then be instantaneously combined in an optimal manner to approximate a desired response. Therefore, ESNs possess a representational recurrent infrastructure that brings the information from the past of the input into the present sample without adapting parameters. The short-term memory of the ESNs is designed a priori, and the neuronal data are processed through a reservoir of recurrent networks that are linked by a random, sparse matrix. Therefore, only the adaptive linear or nonlinear regressor (static mapper) is needed to implement functional mappings. Therefore, the training can be done online in O(N) time, unlike the Wiener or the RMLP that have O(N^2) training algorithms. Figure 3.15 shows the block diagram of an ESN. A set of input nodes denoted by the vector $\mathbf{u}_n \in \Re^{M\times 1}$ is connected to a "reservoir" of N discrete time-recurrent networks by a connection matrix $\mathbf{W}_{in} \in \Re^{M\times N}$. At any time instant n, the readout (state output) from the recurrent neural network (RNN) reservoir is a column vector denoted by $\mathbf{x}_n \in \Re^{N\times 1}$. Additionally, an ESN can have feedback connections from the output to the RNN reservoir. In Figure 3.15, we show two outputs (representing the X–Y Cartesian coordinates of hand trajectory) and the associated feedback connection matrix, $\mathbf{W}_b \in \Re^{N\times 2} = \left[\mathbf{w}_{b1}, \mathbf{w}_{b2}\right]$. The desired outputs form a 2D column vector $\mathbf{d}_n = [d_{xn}; d_{yn}]$. The reservoir states are transformed by a static linear mapper that can additionally receive contributions from the input $\mathbf{u}_n$. Each PE in the reservoir can be implemented as a leaky integrator (first-order gamma memory [44]) and the state output or the readout is given by the difference equation

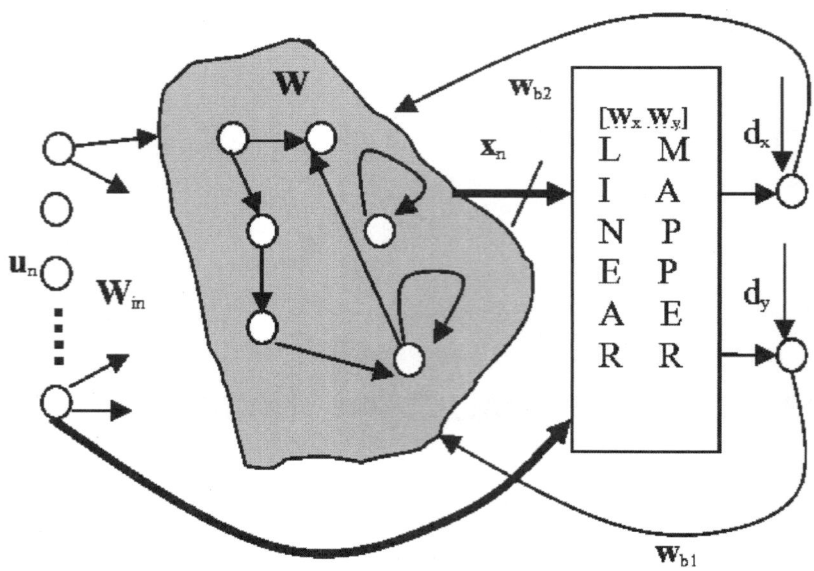

FIGURE 3.15: Block diagram of the echo state networks.

$\mathbf{x}_{n+1} = (1-\mu Ca)\mathbf{x}_n + \mu C(f(\mathbf{W}_{in}^T\mathbf{u}_{n+1} + \mathbf{W}\mathbf{x}_n + \mathbf{W}_b\mathbf{d}_n))$, where $0 < \mu < 1$ is the step-size used for converting a continuous-time leaky integrator into a discrete time difference equation, C is the time constant, and a is the decay rate [42]. The point-wise nonlinear function $f(.)$ is chosen to be the standard tanh sigmoid (i.e., $f(.) = \tanh(.)$). Note that, if μ, C, and a are all equal to unity, the RNNs default to the conventional nonlinear PE [33] without memory. From a signal processing point of view, the reservoir creates a set of bases functions to represent the input, whereas the static mapper finds the optimal projection in this space. There are obvious similarities of this architecture to kernel machines, except that the kernels here are time functions (Hilbert spaces).

We will now give the conditions under which an ESN can be "useful," which Jaeger aptly calls as the "echo state property." Loosely stated, the echo state property says that the current state is uniquely defined by the past values of the inputs and also the desired outputs if there is feedback. A weaker condition for the existence of echo states is to have the spectral radius of the matrix $\hat{\mathbf{W}} = \mu C\mathbf{W} + (1-\mu Ca)\mathbf{I}$ less than unity [42]. Another aspect critical for the success of ESN is to construct a sparse matrix **W**. This will ensure that the individual state outputs have different representations of the inputs and desired outputs, or in other words, the span of the representation space is sufficiently rich to construct the mapping to the desired response (hand trajectory).

3.2.2.2 Design of the ESN. One of the important parameters in the design of the ESN is the number of RNN units in the reservoir. In our experience with BMI simulations, Monte Carlo simulations varying the number of parameters and monitoring performance provided the best technique for selecting the dimensionality of the reservoir. Here, we chose $N = 800$, where increasing N further did not result in any significant improvement in performance. The input weight matrix $\mathbf{W}_{in}$ was fully connected with all the weights fixed to unity. The recurrent connection matrix **W** was sparse, with only 1% of the weights (randomly chosen) being nonzero. Moreover, we fix all the nonzero weights to a value 0.5. Further, each RNN is a gamma delay operator with parameters $[a, C, \mu] = [1, 0.7, 1]$. The next aspect is to set the spectral radius, which is crucial for this problem because it controls the dynamics and memory of the echo states. Higher values are required for slow output dynamics and vice versa [42, 45]. For the experiments in this section, we utilized a single ESN whose spectral radius was tuned to 0.79. Marginal changes (<1%) in performance (both X and Y) were observed when this parameter was altered by ±10%. We also turned off the connections from the output to the RNN reservoir and the direct connections between the inputs and the linear mapper. The network state is set to zero initially. The training inputs were forced through the network, and the states were updated. The first 400 echo state outputs were discarded as transients. The remaining 5000 state outputs were used to train the linear mapper.

Testing of the network on a 2D cursor control task demonstrated that the ESN can perform at the same level as the standard RMLP trained with BPTT. The CC values for the ESN x and y coordinates, respectively, were (0.64 and 0.78), whereas the RMLP produced CC values of (0.66 and 0.79). However, the case of the ESN was trained with far less complexity.

Lastly, we would like to mention the appeal of the ESN as a model for biologically plausible computation. If one thinks of the echo states as neuronal states, we can see how a distributed, recurrent topology is capable of storing information about past inputs into a diffuse set of states (neurons). For the most part, the interconnectivity and the value of the weights (synapses) are immaterial for representation. They become, however, critical for readout (approximation). In this respect, there are very close ties between ESN and liquid state machines, as already mentioned by Maas et al. [43]. These ideas may become useful in developing new distributed paradigms for plasticity and characterization of neuronal response in motor cortex.

3.2.3 Competitive Mixture of Local Linear Models

So far, we have investigated nonlinear decoding models that globally approximate an unknown nonlinear mapping between neural activities and behavior. However, a complex nonlinear modeling task can be elucidated by dividing it into simpler tasks and combining them properly [46]; this is an application of the well-known divide-and-conquer approach extensively used in science and engineering. We will briefly review here the statistical framework called *mixture of experts* [47] that implements this approach, and will also extend it to a more general model called *gated competitive experts* that includes other neural architectures that have been successfully used in time series segmentation [48] and optimal control [49]. We will here summarize the methodology that was originally presented in Reference [48].

Let us consider modeling a vector i.i.d.[4] process that may be multimodal, assuming that the mixing process has memory using a mixture model. To simplify the derivation, a 1D time series is assumed, but the results can easily be generalized to the multidimensional case. Recall that the multivariate PDF of a random process, with an effective memory depth of M, can be decomposed as

$$p(\chi(n) = \prod_{n=1}^{N} p(x(n) \mid \chi(n-1) \qquad (3.31)$$

where $\chi(n-1) = [x(n-1), \ldots x(n-M+1)]^T$. Let us entertain the possibility that the random process is produced by K switching regimes. Therefore, we propose the joint conditional density $p(k, x(n)|\chi(n-1))$, where the discrete variable $k = 1\ldots K$ indicates the regime. We cannot observe

[4]Independent identically distributed.

this PDF directly because we do not know which regime is active at which times, but we can observe its marginal distribution

$$p(x(n)|\chi(n-1)) = \sum_{k=1}^{K} p(k, x(n)|\chi(n-1)) =$$

$$\sum_{k=1}^{K} P(k|\chi(n-1)) \cdot p(x(n)|\chi(n-1), k) = \sum_{k=1}^{K} g_k(\chi(n-1)) \cdot p(x(n)|\chi(n-1), k) \quad (3.32)$$

where $g_k(\chi(n-1)) \equiv P(k|\chi(n-1))$ can be regarded as the a priori probability of the kth regime. If we can find predictors for each of the subprocesses, up to a random term ε_k,

$$x(n) = f(\chi(n-1);\theta_k) + \varepsilon_k(n) \qquad k = 1 \ldots K \quad (3.33)$$

then we can evaluate the conditional PDF in terms of the innovations

$$p(x(n)|\chi(n-1), k) = p_{\varepsilon_k}(\varepsilon_k(n)) \quad (3.34)$$

The residuals are usually modeled as a Gaussian distribution, which for a 1D time series becomes

$$p_{\varepsilon_k}(\varepsilon_k) = \frac{1}{\sqrt{2\pi\sigma_k^2}} \exp\left[\frac{(\tilde{x}_k(n) - x(n))^2}{2\sigma_k^2}\right] \quad (3.35)$$

where we have $\tilde{x}_k(n) \equiv f(\chi(n-1);\theta_k)$ defined to be kth predictor's estimate of the next value of the time series and σ_k, a "nuisance parameter," is the variance of the kth predictor. Taking the expected value of both sides of Eq 3.33 gives the best MMSE prediction of the next value of the time series

$$\tilde{x}(n) \equiv E[x(n) \mid \chi(n-1)] = \sum_{k=1}^{K} g_k(\chi(n-1)) \cdot E(x(n) \mid \chi(n-1), k] = \sum_{k=1}^{K} g_k(\chi(n-1))\tilde{x}_k(n) \quad (3.36)$$

which is a weighted sum of outputs of the individual predictors. We can regard this as the total system output. In the mixture-of-experts algorithm of Jordan and Jacobs [50], this mapping is provided by an adaptable function called the "gate." The particular variation of the algorithm we will examine is Weigend's [51] nonlinear gated experts, which implements the gate using a single MLP. This architecture is shown in Figure 3.16, and note that the individual predictors, called "experts" in this context, and the gates all see the same input. In other implementations, the gate can be trained from the outputs (called therefore output gating).

We now turn to the question of how to train the gate and experts simultaneously, which is the great appeal of this architecutre. In the following development, for ease of reading, we leave out

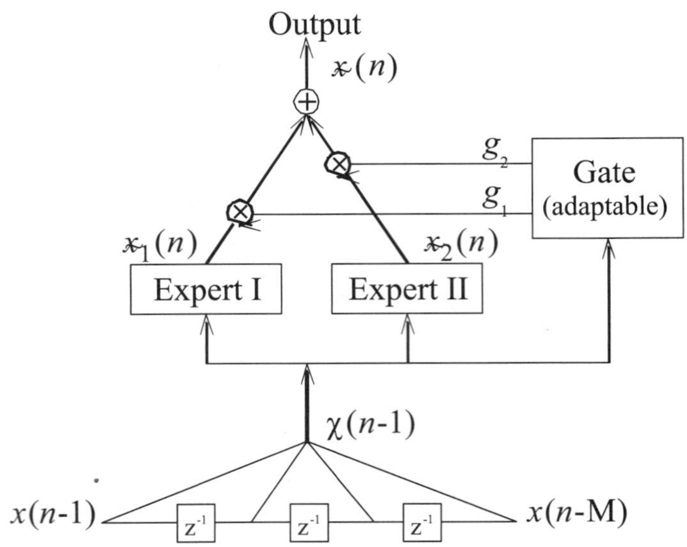

FIGURE 3.16: Mixture of experts during activation for a network with two experts.

explicit time dependence whenever possible. Also, there is an implicit iteration index in all the following equations. Given a time series of length N, we choose the free parameters of the predictors and gate that maximize the process log-likelihood. If the innovations are i.i.d., we can rewrite the process likelihood as

$$p(\chi(n)) = \prod_{n=1}^{N} \sum_{k=1}^{K} g_k(\chi(n-1)) \cdot p_{\varepsilon_k}(\tilde{x}_k(n) - x(n)) \tag{3.37}$$

This is difficult to maximize directly. Therefore, we propose a latent binary indicator, I_k, indicating which expert is valid, allowing the likelihood to be written.

$$L = \prod_{n=1}^{N} \prod_{k=1}^{K} [g_k(\chi(n-1)) \cdot p(x(n) \mid \chi(n-1), k)]^{I_k(n)} \tag{3.38}$$

The indicator variable is "hidden," in the sense that we do not know a priori which expert is valid at any time step. In the E step of the EM algorithm, for a given set of free parameters of the experts and gate, the entire data set is evaluated, holding the free parameters constant, to determine $p(x(n) \mid \chi(n-1), k)$ and $g_k(\chi(n-1))$ for all k and n. We then replace the indicator variables, I_k, at every time step, by their expected value:

$$h_k(n) \equiv E[I_k(n)] = P(k \mid x(n), \chi(n-1) =$$
$$\frac{P(k \mid \chi(n-1)) \cdot p(x(n) \mid \chi(n-1), k)}{p(x(n) \mid \chi(n-1))} = \frac{g_k(\chi(n-1)) \cdot p(x(n) \mid \chi(n-1), k)}{\sum_{k=1}^{K} g_k(\chi(n-1)) p(x(n) \mid \chi(n-1), k)} \tag{3.39}$$

Thus, h_k is the posterior probability of expert k, given both the current value of the time series and the recent past. For the M step, L is maximized or equivalently, the negative log-likelihood,

$$J = \sum_{n=1}^{N}\sum_{k=1}^{K}\left\{ -h_k(n) \cdot \log[g_k(\chi(n-1))] + \frac{1}{2} h_k(n)[\frac{(\tilde{x}_k(n) - x(n))^2}{\sigma_k^2} + \log[\sigma_k^2 \right\} \tag{3.40}$$

is globally minimized over the free parameters. The process is then repeated. If, in the M step, J is only decreased and not minimized, then the process is called the generalized EM algorithm. This is necessary when either the experts or gate is nonlinear, and a search for the global minimum is impractical.

The first term in the summation of (3.40) can be regarded as the cross-entropy between the posterior probabilities and the gate. It has a minimum when only one expert is valid and thus encourages the experts to divide up the input space. To ensure that the outputs of the gate sum to unity, the output layer of the MLP has a "softmax" transfer function,

$$g_k(\chi) = \frac{\exp[s_k(\chi)]}{\sum_{k=1}^{K} \exp[s_j(\chi)]} \tag{3.41}$$

where s_k is the kth input to the softmax. For a gate implemented as an MLP, the cross entropy term in (3.40) cannot be minimized in a single step, and the generalized EM algorithm must be employed. If the gate is trained through gradient descent (backpropagation), the error backpropagated to the input side of the softmax, at each time step is

$$\frac{\partial J}{\partial s_k} = g_k(\chi) - h_k \tag{3.42}$$

This is the same backpropagated error that would result for a MSE with the posterior probabilities acting as the desired signal. Thus, the posterior probabilities act as targets for the gate. For each EM iteration, several training iterations may be required for the gate because it is implemented using a multilayer perceptron.

There is an analytical solution for the experts, at each iteration, when they are linear predictors, $\tilde{x}_k(n) = w_k^T \chi(n-1)$ and $w_k = \mathbf{R}_k^{-1}$ and $\mathbf{p}_k$, where $\mathbf{R}_k$ and $\mathbf{p}_k$ are weighted autocorrelation and cross-correlation matrices, respectively,

$$R_k = \frac{1}{\sum_{n=1}^{N} h_k(n)} \sum_{n=1}^{N} h_k(n)\chi(n-1)\chi^T(n-1) \tag{3.43}$$

$$P_k = \frac{1}{\sum_{n=1}^{N} h_k(n)} \sum_{n=1}^{N} h_k(n)\chi(n-1)x(n) \tag{3.44}$$

Note that the solution represents a weighted Wiener filter, where the data are weighted by the posterior probabilities. If the experts are nonlinear, then the error used to iteratively train each expert is simply interpreted as being weighted by the posterior probabilities, $h_k(n)$. However, Zeevi et al. [52] have shown that the nonlinearity of the gate allows the mixture of experts to be a universal approximator, even if the expert predictors are linear.

No matter whether the experts are linear or nonlinear, there is an exact solution for the Gaussian covariance parameter s $_k$ at the end of each epoch

$$\sigma_k^2 = \frac{1}{\sum_{n=1}^{N} h_k(n)} \sum_{n=1}^{N} h_k(n)[\tilde{x}(n) - x(n)]^2 \tag{3.45}$$

3.2.3.1 Gated Competitive Experts. A gated competitive expert model is characterized by several adaptable "experts" that compete to explain the same data. That is, they all see the same input, and all attempt to produce the same output. Their performance is both monitored and mediated by a gate, whose goal is to determine the relative validity of the experts for the current data and then to appropriately moderate their learning. The history of the gate's decisions can be used to segment the time series.

In the activation phase, the total system output is a simple weighted sum of the experts' outputs

$$y(n) = \sum_{k=1}^{K} g_k(n) \cdot y_k(\chi(n)) \tag{3.46}$$

where $\boldsymbol{y}_k$ is the output of the kth expert with χ as input and g_k is the kth output of the gate. The input, $\chi(n)$ is a vector of delayed versions of the time series, in essence the output of a tapped delay line, repeated here for convenience

$$\chi(n) = \left[x(n)\ x(n-1) \ldots x(n-M+1)\right]^T \tag{3.47}$$

For the mixture to be meaningful, the gate outputs are constrained to sum to one.

$$\sum_{k=1}^{K} g_k(n) = 1 \tag{3.48}$$

Such a linear mixture can represent either a competitive or cooperative system, depending on how the experts are penalized for errors, as determined by the cost function. In fact, it was in the context of introducing their mixture of experts model that Jacobs et al.[53] first presented a cost function that encourages competition among gated expert networks, which we generalize to

$$J(n) = \sum_{k=1}^{N} g_k(n) \cdot f(d(n) - y_k(\chi(n))) \tag{3.49}$$

where $\boldsymbol{d}$ is the desired signal and $f(\boldsymbol{d}(n) - y_k(\chi(n)))$ is a function of the error between the desired signal and the kth expert. Because the desired signal is the same for all experts, they all try to regress the same data and are always in competition. This alone, however, is not enough to foster specialization. The gate uses information from the performance of the experts to produce the mixing coefficients. There are many variations of algorithms that fall within this framework. Let us discuss the important components one at a time, starting with the design of the desired signal.

The formalism represented by (3.49) is a supervised algorithm, in that it requires a desired signal. However, we are interested in a completely unsupervised algorithm. A supervised algorithm becomes unsupervised when the desired signal is a fixed transformation of the input: $\boldsymbol{d} \Rightarrow \boldsymbol{d}(\chi)$. Although many transformations are possible, the two most common transformations involve the delay operator and the identity matrix, resulting in prediction as explained above and auto-association, which yields a generative model for PCA [54].

Gates can be classified into two broad categories, which we designate as input or output based. With input-based gating, the gate is an adaptable function of the input, $g_k \Rightarrow g_k(\chi)$, that learns to forecast which expert will perform the best, as we have seen in the mixture of experts. For output-based gating, the gate is a directly calculated function of the performance, and hence, the outputs, of the experts. The gate in the annealed competition of experts of Pawelzik et al. [55] implements memory in the form of a local boxcar average squared error of the experts. The self-annealing competitive prediction of Fancourt and Principe also uses the local squared error, but using a recursive estimator. The mixture of experts can also keep track of past expert performance, the simplest example of which is the mixture model where the gate is expanded with memory to create an estimate of the average of the posterior probabilities over the data set [51]. Perhaps, the simplest scheme is hard competition, for which the gate chooses the expert with the smallest magnitude error in a winner-take-all fashion, as will be explained below. This method simplifies the architecture and is very appropriate for system identification in a control framework because it simplifies the design of controllers.

The total cost function can be constructed in several ways from the instantaneous cost, depending on the error function $f(\cdot)$. If $f(\cdot)$ is a quadratic function of the error, the total cost over a data set of N samples is the sum of the instantaneous costs.

$$J = \sum_{n=1}^{N} \sum_{k=1}^{N} g_k(n) \cdot [d(n) - y_k(\chi(n))]^2 \tag{3.50}$$

$$J = \prod_{n=1}^{N} \sum_{k=1}^{N} g_k(n) \cdot p(d(n) - y_k(\chi(n))) \tag{3.51}$$

If $f(\cdot)$ is a PDF in the error, the appropriate total cost function is a product of the instantaneous costs and returns us toward the mixture of experts when the gate is input based. We explain now in more detail the hard competitive structure (shown in Figure 3.17). The experts, in this case predictors, are located on a spatial grid, analogous to a Kohonen's self-organizing map [56], and the gate moderates the learning of the experts based on their distance from the best performing expert. The best predictor is the one with the smallest local mean squared error

$$\text{winner}(n) = \arg\min_k [\varepsilon_k(n)] \tag{3.52}$$

where the local MSE is computed using the recursive estimate for each expert, repeated here for convenience

$$\varepsilon_k(n) = \lambda e_k^2(n) + (1-\lambda)\varepsilon_k(n-1) \qquad 0 < \lambda < 1 \tag{3.53}$$

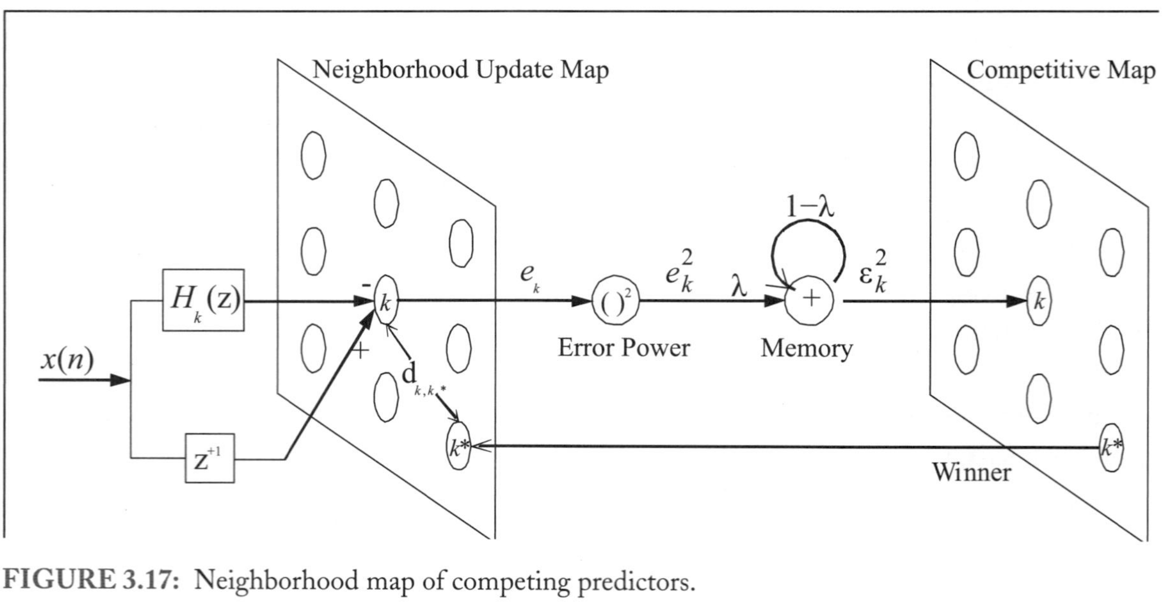

FIGURE 3.17: Neighborhood map of competing predictors.

where e_k is the instantaneous squared error of the kth predictor. The memory term, λ, is identical for all experts, but can be manually adjusted during training. The gating function, which moderates the learning rate of the predictors, is determined by the distance from the winning predictor, k^*, to the other predictors.

$$g_k(n) = \exp\left[\frac{d^2_{k,k^*}(n)}{2\sigma^2(n)}\right] \tag{3.54}$$

where k is the predictor to be updated, k^* is the winning predictor, d_{k,k^*} is the neighborhood distance between them, and σ is an annealing parameter that controls the neighborhood width. This results in the model shown in Figure 3.17. The signal flow graph is shown for only one predictor.

In exact analogy with training a Kohonen map, both the neighborhood width and the global learning rate are annealed during training according to an exponentially decreasing schedule

$$\sigma(t) = \sigma_0 e^{\frac{-i}{\tau}} \quad \mu(t) = \mu_0 e^{\frac{-i}{\tau}} \tag{3.55}$$

where τ is the annealing rate and i is the iteration number. The overall learning rate for the kth predictor at the ith iteration is

$$\eta_k(i) = \mu(i) \cdot g_k(i) \tag{3.56}$$

3.2.3.2 Gated Competitive Experts in BMIs. Let us see how the hard competition of gated experts can be applied to BMIs. Hypothesizing that the neural activity will demonstrate varying characteristics for different localities in the space of the hand trajectories, we expect the multiple model approach, in which each linear model specializes in a local region, to provide a better overall input–output mapping. However, the BMI problem is slightly different from the signal segmentation because the goal is to segment the joint input/desired signal space.

The training of the multiple linear models is accomplished by competitively (hard or soft competition) updating their weights in accordance with previous approaches using the NLMS algorithm. The winning model is determined by comparing the (leaky) integrated squared errors of all competing models and selecting the model that exhibits the least integrated error for the corresponding input [48]. The leaky integrated squared error for the ith model is given by

$$\varepsilon_i(n) = (1-\mu)\varepsilon_i(n-1) + \mu e_i^2(n), i = 1,\cdots,M, \tag{3.57}$$

where M is the number of models and μ is the time constant of the leaky integrator. Then, the jth model wins competition if $\varepsilon_j(n) < \varepsilon_i(n)$ for all $i \neq j$. If hard competition is employed, only the weight vector of the winning model is updated. Specifically, if the *j*th model wins competition, the update rule for the weight vector $\mathbf{w}_j(n)$ of that model is given by

$$\mathbf{w}_j(n+1) = \mathbf{w}_j(n) + \eta \frac{e_j(n)\mathbf{x}(n)}{\gamma + \|\mathbf{x}(n)\|^2}, \tag{3.58}$$

where $e_j(n)$ is the instantaneous error and $\mathbf{x}(n)$ is the current input vector. η represents a learning rate and γ is a small positive constant used for normalization. If soft competition is used, a Gaussian weighting function centered at the winning model is applied to all competing models. Every model is then updated proportional to the weight assigned to that model by this Gaussian weighting function such that

$$\mathbf{w}_i(n+1) = \mathbf{w}_i(n) + \eta \frac{\Lambda_{i,j}(n)e_i(n)\mathbf{x}(n)}{\gamma + \|\mathbf{x}(n)\|^2}, i = 1, \ldots, M, \tag{3.59}$$

where $\mathbf{w}_i$ is the weight vector of the *i*th model. Assuming the *j*th model wins competition, $\Lambda_{i,j}(n)$ is the weighting function defined by

$$\Lambda_{i,j}(n) = \exp\left(-\frac{d_{ij}^2}{2\sigma^2(n)}\right), \tag{3.60}$$

where d_{ij} is the Euclidean distance between indices *i* and *j*, which is equal to $|j - i|$, $\eta(n)$ is the annealed learning rate, and $\sigma^2(n)$ is the Gaussian kernel width decreasing exponentially as *n* increases. The learning rate also exponentially decreases with *n*.

Soft competition preserves the topology of the input space, updating the models neighboring the winner; thus, it is expected to result in smoother transitions between models specializing in topologically neighboring regions (of the state space). However, the empirical comparison using BMI data between hard and soft competition update rules shows no significant difference in terms of model performance (possibly because of the nature of the data set). Therefore, we prefer to utilize hard competition rule because of its simplicity.

With the competitive training procedure, each model can specialize in local regions in the joint space. Figure 3.18 demonstrates the specialization of 10 trained models by plotting their outputs (black dots) with the common input data (40 sec long) in the 3D hand trajectory space. Each model's outputs are simultaneously plotted on top of the actual hand trajectory (red lines) synchronized with the common input. This figure shows that the input–output mappings learned by each model display some degree of localization, although overlaps are still present. These overlaps may be consistent with a neuronal multiplexing effect as depicted in Carmena et al. [26], which suggests that the same neurons modulate for more than one motor parameter (the *x* and *y* coordinates of hand position, velocity, and griping force).

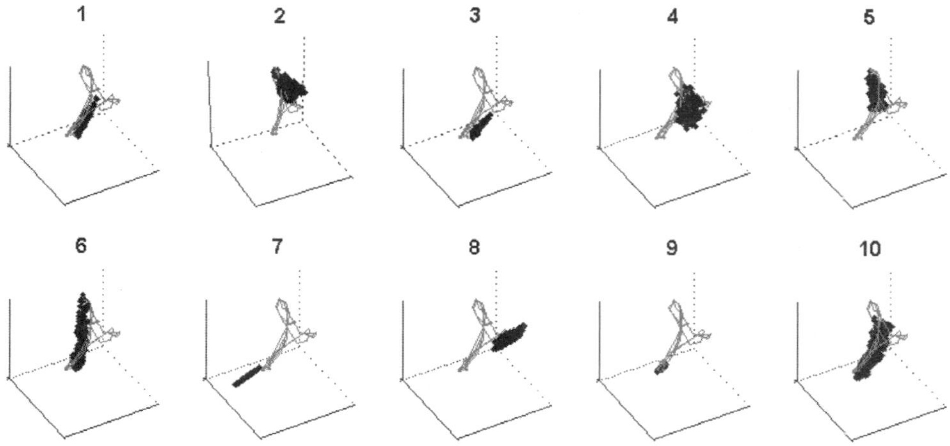

FIGURE 3.18: Demonstration of the localization of competitive linear models.

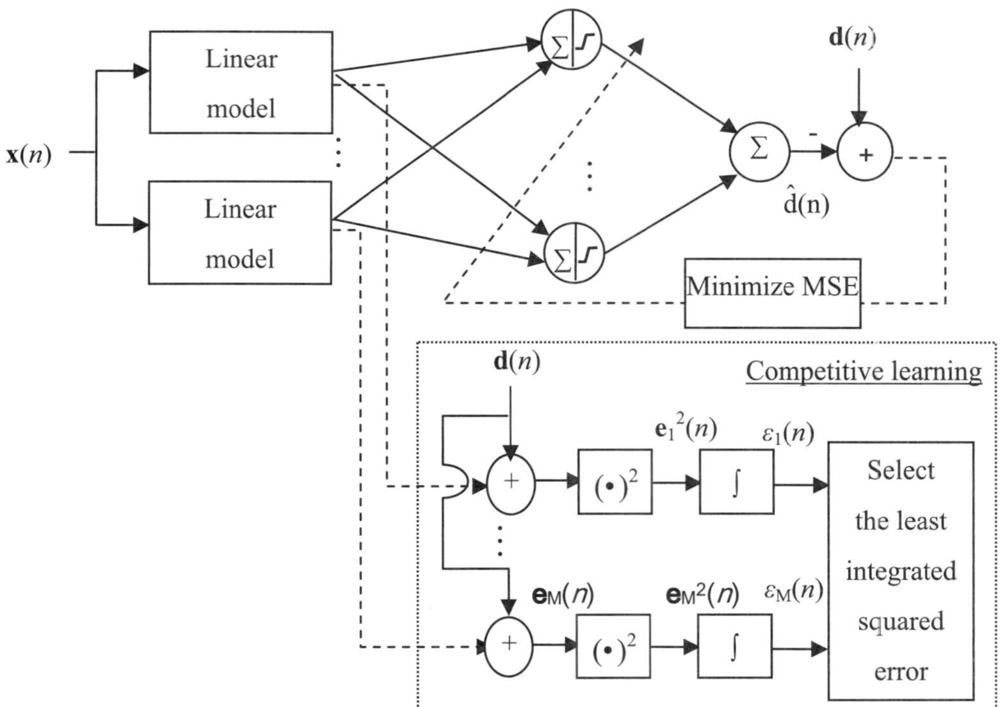

FIGURE 3.19: An overall diagram of the nonlinear mixture of competitive linear models.

The competitive local linear models, however, require additional information for switching when applied to BMIs because the desired signal that is necessary to select a winning model is not available in practice after training. A gate function as in the mixture of experts [47] utilizing input signals needs to be trained to select a local model. Here, we opt for an MLP that directly combines the predictions of all models. Therefore, the overall architecture can be conceived as a nonlinear mixture of competitive linear models [30] and is architecturally a TDNN. This procedure facilitates training of each model compared with the TDNN because only one linear model is trained at a time in the first stage, whereas only a relatively small number of weights are trained by error backpropagation [2] in the second stage.

The overall architecture of nonlinear mixture of competitive linear models is identical to a single hidden-layer TDNN as shown in Figure 3.19. However, the training procedure undertaken here is significantly different. The topology allows a two-stage training procedure that can be performed sequentially in offline training: first, competitive learning for the local linear models and then error backpropagation learning for the MLP. It is important to note that in this scheme, both the linear models and the MLP are trained to approximate the same desired response, which is the hand trajectory.

3.3 SUMMARY

Here, we have presented a comprehensive overview of input–output modeling techniques for BMIs. We would like to address here a few aspects of the modeling techniques that can influence performance. First, the tradeoffs between timing resolution (bin size) and memory depth needed to optimally solve the system identification problem for all models has not been directly addressed [57] and can present as a source of variability in model performance. Second, all of the variations in modeling (linear vs. nonlinear, feedfoward vs. recurrent, global vs. local) and enhancements shown improve the original solution in a marginal but statistically significant way. In more traditional signal-processing applications of these techniques, the differences in modeling performance per topology are often greater. This observation leads one to suspect either that all of the topologies are not optimal for BMI signal processing or there is some common property to the data (for rats, monkeys, and humans) that is influencing all of the results. For all BMI data in the literature, the reported performance ranges in general from 60% to 90% correlation of the model output with the true behavioral trajectories. However, because each research group is using their own data and the performance changes from task to task, animal to animal, and even from day to day, it is very difficult to compare results. A third aspect that complicates the comparison of decoding techniques is the simplicity of the movement tasks studied. The next generation of decoding models will have to assess performance in multipostural reaching, grasping, and holding under the influence of force.

Here, the possible advantages of nonlinear models may be better highlighted. All in all, the viability of input–output decoding the intent of motion contained in multidimensional spike trains collected from the motor cortex has been established by numerous research groups and should be considered an acquired scientific fact.

REFERENCES

1. Soderstrom, T., and P. Stoica, System Identification. 1989, New York: Prentice-Hall.
2. Haykin, S., Adaptive Filter Theory. 3rd ed. 1996, Upper Saddle River, NJ: Prentice-Hall International.
3. Ljung, L. *Black-box models from input–output measurements*, in IEEE Instrumentation and Measurement Technology Conference. 2001. Budapest, Hungary.
4. Kandel, E.R., J.H. Schwartz, and T.M. Jessell, eds. Principles of Neural Science. 4th ed. 2000, New York: McGraw-Hill.
5. Georgopoulos, A.P., A.B. Schwartz, and R.E. Kettner, *Neuronal population coding of movement direction*. Science, 1986. **233**(4771): pp. 1416–1419.
6. Georgopoulos, A., et al., *On the relations between the direction of two-dimensional arm movements and cell discharge in primate motor cortex*. Journal of Neuroscience, 1982. **2**: pp. 1527–1537.
7. Flament, D., and J. Hore, *Relations of motor cortex neural discharge to kinematics of passive and active elbow movements in the monkey*. Journal of Neurophysiology, 1988. **60**(4): pp. 1268–1284.
8. Wessberg, J., et al., *Real-time prediction of hand trajectory by ensembles of cortical neurons in primates*. Nature, 2000. **408**(6810): pp. 361–365.
9. Taylor, D.M., S.I.H. Tillery, and A.B. Schwartz, *Direct cortical control of 3D neuroprosthetic devices*. Science, 2002. **296**(5574): pp. 1829–1832.
10. Serruya, M.D., et al., *Brain–machine interface: Instant neural control of a movement signal*. Nature, 2002. **416**: pp. 141–142.
11. Sanchez, J.C., et al. *A comparison between nonlinear mappings and linear state estimation to model the relation from motor cortical neuronal firing to hand movements*, in SAB Workshop on Motor Control in Humans and Robots: on the Interplay of Real Brains and Artificial Devices. 2002. University of Edinburgh, Scotland.
12. Sanchez, J.C., et al. *Input–output mapping performance of linear and nonlinear models for estimating hand trajectories from cortical neuronal firing patterns*, in International Work on Neural Networks for Signal Processing. 2002. Martigny, Switzerland.
13. Moran, D.W., and A.B. Schwartz, *Motor cortical representation of speed and direction during reaching*. Journal of Neurophysiology, 1999. **82**(5): pp. 2676–2692.

14. Kalaska, J.F., et al., *A comparison of movement direction-related versus load direction-related activity in primate motor cortex, using a two-dimensional reaching task.* Journal of Neuroscience, 1989. **9**(6): pp. 2080–2102.
15. Georgopoulos, A.P., et al., *Mental rotation of the neuronal population vector.* Science, 1989. **243**(4888): pp. 234–236.
16. Thach, W.T., *Correlation of neural discharge with pattern and force of muscular activity, joint position, and direction of intended next movement in motor cortex and cerebellum.* Journal of Neurophysiology, 1978. **41**: pp. 654–676.
17. Scott, S.H., and J.F. Kalaska, *Changes in motor cortex activity during reaching movements with similar hand paths but different arm postures.* Journal of Neurophysiology, 1995. **73**(6): pp. 2563–2567.
18. Todorov, E., *Direct cortical control of muscle activation in voluntary arm movements: A model.* Nature Neuroscience, 2000. **3**(4): pp. 391–398.
19. Gao, Y., et al. *A quantitative comparison of linear and non-linear models of motor cortical activity for the encoding and decoding of arm motions*, in the 1st International IEEE EMBS Conference on Neural Engineering. 2003. Capri, Italy.
20. Wu, W., et al. *Inferring hand motion from multi-cell recordings in motor cortex using a Kalman filter*, in SAB Workshop on Motor Control in Humans and Robots: On the Interplay of Real Brains and Artificial Devices. 2002. University of Edinburgh, Scotland.
21. Kalman, R.E., *A new approach to linear filtering and prediction problems.* Transactions of the ASME Journal of Basic Engineering, 1960. **82**(Series D): pp. 35–45.
22. Wiener, N., Extrapolation, Interpolation, and Smoothing of Stationary Time Series with Engineering Applications. 1949, Cambridge, MA: MIT Press.
23. Haykin, S., Neural Networks: A Comprehensive Foundation. 1994, New York: Macmillan; Toronto, Canada: Maxwell Macmillan.
24. Orr, G., and K.-R. Müller, Neural Networks: Tricks of the Trade. Vol. 1524. 1998, Berlin: Springer.
25. Chapin, J.K., et al., *Real-time control of a robot arm using simultaneously recorded neurons in the motor cortex.* Nature Neuroscience, 1999. **2**(7): pp. 664–670.
26. Carmena, J.M., et al., *Learning to control a brain–machine interface for reaching and grasping by primates.* PLoS Biology, 2003. **1**: pp. 1–16.
27. Shenoy, K.V., et al., *Neural prosthetic control signals from plan activity.* NeuroReport, 2003. **14**: pp. 591–597.
28. Horn, R.A., and C.R. Johnson, Topics in Matrix Analysis. 1991, New York: Cambridge University Press.
29. Shephard, N., *Maximum likelihood estimation of regression models with stochastic trend components.* Journal of the American Statistical Association, 1993. **88**(422): pp. 590–595.

30. Kim, S.P., et al., *Divide-and-conquer approach for brain machine interfaces: Nonlinear mixture of competitive linear models*. Neural Networks, 2003. **16**(5–6): pp. 865–871.
31. Hoerl, A.E., and R.W. Kennard, *Ridge regression: Biased estimation for nonorthogonal problems*. Technometrics, 1970. **12**(3): pp. 55–67.
32. Widrow, B., and S.D. Stearns, Adaptive Signal Processing. Prentice-Hall Signal Processing Series. 1985, Englewood Cliffs, NJ: Prentice-Hall.
33. Príncipe, J.C., N.R. Euliano, and W.C. Lefebvre, Neural and Adaptive Systems: Fundamentals Through Simulations. 2000, New York: Wiley.
34. Sanchez, J.C., et al. *Interpreting neural activity through linear and nonlinear models for brain machine interfaces*, in International Conference of Engineering in Medicine and Biology Society. 2003. Cancun, Mexico.
35. Rao, Y.N., et al. *Learning mappings in brain–machine interfaces with echo state networks*, in International Joint Conference on Neural Networks. 2004. Budapest, Hungary.
36. Sandberg, I.W., and L. Xu, *Uniform approximation of multidimensional myopic maps*. IEEE Transactions on Circuits and Systems, 1997. **44**: pp. 477–485.
37. Todorov, E., *On the role of primary motor cortex in arm movement control*, in Progress in Motor Control III, M. Latash, and M. Levin, eds. 2003, Urbana, IL: Human Kinetics.
38. Puskorius, G.V., et al., *Dynamic neural network methods applied to on-vehicle idle speed control*. Proceedings of the IEEE, 1996. **84**(10): pp. 1407–1420.
39. Werbos, P.J., *Backpropagation through time: What it does and how to do it*. Proceedings of the IEEE, 1990. **78**(10): pp. 1550–1560.
40. Lefebvre, W.C., et al., NeuroSolutions. 1994, Gainesville, FL: NeuroDimension.
41. Vapnik, V., The Nature of Statistical Learning Theory. Statistics for Engineering and Information Science. 1999, New York: Springer-Verlag. 304.
42. Jaeger, H., *The "Echo State" Approach to Analyzing and Training Recurrent Neural Networks*, GMD Report 148. 2001, Sankt Augustin, Germany: GMD-German National Research Institute for Computer Science.
43. Maas, W., T. Natschläger, and H. Markram, *Real-time computing without stable states: A new framework for neural computation based on perturbations*. Neural Computation, 2002. **14**(11): pp. 2531–2560.
44. Principe, J.C., B. De Vries, and P.G. Oliveira, *The gamma filter—A new class of adaptive IIR filters with restricted feedback*. IEEE Transactions on Signal Processing, 1993. **41**(2): pp. 649–656.
45. Ozturk, M.C., D. Xu, and J.C. Principe, *Analysis and Design of Echo State Network for Function Approximation*. Neural Computation, 2006. **19**: pp. 111–138.
46. Farmer, J.D., and J.J. Sidorowich, *Predicting chaotic time series*. Physical Review Letters, 1987. **50**: pp. 845–848.

47. Jacobs, R., et al., *Adaptive mixture of local experts*. Neural Computation, 1991. **3**: pp. 79–87.
48. Fancourt, C., and J.C. Principe. *Temporal self-organization through competitive prediction*, in ICASSP. 1996.
49. Cho, J., et al., *Self-organizing maps with dynamic learning for signal reconstruction*. Neural Networks, 2007. **20**(2): pp. 274–284.
50. Jordan, M.I., and R.A. Jacobs, *Hierarchical mixtures of experts and the EM algorithm*. Neural Computation, 1994. **6**: pp. 181–214.
51. Weigend, A.S., M. Mangeas, and A.N. Srivastava, *Nonlinear gated experts for time-series—Discovering regimes and avoiding overfitting*. International Journal of Neural Systems, 1995. **6**(4): pp. 3773–399.
52. Zeevi, A.J., T. Meir, and V. Maiorov, *Error bounds for functional approximation and estimation using mixtures of experts*. IEEE Transactions on Information Theory, May 1998. **44**: pp. 1010–1025.
53. Jacobs, R.A., et al., *Adaptive mixtures of local experts*. Neural Computation, 1991. **3**: pp. 1–12.
54. Fancourt, C., and J.C. Principe, *Competitive principal component analysis for locally stationary time series*. IEEE Transactions on Signal Processing, 1998. **46**(11): pp. 3068–3081.
55. Pawelzik, K., J. Kohlmorgen, and K.-R. Muller, *Annealed competition of experts for a segmentation and classification of switching dynamics*. Neural Computation, 1996. **8**(2): pp. 340–356.
56. Kohonen, T., *Analysis of a simple self-organizing process*. Biological Cybernetics, 1982. **44**(2): pp. 135–140.
57. Wu, W., et al., *Bayesian population decoding of motor cortical activity using a Kalman filter*. Neural Computation, 2005. **18**: pp. 80–118.

• • • •

CHAPTER 4

Regularization Techniques for BMI Models

Additional Contributor: Sung-Phil Kim

In Chapter 3, we have demonstrated the design of linear and nonlinear filters which can be adapted for BMI applications. Despite the intrinsic sophistication of the BMI system, early tests with the simple linear filter (which merely combines the weighted bin count inputs) showed reasonable estimation of hand position (HP) from neuronal modulation. Therefore, it makes sense to fine-tune these linear models by means of advanced learning techniques. There are three major issues in the design of BMIs: the large number of parameters, irrelevant inputs, and the noise in the data due to spike sorting and nonstationarities. They all affect model performance in different ways. The first two can be handled using sophisticated signal processing techniques during offline analysis of the data. The latter issue of spike sorting is usually handled during the time of data collection, and is considered as a preprocessing step. Here we assume that the data have been appropriately spike sorted, and will focus on the signal processing techniques used in offline analysis.

When training BMI models the first challenge one encounters is how to deal with model overfitting when hundreds of neurons are used as model inputs. In the examples, we showed that the introduction of extra degrees of freedom not related to the mapping can result in poor generalization, especially in topologies where tap-delay memory structures are implemented in the neural input layer (i.e., the TDNN topology). The problem occurs with each additional memory delay element that scales the number of free parameters by the number of input neurons. This explosion in the number of free parameters also puts a computational burden on computing an optimal solution especially when the goal is to implement the BMI in low-power, portable hardware. A table of the number of free parameters for the topologies is described in Chapter 3.

The generalization of the model can be explained in terms of the bias-variance dilemma of machine learning [1], which is related to the number of free parameters of a model. The MIMO structure of BMIs built for the neuronal data presented here can have as few as several hundred to as many as several thousand free parameters. On one extreme, if the model does not contain enough parameters, there are too few degrees of freedom to fit the to be estimated, which results in bias errors. On the other extreme, models with too many degrees of freedom tend to overfit the function to be estimated. In terms of BMIs, models tend to err on the latter because of the large dimensionality of the input.

The first approach we presented for reducing the number of free parameters involved modifications to the model topology itself. More parsimonious models such as the recurrent multilayer perceptron (RMLP) [2–4] were studied. The topology of this model significantly reduced the number of free parameters by implementing feedback memory structures in hidden network layers instead of in the input where the dimensionality is large.

Besides modifying the model topology, a second statistical approach can be taken during model training. Regularization techniques [5] attempt to reduce 4the value of unimportant weights to zero, and effectively prune the size of the model topology. Regularized least squares (LS), subspace projections using partial least squares (PLS), or special memory structures can be used to reduce the number of free parameters [6]. These approaches are strictly statistical, and require lots of data and computation, are not trivial to use, and do not necessarily provide information about the importance of neurons. As an alternative, the number of inputs given to the models could be manually pruned using neurophysiological analysis of the correlation of neuronal function with behavior; however, it is difficult to know how it will affect BMI model performance. To overcome this issue, neural selection has also been attempted using sensitivity analysis and variable selection procedures [4, 7].

This chapter presents our efforts to deal with these problems which are basically divided between regularization of the optimal solution and channel selection. Moreover, once we have a well-tuned model, then the model itself can help us learn about the neurophysiology of the motor cortex. We also show the quantification of neural responses that is possible by "looking inside" the trained models.

4.1 LEAST SQUARES AND REGULARIZATION THEORY

The nature of the neural systems used in BMI model architectures creates MIMO systems with a large input space. Even with the LS solution, an issue of well-posedness follows immediately because the algorithm is formulated in very high dimensional spaces with finite training data available. The concept of well-posedness was proposed by Hadamard [8]. Regularization as a remedy for ill-posedness became widely known because of the work of Tikhonov [9], and also from Bayesian Learning for Neural Networks [10]. In solving LS problems, the Tikhonov regularization is essentially a trade-off between fitting training data and reducing solution norms. Consequently, it reduces the sensitivity of the solution to small changes in the training data and imposes stability on the ill-posed problem. Moreover, the significance of well-posedness related to generalization ability has been also revealed recently in statistical learning theory [11].

Recall that in LS, the multiple linear regression model is hypothesized as

$$y_n = W^{\circ}(\Phi_n) + v(n) \tag{4.1}$$

where W represents the model weights, Φ_n is the identity or a nonlinear function depending on whether the model is linear or nonlinear, and $v(n)$ is the modeling uncertainty. We will use Φ_n to mean $\Phi(x_n)$, where x is the neuronal input.

Denote the transformed data matrix as $D_W^{\mathrm{T}} = [W(x_1), W(x_2), \ldots, W(x_N)]$, the correlation matrix by

$$R_W = D_W^V D_W / N \tag{4.2}$$

the cross-correlation vector $p_\Phi = \sum_{i=1}^{N} \Phi(x_i) y_i / N = D_\Phi^{\mathrm{T}} \vec{y} / N$, and the Gram matrix

$$G_\Phi = D_\Phi D_\Phi^{\mathrm{T}} = [\kappa(x_i, x_j)]_{N \times N} \tag{4.3}$$

Then, the LS solution is known to satisfy the following normal equations [8]

$$R_\Phi \hat{W} = p_\Phi \tag{4.4}$$

Unfortunately, the LS problem is not always well posed. According to Hadamard [8], a problem is well posed if the solution exists, is unique, and smoothly depends on data and parameters. More specifically, the following theorems demonstrate how to get a well-posed LS algorithm.

Theorem 4.1: *(uniqueness theorem of LS [12]) The least-squares estimate $\hat{W}$ is unique if and only if the correlation matrix R_Φ is nonsingular and the solution is given by*

$$\hat{W} = R_\Phi^{-1} p_\Phi \tag{4.5}$$

Theorem 4.2: *(pseudo-inverse [12]) If R_Φ is singular, there are infinitely many solutions to (4.5). Of them, there is a unique minimum-norm solution given by*

$$\hat{W} = D_\Phi^{+} \vec{y} \tag{4.6}$$

Here, D_Φ^{+} is the general pseudo-inverse of D_Φ given by

$$D_\Phi^{+} = p \begin{bmatrix} S^{-1} & 0 \\ 0 & 0 \end{bmatrix} Q^{\mathrm{T}} \tag{4.7}$$

and matrices P, Q, and S are given by the singular value decomposition (SVD) of D_Φ [13]

$$D_\Phi = Q\begin{bmatrix} S & 0 \\ 0 & 0 \end{bmatrix}P^{\mathrm{T}} \tag{4.8}$$

where P, Q are orthogonal matrices, and $S = \mathrm{diag}\{s_1, s_2, \ldots, s_r\}$ assuming the rank of D_Φ is r. s_i denotes the singular values of D_Φ and assumed $s_1 \geq s_2 \geq \cdots \geq s_r > 0$.

Remarks: The solution (4.6) includes (4.5) as special case, and together can be expressed as

$$\hat{W} = R_\Phi^+ p_\Phi \tag{4.9}$$

In an ideal world, the designer's job is finished with (4.9), but in the real world more caution should be taken. The fundamental issues are the finite precision of our computers and the noise—universal thermal noise, measurement errors, modeling uncertainty, etc. For example, when the modeling uncertainty $v(n)$ in (4.1) is white with zero mean and constant variance s ^{2}, the covariance matrix of the LS estimate (4.6) is [14]

$$\mathrm{cov}[\hat{W}] = \sigma^2 P\begin{bmatrix} S^{-2} & 0 \\ 0 & 0 \end{bmatrix}P^{\mathrm{T}} \tag{4.10}$$

If some singular values are too small, the variance of the estimate is pessimistically large. On the other hand, too small singular values also cause catastrophic numerical errors because of the finite precision [13]. The typical symptoms of this type of ill-posedness are large norm solutions and extreme sensitivity to small changes in training data.

The Tikhonov regularization is widely used to address this problem. Because small norm solutions are desirable, a regularization term is introduced in the LS cost function which penalizes the solution norm

$$\min_W R_{\mathrm{reg}}\left[W \in H_2, Z^N\right] = \|\vec{y} - D_\Phi W\|^2 + \lambda\|W\|_F^2 \tag{4.11}$$

In the Bayesian interpretation, the error square term is the likelihood and the regularized term is some priori knowledge on the norm of the solution, which corresponds to a Gaussian prior for the L_2 norm [14].

Theorem 4.3: *(Tikhonov regularization [9]) The Tikhonov regularized LS solution is*

$$\hat{W}_{\mathrm{TR}} = (D_\Phi^T D_\Phi + \lambda I)^{-1} D_\Phi^{\mathrm{T}} \vec{y} \tag{4.12}$$

Remarks: More insights can be obtained through the SVD,

$$\hat{W}_{\mathrm{TR}} = P\mathrm{diag}\left(\frac{s_1}{s_1^2+\lambda},\ldots,\frac{s_r}{s_r^2+\lambda},0,\ldots,0\right)Q^{\mathrm{T}}\vec{y} \tag{4.13}$$

Comparing it with the pseudo-inverse solution

$$\hat{W} = P\mathrm{diag}(s_1^{-1},\ldots,s_r^{-1},0,\ldots,0)Q^{\mathrm{T}}\vec{y} \tag{4.14}$$

we see that the Tikhonov regularization modifies the singular values through the following regularization function:

$$H_{\mathrm{TR}}(x) = x^2/(x^2+\lambda) \tag{4.15}$$

Note that $H_{\mathrm{TR}}(x) \to 1$ when x is large and $H_{\mathrm{TR}}(x) \to 1$ when x is small. In this sense, the Tikhonov regularization smoothly filters out the singular components that are small (relative to λ). This viewpoint actually brings new understanding into all kinds of regularization techniques as will be described in this chapter.

4.1.1 Ridge Regression and Weight Decay

When building linear models for BMI with neuronal activity that is highly variable with a large dynamic range of bin counts, the condition number of an input correlation matrix may be relatively large as shown in Chapter 3. In this situation, the performance of the model for reconstructing the movement trajectories from neural activity is highly variable because of the extra degrees of freedom that are not constrained. To reduce the condition number, we can, according to (4.12), add an identity matrix multiplied by a white noise variance to the correlation matrix, which is known as ridge regression (RR) in statistics [15] when LS are being solved as in the case of the Wiener solution

We can also use a similar regularizer in the iterative updates based on the least mean squares (LMS), or normalized LMS (NLMS) algorithm, which is called weight decay [16]. Rewriting (4.11), the criterion function of the regularized solution is

$$J(\mathbf{w}) = E[\|\mathbf{e}\|^2] + \delta\|\mathbf{w}\|^2 \tag{4.16}$$

where the additional term $\delta||\mathbf{w}||^2$ smoothes the cost function. The instantaneous gradient that corresponds to this equation can be written as

$$\mathbf{w}(n+1) = \mathbf{w}(n) + \eta\hat{\nabla}\zeta(n) - \delta\mathbf{w}(n) \tag{4.17}$$

where $\hat{\nabla}\zeta(n) = \partial E\left[\|\mathbf{e}(n)\|^2\right]/\partial\mathbf{w}(n)$, which is called the weight decay gradient update. Both RR and weight decay can be viewed as implementations of a Bayesian approach to complexity control in supervised learning using a zero-mean Gaussian prior [10].

The choice of the amount of regularization (δ) plays an important role in the generalization performance, because there is a trade-off between the condition number and the achievable MSE for a particular δ. A larger δ can decrease the condition number at the expense of increasing the MSE, whereas a smaller δ can decrease the MSE but also increase the condition number. Larsen et al. [17] proposed that δ can be optimized by minimizing the generalization error with respect to δ. Following this procedure, we utilize the *K*-fold cross-validation [18], which divides the data into *K* randomly chosen disjoint sets, to estimate the average generalization error empirically,

$$\hat{\xi} = \frac{1}{K}\sum_{k=1}^{K}\varepsilon_k \tag{4.18}$$

where ε_k is MSE of the validation for the *k*th set. Then, the optimal regularization parameter is learned by using gradient descent,

$$\delta(n+1) = \delta(n) - \eta\frac{\partial\hat{\xi}(n)}{\partial\delta} \tag{4.19}$$

where $\hat{\xi}(n)$ is an estimate computed with $\delta(n)$, and $\eta > 0$ is a learning rate. See Reference [17] for the procedure of estimation of $\partial\hat{\xi}(n)/\partial\delta$ using weight decay. Once the procedure is applied, the following change in weight distribution can be obtained as in Figure 4.1. Here, a linear BMI was trained with both weight decay and standard NLMS. As we can see, the solid line has many more weights with the value of zero. This means that during adaptation, the extra degrees of freedom in the model were effectively eliminated. The elimination of the influence of many weights resulted in an increase in the testing correlation coefficient, which also had a corresponding decrease in variance as shown in Table 4.1.

4.1.2 Gamma Filter

The large number of parameters in decoding models is caused not only by the number of neurons but by the number of time delays required to capture the history of the neuron firings over time. This attribute of the neural input topology was especially evident with the analysis of the TDNN. This problem is compounded in the context of neural decoding we showed that the time resolution of the neural rate coding can influence performance. If smaller time bins are used in decoding, additional numbers of parameters could be introduced. For example, although we use a 10-tap delay line for 100-msec bin sizes, the size of the delay line is variable depending on the bin size (e.g., if we use a 50-msec time bin, then the number of time lags doubles for an equivalent memory depth).

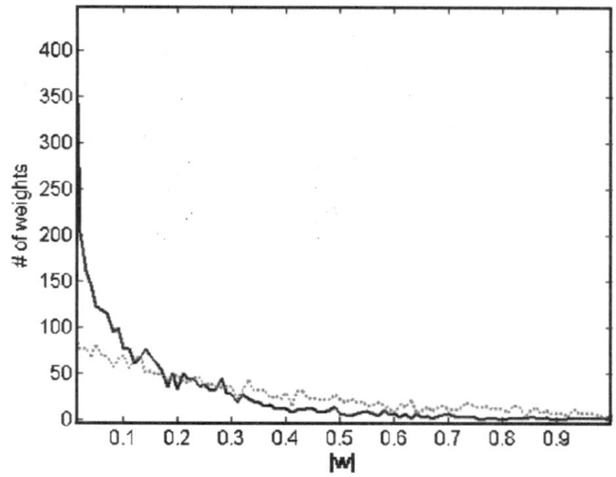

FIGURE 4.1: Histogram of the magnitudes of weights over all the coordinates of hand position (HP), trained by weight decay (solid line), and NMLS (dotted line).

A generalized feedforward filter (GFF) provides a signal processing framework to incorporate both finite (FIR) and infinite impulse response (IIR) characteristics into a single linear system by using a local feedback structure in the delay line as a generalized delay operator [19]. As shown in Figure 4.2, an input signal is delayed at each tap by a delay operator defined by specific transfer function $G(z)$. Note that when $G(z) = z^{-1}$, it becomes an FIR filter. The transfer function of an overall system $H(z)$ is stable when $G(z)$ is stable because

$$H(z) = \sum_{k=0}^{K} w_k \left(G(z)\right)^k \tag{4.20}$$

where K is the number of taps. It has been shown that a GFF can provide trivial stability checks if $G(z)$ is a first- or second-order polynomial in z, and easy adaptation while decoupling the memory

TABLE 4.1: Comparison of testing performance for weight decay and NLMS

	NLMS	WEIGHT DECAY
Testing a correlation coefficient for reaching task	0.75 ± 0.20	0. 77 ± 0.18

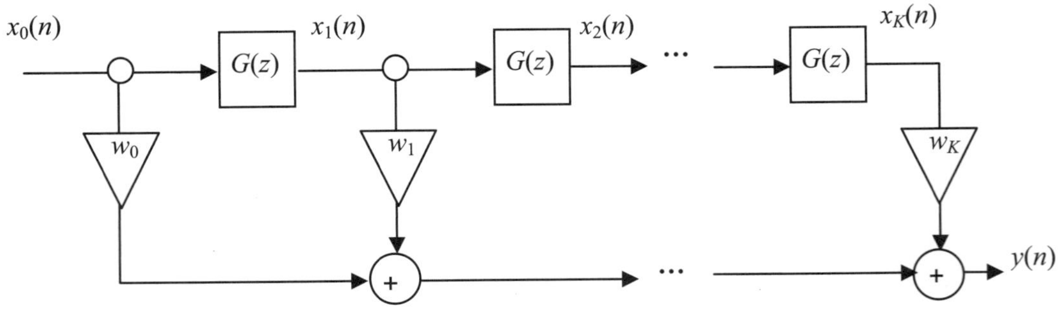

FIGURE 4.2: An overall diagram of a generalized feedforward filter [19]. $x_0(n)$ is an instantaneous input and $y(n)$ is a filter output.

depth from the filter order [19]. Memory depth in the present context is defined as the center of mass of the impulse response of the last tap of the GFF filter [19].

The gamma filter is a special case of the GFF with $G(z) = \mu/(z - (1 - \mu))$, where μ is a feedback parameter. The impulse response of the transfer function from an input to the kth tap, denoted as $g_k(n)$, is given by

$$g_k(n) = Z^{-1}(G_k(z)) = Z^{-1}\left(\left(\frac{\mu}{z-(1-\mu)}\right)^k\right) = \binom{n-1}{k-1}\mu^k(1-\mu)^{n-k}u(n-k) \tag{4.21}$$

where $Z^{-1}(\cdot)$ indicates the inverse z transform and $u(n)$ the step function. When $\mu = 1$, the gamma filter becomes an FIR filter. The stability of the gamma filter in adaptation is guaranteed when $0 < \mu < 2$ because of a local feedback structure.

The memory depth D with a feedback parameter μ in the Kth-order gamma filter is given by

$$D = \frac{K}{\mu} \text{ for } \mu < 1, \text{ or } D = \frac{K}{2-\mu} \text{ for } \mu > 1. \tag{4.22}$$

If we defined the resolution $R \equiv \mu$, the property of the gamma delay line can be described as

$$K = D \times R \text{ for } \mu < 1, \text{ or } K = D \times (2 - R) \text{ for } \mu > 1. \tag{4.23}$$

This property shows that the gamma filter decouples the memory depth from the filter order by adjusting a feedback parameter (μ). In the case of $\mu = 1$ (i.e., the FIR filter), the resolution is maximized, whereas the memory depth is minimized for a given filter order. But this choice sometimes results in overfitting when a signal to be modeled requires more time delays than the number of descriptive parameters. Therefore, the gamma filter with the proper choice of a feedback parameter can avoid overfitting by reducing the number of parameters required to span a certain memory depth. Sandberg and Xu showed that the gamma neural model, which is obtained by substituting the tap delay line in the TDNN by a gamma memory, is still a universal mapper in functional spaces [20].

The tap weights can be either computed by LS or updated using NLMS, and therefore the computational complexity is of the same order of FIR filters. The feedback parameter μ can also be adapted from the data using an LMS type of update [21]. However, instead of adaptively finding μ, we can search the best combination of K and μ through cross-validation.

In Table 4.2, we can see that by using the Gamma architecture the number of model parameters can be reduced by 40%. The reduction yielded a 3% increase in the testing correlation coefficient with a reduction in the variance.

4.1.3 Subspace Projection

In addition to unnecessary weights in the model, one of the challenges in the design of decoding models for BMIs is that some neuron firings are not substantially modulated during task performance, and they only add noise to the solution. In the overall representation contained within these neural systems, the superfluous neurons may be coding other kinematics, sensory information, or are involved in internal signaling within the network. Although these neurons are important to the overall function of the network, they may not be correlated with the particular behavior

TABLE 4.2 Comparison of testing performance and number of parameters for the Gamma architecture and NLMS (1-sec memory depth)

	NLMS	GAMMA
Testing correlation coefficient for a reaching task	0.75 ± 0.20	0.78 ± 0.19
Number of parameters	2973	1191

studied during the BMI experiment.[1] In addition, some neuron firings are correlated with each other, which creates problems in the estimation of LS (R may become rank deficient); thus, it may be advantageous to blend these inputs to improve model performance. Subspace projection, which can at the same time reduce the noise, and blend correlated input signals together, may be very beneficial. Moreover, it also reduces the number of degrees of freedom in the multichannel data, and consequently decreases the variance of the model for the same training data size. Here, we introduce a hybrid subspace projection method that is derived by combining the criteria of PCA and PLS. Then, we will design the subspace Wiener filter based on this hybrid subspace projection for BMIs.

PCA, which preserves maximum variance in the subspace data, has been widely adopted as a projection method [14]. The projection vector w_{PCA} is determined by maximizing the variance of the projection outputs as

$$w_{\text{PCA}} = \arg\max_{w} \quad J^{\text{PCA}}(w) = E\left[\left\|x^{\text{T}} w\right\|^2\right] = w^{\text{T}} \mathbf{R}_s w \tag{4.24}$$

where $\mathbf{R}_s$ is the input covariance matrix computed over the neuronal space only (it is an $M \times M$ matrix where M is the number of neurons). x is an $M \times 1$ instantaneous neuronal bin count vector. It has been well known that w_{PCA} turns out to be the eigenvector of $\mathbf{R}_s$ corresponding to the largest eigenvalues. Then an $M \times S$ projection matrix that constructs an S-dimensional subspace consists of S eigenvectors corresponding to the S largest eigenvalues. However, PCA does not exploit information in the joint space of the input (neuronal) and desired (behavioral) response. This means that there may be directions with large variance that are not important to describe the correlation between input and desired response (e.g., some neuronal modulations related to the BMI patient's anticipation of reward might be substantial, but less useful for the direct estimation of movement parameters), but will be preserved by the PCA decomposition.

One of the subspace projection methods to construct the subspace in the joint space is PLS, which seeks the projection maximizing the cross-correlation between the projection outputs and desired response [22]. Given an input vector x, and a desired response d, a projection vector of PLS, w_{PLS}, maximizes the following criterion,

$$w_{\text{PLS}} = \arg\max_{w} \quad J^{\text{PLS}}(w) = E\left[\left(x^{\text{T}} \mathbf{w}\right) d\right] = E\left[w^{\text{T}} (xd)\right] = w^{\text{T}} p \tag{4.25}$$

[1]At the time of surgery, the targeting of neurons is related to the broad functional specialization of each area of cortex. The ability to target specific neurons can be achieved through sensory, motor, or electrical stimulation. However, it is common to obtain the activity of neurons not related to the motor task of interest.

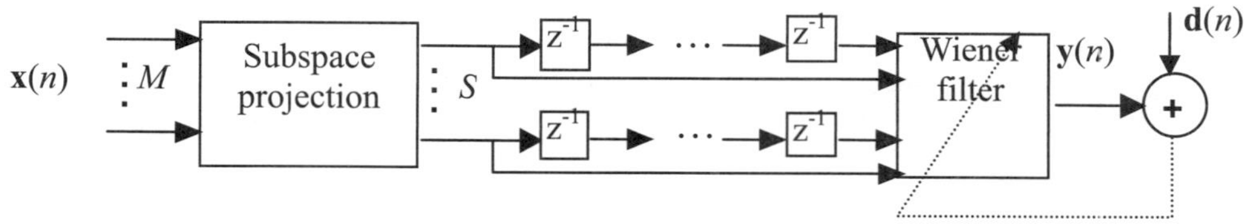

FIGURE 4.3: The overall diagram of the subspace Wiener filter. $y(n)$ denotes the estimated HP vector. There are $L - 1$ delay operators (z^{-1}) for each subspace channel.

where $\boldsymbol{p}$ is defined as an $M \times 1$ cross-correlation vector between $\boldsymbol{x}$ and d. The consecutive orthogonal PLS projection vectors are computed using the deflation method [14].

There have been efforts to find a better projection that can combine the properties of PCA and PLS. The continuum regression (CR), introduced by Stone and Brooks [23], attempted to blend the criteria of LS, PCA, and PLS. Recently, we proposed a hybrid criterion function similar to the CR, together with a stochastic learning algorithm to estimate the projection matrix [24]. The learned projection can be either PCA, PLS, or combination of the two. A hybrid criterion function combining PCA and PLS is given by

$$J(\boldsymbol{w},\lambda) = \frac{(\boldsymbol{w}^{\mathrm{T}}\boldsymbol{p})^{2\lambda}(\boldsymbol{w}^{\mathrm{T}}\mathbf{R}_s\boldsymbol{w})^{1-\lambda}}{\boldsymbol{w}^{\mathrm{T}}\boldsymbol{w}} \tag{4.26}$$

where λ is a balancing factor between PCA and PLS. This criterion covers the continuous range between PLS ($\lambda = 1$) and PCA ($\lambda = 0$).[2] Because the log function is monotonically increasing, the criterion can be rewritten as,

$$\log(J(\boldsymbol{w},\lambda)) = \lambda \log\left(\boldsymbol{w}^{\mathrm{T}}\boldsymbol{p}\right)^2 + (1-\lambda)\log(\boldsymbol{w}^{\mathrm{T}}\mathbf{R}_s\boldsymbol{w}) - \log\left(\boldsymbol{w}^{\mathrm{T}}\boldsymbol{w}\right) \tag{4.27}$$

We seek to maximize this criterion for $0 \le \lambda \le 1$. There are two learning algorithms derived in [24] to find $\boldsymbol{w}$ (one is based on gradient descent and the other is based on the fixed-point algorithm), but we opted to use the fixed-point learning algorithm here because to its fast convergence and independence of learning rate. The estimation of $\boldsymbol{w}$ at the k+1th iteration in the fixed-point algorithm is given by

$$\boldsymbol{w}(k+1) = (1-T)\boldsymbol{w}(k) + T\left[\frac{\lambda\boldsymbol{p}}{\boldsymbol{w}(k)^T\boldsymbol{p}} + \frac{(1-\lambda)\mathbf{R}_s\boldsymbol{w}(k)}{\boldsymbol{w}(k)^T\mathbf{R}_s\boldsymbol{w}(k)}\right] \tag{4.28}$$

[2]The CR covers LS, PLS, and PCA. However, because we are only interested in the case when subspace projection is necessary, LS can be omitted in our criterion.

TABLE 4.3: Comparison of testing performance for PLS and NLMS

	NLMS	PLS
Testing correlation coefficient for a reaching task	0.75 ± 0.20	0.77 ± 0.18

with a random initial vector $\boldsymbol{w}(0)$. T $(0 < T < 1)$ is a balancing parameter to remove the oscillating behavior near convergence. The convergence rate is affected by T that produces a trade-off between the convergence speed and the accuracy. Note that the fastest convergence can be obtained with $T = 1$. The consecutive projection vectors are also learned by the deflation method to form in each column of a projection matrix $\mathbf{W}$. After projection onto the subspace by $\mathbf{W}$, we embed the input signal at each channel with an L-tap delay line and design the Wiener filter to estimate the HP. Figure 4.3 illustrates the overall diagram of the subspace Wiener filter. For this architecture, the testing performance of the model also improved with a decrease in CC variance as shown in Table 4.3.

4.2 CHANNEL SELECTION

Because our ultimate goal for BMIs is to design the most accurate reconstructions of hand kinematics from cortical activity using adaptive signal processing techniques, it seems natural to equate here neural importance (selection of important neurons) to model fitting quality. Moreover, these measures should be compared with the available neurophysiological knowledge, with the hope that we can understand better the data, and enhance our methodologies and, ultimately, the performance of BMIs. Therefore, the importance of neurons will be ascertained using three techniques:

- Sensitivity analysis through trained linear and nonlinear models
- L_1-norm penalty pruning
- Real-time input selection

Given a set of data, we would like to evaluate how well these three methodologies are able to find important neurons for building BMI models. Second, we would like to use this information to tackle the model generalization issues encountered in BMIs. The goals of the study are formulated in the following questions:

- Can our methods automatically indicate important neurons for the prediction of the kinematic variables in the tasks studied?

- In this model-based framework, can better BMIs be built using a subset of important neurons?

It is well known that neurons vary in their involvement in a given task [25]. However, quantifying neuronal involvement for BMI applications is still an ongoing area of research. This is where BMI modeling is an asset, because once trained, the model implicitly contains the information of how cells contribute for the mapping. The difficulty is that the assessment is in principle dependent on the type of model chosen to predict the kinematic variables and its performance (model-dependent). We will first compare the linear FIR model and nonlinear RMLP. Our second question quantifies the change in performance when only a small subset of cells is used to build the BMI. In principle, one could think that any subset of cells will perform worse than the whole ensemble, but because of the poor generalization of large models, performance may in fact be better in a test set with a reduced number of important cells. Of course, this also makes BMIs more dependent on the stability over time of these cells, and in the long run we have shown that performance can either worsen or improve. The ultimate goal is to improve understanding of how cells encode kinematic parameters so that better "gray-box" models can be built using the underlying information in neural recordings.

Assumptions for ranking the importance of a neuron. We would like to obtain an automatic measure of each cell's contribution to encoding motor parameters for a given task, which we call the cell importance. For this reason, a structured approach is taken to ascertaining the importance of neurons with the three methods described earlier. Our methodological choices, however, are not free from assumptions. First, the methods assume stationarity in the data. A snapshot of neural activity is taken and importance is ascertained without addressing time variability in the recordings, which is a shortcoming. Second, despite the highly interconnected nature of neural structures, importance is often computed independently for each individual neuron. With this independence assumption, it is difficult to quantify the importance of pairs, triples, etc., of cells. In contrast, the model sensitivity analysis considers covariations in firing rate among groups of cells in the neural ensemble, but depends on the type of model utilized. Third, a technique may only consider the instantaneous neural activity, whereas other techniques include memory structures (tap delay lines). Finally, each technique for ascertaining importance focuses on different neuronal firing features.

4.2.1 Sensitivity-Based Pruning

With the weights of the trained linear and nonlinear networks, we have a tool with which we can identify the neurons that affect the output the most. A sensitivity analysis, using the Jacobian of the output vector with respect to the input vector, tells how each neuron's spike counts affect the output given the data of the training set. Because the model topology can affect the interpretation of sensitivity, we will first examine sensitivities through the FIR filter that will serve as the "control model" throughout this book. The procedure for deriving the sensitivity for a feedforward topology is an application of the chain rule [26]. For the case of the FIR filter, differentiating the output with respect to the input [see (3.8)] directly yields a sensitivity with respect to each neuronal input i in (4.29).

$$\frac{\partial \mathbf{y}_j}{\partial \mathbf{x}_i} = \mathbf{W}_{10(i-1)+1:10(i-1)+10,j} \tag{4.29}$$

Hence, a neuron's importance can be determined by simply reading the corresponding weight value[3] in the trained model, if the input data for every channel is power-normalized. Because this is not the case for neural data, the neuron importance is estimated in the vector Wiener filter by multiplying the absolute value of a neuron's sensitivity with the standard deviation of its firing computed over the data set[4] as in (4.30). To obtain a scalar sensitivity value for each neuron, the weight values are also averaged over the 10 tap delays and three output dimensions.

$$\text{Sensitivity}_i = \sigma_i \frac{1}{2}\sum_{j=1}^{3}\frac{1}{10}\sum_{k=1}^{10}\left|\mathbf{W}_{10(i-1)=k,j}\right| \tag{4.30}$$

The procedure for deriving the sensitivity for a feedfoward multilayer perceptron (MLP), also discussed in [26], is again a simple application of the chain rule through the layers of the network topology as in (4.31):

$$\frac{\partial y_2(t)}{\partial x(t)} = \frac{\partial y_2(t)}{\partial y_1(t)}\frac{\partial y_1(t)}{\partial x(t)} \tag{4.31}$$

In the case of a nonlinear, dynamical system like the RMLP, the formulation must be modified to include time. Because the RMLP model displays dependencies over time that results from feedback in the hidden layer, we must modify this procedure [26]. Starting at each time t, we compute the sensitivities in (4.31) as well as the product of sensitivities clocked back in time. For example, using the RMLP feedforward equations [see (3.29) and (3.30)], we can compute at $t = 0$, the chain rule shown in (4.32). D_t is the derivative of the hidden layer nonlinearity evaluated at the operating point shown in (4.33). Notice that at $t = 0$ there are no dependencies on y_1. If we clock back one cycle, we must now include the dependencies introduced by the feedback, which is shown in (4.34). At each

[3]In this analysis, we consider the absolute values of the weights averaged over the output dimensions and the 10-tap delays per neuron.

[4]By multiplying the model weights by the firing standard deviation, we have modified the standard definition of sensitivity; however, for the remainder of this analysis we will refer to this quantity as the model sensitivity.

clock cycle back in time, we simply multiply an additional $\mathbf{W}_f^{\mathrm{T}} D_{t-i}$. The general form of the sensitivity calculation is shown in (4.35). Experimentally, we determined that the effect of an input decays to zero over a window of 20 samples (Figure 4.4). At each time t, the sensitivity of the output with respect to the input is represented as the sum of the sensitivities over the 20-sample window.

$$\frac{\partial y_2(t)}{\partial x(t)} = \mathbf{W}_2^{\mathrm{T}} D_t \mathbf{W}_1^{\mathrm{T}} \tag{4.32}$$

$$\mathbf{D} = \begin{bmatrix} f'(z_1^1) & 0 & \cdots & 0 \\ 0 & f'(z_1^2) & \ddots & \vdots \\ \vdots & \ddots & \ddots & 0 \\ 0 & \cdots & 0 & f'(z_1^n) \end{bmatrix} \tag{4.33}$$

$$\frac{\partial y_2(t)}{\partial x(t-1)} = \mathbf{W}_2^{\mathrm{T}} D_t \mathbf{W}_f^{\mathrm{T}} D_{t-1} \mathbf{W}_1^{\mathrm{T}} \tag{4.34}$$

$$\frac{\partial y_2(t)}{\partial x(t-\Delta)} = \mathbf{W}_2^{\mathrm{T}} D_t \prod_{i=1}^{\Delta} \left(\mathbf{W}_f^{\mathrm{T}} D_{t-i} \right) \mathbf{W}_1^{\mathrm{T}} \tag{4.35}$$

Compared to the FIR model, which produces a static measure of sensitivity, the RMLP produces a time-varying sensitivity that we will now use to analyze three similar movements from

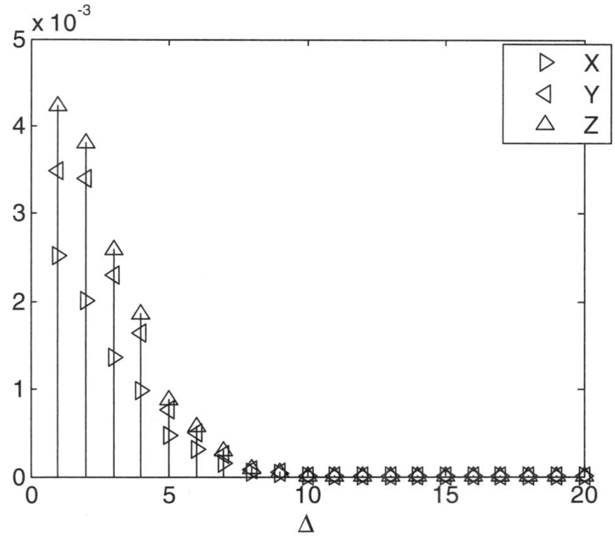

FIGURE 4.4: Sensitivity at time t for a typical neuron as a function of Δ.

the testing trajectory; specifically, reaching movements shown in the testing plots of Figure 4.5. Notice that movements are similar except that movement 3 does not have a decrease in the y coordinate. To visualize trends in how the input affects the output, we plot the neuronal activity (input) along with the computed sensitivity (averaged over 104 neurons and 20 delays) in Figure 4.5b–c. We take first a macroscopic view of both the neuronal activity and sensitivity by summing over 104 neurons at each time bin. Comparing Figure 4.5a with Figure 4.5b, we see that it is difficult to visually extract features that relate neuronal firing counts to hand movement. Despite this fact, a few trends arise:

- Neuronal activity consists of a time-varying firing rate around some mean firing rate.
- Movements 1 and 3 seem to show increased neuronal activity at the beginning and end of a movement, whereas 2 does not.
- All three movements contain a decrease in neuronal activity during the peak of the movement.

With these three trends in mind, we now include the sensitivity plot estimated for the RMLP in Figure 4.5c. We can observe the following:

- In general, the network becomes more sensitive during all three movements.
- Sensitivity is large when velocity is high.
- The sensitivity shows large, sharp values at the beginning and end of the movement, and a plateau during the peak of the movement.

From sensitivity peaks during the movements in Figure 4.5c, we ascertain that the sensitivity analysis is a requirement for relating neuronal activity to behavior. Without the model-based sensitivity, finding relationships in the raw data is difficult. Now that we have a mesoscopic view of how the ensemble of sampled neuronal activity affects the output of the network, we change our focus, and use the model-based sensitivity to "zoom in" on the important individual neurons. We are interested in learning why the individual neurons in a given cortical region are necessary for constructing the network output movement. Using the sensitivity analysis, we can select neurons that are most closely associated with the reaching movements (i.e., neurons that elicit large perturbations in the output with small perturbations in firing).

Neurons for a particular movement are selected for the RMLP by choosing the maximum of a sorted list of neuronal sensitivities computed by averaging the time-varying sensitivity in (4.35) over the interval of a movement and all three coordinate directions. For comparison, the neurons from the FIR filter are sorted directly using (4.30). The sorted ensemble neuronal sensitivities for

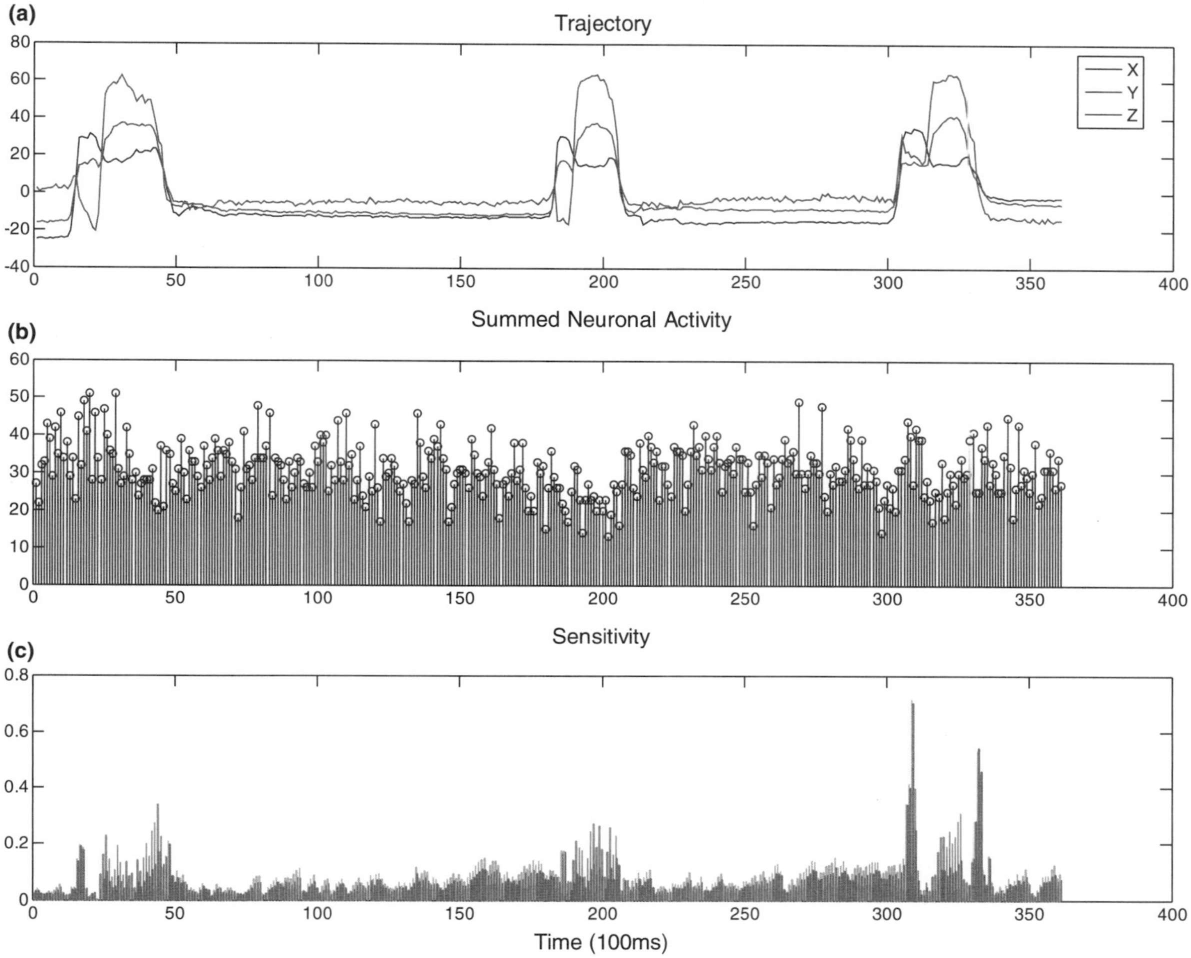

FIGURE 4.5: RMLP time-varying sensitivity. (a) *X*, *Y*, and *Z* desired trajectories for three similar movements. (b) Neuronal firing counts summed (at each time bin) over 104 neurons. (c) Sensitivity (averaged over 104 neurons) for three coordinate directions.

both the FIR and RMLP models (Figure 4.6) shows an initially sharp decrease from maximum indicating that only a few neurons are required for outlining the movement reconstruction. Of the 104 neurons from the animal used in this study, the 10 highest sensitivity neurons for a given movement are presented in the image plot in Figure 4.6, and are grayscale coded by cortical area (PP, posterior parietal; M1, primary motor; PMd, premotor dorsal; M1/PMd, ipsilateral). The highest sensitivity neurons are primarily distributed over the PP and M1. Of the 104 neurons, 7 of the 10 highest-ranking neurons are common for the WF and RMLP. By experimentally verifying the

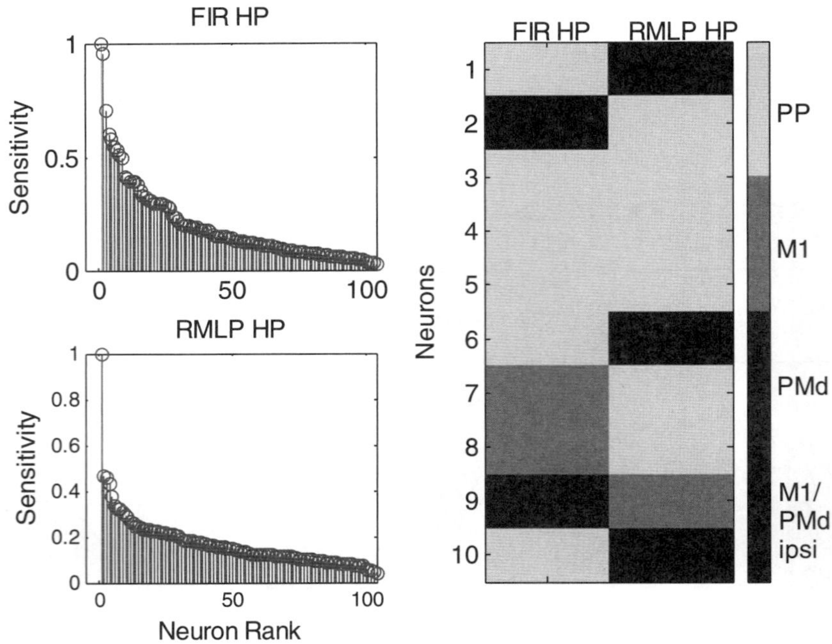

FIGURE 4.6: Reaching task neuronal sensitivities sorted from minimum to maximum for a movement. The 10 highest sensitivities are labeled with the corresponding neuron.

effect of computing the sensitivities through both WF and RMLP topologies, we have found that sensitivity-based selection of neurons is not heavily dependent on the model topology, even with two distinct topologies (linear-feedforward vs. nonlinear-feedback) are utilized [4].

As a further check of the sensitivity importance measure, we trained three identical RMLP networks of five hidden PEs, one with the 10 highest sensitivity neurons, one with the 84 intermediate sensitivity neurons, and one with the 10 lowest sensitivity neurons. A plot of the network outputs in Figure 4.7a, b, and c shows that the reduced set of highest sensitivity neurons does a good job of capturing the peaks of the movement, whereas the lowest sensitivity neurons are unable to discover the mapping from spike trains to hand movements. The remaining intermediate sensitivity neurons contain information about the movement but it is not enough to capture the full detail of the trajectory, although it does improve the overall trajectory fitting. Using the cumulative error metric (CEM)[5] for BMI performance in movements, we see that the curves approach the bisector of the space as sensitive neurons are dropped from the mapping, indicating a decrease in performance.

[5]CEM(r) is the estimated probability that the radius of the error vector is less than or equal to a certain value r.

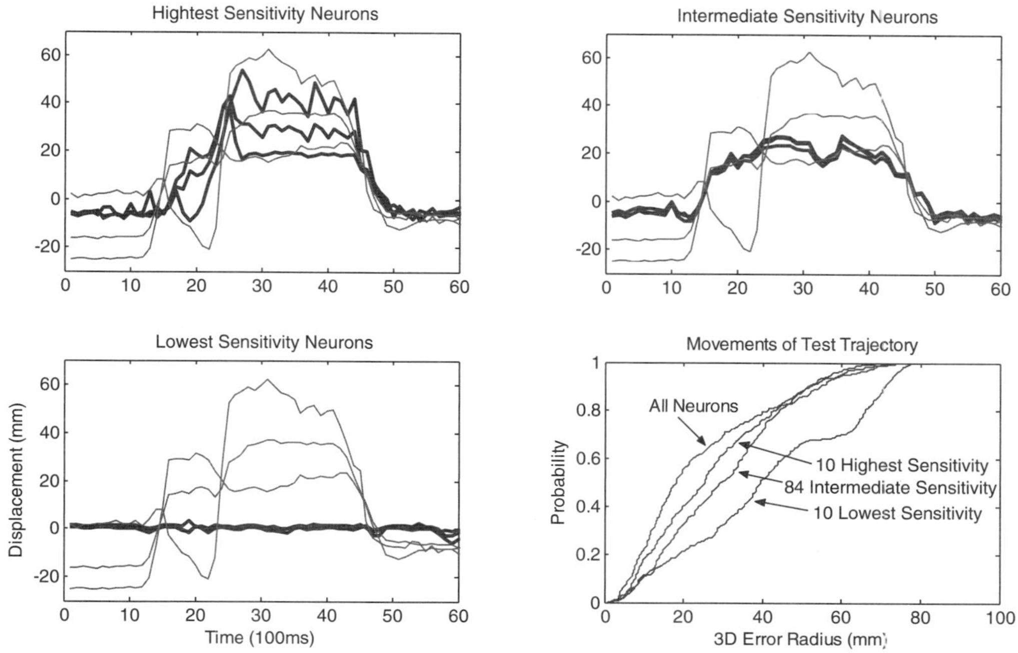

FIGURE 4.7: Testing outputs for RMLP models trained with subsets of neurons. (a–c) x, y, and z trajectories (bold) for one movement (light) from three RMLPs trained with the highest, intermediate, and lowest sensitivity neurons. (d) Cumulative error metric decreases as sensitive neurons are dropped.

This result verifies that the method of RMLP model-based sensitivity analysis for BMIs is able to produce a graded list of neurons involved in the reconstructing a hand trajectory.

We now need to address why the neurons with the highest sensitivity are important for the production of the output movement. The answer is quite simple and can be obtained by plotting the binned firing counts for highest and lowest sensitivity neurons in the vicinity of a reaching movement (Figure 4.8). In the top subplot of each column, we have included the first reaching movement as in the above simulations. For the highest sensitivity neurons, we can see some strong correlations between the neuronal firing and the rest/food, food/mouth, and mouth/rest movements. Neurons 4, 5, 7, 19, and 26 (PP) display increased firing during the rest/food movement, and are zero otherwise. Neurons 93 and 104 (M1) show a firing increase during the food/mouth movement. Neuron 45 (M1) does not correlate with the movement directly but seems to fire right before the movement, indicating that it may be involved in movement initiation. Neuron 84 (M1) fires only single pulses during the transitions of the movement. In the firing patterns of these neurons, we see the effect of the decay of the time-varying sensitivity at time t, which can be influenced by samples 2 sec in the past.

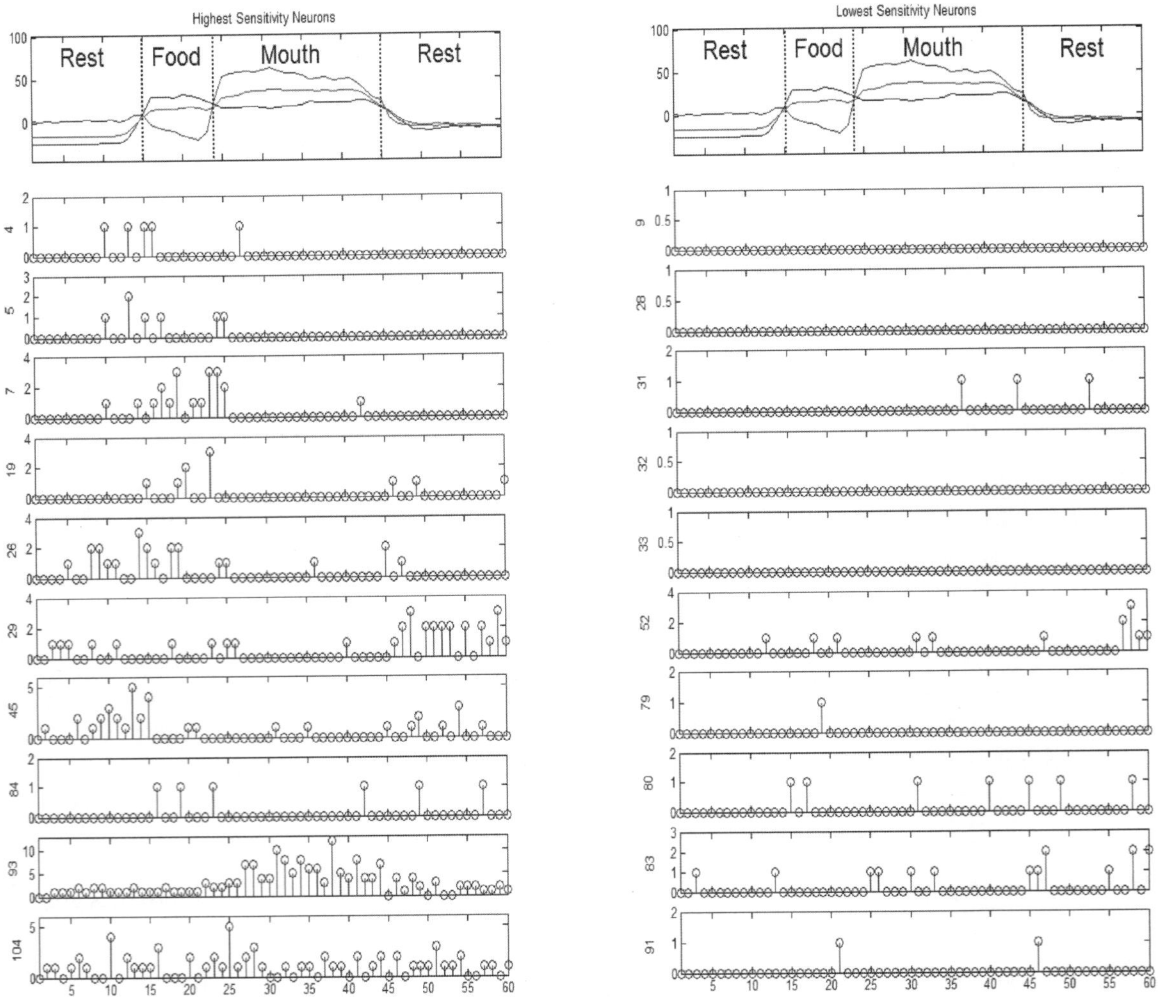

FIGURE 4.8: Neuronal firing counts from the 10 highest and lowest sensitivity neurons time-synchronized with the trajectory of one reaching movement.

All of the lowest sensitivity neurons primarily contain single firings, which do not display any measurable correlations to the movement. The data from these neurons are relatively sparse when compared to the highest sensitivity neurons, indicating that the network is not using them for the mapping.

Cellular Importance for Cursor Control Tasks. We extend our analysis methodology for selecting important neurons for constructing the BMI mapping in a cursor control task. This time, however, the analysis will include all the kinematic variables available, HP, hand velocity (HV), and grip force (GF) from a rhesus monkey's neural recordings. For each variable, separate RMLPs were trained, and the sensitivity was computed as in (4.35). For the two-dimensional (2D) cursor task, sensitivities were averaged over the x/y coordinates and an arbitrary 3000 sample window because no clear marker of movement/nonmovement could be discriminated. The sorted position, velocity, and gripping force neuronal sensitivities for each session's ensemble of neurons is plotted in Figure 4.9a and b. Like the reaching task cellular analysis, an initially sharp decrease from maximum

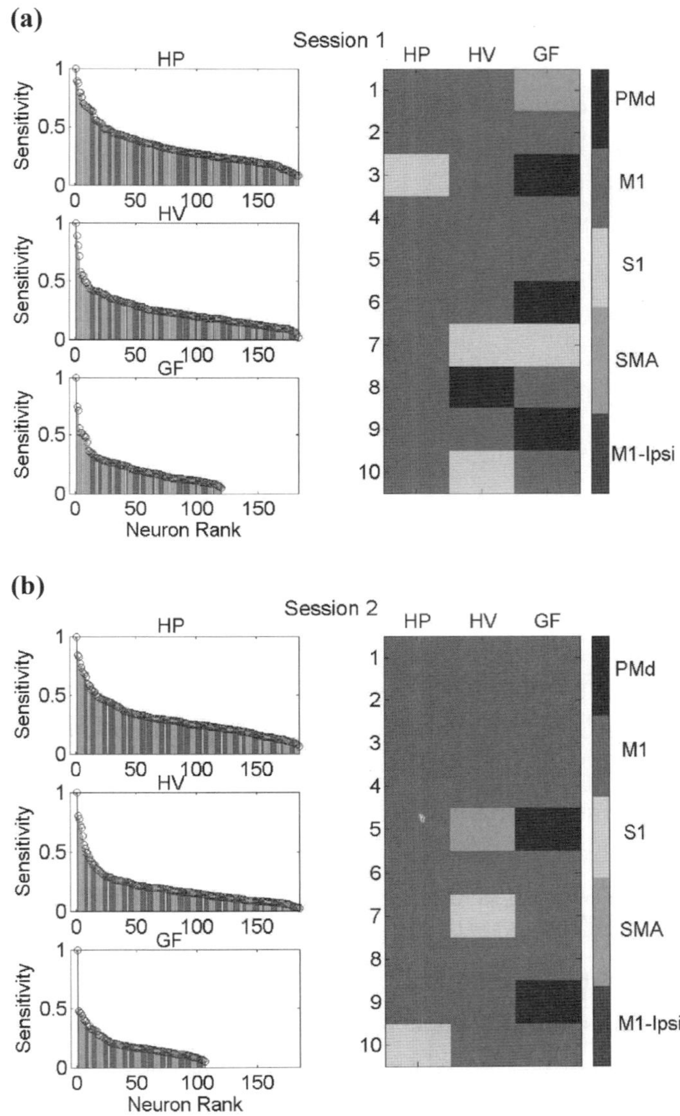

FIGURE 4.9: Sensitivity-based neuronal ranking for HP and HV for two recording sessions using an RMLP: (a) session 1 and (b) session 2. The cortical areas corresponding to the 10 highest ranking HP, HV, and GF neurons are given by the color map. HV, hand velocity; GF, grip force.

indicates that there are a few neurons that are significantly more sensitive than the rest of the ensemble. For HP and HV, the most important neurons are again mostly located in the primary motor cortex (M1). For the GF, we can also see that initially during first recording session multiple brain areas are contributing (30% premotor dorsal, 10% supplementary motor associative, 10% somatosensory), but many of these cells are replaced during session two with cells from M1.

4.2.2 L_1-norm Penalty Pruning

Recent studies on statistical learning have revealed that the L_1-norm penalty sometimes provides a better solution as a regularization method in many applications than the L_2-norm penalty [27]. Basically, the L_1-norm based regularization methods can select input variables more correlated with outputs as well as provide sparser models, compared to the L_2-norm based methods. Linear Absolute Shrinkage and Selection Operator (LASSO) has been a prominent algorithm [28] among the L_1-norm based regularization methods. However, its implementation is computationally complex. Least Angle Regression (LAR), which was recently proposed by Efron et al., provides a framework to incorporate LASSO and forward stagewise selection [29]. With LAR, the computational complexity in the learning algorithm can be significantly reduced.

The LAR algorithm has been recently developed to accelerate computation and improve performance of forward model selection methods. It has been shown in Efron et al. that simple modifications to LAR can implement the LASSO and the forward stagewise linear regression [29]. Essentially, the LAR algorithm requires the same order of computational complexity as ordinary least squares (OLS).

The selection property of LAR, which leads to zeroing model coefficients, is preferable for sparse systems identification when compared to regularization methods based on the L_2-norm penalty. Also, the analysis of the selection process often provides better insights into the unknown system than the L_2-norm based shrinkage methods.

The LAR procedure starts with an all-zero coefficients initial condition. The input variable having the most correlation with desired response is selected. We proceed in the direction of the selected input with a step size which is determined such that some other input variable attains as much correlation with the current residual as the first input. Next, we move in the equiangular direction between these two inputs until the third input has the same correlation. This procedure is repeated until either all input variables join the selection, or the sum of coefficients meets a preset threshold (constraint). Note that the maximum correlation between inputs and the residual decreases over successive selection step in order to decorrelate the residual with inputs. Table 4.4 summarizes the details of the LAR procedure [29].

The illustration in Figure 4.10 (taken from Efron et al. [29]) will help to explain how the LAR algorithm proceeds. In this figure, we start to move on the first selected input variable x_1 until the next variable (x_2 in this example) has the same correlation with the residual generated by x_1. $\boldsymbol{\mu}_1$ is the unit

TABLE 4.4: Procedure of the LAR algorithm

Given an $N \times M$ input matrix $\mathbf{X}$ (each row being M-dimensional sample vector), and an $N \times 1$ desired response matrix $\mathbf{Y}$, initialize the model coefficient $\beta_i = 0$, for $i = 1,\ldots, M$, and let $\boldsymbol{\beta} = [\beta_1, \ldots, \beta_M]^T$

Then the initial LAR estimate becomes $\hat{\mathbf{Y}} = \mathbf{X}\boldsymbol{\beta} = 0$.

Transform $\mathbf{X}$ and $\mathbf{Y}$ such that $\frac{1}{N}\sum_{i=1}^{N} x_{ij} = 0$, $\frac{1}{N}\sum_{i=1}^{N} x_{ij}^2 = 1$, $\frac{1}{N}\sum_{i=1}^{N} y_i = 0$, for $j = 1,\ldots, M$.

a) Compute the current correlation $c = \mathbf{X}^T(\mathbf{Y} - \hat{\mathbf{Y}})$

b) Find $C_{max} = \max_j \{|c_j|\}$, and a set $A = [j: |c_j| = C_{max}]$.

c) Let $\mathbf{X}_a = [\ldots, \text{sign}(c_j)\mathbf{x}_j, \ldots]$ for $j \in A$.

d) Let $\Phi = \mathbf{X}_a^T\mathbf{X}_a$, and $\alpha = (\mathit{1}_a^T\Phi^{-1}\mathit{1}_a)^{-1}$, where $\mathit{1}_a$ is a vector of 1's with a length equal to size of A.

e) Compute the equiangular vector $\mu = \mathbf{X}_a(\alpha\Phi^{-1}\mathit{1}_a)$ that has the unit length. Note that that $\mathbf{X}_a\mu = \alpha\mathit{1}_a$ (angles between all inputs in A and μ are equal).

f) Compute the step size, $\gamma = \min^{+}_{j \in A^C}\left\{\frac{C_{max} - c_j}{\alpha - \theta_j}, \frac{C_{max} + c_j}{\alpha + \theta_j}\right\}$

where $\min^{+}$ indicates considering only positive minimum values over possible j.

g) Compute θ_j, which is defined as the inner product between all inputs and μ such as,

$\theta_j = \mathbf{X}^T\mu$

h) Update $\hat{\mathbf{Y}}_+ = \hat{\mathbf{Y}} + \gamma\mu$.

Repeat a–h until all inputs join the active set A, or $\sum_j |\beta_j|$ exceeds the given threshold.

vector in this direction as computed in Table 4.4(e). The amount of movement along $\boldsymbol{\mu}_1$, denoted as γ_1, is computed by equation in Table 4.4(f). y_1 denotes the OLS estimate of desired response y with input x_1. Note that the estimate by LAR ($\hat{y}_1$) moves toward y_1, but does not reach it. The next direction $\boldsymbol{\mu}_2$ intersects the angle between x_1 and x_2 (equiangular vector in the 2D space of x_1 and x_2) such that the angle between x_1 and the updated residual ($r_1 = y_2 - \gamma_1\boldsymbol{\mu}_1$) is same as the one between x_2 and r_1. Because every input is standardized such that the correlation that is measured by the inner product of x_1 and x_2 can be estimated by the angle between x_1 and x_2, these two variables have the same absolute correlation with r_1 following the equation in Table 4.4(a). The coefficient γ_2 is computed again following Table 4.4(f), such that x_3 has the same absolute correlation with the next residual $r_2 = y_3 - (\gamma_1\mu_1 + \gamma_2\mu_2)$ as x_1 and x_2. So, the next direction μ_3 intersects the angle between x_1, x_2, and x_3. This procedure is repeated until the L_1 norm of coefficients reaches a given threshold.

LAR can be easily modified to implement LASSO; when some coefficients cross zero at a given step, those are forced to be zero. And the corresponding inputs are removed from the selected

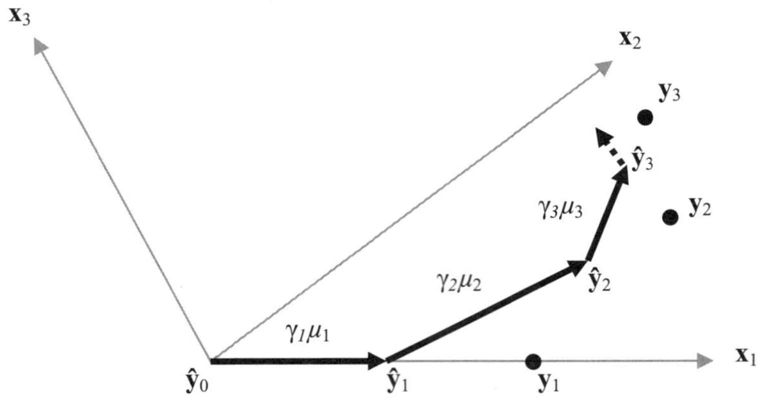

FIGURE 4.10: An illustration of the LAR procedure.

joint set. The LAR procedure can be continued with the remaining inputs because they still have the same absolute correlation with the current residual.

There are three major considerations in the implementation of LAR. First, LAR assumes the linearly independence between input variables. Second, the determination of threshold for the L_1 norm of coefficients is an open data-dependent problem. The performance of the linear model learned by LAR can be greatly influenced by a choice of this threshold. Finally, LAR was originally derived for static data. If we attempt to apply LAR to a linear model in BMIs, we have to cope with these three difficulties. Indeed, the embedded inputs are likely to be correlated with each other (although they might be linearly independent), so that LAR might not be able to operate optimally. Also, finding an optimal threshold will be a nontrivial task, but we may devise a surrogate data method to deal with this difficulty. The third issue is critical because the procedure has to be applied online for channel selection, and we will start by addressing our implementation [6] that can have wide applicability to time series modeling (not only BMIs).

Online Variable Selection. The LAR algorithm has been applied to variable selection in multivariate data analysis (static problems), but because LAR selects input variables by computation of correlation using the entire data set, it requires the assumption of stationary statistics when applied to time series. Therefore, a modified version of LAR, which selects a subset of input variables locally in time without the stationary assumption, has been proposed in Kim et al. [6]. Moreover, this algorithm provided can be computed online, and provides a real-time variable selection tool for the time-variant system identification problems.

Correlation between neuronal inputs and the desired response (hand trajectory) can be accomplished by recursively updating the correlation vector. The input covariance matrix can also be estimated recursively. By decoupling the variable selection part from the model update part in LAR, the input variables can be selected locally with recursive estimates of correlations. The modification to the LAR procedure using these recursions is described next.

The correlation in step a in Table 4.4 at a given step can be simply updated without computing residuals,

$$c_j(k) = c_j(k-1) - \gamma\theta_j \tag{4.36}$$

for the *k*th step of variable selection. Hence, the update procedure of step h of Table 4.4 can be removed. Instead of computing the correlation with entire data, we can recursively estimate the correlation using a forgetting factor ρ, given by

$$\boldsymbol{p}(n) = \rho\boldsymbol{p}(n-1) + d(n)\boldsymbol{x}(n) \tag{4.37}$$

where $\boldsymbol{x}(n)$ is an $1 \times M$ input vector at time instance n. $\boldsymbol{p}(n)$ is utilized by the LAR routine such that $c_j(0) = p_j(n)$. For the computation of the covariance matrix Φ in step d in Table 4.4, we also estimate the input covariance matrix as

$$\mathbf{R}(n) = \rho\mathbf{R}(n-1) + \boldsymbol{x}(n)^T\boldsymbol{x}(n) \tag{4.38}$$

This matrix is not directly used because Φ is the covariance of only a subset of inputs. Also, the input vectors are multiplied by the sign of correlations before computing Φ. Therefore, we need to introduce a diagonal matrix $\mathbf{S}$ whose elements are signs of $c_j(k)$ for $j \in A$. Then Φ can be computed using $\mathbf{R}(n)$ and $\mathbf{S}$ as,

$$\Phi = \mathbf{S}\mathbf{R}_a\mathbf{S} \tag{4.39}$$

where $\mathbf{R}_a$ is an $L_a \times L_a$ (L_a is the length of A) matrix representing covariance among the selected input variables. $\mathbf{R}_a$ can be given by the elements of $\mathbf{R}(n)$, that is, r_{ij} for $i, j \in A$. To remove the computation of the equiangular vector μ that requires a batch operation, we incorporate step e in Table 4.4 into step g in Table 4.4 such that

$$\theta_j = \mathbf{X}^T\mu = \mathbf{X}^T\mathbf{X}_a(\alpha\Phi^{-1}\boldsymbol{1}_a) = \alpha\,\mathbf{X}^T\mathbf{X}_a\,\Phi^{-1}\boldsymbol{1}_a \tag{4.40}$$

By noting that $\mathbf{X}^T\mathbf{X}_a$ is the *j*th columns of $\mathbf{R}(n)$, for $j \in A$ followed by multiplication with $\mathbf{S}$, we define $\mathbf{R}_{acol}$ to be a submatrix of $\mathbf{R}(n)$ consisting of the *j*th columns for $j \in A$. Then step g in Table 4.4 can be computed as

$$\theta_j = \alpha\,\mathbf{R}_{acol}\,\mathbf{S}\Phi^{-1}\boldsymbol{1}_a \tag{4.41}$$

TABLE 4.5: Online variable selection algorithm (taken from [30])

Given an $N \times M$ input matrix $\mathbf{X}$ (each row being M-dimensional sample vector), and an $N \times 1$ desired response matrix $\mathbf{Y}$, initialize $p(0) = 0$ and $\mathbf{R}(0) = 0$ Transform $\mathbf{X}$ and $\mathbf{Y}$ such that $\frac{1}{N}\sum_{i=1}^{N} x_{ij} = 0$, $\frac{1}{N}\sum_{i=1}^{N} x_{ij}^2 = 1$, and $\frac{1}{N}\sum_{i=1}^{N} y_i = 0$ for $j = 1, \ldots, M$. Update the correlation: $p(n) = (1 - \rho)p(n - 1) + \rho(d(n)x(n))$ Update the input covariance: $\mathbf{R}(n) = (1-\delta)\mathbf{R}(n - 1) + \delta(x(n)^T x(n))$
Let $\mathbf{c}(0) = p(n)$. For $k = 0, \ldots, M - 1$ Let $C_{max} = \max_j \{\lvert c_j \rvert\}$, and $A = [j: \lvert c_j(k) \rvert = C_{max}]$. Compute a diagonal matrix $\mathbf{S}$ with elements of sign of $c_j(k)$ for $j \in A$. $\Phi = \mathbf{S}\mathbf{R}_a\mathbf{S}$, where $\mathbf{R}_a$ is submatrix of $\mathbf{R}(n)$ with jth rows and jth columns for $j \in A$. $\alpha = (\mathbf{1}_a^T \Phi^{-1} \mathbf{1}_a)^{-1}$ $\theta_j = \alpha \mathbf{R}_{acol} \mathbf{S} \Phi^{-1} \mathbf{1}_a$, where $\mathbf{R}_{acol}$ is a matrix consisting of jth columns of $\mathbf{R}(n)$ for $j \in A$. Compute the step size, $\gamma = \min^{+}_{j \in A^C} \left\{ \frac{C_{max} - c_j}{\alpha - \theta_j}, \frac{C_{max} + c_j}{\alpha + \theta_j} \right\}$. Update correlation: $c_j(k) = c_j(k) - \gamma\theta_j$.

Hence, using Φ obtained by $\mathbf{R}(n)$ and $\mathbf{S}$, we can compute α and consecutively θ_j for $j \in A$. This modification removes the computation of the equiangular vector in step e in Table 4.4, which is not directly required for computing θ_j or γ.

In this way, $\mathbf{R}(n)$ and $\mathbf{p}(n)$ are estimated with the current input–output samples and fed into the modified LAR routine. Through this routine, a subset of input variables is selected with certain threshold for the L_1 norm of the coefficient vector. Therefore, it is possible to estimate which input variables are more correlated with the current desired response at every time instance. The procedure of this online variable selection is described in Table 4.5.

4.2.3 Real-Time Input Selection for Linear Time-variant MIMO Systems

In the linear MIMO system considered in this section, system outputs are assumed to be linearly and causally correlated with the neuronal spatiotemporal pattern of input channels. The temporal pattern of each input channel is represented by embedding input time series using a time delay line. In our representation, only a discrete time series is considered. Hence, an input sample is delayed by a discrete time delay operator, denoted by z^{-1} in the Z-domain. Let $x_j(n)$ be an input sample for the jth channel at time instance n. The temporal pattern of the jth channel is represented in the embedded vector space as $\boldsymbol{x}_j(n) \in \Re^L$, where L is the number of taps in the delay line. The linear approximation of the time-varying input–output mapping between all input channels and desired output signals, denoted by $y(n)$, is given by

$$y(n) = \sum_{j=1}^{M} \boldsymbol{w}_j(n)^T \boldsymbol{x}_j(n) + b(n) \tag{4.42}$$

where $\boldsymbol{w}_j(n) \in \Re^L$ is the coefficient vector for the jth channel at time instance n, $b(n)$ is bias, and M is the number of channels. Note that the operation of $\boldsymbol{w}_j(n)^T\boldsymbol{x}_j(n)$ can be considered as filtering the jth channel input signal by a linear time-variant FIR filter. Therefore, $\sum_j \boldsymbol{w}_j(n)^T\mathbf{x}_j(n)$ can be the sum of the FIR filter outputs from every channel. We can also remove the bias term $b(n)$ by normalizing input and output, forcing them to have zero-mean.

The coefficient vector $\boldsymbol{w}_j(n)$ can be adapted by a number of methods, among which the LMS algorithm plays a central role because of its computational simplicity and tracking capability. With the LMS algorithm, the coefficient vector is updated as

$$\boldsymbol{w}_j(n+1) = \boldsymbol{w}_j(n) + \eta e(n)\boldsymbol{x}_j(n) \tag{4.43}$$

for $j = 1,\ldots, M$. η is the learning rate, controlling convergence speed and the misadjustment, and $e(n)$ is the instantaneous error such that

$$e(n) = y(n) - \hat{y}(n) = y(n) - \sum_{j=1}^{M} \boldsymbol{w}_j(n)^T \boldsymbol{x}_j(n)\,. \tag{4.44}$$

See [31] for the review of the tracking linear time-variant systems using LMS.

Assuming linear independence between input channels in MIMO systems, we can apply online variable selection to channels instead of each tap output. To do so, we need to identify a variable that can represent the temporal patterns of input time series at each channel. If we consider learning a linear MIMO system, the estimation of desired signal is simply the sum of the FIR filters outputs from each channel. These individual filtered outputs indicate the relationship between desired outputs and the filtered channel input. Because online variable selection operates based on correlation, the filter output is hypothesized to be a sufficient variable to provide the correlation information between desired outputs and input temporal patterns

Hence, we choose the filter outputs as input to the online variable selection procedure. Then, the remaining question is how to learn each channel filter in real time, and here we choose to utilize LMS because of its simplicity and reasonable tracking performance in nonstationary environments [14].

Figure 4.11 depicts the overall architecture of the proposed online input channel selection approach. The embedded input vector $\boldsymbol{x}_j(n)$ at the jth channel is filtered by an FIR filter with a coefficient vector of $\boldsymbol{w}_j(n)$, yielding the filter output vector $\boldsymbol{z}(n) = [z_1(n), \ldots, z_M(n)]^T$. The autocovariance matrix $\mathbf{R}(n)$ of $\boldsymbol{z}(n)$, and the cross-correlation vector $\boldsymbol{p}(n)$ between $\boldsymbol{z}(n)$ and desired output $y(n)$, are recursively estimated by (4.37) and (4.38), respectively. Then, the online variable selection algorithm receives $\mathbf{R}(n)$ and $\boldsymbol{p}(n)$ to yield an LAR coefficient vector $\boldsymbol{g}(n) = [g_1(n), \ldots, g_M(n)]^T$. Note that some of elements in $\boldsymbol{g}(n)$ can be equal to zero because of the L_1-norm constraint in the LAR procedure (Figure 4.12). Because the estimate of desired output is here the weighted sum of the channel filter outputs, an instantaneous error becomes

$$e(n) = y(n) - \hat{y}(n) = y(n) - \mathbf{g}(n)^T \mathbf{z}(n) \tag{4.45}$$

The update of $\boldsymbol{w}_j(n)$ can then be accomplished by

$$\mathbf{w}_j(n+1) = \mathbf{w}_j(n) + \eta e(n) g_j(n) \mathbf{x}_j(n) \tag{4.46}$$

Note the difference between this update rule and the one in (4.43) by additional weights on the filter outputs.

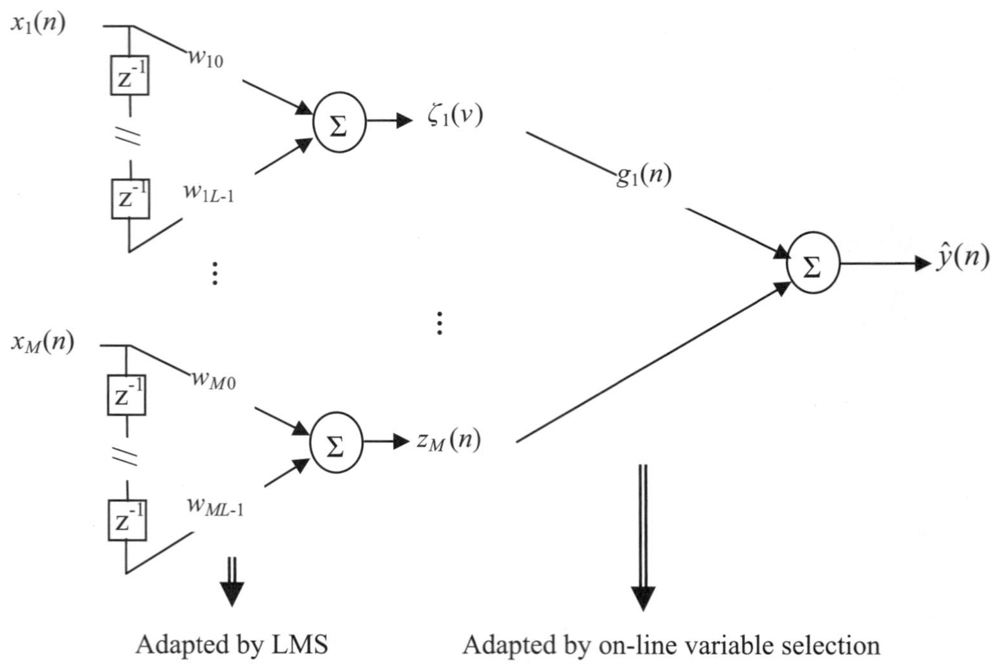

FIGURE 4.11: A diagram of the architecture of online channel selection method.

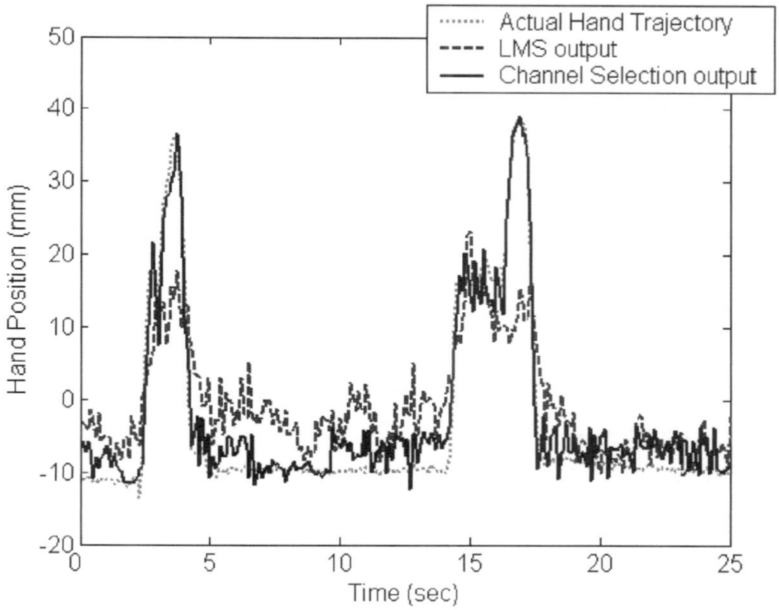

FIGURE 4.12: An example of the outputs of two tracking systems, SYS1 (dashed line) and SYS2 (solid line), on top of actual hand trajectory (dotted line).

Selection of the Threshold. The constraint on the L_1-norm of the LAR coefficients impacts the neuronal subset selection as follows: a smaller threshold of the constraint will result in a large subset, increasing a chance to select irrelevant channels, and a larger threshold will result in a smaller subset, increasing a chance to miss relevant channels. Hence, a careful approach to determine this threshold is critical. Towards this end, we propose using surrogate data to select the criterion parameters.

The surrogate data are generated to uncouple the hand movements from the neural activity using two different procedures: 1) shifting either the kinematics or the neural activity, or 2) shuffling the phase of the kinematic signals while preserving its power spectral density (PSD) to ensure the same second-order statistics after perturbation [32]. The hypothesis is that there is no correlation between the neural data and the hand movements in the surrogates. Therefore, the threshold should be set such that no subsets are found in the surrogate data sets.

The online variable selection includes two components: a threshold for the correlation between the model output $\hat{d}_{\text{LMS}}(n)$ and desired signal $d(n)$; and a threshold for the maximum correlation in the variable selection algorithm, that is $C_{\max}(j)$. The first threshold, δ_1, plays a role such that the variable selection algorithm is not activated when a correlation between $\hat{d}_{\text{LMS}}(n)$ and $d(n)$ is lower than δ_1. In this case, no subset is selected at time n. This happens when the FIR filters are not sufficiently adapted to track the target system. Accordingly, unreliable subsets may result if the selection is performed on such inadequate filter outputs. On the other hand, when the correlation exceeds δ_1, the variable selection algorithm runs until $C_{max}(j)$ becomes lower than the second threshold, δ_2. We empirically search certain values of $[\delta_1, \delta_2]$ with which the probability of selecting at least one channel is very low for both surrogate data sets, but reasonably high for the original data.

To determine δ_1, a correlation between $\hat{d}_{\mathrm{LMS}}(n)$ and $d(n)$ is recursively estimated such as

$$\xi(n) = \mu\xi(n-1) + \frac{\hat{d}_{\mathrm{LMS}}(n)d(n)}{\sqrt{p(n)q(n)}}, \tag{4.47}$$

where μ is a forgetting factor that is usually defined for the recursive least squares (RLS) [14]. $p(n)$ and $q(n)$ represent the power estimates for $\hat{d}_{\mathrm{LMS}}(n)$ and $d(n)$, respectively. Normalization by the square root of $p(n)q(n)$ prevents $\xi(n)$ from being biased to a large magnitude of $d(n)$. The power is estimated through similar recursions:

$$\begin{aligned} p(n) &= \mu p(n-1) + \hat{d}^2_{\mathrm{LMS}}(n) \\ q(n) &= \mu q(n-1) + d^2(n) \end{aligned}. \tag{4.48}$$

If $\xi(n) \geq \delta_1$, the online variable selection algorithm is activated. If $\xi(n) < \delta_1$, an empty subset is yielded and $\hat{d}_{\mathrm{LAR}}(n)$ is set equal to $\hat{d}_{\mathrm{LMS}}(n)$.

Once the online variable selection algorithm is started, δ_2 is used to stop the algorithm until $C_{max}(j) < \delta_2$ at the jth iteration. Here, we describe how $C_{max}(j)$ represents the correlation of inputs with a desired output. In the LAR, if two successively selected inputs, x_{j-1} and x_j, have similar correlations with a desired output, d, the decrease from $C_{max}(j-1)$ to $C_{max}(j)$ will be small. On the other hand, if x_{j-1} has more correlation than x_j, $C_{max}(j)$ will be much smaller than $C_{max}(j-1)$. This is illustrated in Figure 4.13. Consider the data $[\mathbf{X}, \boldsymbol{d}]$, where $\mathbf{X}$ is an input matrix whose rows are input samples, and $\boldsymbol{d}$ is an output vector. Suppose that $\mathbf{X}$ has two columns, x_1 and x_2. We assume that x_1, x_2, and $\boldsymbol{d}$ are standardized with zero mean and unit variance. Suppose x_1 has more correlation with $\boldsymbol{d}$. The LAR starts to move in the direction of x_1. It finds the coefficient β_1 for x_1 such that $|x_1^T r_1| = |x_2^T r_1|$, where $r_1 = \boldsymbol{d} - \beta_1 x_1 \equiv \boldsymbol{d} - y_1$. Then, the maximum correlation changes from $C_{max}(0) = |x_1^T \boldsymbol{d}|$ to $C_{max}(1) = |x_1^T r_1| = |x_2^T r_1|$. From this, we can see that the angle between x_1 and r_1 is equal to the angle between x_2 and r_1—the equiangular property of the LAR. $C_{max}(j)$ is related with the angles such that $C_{max}(j) \approx \cos\theta_j$, where θ_j represents the angle at the jth step. The diagram on the left side of Figure 4.13 illustrates the case when x_1 and x_2 have similar correlations with $\boldsymbol{d}$. In this case, a small difference between θ_0 and θ_1 leads to a small decrease from $C_{max}(0)$ to $C_{max}(1)$. On the other hand, the diagram on the right side illustrates the case when x_2 is considerably less correlated with $\boldsymbol{d}$ than x_1. In this case, a large difference between θ_0 and θ_1 causes $C_{max}(1)$ to decrease significantly.

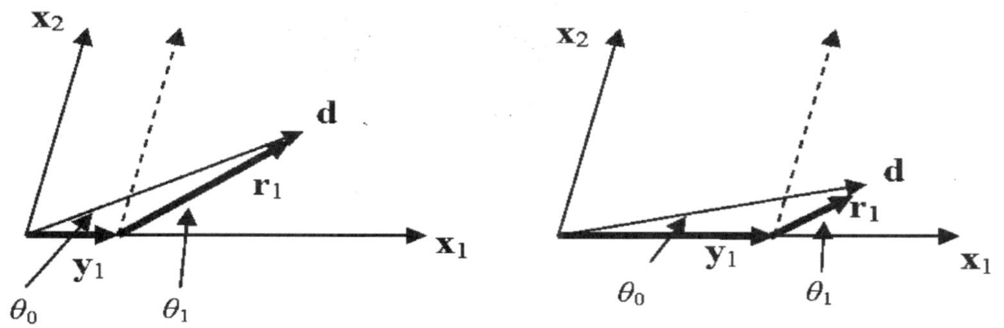

FIGURE 4.13: Illustration of the maximum correlation in the online variable selection algorithm. Note that **d** represents a desired output projected in the input space.

Therefore, the curve of $C_{max}(j)$ over iterations can represent the correlation between each selected input and a desired output.

4.3 EXPERIMENTAL RESULTS

We first determined two thresholds, δ_1 and δ_2, by applying the linear adaptive systems to both the original and surrogate data sets. The first surrogate data are composed of neuronal inputs and the shifted HPs. The HP data are shifted by 5 sec for a hand reaching task because successive reaching movements has an interval of approximately 10 sec. For the 2D cursor control task, the HP data are shifted by 10 sec to sufficiently destroy synchronization. The second surrogate data consist of neuronal inputs and the shuffled hand trajectory data in which the phase of the HP signals are shuffled while preserving PSD.

Figure 4.14 shows an example of the curve of $C_{max}(j)$ for the case when we use the original data and the surrogate data, respectively. In this example, $C_{max}(j)$ for both the original and the surrogate data were evaluated at the same time instance. A large difference between the C_{max} curves can be observed; for example, three channels were selected for the original data whereas only one channel was selected for the surrogate data. The dotted line indicates the threshold of selection.

The linear adaptive systems were implemented for 2000-sec (20 000 samples three-dimensional (3D) data and 1600-sec (16 000 samples) 2D data sets. The online variable selection algorithm starts after 100 sec for the 3D data, and 400 sec for the 2D data, in order to allow the LMS to sufficiently adapt filter weights in the beginning. A learning rate of LMS is set such that the sum of the FIR filter outputs without subset selection can track the hand trajectory reasonably well in the original data. Note that it must not be set too large, otherwise the filter weights change too rapidly over time, which will lead to a large misadjustment; therefore, the estimation of the correlation of individual neuronal channels from the filter outputs becomes unreliable.

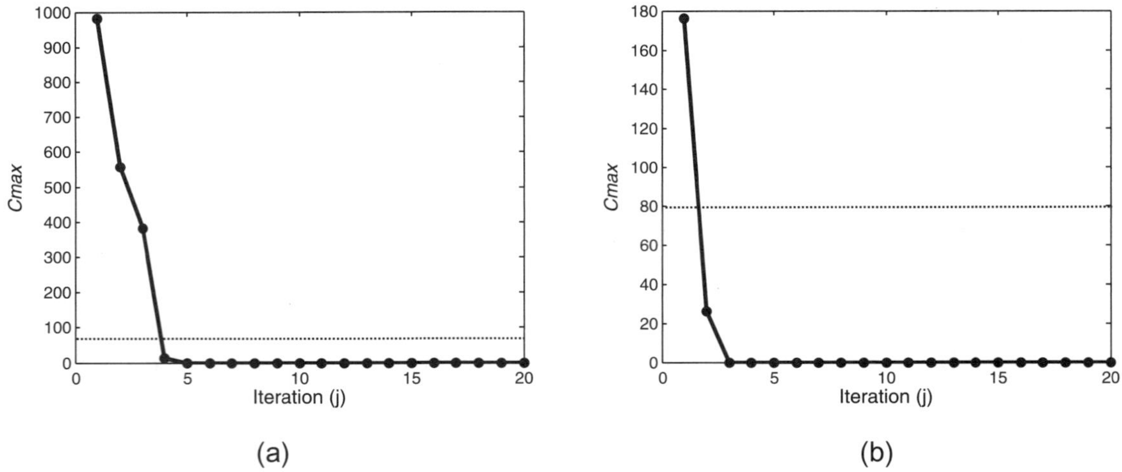

FIGURE 4.14: Examples of the maximum absolute correlation curves over variable selection iterations for (a) the original data and (b) the surrogate data.

From experiments with the surrogate data sets, the first threshold, δ_1, is determined such that few neuronal channels are selected in the surrogate data. The model parameters and the thresholds are summarized in Table 4.6. Note that δ_1 is determined much lower for more complex target hitting data, indicating that the correlation of the FIR filter outputs with kinematics tends to be lower for these data.

We show examples of neuronal subset selection results for the original and surrogate hand reaching data, with the thresholds setting as described earlier, in Figure 4.15. Figure 4.15a shows

TABLE 4.6: The model parameters and neuronal subset selection thresholds

PARAMETER	3D	2D
A learning rate for NLMS; η in (4.46)	0.1	0.3
A forgetting factor for FIR filter correlation; μ in (4.47)	0.95	0.95
A forgetting factor for channel covariance; ρ in (4.38)	0.8	0.9
A threshold for activation of variable selection; δ_1	0.7	0.3
A threshold for correlation in variable selection; δ_2	80	80

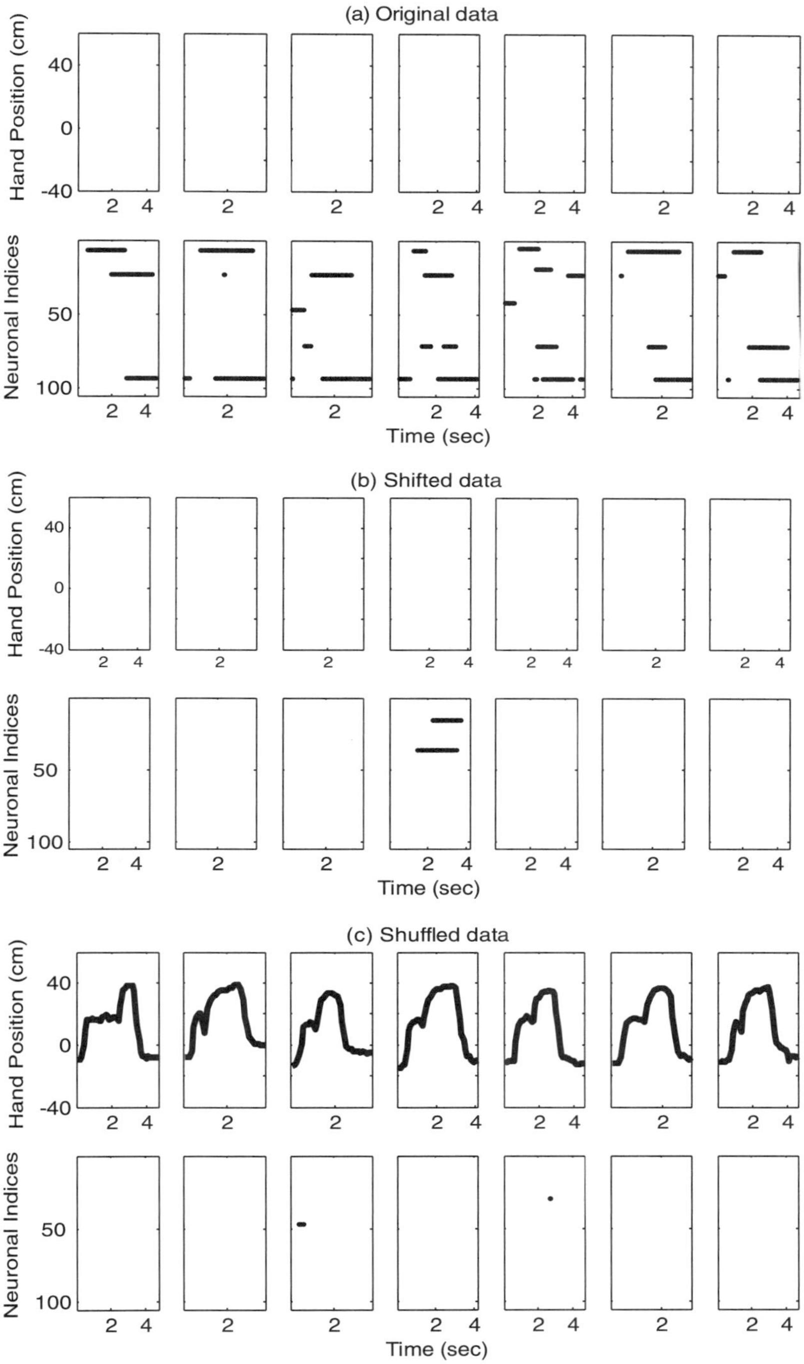

FIGURE 4.15: Neuronal subset examples for the identical 3D reaching movements using different data: (a) the original data; (b) the shifted data by 5 sec; (c) the shuffled data. Black dots in the below figures denote the selected neuronal channels.

the examples of neuronal subset selection for the original data. Figure 4.15b and c show the examples for the shifted data and the shuffled data, respectively. The neuronal subsets are presented along with the corresponding HPs (z coordinate) of seven reaching movement examples. It is clear that very few subsets are selected for the surrogate data. To quantify these subset selection results, we define a selection rate as the average number of neuronal channels per time instance. From the experimental results, the selection rates were 0.006 ± 0.006 for the original hand reaching data, 0.001 ± 0.003 for the shifted data, and $9.1 \times 10^{-4} \pm 0.004$ for the shuffled data, respectively. For 2D cursor control data, the selection rates were 0.015 ± 0.009 for the original data, and 0.002 ± 0.006 for the shifted data (the selection rate for the shuffled data is considerably lower). Therefore, these results demonstrate that we can determine the thresholds using the surrogate data with which neuronal subsets selected from the synchronized data is determined to represent real-time correlations of neuronal activities and kinematics.

We examined if the neuronal subsets showed dependency on the initial condition of the linear adaptive system. We first define a selection vector as

$$\mathbf{s}(n) = [s_1(n), s_2(n), \cdots, s_M(n)]^T \tag{4.49}$$

where $s_j(n) = 1$ if the jth neuronal channel was selected at the nth bin, and $s_j(n) = 0$ otherwise. We repeated the neuronal subset selection analysis 100 times for different initial filter weights. Then, an average of $\mathbf{s}(n)$, denoted as $\underline{\mathbf{s}}(n)$, over 100 repetitions was computed. If the linear adaptive system is robust to initial conditions, the nonzero elements of $\underline{\mathbf{s}}(n)$ should be close to 1. Hence, we examined the statistics of those nonzero elements. For 3D food reaching data, the average and standard deviation were 0.85 and 0.26, respectively. However, if we are only concerned with the subset selection for movements, the average and standard deviation estimated only during movements were 0.94 and 0.15, respectively. Because we are typically more interested in neuronal subsets during given movements, the latter result demonstrates that the linear system is reasonably robust to initial conditions. For 2D cursor control data, the average and standard deviation of the nonzero elements of $\underline{\mathbf{s}}(n)$ were 0.91 and 0.19, respectively, which again demonstrates the robustness to initial conditions.

To find neuronal subsets without separation into each spatial coordinate, we performed the analysis for individual coordinates and combined the resulting subsets into one. To that end, we performed a Boolean OR operation with $\mathbf{s}(n)$ resulted from every coordinate analysis. This operation yields a single joint neuronal subset at each time instance. In other words, if a channel is selected for at least one coordinate of HP, it belongs to the joint subset.

To investigate the temporal variation of neuronal subsets, we tracked the joint subsets over the entire data. To focus the analysis only on the movement portion of the 3D data, we divided the entire neuronal subsets into individual segments corresponding to each reaching movement. This resulted in 149 movement segments. In each segment, we collected neuronal channels that were selected at least for two consecutive bins. Figure 4.16 displays this analytical result. From this figure, we can observe which neurons were consistently selected through the entire set of reaching movements. For instance, neurons indexed by 5, 7, 23, 29, and 93 are shown to be selected in most segments. On the other hand, neurons indexed by 19, 45, 70, 71, and 84 are partially selected in some segments. In particular, neuron 71 seems to be consistently selected in the late segments.

This observation is further demonstrated in Figure 4.17. In this figure, we displayed neuronal subset examples in the early movement segments (485, 509, and 600 sec from the beginning of subset selection) and the late movement segments (1666, 1713, and 1781 sec). It can be observed from this figure that there are neurons that are consistently selected such as 5, 7, and 93 for both early and late segments. However, some neurons selected in the early segments are not selected in the late segments. For instance, neuron 70 is no longer selected in the late segments. Also, other neurons (such as 71) are not selected in the early segments but selected in the late segments. It is interesting to see this transition of a correlation with kinematics from the neuron 70 to 71 through time, because the activities of those neurons are collected from the adjacent electrodes in the PMd area.

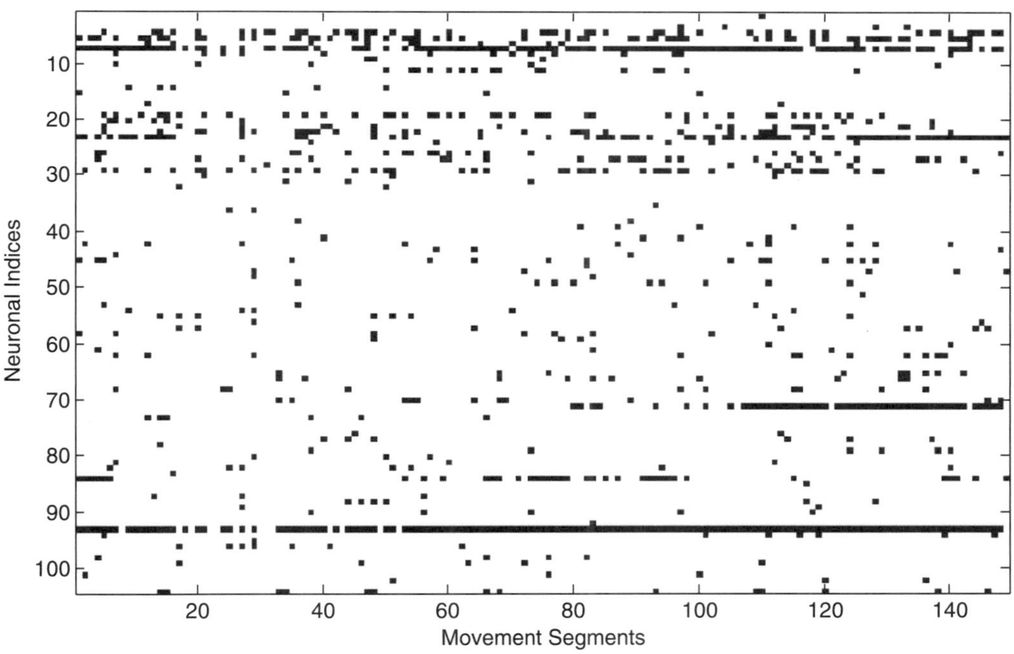

FIGURE 4.16: Temporal patterns of neurons selected during 3D food reaching movements. Black dots represent selection of neurons for each movement segment (see text). The segment indices are sequentially ordered in time.

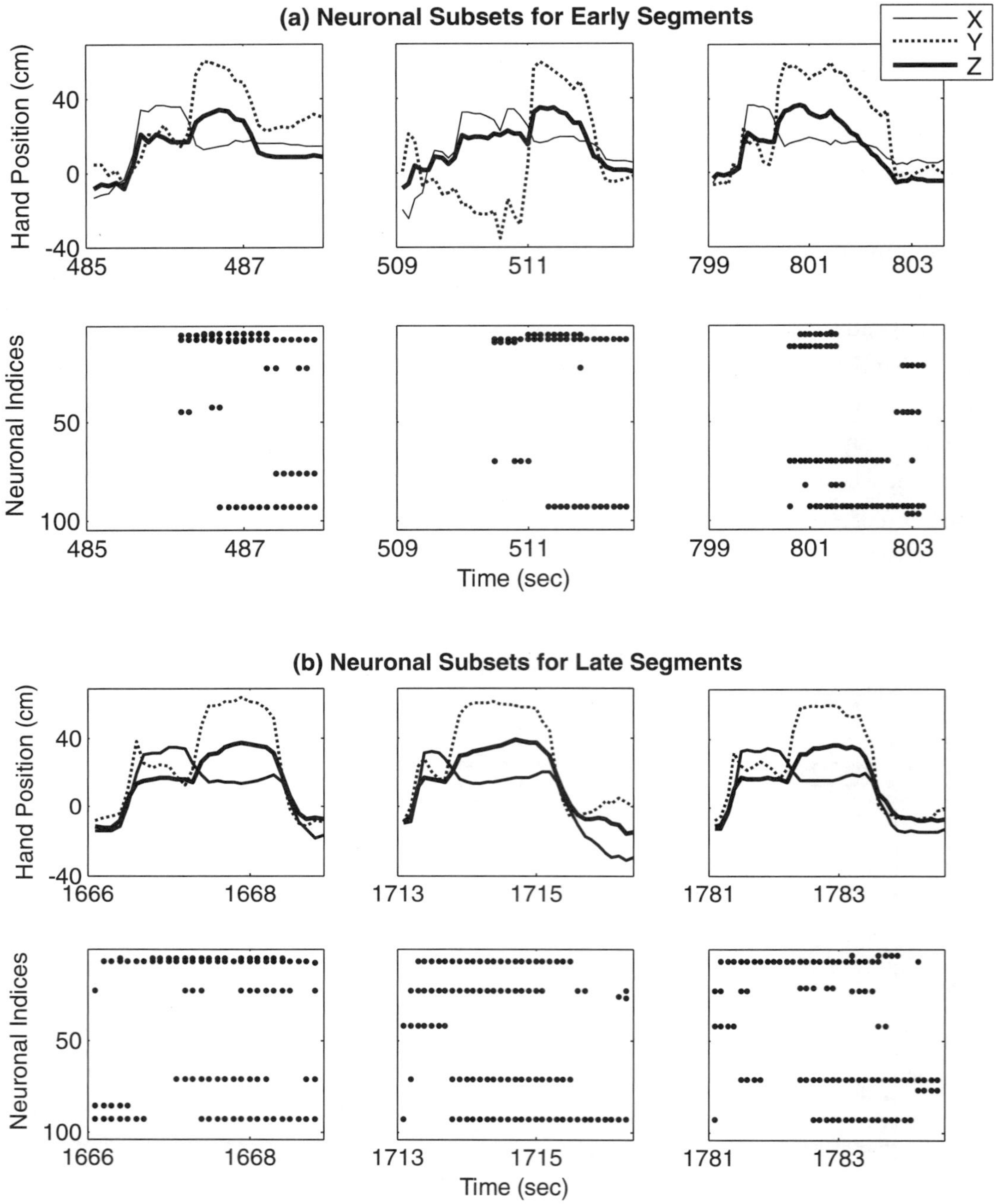

FIGURE 4.17: Neuronal subset examples for the entire 3D food reaching movement segments.

Similar observations were made for the 2D cursor control data. In Figure 4.18, neuronal subset examples for five segments with specific hand movement are displayed. In these segments, the animal moved the hand in the positive diagonal direction; that is, the hand was moved back and forth in 270° and 45° for about 2.5–5 sec. From these examples, we can observe neurons that are consistently selected such as 69, 80, 84, 92, 99, 108, and 110. However, there are a number of neurons that are selected only in particular segments such as 45, 54, 67, 149, etc. Most of selected neurons were collected from the M1 area, yet some neurons from the PMd area join the subsets in the last two segments (late parts of the data set).

When we compared neuronal subsets from our analysis to the sensitivity analysis obtained from the trained models in the same data set [7], we were able to find consistency among neuronal contributions. For instance, neurons 5, 7, 23, 71, and 93 in the 3D food reaching data were observed in the top-ranked group of neurons sorted by sensitivity analysis. Also, neurons 54, 69, 80, 84, 92,

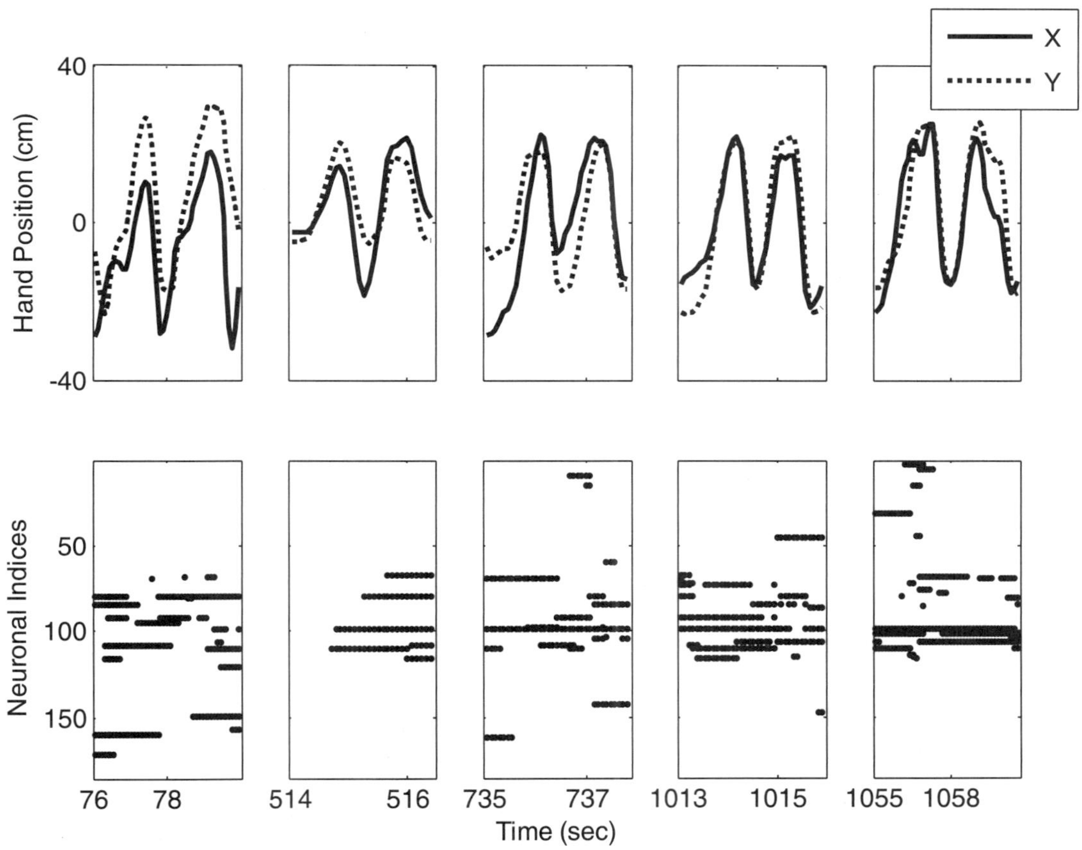

FIGURE 4.18: Neuronal subset examples for the 2D cursor control task.

99, 108, 110, 149, and 167 in the2D target hitting data were similarly observed in the top-ranked sensitivity group. These comparisons show that neuronal subsets selected in our analysis significantly contribute to building decoding models (nonlinear as well as linear models). The advantage of this method would be the capability of detecting time-varying changes of composition of subsets.

To demonstrate the effect of the time-variant relationship between the neural activity and kinematics to the decoding, we performed a brief statistical test on the decoding performance using the linear filter. We first divided the entire data into three disjoint sequential segments, say, SEG1, SEG2, and SEG3 (such that SEG1 preceded SEG2 and SEG3). Then, the first linear filter was built using samples randomly drawn from SEG1, where the Wiener–Hopf solution was used to learn the filter coefficients [14]. The second linear filter was built using samples randomly drawn from SEG2. After building these filters, we drew samples randomly from the last segment SEG3, and used them to test each filter's decoding accuracy. The accuracy was measured by the mean absolute error (MAE), which indicates the mean distance between the actual hand trajectory and the estimated trajectory. We hypothesized that the decoding accuracy of two filters should be similar to each other if the relationship of neural activity to kinematics is unchanged. We repeated this building and decoding procedure multiple times, and obtained a set of MAE values for each filter. Then, a nonparametric statistical method (the Kolmogorov–Smirnov (KS) test) was applied to test if two empirical distributions of the MAE from each filter were statistically equal. Specifically, the null hypothesis of the KS test was given by,

$$H_0 : F_1(|\mathbf{e}|) \geq F_2(|\mathbf{e}|), \tag{4.50}$$

where $F_i(|\mathbf{e}|)$ is an empirical cumulative density function (CDF) of the MAE from SEG1 (i = 1) or SEG2 (i = 2). If H_0 is rejected, the alternative hypothesis is then $F_1(|\mathbf{e}|) < F_2(|\mathbf{e}|)$, which indicates that the second linear filter yielded less errors.

The size of SEG1 or SEG2 was 800 sec long for the 3D data and 700 sec for the 2D data. We randomly drew 4000 samples from each segment to build the linear filter. The size of SEG3 was 400 sec long for the 3D data and 200 sec for the 2D data. We randomly drew 500 samples from SEG3 to test the linear filter. The procedure was repeated 1000 times. The empirical statistics are

TABLE 4.7: Average MAE using two linear filters (in centimeters)

	3D	2D
First filter (SEG1)	32.83 ± 3.43	19.39 ± 0.47
Second filter (SEG2)	17.94 ± 0.61	18.64 ± 0.37

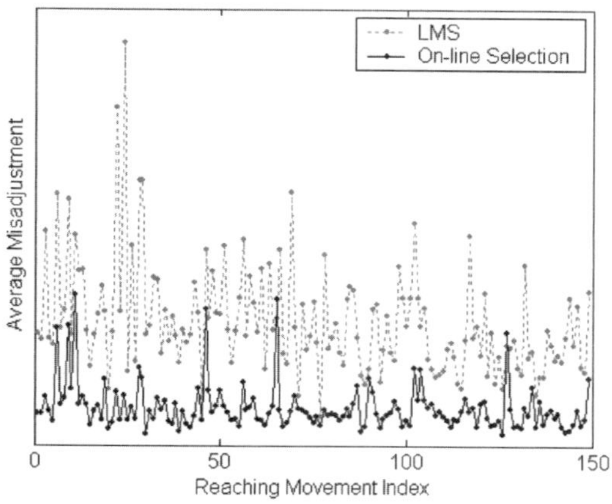

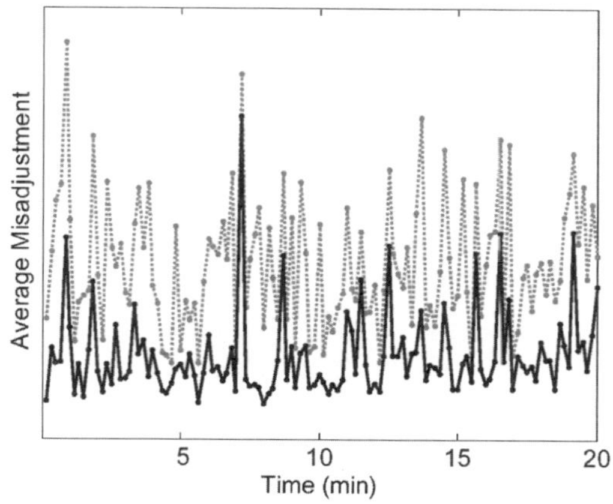

FIGURE 4.19: A comparison of misadjustments between the online neuronal channel selection and the NLMS: (a) 3D data, (b) 2D data.

described in Table 4.7, in which it is clearly shown that the second linear filter decoded kinematics from neural activity for samples in SEG2 better than the first filter. The statistical test showed that the null hypothesis was rejected for both 3D and 2D data sets ($P < 10^{-12}$). These results imply that we may need to rebuild the linear filter during the session to track the change of the relationship between neural activity and kinematics.

Finally, we investigated the misadjustment of two linear adaptive systems, one of which was adapted using the NLMS and the online subset selection algorithm, whereas the other was adapted using only the NLMS. For the 3D data, we computed the average misadjustment for each movement segment. For the 2D data in which the continuous movement was recorded, we arbitrarily divided the data into 10-sec segments and computed the average misadjustment for each segment. Figure 4.19 shows the results of these average misadjustments for each data set. In this figure, the linear adaptive system with the online subset selection exhibits smaller misadjustments than the normal linear adaptive system for almost every segment. This result is consistent with a previous demonstration of the superior tracking performance of the online subset selection algorithm, as reported in [33].

4.4 SUMMARY

A reduction in the number of free parameters without affecting performance leads directly to BMI systems that require less power, bandwidth, and computational demands. Solving these challenges will bring us one step closer to real, portable BMIs. Regularization and analysis of the most important neurons in the model-dependent methods has opened up directions to better understand how neural activity can be effectively used for BMI design. Since the work of Wessberg et al, it is known that the performance of linear models for BMIs improves as more neurons are recorded, but that the performance improvement must be coupled with the *right* neurons. Irrelevant neurons can add to the model bias and variance, thus reducing performance. One must also remember that if channel selection is used the reduced number of cells makes BMIs more sensitive to the instability of neuronal firings over time. Recent studies are showing that the activity of individual neurons and cortical areas used in BMI experiments can vary considerably from day to day [34]; therefore, the variance over time of the importance of neurons must be quantified in future studies. It is therefore not clear how practical the neuron selection techniques will be in the surgery stage. For these reasons, we advocate the use of a higher sampling of the cortical activity to help improve this ratio until other models are proposed that take advantage of the information potentially contained in the spike trains and not exploited by linear models.

In this model-based framework, it is clear that better BMIs can be built by combining regularization techniques with a subset of important cells. This question is rooted in our goal to build models for BMIs that generalize optimally. It should not be forgotten that regularization and channel selection is biased by the type of model chosen, its performance level, and by noise in the data. Therefore, it is important to pursue a model-independent approach to preprocess the data. We hypothesized from a combined neurophysiological and modeling point of view that highly modulated neurons spanning the space of the kinematic parameter of interest should be chosen. Intuitively, these constraints make sense for the following reasons: If a cell is modulated to the kinematic parameter, the adaptive filter will be able to correlate this activity with the behavior through training. Otherwise, neural firing works as a broadband excitation that is not necessarily related to better performance. If a group of cells are highly modulated to only a part of the space, the adaptive filter may not be able to reconstruct data points in other parts of the space.

A final comment goes to the overall performance of the BMI system built from adaptive models. Although it is impressive that an optimal linear or nonlinear system is able to identify the complex relations bnetween spike trains in the motor cortex and hand movements/gripping force, a correlation coefficient of ~0.8 may not be sufficient for real-world applications of BMIs. Therefore, further research is necessary to understand what is limiting the performance of this class of adaptive linear and nonlinear systems. Another issue relates to the unrealistic assumption of stationarity in neural firings over recording sessions that is used to derive results presented in this. In future studies, it will be necessary to assess the time variability of the neuronal rankings and determine its effect on model generalization.

REFERENCES

1. GemanS., E. Bienenstock, E., and R. Doursat, *Neural networks and the bias/variance dilemma.* Neural Computation, 1992. **4**: pp. 1–58.

2. Sanchez, J.C., et al., *Input–output mapping performance of linear and nonlinear models for estimating hand trajectories from cortical neuronal firing patterns*, in International Work on Neural Networks for Signal Processing. 2002. IEEE, Martigny, Switzerland.
3. Sanchez, J.C., et al., *A comparison between nonlinear mappings and linear state estimation to model the relation from motor cortical neuronal firing to hand movements*, in SAB Workshop on Motor Control in Humans and Robots: on the Interplay of Real Brains and Artificial Devices. 2002. University of Edinburgh, Scotland.
4. Sanchez, J.C., et al., *Interpreting neural activity through linear and nonlinear models for brain machine interfaces*, in International Conference of Engineering in Medicine and Biology Society. 2003. Cancun, Mexico.
5. Wahba, G., Spline Models for Observational Data. 1990, Montpelier: Capital City Press.
6. Kim, S.-P., J.C. Sanchez, and J.C. Principe, *Real time input selection for linear time-variant MIMO systems.* Optimization Methods and Software, 2007. **22**: pp. 83–98.
7. Sanchez, J.C., et al., *Ascertaining the importance of neurons to develop better brain machine interfaces.* IEEE Transactions on Biomedical Engineering, 2003. **61**(6): pp. 943–953.
8. Hadamard, J.P.U.B., *Sur les problèmes aux dérivées partielles et leur signification physique*. Princeton University Bulletin, 1902: pp. 49–52.
9. Tikhonov, A. and V. Arsenin, Solution of Ill-Posed Problems. 1977, Washington: Winston.
10. Neal, R., Bayesian Learning for Neural Networks. 1996, Cambridge: Cambridge University Press.
11. Vapnik, V., The Nature of Statistical Learning Theory. Statistics for Engineering and Information Science. 1999, New York: Springer-Verlag.
12. Stewart, G.W., Introduction to Matrix Computations. 1973, New York: Academic Press.
13. Klema, V.C. and A.J. Laub, *The singular value decomposition: Its computation and some applications.* IEEE Transactions on Automatic Control, 1980. **AC-25**: pp. 164–176.
14. Haykin, S., Adaptive Filter Theory. 3rd ed. 1996, Upper Saddle River, NJ: Prentice-Hall International.
15. Hoerl, A.E. and R.W. Kennard, *Ridge regression: Biased estimation for nonorthogonal problems.* Technometrics, 1970. **12**(3): pp. 55–67.
16. Weigend, A.S., D.E. Rumelhart, and B.A. Huberman, *Generalization by weight-elimination with application to forecasting*. Advances in Neural Information Processing Systems 3. R.P. Lippmann, J. Moody, and D.S. Touretzky, eds. 1991. pp. 875–882, Morgan Kaufmann: San Mateo, CA.
17. Larsen, J., et al., *Adaptive regularization in neural network modeling*, Neural Networks: Tricks of the Trade, G.B. Orr and K. Muller, eds., 1996, Germany: Springer, pp. 113–132.
18. Geisser, S., *The predictive sample reuse method with applications.* Journal of the American Statistical Association, 1975. **50**: pp. 320–328.

19. Principe, J.C., B. De Vries, and P.G. Oliveira, *The gamma filter—a new class of adaptive IIR filters with restricted feedback.* IEEE Transactions on Signal Processing, 1993. **41**(2): pp. 649–656.
20. Sandberg, I.W. and L. Xu, *Uniform approximation of multidimensional myopic maps.* IEEE Transactions on Circuits and Systems, 1997. **44**: pp. 477–485.
21. De Vries, B. and J.C. Príncipe, *The gamma model: a new neural network model for temporal processing.* Neural Networks, 1993. **5**: pp. 565–576.
22. de Jong, S., *SIMPLS: An alternative approach to partial least squares regression.* Chemometrics and Intelligent Laboratory Systems, 1993. **18**: pp. 251–263.
23. Stone, M. and R.J. Brooks, *Continuum regression: cross-validated sequentially constructed prediction embracing ordinary least squares, partial least squares and principal components regression (with discussion).* Journal of Royal Statistical Society, Series B, 1990. **52**: pp. 237–269.
24. Kim, S.P., et al. *A hybrid subspace projection method for system identification*, in Proceedings of the International Conference on Acoustics, Speech, and Signal Processing. 2003.
25. Fetz, E.E., *Are movement parameters recognizably coded in the activity of single neurons.* Behavioral and Brain Sciences, 1992. **15**(4): pp. 679–690.
26. Fu, L. and T. Chen. *Sensitivity analysis for input vector in multilayer feedforward neural networks*, in IEEE International Conference on Neural Networks. 1993. San Francisco, CA.
27. Hastie, T., R. Tibshirani, and J. Friedman, Elements of statistical learning: data mining, inference and prediction. 2001, New York: Springer-Verlag.
28. Tibshirani, R.J., *Regression shrinkage and selection via the lasso.* Royal Statististical Society B, 1996. **58**(1): pp. 267–288.
29. Efron, B., et al., *Least angle regression.* Annals of Statistics, 2004. **32**: pp. 407–499.
30. Kim, S.-P., et al., *Tracking multivariate time-variant systems based on on-line variable selection*, in IEEE Workshop on Machine Learning for Signal Processing. 2004: Sao Luis, Brazil.
31. Haykin, S., et al., *Tracking of Linear Time-Variant Systems,* in IEEE MILCOM. 1995.
32. Palus, M. and D. Hoyer, *Detecting nonlinearity and phase synchronization with surrogate data.* IEEE Engineering in Medicine Biology Magazine, 1998. **17**(6): pp. 40–45.
33. Kim, S.P., J.C. Sanchez, and J.C. Principe, *Real time input subset selection for linear time-variant MIMO systems.* Optimization Methods and Software, 2007. **22**(1): pp. 83–98.
34. Carmena, J.M., et al., Learning to control a brain–machine interface for reaching and grasping by primates. PLoS Biology, 2003. **1**: pp. 1–16.

• • • •

CHAPTER 5

Neural Decoding Using Generative BMI Models

Additional Contributors: Yiwen Wang and Shalom Darmanjian

This chapter will address generative models for BMIs, which are a more realistic modeling approach because they take into consideration some of the known features of motor cortex neurons. This chapter still uses neuronal firing rates as inputs to the model, but the mathematical foundations can also be applied, with appropriate adaptations, to spike trains as will be discussed in the next chapter. Generative models are more in tune with the physical neural systems that produce the data and therefore are examples of "gray box" models. Three examples of "gray box" models can be found in the BMI literature. One of the most common examples is Georgopoulos population vector algorithm (PVA) [1] as we briefly mentioned in Chapter 3. Using observations that cortical neuronal firing rates were dependent on the direction of arm movement, a model was formulated to incorporate the weighted sum of the neuronal firing rates. The weights of the model are then determined from the neural and behavioral recordings. A second example is given by Todorov's work who extended the PVA by observed multiple correlations of M1 firing with movement position, velocity, acceleration, force exerted on an object, visual target position, movement preparation, and joint configuration [2–13]. With these observations, Todorov proposed a minimal, linear model that relates the delayed firings in M1 to the sum of many mechanistic variables (position, velocity, acceleration, and force of the hand) [14]. Todorov's model is intrinsically a *generative model* [15, 16]. Using knowledge about the relationship between arm kinematics and neural activity, the *states* (preferably the feature space of Todorov) of linear or nonlinear *dynamical* systems can be assigned. This methodology is supported by a well known training procedure developed by Kalman for the linear case [17], and has been recently extended to the nonlinear case under particle filters [18] and other graphical models or Bayesian network frameworks [19]. Because the formulation of generative models is recursive in nature, the model is well suited for learning about motor systems because the states are all intrinsically related in time. The third modeling approach is the hidden Markov model (HMM), a graphical model where the dependencies over time are contained in the present state, and the observations are assumed to be discrete. HMM models are the leading technology in speech recognition because they are able to capture very well the piecewise nonstationary of speech [20]. It turns out that speech production is ultimately a motor function, so HMMs can potentially be also very useful for motor BMIs. One of the major differences is that there is a well established finite repertoire of speech atoms called the phonemes, which allow for the use of the HMM framework. Research into animal motor control is less advanced, but recently there has been interest in studying complex motor actions as a succession of simpler movements that can be properly called "movemes" [21]. If this research direction meets the expectation, then HMMs would play a more central role in motor BMIs because the model building problem would be similar to speech recognition.

5.1 POPULATION VECTOR CODING

As presented in the Introduction, the population vector algorithm is a physiologically based model that assumes that a cell's firing rate is a function of the velocity vector associated with the movement performed by the individual. Population vector coding is based on the use of tuning curves [22], which in principle provide a statistical relationship between neural activity and behavior. The tuning or preferred direction [2] of each cell in the ensemble convey the expected value of a probability density function (PDF) indicating the average firing a cell will exhibit given a particular movement direction. The PVA model relating the tuning to kinematics is given by (5.1)

$$s_n(\mathbf{V}) = b_0^n + b_x^n v_x + b_y^n v_y + b_z^n v_z = \mathbf{B} \cdot \mathbf{V} = |\mathbf{B}||\mathbf{V}|\cos\theta\ , \tag{5.1}$$

where the firing rate s for neuron n is a weighted ($b^n_{x,y,z}$) sum of the components ($v_{x,y,z}$) of the unit velocity vector $\mathbf{V}$ of the hand plus mean firing rate b^0_n. The relationship in (5.1) is the inner product between the velocity vector of the movement and the weight vector for each neuron, that is, the population vector model considers each neuron independently of the others (see Figure 5.1). The inner product (i.e., spiking rate) of this relationship becomes maximum when the weight vector $\mathbf{B}$ is collinear with the velocity vector $\mathbf{V}$. At this point, the weight vector $\mathbf{B}$ can be thought as the neuron's preferred direction for firing because it indicates the direction for which the neuron's activity will be maximum. The weights b_n can be determined by least squares techniques [1]. Each neuron makes a vector contribution w in the direction of P_i with magnitude given in (5.2), where b_0 is the mean firing rate for neuron n. The resulting population vector or movement is given by (5.3), where the reconstructed movement at time t is simply the sum of each neuron's preferred direction weighted by the firing rate.

$$w_n(\mathbf{V},t) = s_n(\mathbf{V}) - b_0^n, \tag{5.2}$$

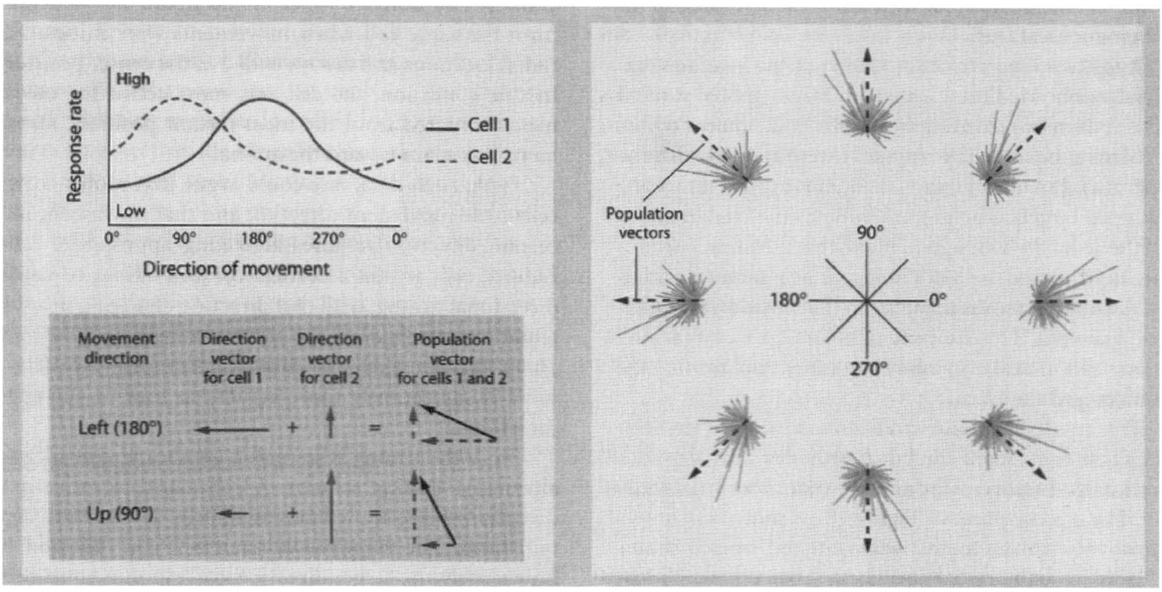

FIGURE 5.1: Illustration of the mechanisms of population vector approach.

$$\mathbf{P}(\mathbf{V},t) = \sum_{n=1}^{N} w_n(\mathbf{V},t)\frac{\mathbf{B}_n}{\|\mathbf{B}_n\|}, \tag{5.3}$$

The population vector can therefore be thought of as a local linear model, where the contributions of each neuron are weighted appropriately from the trajectory. It should be noted here that the PVA approach includes several assumptions whose appropriateness in the context of neural physiology and motor control will only be briefly considered here (see Ref. [23] for a full discussion). First, each cell is considered independently in their contribution to the kinematic trajectory. The formulation does not consider feedback of the neuronal firing patterns; a feature found in real interconnected neural architectures. Second, neuronal firing counts are linearly combined to reproduce the trajectory. At this moment it is unclear how the neural activation of nonlinear functions can create arbitrarily complex movement trajectories. The PV model determines the movement direction from neural activity, but the reconstruction of hand trajectory also requires the estimation of the speed. Georgopoulos et al [1] directly used the magnitude of the population vector, $|\mathbf{P}(n)|$, for estimating the instantaneous speed, and Moran and Schwartz [24] extended the PVA model to include the hand speed. Then, the trajectory of the hand position was reconstructed by a time series of population vectors that were connected tip to tail as time progresses.

5.2 SEQUENTIAL ESTIMATION

A second class of models that has been used for BMIs uses a generative, state space approach instead of the input-output modeling described in Chapters 3 and 4 [5, 16]. Generative models utilize a Bayesian formulation and they offer a general framework to estimate goal directed behavior encoded in multichannel spike trains. The movement kinematics are defined as state variables of a neural dynamical system from a sequence of noisy neuronal modulation observations. In this approach, the data are analyzed in the neural space instead of using an input-output model to map the neural inputs and desired kinematics. The generative approach creates an observation model that incorporates information from measurements using a recursive algorithm to construct the posterior PDF of the kinematics given the observations at each time. By estimating the expectation of the posterior density (or by maximum likelihood estimation), the movement kinematic estimate can be recovered probabilistically from the multichannel neural recordings.

Bayesian learning assumes that all forms of uncertainty should be expressed and measured by probabilities. At the center of the Bayesian formulation is a simple and very rich expression known as Bayes' rule. Given a data set $D = [u_{1:N}, z_{1:N}]$, and a set of possible descriptive data models M_k, $k = 0, \ldots, K$, the Bayes' rule becomes

$$p(M_i \mid D) = \frac{p(D \mid M_i)}{\sum_k p(D \mid M_k) p(M_k)} p(M_i) \tag{5.4}$$

This rule simply states that the posterior distribution ($p(M|D)$) can be computed as the ratio of the likelihood ($p(D|M)$) over the evidence ($p(D) = \sum_k p(D \mid M_k) p(M_k)$) times the prior probability ($p(M)$). The resulting posterior distribution incorporates both a priori knowledge and the information conveyed by the data.

This approach can be very useful in parametric model building ($f(u,w)$), where u is the input and w are system parameters because one can assign prior probabilities not only to the parameters of the system but also to model order k as $p(w,k)$ and even to the noise. Once the data are passed thru the system we can build the posterior distribution $p(w,k| u_{1:N}, z_{1:N})$ using Bayes' rule, where z are the system outputs. Because the posterior embodies all the statistical information about the parameters and their number given the measurements and the prior, one can theoretically obtain all features of interest (marginal densities, model selection, and parameter estimation) by standard probability marginalization and transformation techniques. For instance, we can predict the value of the next system output by (5.5)

$$E(z_{N+1} \mid u_{1:N+1}, z_{1:N}) = \int \hat{f}(k, w, u_{N+1}) p(k, w \mid u_{1:N}, z_{1:N}) \mathrm{d}k\mathrm{d}w \tag{5.5}$$

Note that the predictions must be based on *all* possible values of the network parameters weighted by their probability in view of the training data. Moreover, note that Bayesian learning is fundamentally based on *integration* over the unknown quantities. In (5.5), because we want to find the expected value of the output, which is a function of the model parameters and model order, we will have to integrate over these unknown quantities using all the available data. Because these integrals are normally multidimensional, this poses a serious problem that was only recently mitigated through the development of very efficient *sampling-based evaluation of integrals* called Monte Carlo sampling [25]. From a Bayesian perspective, Monte Carlo methods allow one to compute the full posterior probability distribution.

In BMIs we are interested in problems where the noisy neuronal recordings can be thought as the output of a system controlled by its state vector $\boldsymbol{x}$, which could include kinematics such as position, velocity, acceleration, and force. Moreover, we assume that the system state vector may change over time as shown in Figure 5.2.

The state space representation of a discrete stochastic dynamical system is given by

$$\begin{cases} x_{t+1} = x_t + v_t \\ z_t = \hat{f}_t(u_t, x_t) + n_t \end{cases} \tag{5.6}$$

where the noise term v_t is called the process noise and n_t the measurement noise. The first equation defines a first-order Markov process $p(x_{t+1}|x_t)$ (also called the system model), whereas the second defines the likelihood of the observations $p(z_t|x_t)$ (also called the measurement model). The

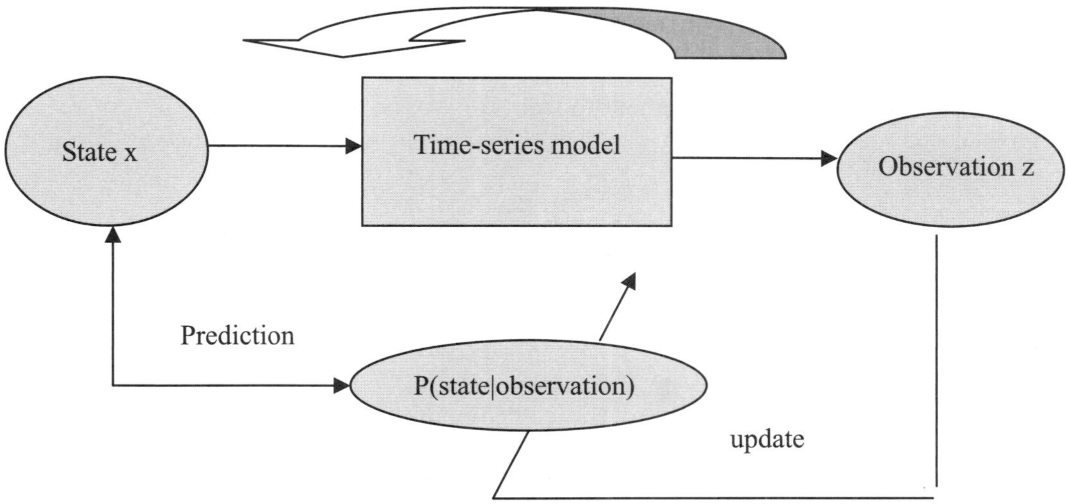

FIGURE 5.2: Block diagram of the state machine for BMIs.

problem is completely defined by specifying the prior distribution $p(x_0)$. The posterior distribution $p(x_{0:t}|u_{1:t},z_{1:t})$ constitutes the complete solution to the sequential estimation problem, however many times we are interested in one of the marginals, such as the *filtering density* $p(x_t|u_{1:t}, z_{1:t})$, because one does not need to keep track of all the model parameters, and still can answer many important modeling questions.

According to (5.6), there are two models required to analyze and infer the state of a dynamical system: the system model, which describes the evolution of the state with time, and the continuous observation measurement model, which relates the noisy measurements to the state. There are two stages to adapt the filtering density: prediction and update. The prediction stage uses the system model $p(x_{t+1}|x_t)$ to propagate into the future the posterior probability density of the state given the observation, called the Chapman–Kolmogorov equation $p(x_{t-1}|u_{1:t-1},z_{1:t-1})$ as follows

$$p(x_t \mid u_{1:t-1}, z_{1:t-1}) = \int p(x_t \mid x_{t-1}) p(x_{t-1} \mid u_{1:t-1}, z_{1:t-1}) dx_{t-1}, \tag{5.7}$$

The update stage applies Bayes rule when new data (u_t,z_t) is observed

$$p(x_t \mid u_{1:t}, z_{1:t}) = \frac{p(z_t \mid x_t, u_t) p(x_t \mid x_{1:t-1}, z_{1:t-1})}{p(u_t \mid u_t, z_{1:t-1})} \tag{5.8}$$

To evaluate (5.6) and (5.7), one needs to still compute the system model from the process noise v_{t-1} as

$$p(x_t \mid x_{t-1}) = \int p(x_t \mid v_{t-1}, x_{t-1}) p(v_{t-1} \mid x_{t-1}) dv_{t-1} = \int \Xi(x_t - v_{t-1} - x_{t-1}) p(v_{t-1}) dv_{t-1} \tag{5.9}$$

where the notation $\Xi(.)$ means that the computation is deterministic, and the conditional probability reduces to $p(v_{t-1})$ because of the assumed independence of the noise. The likelihood density function is determined by the measurement model

$$p(z_t \mid x_t, u_t) = \int \Xi(z_t - \hat{f}(u_t, x_t) - n_t) p(n_t) dn_t \tag{5.10}$$

Although the evidence is estimated by

$$p(z_t \mid z_{1:t-1}, u_t) = \int p(z_t \mid x_t, u_t) p(x_t \mid z_{1:t-1}, u_{1:t-1}) dx_t \tag{5.11}$$

When one closely examines the procedure of computing these expressions, it is clear that the difficulty is one of integrating PDFs. The most popular approach is to assume a Gaussian distribution for the PDFs and proceed with closed form integration, which provides the Kalman filter when the model is linear [17], and its variants (extended Kalman filter (EKF) and the unscented

Kalman filter). The most general case is called the particle filter where the system and observation models can be nonlinear and there are no modeling constrains imposed on the posterior density. The particle filter became very popular recently with the availability of fast processors and efficient algorithms, such as Monte Carlo integration, where a set of samples drawn from the posterior distribution of the model parameters is used to approximate the integrals by sums [25]. Alternatively, the sum of Gaussian models can be used, or more principally graphical models, which take advantage of known dependencies in the data to simplify the estimation of the posterior density.

5.3 KALMAN FILTER

Our discussion of generative models will begin with the most basic of the Bayesian approaches: the Kalman filter. The Kalman filter has been applied in BMI experimental paradigms by several groups [7, 16, 27–29]. This approach assumes a linear relationship between hand motion states and neural firing rates (i.e., continuous observations obtained by counting spikes in 100-msec windows), as well as Gaussian noise in the observed firing activity. The Kalman formulation attempts to estimate the state, $\mathbf{x}(t)$, of a linear dynamical system as shown in Figure 5.3. For BMI applications, we define the states as the hand position, velocity, and acceleration, which are governed by a linear dynamical equation as shown in (5.12)

$$\mathbf{x}(t) = [\mathbf{HP}(t) \quad \mathbf{HV}(t) \quad \mathbf{HA}(t)]^{\mathrm{T}} \tag{5.12}$$

where **HP**, **HV**, *and* **HA** are the hand position, velocity, and acceleration vectors,[1] respectively. The Kalman formulation consists of a generative model for the data specified by a linear dynamic equation for the state in (5.13)

$$\mathbf{x}(t+1) = \mathbf{A}\mathbf{x}(t) + \mathbf{u}(t) \tag{5.13}$$

where $\mathbf{u}(t)$ is assumed to be a zero-mean Gaussian noise term with covariance $\mathbf{U}$. The observation model also called the output mapping (from state to spike trains) for this BMI linear system is simply

$$\mathbf{z}(t) = \mathbf{C}\mathbf{x}(t) + \mathbf{v}(t) \tag{5.14}$$

where $\mathbf{v}(t)$ is the zero-mean Gaussian measurement noise with covariance $\mathbf{V}$ and $\mathbf{z}$ is a vector consisting of the neuron firing patterns binned in nonoverlapping (100 msec) windows. In this specific formulation, the output-mapping matrix $\mathbf{C}$ has dimension $N \times 9$. Alternatively, we could have

[1]The state vector is of dimension $9 + N$; each kinematic variable contains an x, y, and z component plus the dimensionality of the neural ensemble.

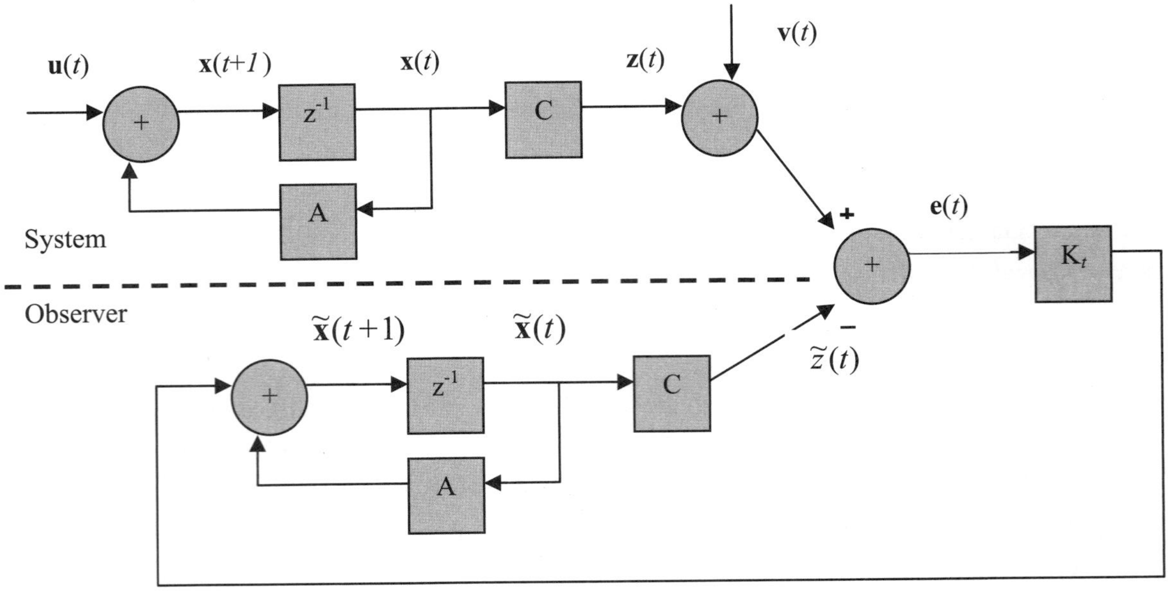

FIGURE 5.3: Kalman filter block diagram.

also included the spike counts of N neurons in the state vector as $\mathbf{f}_1, \ldots, \mathbf{f}_N$. This specific formulation would exploit the fact that the future hand position is not only a function of the current hand position, velocity, and acceleration, but also the current cortical firing patterns. However, this advantage comes at the cost of large training set requirements, because this extended model would contain many more parameters to be optimized. To train the topology given in Figure 5.3, L training samples of $\mathbf{x}(t)$ and $\mathbf{z}(t)$ are utilized, and the model parameters $\mathbf{A}$ and $\mathbf{U}$ are determined using least squares. The optimization problem to be solved is (5.15).

$$\mathbf{A} = \underset{\mathbf{A}}{\arg\min} \sum_{t=1}^{L-1} \|\mathbf{x}(t+1) - \mathbf{A}\mathbf{x}(t)\|^2 \tag{5.15}$$

The solution to this optimization problem is found to be (5.16)

$$\mathbf{A} = \mathbf{X}_1\mathbf{X}_0^{\mathrm{T}}(\mathbf{X}_1\mathbf{X}_1^{\mathrm{T}})^{-1} \tag{5.16}$$

where the matrices are defined as $\mathbf{X}_0 = [x_1 \; \cdots \; x_{L-1}]$, $\mathbf{X}_1 = [x_2 \; \cdots \; x_L]$. The estimate of the covariance matrix $\mathbf{U}$ can then be obtained using (5.13).

$$\mathbf{U} = (\mathbf{X}_1 - \mathbf{A}\mathbf{X}_0)(\mathbf{X}_1 - \mathbf{A}\mathbf{X}_0)^{\mathrm{T}} / (L-1) \tag{5.17}$$

Once the system parameters are determined using least squares on the training data, the model obtained (**A**, **C**, **U**) can be used in the Kalman filter to generate estimates of the hand positions from neuronal firing measurements. Essentially, this model assumes a linear dynamical relationship between current and future trajectory states.

The Kalman filter is an adaptive state estimator (observer) where the observer gain is optimized to minimize the state estimation error variance. In real-time operation, the Kalman gain matrix **K** (5.19), is updated using the projection of the error covariance in (5.18) and the error covariance update in (5.21). During model testing, the Kalman gain correction is a powerful method for decreasing estimation error. The state in (5.20) is updated by adjusting the current state value by the error multiplied with the Kalman gain.

$$\mathbf{P}^{-}(t+1) = \mathbf{AP}(t)\mathbf{A}^{\mathrm{T}} + \mathbf{U} \tag{5.18}$$

$$\mathbf{K}(t+1) = \mathbf{P}^{-}(t+1)\mathbf{C}^{\mathrm{T}}(\mathbf{CP}^{-}(t+1)\mathbf{C}^{\mathrm{T}})^{-1} \tag{5.19}$$

$$\tilde{\mathbf{x}}(t+1) = \mathbf{A}\tilde{\mathbf{x}}(t) + \mathbf{K}(t+1)(\mathbf{Z}(t+1) - \mathbf{CA}\tilde{\mathbf{x}}(t)) \tag{5.20}$$

$$\mathbf{P}(t+1) = (\mathbf{I} - \mathbf{K}(t+1)\mathbf{C})\mathbf{P}^{-}(t+1) \tag{5.21}$$

where the notation $P^{-}(t+1)$ means an intermediate (prior) value of P at time $t+1$. Using the propagation equations above, the Kalman filter approach provides a recursive and on-line estimation of hand kinematics from the firing rate, which is more realistic than the traditional linear filtering techniques and potentially better. The testing outputs for the Kalman BMI are presented in Figure 5.4. The Kalman filter performs better than the linear filter in peak accuracy (CC = 0.78 ± 0.20) but suffers in

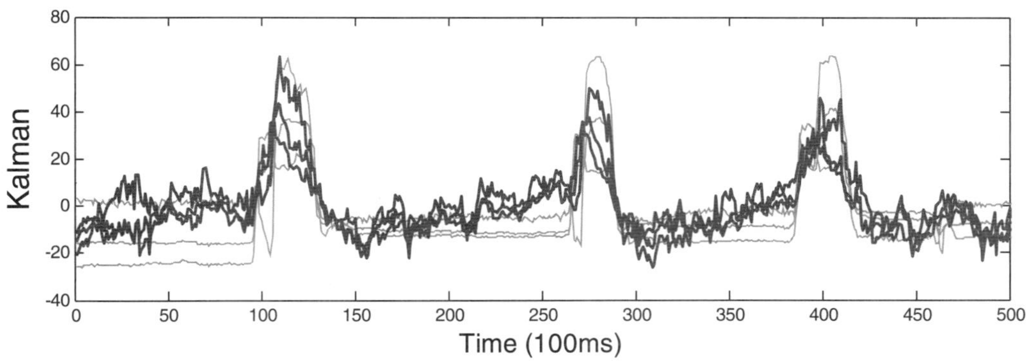

FIGURE 5.4: Testing performance for a Kalman filter for a reaching task. Here, the red curves are the desired *x*, *y*, and *z* coordinates of the hand trajectory, whereas the blue curves are the model outputs.

smoothness because of the number of free parameters (12 073 for a system with 104 neuronal inputs and 9 outputs of 3D position, velocity, and acceleration).

5.4 PARTICLE FILTERS

Although the Kalman filter provides a closed-form decoding procedure for linear Gaussian models, we have to consider the fact that the relationship between neuronal activity and behavior may be nonlinear. Moreover, measured neuronal firing often follows Poisson distributions, and even after binning, there is reason to believe that the Gaussian assumption is too restrictive. The consequences for such a mismatch between model and the real system will be expressed as additional errors in the final position estimates. To cope with this problem, we need to go beyond the linear Gaussian model assumption. In principle, for an arbitrary nonlinear dynamical system with arbitrary known noise distributions, the internal states (**HP**, **HV**, and **HA**) can be still estimated from the measured outputs (neuronal activity) using the sequential estimation framework presented. In the BMI literature, researchers have already implemented the most general of these models called the particle filter [15].

The particle filter framework alleviates the restrictions of the Kalman filter (linearity and Gaussianity) but substantially complicates the computations because, as a result of the generality of the model, there is no closed form solution and therefore the posterior distribution has to be estimated by probing. To help create at the output an estimate of the posterior density, a set of samples drawn from a properly determined density that is estimated at each step, is sent through the system with the present parameters. The peak of this posterior (or another central moment) is considered as the state estimate. Particle filters have also been applied to BMIs [26] where the tuning function has been assumed exponential on linear filtered velocities [32].

In this most general framework, the state and output equations can include nonlinear functions as given in (5.22) and (5.23).

$$x_k = F_k(x_{k-1}, v_{k-1}). \tag{5.22}$$

$$z_k = H_k(x_k, n_k) \; . \tag{5.23}$$

where $F_k : \Re^{n_x} \times \Re^{n_v} \rightarrow \Re^{n_x}$ and $H_k : \Re^{n_x} \times \Re^{n_n} \rightarrow \Re^{n_z}$ are known, possibly *nonlinear*, functions of the state $\mathbf{x}_{k-1}$, $\{v_{k-1}, k \in N\}$, and $\{n_k, k \in N\}$ are both independent and identically distributed (i.i.d.) process noise; n_x, n_v, n_z, and n_n are dimensions of the state $\mathbf{x}_k$, noise vector $\mathbf{v}_{k-1,}$ the measurement $\mathbf{z}_k$, and measurement noise vector $\mathbf{n}_k$, respectively; and N is the set of natural numbers. Note that $\mathbf{v}_k$ and $\mathbf{n}_k$ are assumed *non-Gaussian* in the particle filter.

To deal with these general distributions in a probabilistic way, the integrals have to be evaluated in a numeric way. Particle filtering is implemented to propagate and update the posterior density of the state $\mathbf{x}_k$ given the measurement $\mathbf{z}_k$ recursively over time. Particle filtering uses Sequential Importance Sampling [33] to discretely approximate the posterior distribution. The key idea is to represent the required posterior density function by a set of random samples, which are called particles, with associated weights, and to estimate the posterior density of the state given measurement based on these samples and weights. This Monte Carlo characterization becomes an equivalent representation to the functional description of the posterior PDF of the states when the number of the samples is large enough, and the particle filter solution will also approach the optimal Bayesian estimation.

Let $\{x_{0:k}^i, w_k^i\}_{i=1}^N$ denote a random measure of the posterior PDF $P(x_{0:k}|z_{1:k})$, where $\{x_{0:k}^i\}_{i=1}^N$ is a set of N states up to time k with associated normalized weights $\{w_k^i\}_{i=1}^N$, $\sum_{i=1}^N w_k^i = 1$. Then, the posterior density at time k can be represented by a discrete weighted approximation,

$$p(x_{0:k} \mid z_{1:k}) \approx \sum_{i=1}^{N} w_k^i \delta(x_{0:k} - x_{0:k}^i). \tag{5.24}$$

The weights are chosen according to the principle of Importance Sampling, which generates samples easily from a proposal Importance Density defined as $q(x_{0:k}|z_{1:k})$ [34, 35],

$$\begin{aligned} w_k^i &\propto \frac{p(x_{0:k}^i \mid z_{1:k})}{q(x_{0:k}^i \mid z_{1:k})} \\ &= \frac{p(z_k \mid x_{0:k}^i, z_{1:k-1}) p(x_k^i \mid x_{0:k-1}^i, z_{1:k-1}) p(x_{0:k-1}^i \mid z_{1:k-1})}{p(z_k \mid z_{1:k-1}) q(x_{0:k}^i \mid z_{1:k})} \\ &= \frac{p(z_k \mid x_k^i) p(x_k^i \mid x_{k-1}^i) p(x_{0:k-1}^i \mid z_{1:k-1})}{p(z_k \mid z_{1:k-1}) q(x_{0:k}^i \mid z_{1:k})} \\ &\propto \frac{p(z_k \mid x_k^i) p(x_k^i \mid x_{k-1}^i) p(x_{0:k-1}^i \mid z_{1:k-1})}{q(x_{0:k}^i \mid z_{1:k})} \end{aligned} \tag{5.25}$$

If the importance density is only dependent on x_{k-1} and z_k, we can discard the path $x_{0:k-1}^i$ and the observation history $z_{1:k}$ to simply modify the weight by

$$\begin{aligned} w_k^i &\propto \frac{p(z_k \mid x_k^i) p(x_k^i \mid x_{k-1}^i) p(x_{0:k-1}^i \mid z_{1:k-1})}{q(x_{0:k}^i \mid z_{1:k})} \\ &\propto \frac{p(z_k \mid x_k^i) p(x_k^i \mid x_{k-1}^i) p(x_{0:k-1}^i \mid z_{1:k-1})}{q(x_k^i \mid x_{0:k-1}^i, z_{1:k}) q(x_{0:k-1}^i \mid z_{1:k-1})} \\ &= w_{k-1}^i \frac{p(z_k \mid x_k^i) p(x_k^i \mid x_{k-1}^i)}{q(x_k^i \mid x_{k-1}^i, z_k)} \end{aligned} \tag{5.26}$$

Usually, the importance density $q(x_k^i|x_{k-1}^i,z_k)$ is often chosen as the prior density $p(x_k^i|x_{k-1}^i)$ for convenience. The importance density method requires the generation of new samples from $p(x_k^i|x_{k-1}^i)$. The new sample can be generated by the system model with a process noise sample $\mathbf{v}_{k-1}^i$ generated according to the PDF of $\mathbf{v}_{k-1}$. For the importance density that we choose the posterior filtered density $p(x_k|z_{1:k})$ can be approximated as

$$p(x_k \mid z_{1:k}) \approx \sum_{i=1}^{N} w_k^i \delta(x_k - x_k^i) \tag{5.27}$$

where

$$w_k^i \propto w_{k-1}^i p(z_k \mid x_k^i) \tag{5.28}$$

It can be shown that when the number of the samples is very large, (5.27) approaches the true posterior density $p(x_k|z_{1:k})$.

The particle filter based on important sampling displays a phenomenon called degeneracy [36]. All but a few particles have negligible weight after several iterations, which implies a large computational effort to update the particles with a minute contribution in the estimation of the posterior density and results in a loss of diversity in the particle pool. When a significant degeneracy appears, *resampling* is implemented to reduce degeneracy [36]. The basic idea of resampling is to eliminate the particles with small weight and to concentrate on particles with large weights. Resampling is applied at every time index, so that the samples are i.i.d. from the discrete uniform density with $w_{k-1}^i = 1/N$. The weight then changes proportionally given by

$$w_k^i \propto p(z_k \mid x_k^i) \tag{5.29}$$

The weights given by (5.29) are normalized every time before resampling. The assumptions behind resampling are weak. The state dynamics $f_k(x_{k-1}, v_{k-1})$ and measurement function $h_k(x_k, n_k)$ are required to be known. The realization of the posterior density of the state $\mathbf{x}_k$ given the measurement $\mathbf{z}_k$ is sampled from the process noise distribution $\mathbf{v}_k$ and the prior. Finally, the likelihood function $p(\mathbf{z}_k|\mathbf{x}_k)$ is also required to be available to generate new weights. The new sample $x_k^i \sim p(x_k \mid x_{k-1}^i)$ is generated by setting $x_k = f_k(x_{k-1}, v_{k-1})$ with a process noise sample $v_{k-1}^i \sim p_v(v_{k-1})$, where $p_v(v_{k-1})$ is the PDF of v_{k-1}.

The particle filter can be used as a statistical learning and probabilistic inference technique to infer the hand position of a subject from multielectrode recording of neural activity in motor cortex [26].

The hand movement (position, velocity, and acceleration) can be modeled as the system state $\mathbf{x}_k$. The firing rate of the neurons can be modeled as the observation (measurement) $\mathbf{z}_k$. A non-Gaussian model of "tuning" that characterizes the response of the cell firing $\mathbf{z}_k$ conditioned on hand velocity $\mathbf{x}_k$

is explored by adopting particle filter to estimate the posterior distribution recursively. When there is enough neural recordings, the particle filtering method can generate a more complete coverage of the hand movement space. The algorithm can be described as follows:

Initialization N state samples $\mathbf{x}_0^i$, $i = 1, \ldots, N$

For all the time index $k = 1{:}T$

Draw new samples from $x_k^i \sim p(x_k \mid x_{k-1}^i)$ by setting $x_k = f_k(x_{k-1}, v_{k-1})$ with a process noise sample $v_{k-1}^i \sim p_v(v_{k-1})$

Calculate weights $w_k^i = p(z_k \mid x_k^i)$

Normalize the weights $w_k^i = p(z_k \mid x_k^i) / \sum_{i=1}^{N} p(z_k \mid x_k^i)$

Resample $[\mathbf{x}_k^i, \mathbf{w}_k^i]$ to reduce the degeneracy

Pick out $x_k^{j^*}$ as the estimation state that maximize the posterior density $p(x_k^j \mid z_k)$.

5.5 HIDDEN MARKOV MODELS

The final generative model discussed in this chapter is the HMM. HMMs are famous for capturing the statistical dependencies in time series under the mild assumption that the past of the data is fully described by the present state. For this important reason, HMMs are the leading technology to recognize speech and have even been applied to modeling open-loop human actions, and analyzing similarity between human control strategies. In past efforts, HMMs were trained on neural data to determine the state transition of animal behavior rather than using the likelihood of particular actions [37, 38]. By using HMMs to find likelihoods of simpler movement actions or "movemes" to construct complex motor actions, an unsupervised way to model neural recordings is possible. Here we demonstrate that HMMs are capable of recognizing neural patterns corresponding to two diverse kinematic states (arm at rest or moving) just by analyzing the spatiotemporal characteristics of neural ensemble modulation (i.e., without using a desired signal).

An HMM is a probabilistic model of the joint probability of a collection of random variables $[O_1, \ldots, O_T, X_1, \ldots, X_T]$ (we are going to substitute O for Z to adhere to the notation in the HMM literature). The O_i variables are either continuous or discrete observations, and the X_i variables are *hidden* and *discrete*. Under an HMM, there are two conditional independence assumptions made about these random variables that make associated algorithms tractable. These independence assumptions are:

1. The tth hidden variable, given the $(t-1)^{\text{st}}$ hidden variable, is independent of previous variables

$$P(X_t \mid X_{t-1}, \ldots, X_1) = P(X_t \mid X_{t-1}) \tag{5.30}$$

2. The tth observation, given the tth hidden variable, is independent of other variables

$$P(O_t \mid X_T, O_T, X_{T-1}, O_{T-1}, \ldots, X_t, O_{t-1}, X_{t-1}, \ldots, X_1, O_1) = P(O_t \mid X_t) \tag{5.31}$$

Although continuous and semicontinuous HMMs have been developed, discrete output HMMs are often preferred in practice because of their relative computational simplicity and reduced sensitivity to initial parameter settings during training [39]. A discrete hidden Markov chain depicted in Figure 5.5 consists of a set of n states, interconnected through probabilistic transitions, and it is completely defined by the triplet, $\lambda = [\mathbf{A}, \mathbf{B}, \pi]$, where $\mathbf{A}$ is the probabilistic $n \times n$ state transition matrix, $\mathbf{B}$ is the $L \times n$ output probability matrix (with L discrete output symbols), and π is the n-length initial state probability distribution vector [39, 40].

With respect to the Kalman filter, the HMMs have discrete hidden state variables (as opposed to continuous). Additionally, the HMM has a multistage stochastic state transition, whereas the Kalman filter follows a *single-step* Markov model, following a known deterministic rule, with the randomness appearing through the additive process noise. A graphical model representation of the two topologies in Figure 5.5 makes this similarity even more apparent.

In contrast to the Gaussian noise model that is used for the Kalman filter, the HMM is similar to the particle filter in that it can represent an arbitrary distribution for the next value of the state variables. The differences between the two models become more apparent in the training because the Kalman filter updates are performed sample by sample, whereas in HMMs batch updates are normally used. For an observation sequence O, we locally maximize $P(O|\lambda)$ (i.e., the probability of the observation sequence O given the model λ) by finding the maximum likelihood estimate using an expectation maximization (EM) algorithm called the Baum–Welch algorithm [40].

Recall the definition of the maximum-likelihood estimation problem. We have a density function $p(O|\lambda)$ that is governed by the set of parameters λ and a data set of size N, supposedly drawn from this distribution, that is, $O = [O_1, \ldots, O_N]$. That is, we assume that these data vectors are i.i.d. with distribution p. Therefore, the resulting density for the samples is

$$p(O|\lambda) = \prod_{i=1}^{N} p(O_i|\lambda) = L(\lambda|O) \tag{5.32}$$

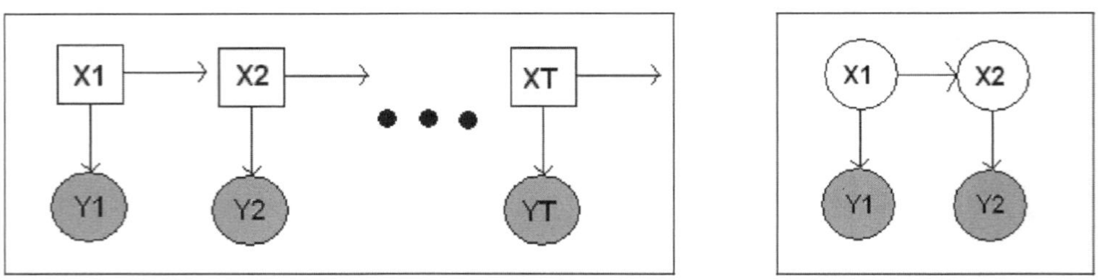

FIGURE 5.5: HMM (left) and Kalman filter (right). Squares denote discrete quantities, circles continuous quantities. White denotes unobservable quantities, and gray observable quantities. Arrows mean dependencies.

This function $L(\lambda | O)$ is called the likelihood of the parameters given the data, or just the likelihood function. In the maximum likelihood problem, our goal is to find the λ that maximizes L, that is, find λ^* such that

$$\lambda^* = \underset{\lambda}{\operatorname{argmax}} L(\lambda | O) \tag{5.33}$$

Often, we maximize log $(L(\lambda | O))$ because it is analytically easier. The EM algorithm is particularly useful in situations where one seeks to maximize the likelihood function and the formulation has hidden parameters, which creates uncertainty. Therefore, the observed data do not tell the full story about the problem, and we assume that a complete data set exists $Z = (O, X)$, where X takes into consideration the hidden variables, and specify a joint density function

$$p(z|\lambda) = p(O, X|\lambda) = p(X|O, \lambda)p(O|\lambda) \tag{5.34}$$

From this joint density, we can define the complete data likelihood

$$L(\lambda|Z) = L(\lambda|O, X) = p(O, X|\lambda) \tag{5.35}$$

Note that $L(\lambda|O, X) = h_{\Lambda,O}(X)$ where $h_{\Lambda,O}(.)$ is some function, λ and O are constants, and X is a random variable (the missing information due to the hidden variable). The original likelihood is referred to as incomplete data likelihood.

The EM algorithm is an iterative algorithm with two steps. Let us call the optimization function we wish to maximize $Q(\lambda, \lambda^{(i-1)})$, where λ stands for the new optimized parameters and $\lambda^{(i-1)}$ the actual parameter set at iteration $i-1$. First (expectation step), EM finds the expected value of the complete data likelihood log $p(O, X|\lambda)$ with respect to the unknown data X given the observed data O and the current parameter estimates.

$$Q(\lambda, \lambda^{(i-1)}) = E[\log p(O, X | \lambda) | O, \lambda^{(i-1)}] = \int \log p(O, X | \lambda) f(X | O, \lambda^{(i-1)}) \mathrm{d}X \tag{5.36}$$

where $f(X|O, \lambda^{(i-1)})$ is the marginal distribution of the unobserved data, which is dependent on the observed data and current parameters.

The second step (M step) of the EM algorithm maximizes the expectation computed in the first step, that is,

$$\lambda^{(i)} = \underset{\lambda}{\operatorname{arg\,max}} Q(\lambda | \lambda^{(i-1)}) \tag{5.37}$$

The beauty of this algorithm is that it has been shown to increase the log likelihood at each step, which means that convergence to a local maximum is guaranteed [40].

The Baum–Welch algorithm is an application of the EM algorithm for HMMs. In an HMM, X_t is a discrete hidden random variable with possible states $[1, \dots, N]$. We further assume that the parameters of the hidden Markov chain are time-independent, that is, $P(X_t|X_{t-1})$ is time homogeneous. Therefore, $P(X_t|X_{t-1})$ can be represented by a stochastic matrix $A[a_{ij}] = p(X_t = j|X_{t-1} = i)$ that does not change over time. The special case $t = 1$ is specified by the initial distribution $\pi_i = p(X_1 = i)$. A particular observation sequence O is described by $O = (O_1 = o_1, \dots, O_T = o_T)$. The probability of a particular observation vector at time t for state j is described by $b_j(o_t) = p(O_t = o_t | X_t = j)$. The complete collection of parameters is represented by $B=[b_j(.)]$.

There are three basic problems in HMM design but here we will only focus on the following two:

Find $p(O|\lambda)$ for some $O=(o_1, \dots, o_T)$. We will use the forward procedure because it is much more efficient than direct evaluation.

Find $\lambda^* = \arg\max_{\lambda} p(O|\lambda)$ using the Baum–Welch algorithm.

For the forward algorithm, let us define a forward variable $\alpha_t(i)$:

$$\alpha_t(i) = P(O_1, O_2, \dots, O_t, q_t = X_i | \lambda) \tag{5.38}$$

which refers to the probability of the partial observation sequence $[O_1, \dots, O_t]$ while being in state X_i at time t, given the model λ [39]. The $\alpha_t(i)$ variables can be computed inductively, and from them, $P(O|\lambda)$ is easily evaluated as explained by Rabiner [39] and Baum et al. [40]. The forward algorithm is defined below:

1. Initialize:

$$\alpha_1(i) = P(O_1, q_1 = X_i | \lambda) \tag{5.39}$$

$$\alpha_1(i) = P(O_1 | q_1 = X_i, \lambda) P(q_1 = X_i | \lambda) \tag{5.40}$$

$$\alpha_1(i) = \pi_i b_i(O_1), i = \{1, \dots, N\} \tag{5.41}$$

2. Induction:

$$\alpha_{t+1}(j) = P(O_1, \dots, O_{t+1}, q_{t+1} = X_j | \lambda) \tag{5.42}$$

$$\alpha_{t+1}(j) = \left[\sum_{i=1}^{N} P(O_1, \dots, O_t, q_t = X_i | \lambda) P(q_{t+1} = X_j | q_t = X_i, \lambda) \right] P(O_{t+1} | q_{t+1} = X_j, \lambda)$$

$$\alpha_{t+1}(j) = \left(\sum_{i=1}^{N} \alpha_t(i) a_{ij} \right) b_j(O_{t+1}), t \in \{1, \dots, T\}, j \in \{1, \dots, T) \tag{5.43}$$

3. Completion:

$$P(O \mid \lambda) = \sum_{i=1}^{N} \alpha_T(i) \tag{5.44}$$

The computation of equation (5.42) in the induction step above (step 2) accounts for all possible state transitions from time step t to time step $t + 1$, and the observable at time step $t + 1$. Figure 5.6 below illustrates the induction step graphically.

The second problem is the optimization. To maximize the probability of the observation sequence O, we must estimate the model parameters (**A**, **B**, π) for λ with the iterative Baum–Welch method [40].

Specifically, for the Baum–Welch method, we provide a current estimate of the HMM λ = [**A**, **B**, π] and an observation sequence $O = [O_1, \ldots, O_T]$ to produce a new estimate of the HMM given by $\bar{\lambda} = \{\bar{\mathbf{A}}, \bar{\mathbf{B}}, \bar{\pi}\}$, where the elements of the transition matrix $\bar{\mathrm{A}}$ are,

$$\bar{a}_{ij} = \left(\frac{\sum_{t=1}^{T-1} \zeta_t(i,j)}{\sum_{t=1}^{T-1} \gamma_t(i)} \right), \qquad i, j \in \{1, \ldots, N\} \tag{5.45}$$

Similarly, the elements for the output probability matrix $\overline{\mathbf{B}}$ are given by,

$$\bar{b}_j(k) = \left(\frac{\sum_t \gamma_t(j)(\text{where } \ \forall O_t = v_k)}{\sum_{t=1}^{T} \gamma_t(j)} \right), \qquad j \in \{1, \ldots, N\}, k \in \{1, \ldots, L\} \tag{5.46}$$

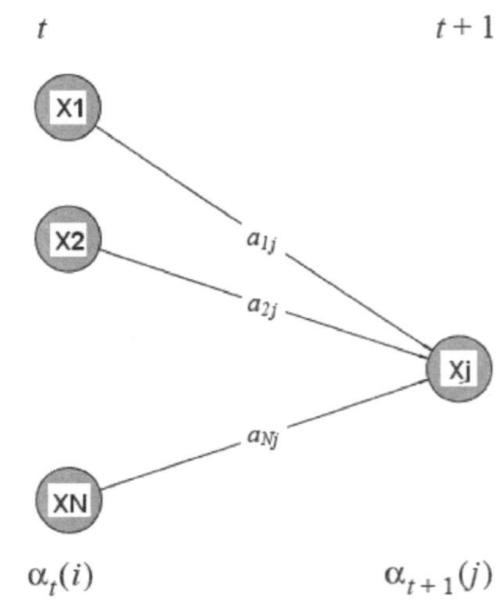

FIGURE 5.6: Forward update.

and finally the $\bar{\pi}$ vector,

$$\bar{\pi} = \gamma_1(i), i \in \{1,\ldots,N\}. \tag{5.47}$$

where,

$$\zeta_t(i,j) = \frac{\alpha_t(i) a_{ij} b_j(O_{t+1}) \beta_{t+1}(j)}{P(O|\lambda)} \tag{5.48}$$

and

$$\gamma_t(i) = \sum_{j=1}^{N} \zeta_t(i,j). \tag{5.49}$$

Note that β is the backward variable, which is similar to the forward variable α except that now we propagate the values back from the end of the observation sequence, rather than forward from the beginning of O [40]. First, we defined the backward variables (similar to the forward variables) to,

$$\beta_t(i) = P(O_{t+1},\ldots,O_T \mid q_t = X_i, \lambda), \tag{5.50}$$

which refers to probability of the partial observation sequence $[O_{t+1}, \ldots, O_T]$ and being in state X_i at time t, given the model λ [39]. Just like the forward variables, the backward variables can be computed inductively.

1. Initialize:

$$\beta_T(i) = 1, i \in \{1,\ldots,N\} \tag{5.51}$$

2. Induction:

$$\beta_t(i) = \sum_{j=1}^{N} a_{ij}\, b_j(O_{t+1}) \beta_{t+1}(j), t \in \{T-1, T-2, \ldots, 1\}, i \in \{1,\ldots,N\} \tag{5.52}$$

5.5.1 Application of HMMs in BMIs

The goal of applying HMM to BMIs is different from the previously discussed generative models. With HMMs the hypothesis is that the emerging states correspond to some underlying organization of the local cortical activity that is specific to external events and to their behavioral significance. Specifically, the HMM can be used to recognize a specific pattern (class) of motor neural activity. If we assume that each class of neural data is associated with a given motor behavior, then a

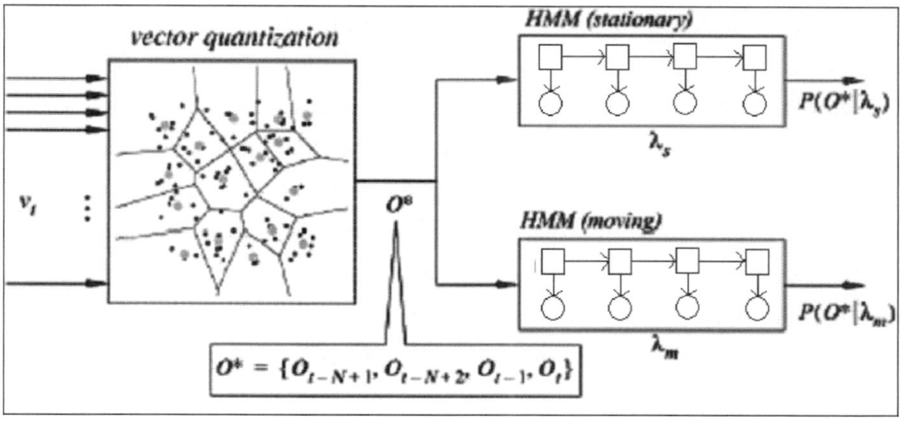

FIGURE 5.7: Stationary/moving classifiers.

forward model can be derived by the techniques discussed in Chapter 3, specifically the gated competitive experts. This allows multiple HMMs to map discrete portions of the neural data to complex trajectories. Consequently, the individual forward models only learn a segment of the trajectory, outperforming single forward model that must generalize over the full trajectory [41].

HMMs were utilized as gates in a gated competitive mixture of linear experts to differentiate and model arm at rest from arm moving in a real-time scenario solely by analysis of the neural recordings. Each one of these two possible outcomes are modeled by an HMM defined by parameters λ_m (movement) and λ_s (stationary), and the goal is to find which one is more likely given that a sequence of neuronal firing rates is observed (Figure 5.7).

There were several difficulties that had to be conquered to apply HMMs to motor BMI data. First, the input data are very high dimensional and discrete, that is 104 neurons of 100-msec binned neural recordings. To decrease the number of parameters of an HMM that would model the posterior density, we created a vector quantization (VQ) preprocessor using the Linde-Buzo-Gray (LBG) algorithm [42] to decrease the training requirements for the HMM and achieve better results. A second method to decrease the number of HMM parameters is to invoke the channel independence assumption and implement an ensemble of single neural-channel HMM chains to form an independently coupled hidden Markov model (ICHMM). Consequently, this classifier takes advantage of the neural firing properties and removes the distortion associated with VQ while jointly improving the classification performance and the subsequent linear prediction of the trajectory.

Vector Quantizing HMM. In this section, we broadly describe the VQ-HMM-based classifier illustrated in Figure 5.7. We demonstrate an experiment in which arm movement is being classified as moving or nonmoving based solely on neural recordings using two HMMs and a winner-take-all strategy. The HMMs act like a switch and allow the neural data belonging to each class to be fed into a feed-forward model to fit it to the respective arm trajectory. To train such a system, we partition the neural recordings into two groups: the first group contains data where the arm appears to be stationary, whereas the second group should contain data where the arm appears to be moving. We use a simple threshold to achieve this grouping: if the instantaneous velocity of the arm is below the noise threshold of the sensor (determined by inspecting the velocity data visually), the corresponding neural data are classified as *stationary*; otherwise, the neural data are classified as *moving* [43]. More specifically, the classifier works as follows:

1) At time index t, the neural binned vector $\mathbf{v}_t$ of length 104 (equal to number of neurons) is converted into a discrete symbol O_t in preparation for discrete output HMM evaluation.

The multidimensional neural recording must be converted to a sequence of discrete symbols. This process involves vector quantizing the input-space vectors to discrete symbols in order to use discrete-output HMMs. We choose the well-known LBG VQ algorithm [42], which iteratively generates vector codebooks of size $L = 2^m, m \in \{0,1...\}$, and can be stopped at an appropriate level of discretization (represented by m), as determined by the amount of available data. For our example experiment, we varied L from 8 prototype vectors to 256. This optimization seemed to be a good trade off given the 10 000 samples available for training and the 104 dimensional input vector (with each neural bin having about 20 possible bin counts or 20^{104} possible vectors). By optimizing the vector codebook on the neural recordings, we seek to minimize the amount of distortion introduced by the VQ process.

2) Next, the conditional probabilities $P(O|\lambda_s)$ and $P(O|\lambda_m)$ are evaluated where

$$O = [O_{t-N+1}, O_{t-N+2}, O_{t-1}, O_t],\ N > 1, \tag{5.53}$$

and λ_s and λ_m denote HMMs that correspond to the two possible states of the arm (stationary vs. moving).

The vector quantized input generates discrete symbols as input to a left-to-right (or Bakis) HMM chain. Given that we expect the monkey' arm movement to be dependent not only on current neural firings, but also on a recent time history of firings, we train each of the HMM models on observation sequences of length N. During run-time evaluation of $P(O|\lambda_s)$ and $P(O|\lambda_m)$, we use the same value of N as was used during training. Based on other experimental paradigms [44], we varied N from 5 to 10 to correspond to a half of second of data to a second of data. The HMM was trained with the Baum–Welch method on average with five iterations despite our convergence criterion of 0.000001 being met much earlier (because we set the minimum number of iterations to be five). The number of hidden states for the HMM were varied from 2 to 8 as to not exceed the observation sequence length.

3) Finally, the arm is stationary if,

$$P(O|\lambda_s) > P(O|\lambda_m) \tag{5.54}$$

and is moving if,

$$P(O|\lambda_m) > P(O|\lambda_s) \tag{5.55}$$

To explicitly compute $P(O|\lambda)$, we use the practical and efficient forward algorithm [40] described earlier in this section.

The classification decision in (5.53) and (5.54) is too simplistic because it does not optimize for overall classification performance, and does not account for possible desirable performance metrics. For example, it may be very important for an eventual modeling scheme to err on the side of predicting arm motion (i.e., moving class). Therefore, our previous classification decision to include the following classification boundary is modified to :

$$\frac{P(O|\lambda_m)}{P(O|\lambda_s)} = y \tag{5.56}$$

where y now no longer has to be strictly equal to 1. Note that by varying the value of y, we can essentially tune classification performance to fit our particular requirements for such a classifier. Moreover, optimization of the classifier is now no longer a function of the individual HMM evaluation probabilities, but rather a function of overall classification performance.

In addition, we have determined previously through first- and second-order statistical analysis that some of the 104 recorded neural channels contribute only negligibly to arm movement prediction [41]. Therefore, in the experiments we not only vary the four parameters listed above, but also the subset of neural channels that are used as temporal features in the segmentation process. Table 5.1 lists the seven different subsets of neural channels that were used in our experiments.

Tables 5.2 and 5.3 report experimental results for different combinations of the four parameters and subsets of neural channels. These tables present representative segmentation results for a large number of experiments.

From Table 2, we see that the VQ-HMM can reasonable classify the arm gross dynamics (i.e., moving or not). Unfortunately, even after exhausting many parameter combinations the results were only able to reach a plateau in performance at about 87%. We also note that because a subset of neural channels at the input yielded the best performance, some of the measured neural channels may offer little information about the arm movement. These two observations help to motivate the next section.

TABLE 5.1: Subsets of neural channels

SUBSET NUMBER	EXPLANATION
1	All 104 neural channels
2–5	Different combinations of neural channels with statistically significant correlations between arm movement/nonmovement [3]
6	Neural channels determined to be significant in artificial neural network mappings of neural activity to 3D arm movement
7	51 neural probe channels[a]

[a]The 104-channel neural data through spike sorting of data from 51 neural probes.

In terms of the final trajectory reconstruction, each HMM will be gating a linear model for the particular pieces of the trajectory where it wins the competition (i.e., rest or movement). Each linear model is a multiinput–multioutput finite impulse response (FIR) filter that is trained using normalized least mean square:

$$W_{\text{winner}}(n+1) = W_{\text{winner}}(n) + \frac{\eta e_{\text{winner}}(n) X(n)}{\gamma + \|X(n)\|^2} \tag{5.57}$$

TABLE 5.2: Results from varying the different parameters and neural subsets

SUBSET NO.	STATIONARY %	MOVING %	L	N	n
1	82.1	85.6	8	7	7
4	83.5	87.5	128	10	6
5	84.0	87.5	256	10	6
6	75.6	81.3	256	10	4
7	87.3	86.1	256	10	5

where $\mathbf{w}_{\text{winner}}$ is the winning filter's weight vector, $\mathbf{x}(n)$ is the present input, $e_{\text{winner}}(n)$ is the error produced from the winning filter, η is the learning rate, and γ is the small positive constant. The data for training is obtained from segments selected by the corresponding HMM (to avoid discontinuities, one second of data is used to fill the FIR memory).

In terms of the final trajectory reconstruction, we see in Figure 5.8 that qualitatively the bimodel system performs well in terms of reaching targets; this is especially evident for the first, second, and the seventh peaks in the trajectory. Overall, prediction performance of the bimodel system is approximately similar to the recurrent neural network (RNN), and superior to the single moving average (MA) model (Wiener filter). The mean of the signal to error ratio (SER) averaged over all dimensions for the single MA model, the RNN, and the bimodal system are -20.3 ± 1.6 (SD), -12.4 ± 16.5, and -15.0 ± 18.8 dB, respectively, whereas the maximum SERs achieved by each model are 10.4, 30.1, and 24.8dB, respectively.

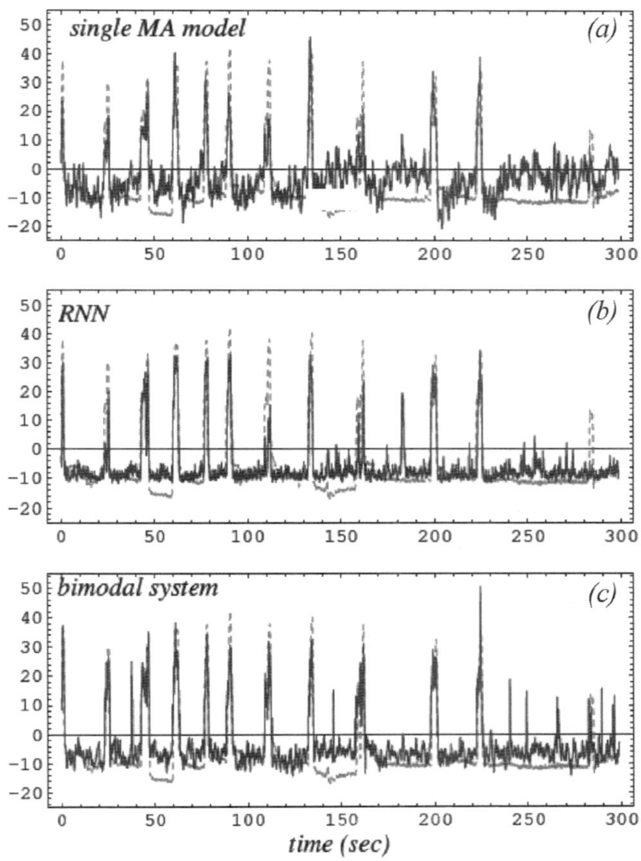

FIGURE 5.8: 3D predicted (blue) and actual arm trajectory (dashed red) for (a) single MA model, (b) RNN, and (c) bimodal system over the test data.

Independently Coupled HMM. One of the difficulties of the VQ-HMM is the distortion associated with VQ, which limited the performance of the classification. In addition, because different neurons generally have different response properties, dynamic ranges and preferred stimuli, synchronous neural operation is not a good assumption (but rather piecewise stationary). It is difficult, however, to train a HMM with a 104 input vector with the amount of data available in BMI experiments. Therefore, our next implementation invoked an independence assumption among the channels to create an ICHMM.

There is some evidence for this assumption. The literature explains that different neurons in the brain modulate independently from other neurons [45] during the control of movement. Specifically, during movement, different muscles may activate for synchronized directions and velocities, yet, are controlled by independent neural masses or clusters in the motor cortex [2, 12, 45]. Conversely, within the neural clusters themselves, temporal dependencies (and coactivations) have been shown to exist [45]. Therefore, for our HMM classifier, we make the assumption that enough neurons are sampled from different neural clusters to avoid overlap or dependencies. We can further justify this assumption by looking at the correlation coefficients (CC) between all the neural channels in our data set.

The best CCs (0.59, 0.44, 0.42, 0.36) occurred between only four out of thousands of possible neural pairs, whereas the rest of the neural pairs were a magnitude smaller. We believe this indicates weak dependencies between the neurons in our particular data set. In addition, despite these possible weak underlying dependencies, there is a long history of making such independence assumptions to create models that are tractable or computationally efficient. The factorial hidden Markov model is one example among many [46].

By making an independence assumption between neurons, we can treat each neural channel HMM independently. Therefore the joint probability

$$P(O_T^{(1)}, O_T^{(2)}, \ldots O_T^{(D)}, |\lambda_{\text{full}}) \tag{5.58}$$

becomes the product of the marginals

$$\prod_{i=1}^{D} P(O_T^{(i)} \,|\, \lambda^{(i)}) \tag{5.59}$$

of the observation sequences (each length T) for each dth HMM chain I . Because the marginal probabilities are independently coupled, yet try to model multiple hidden processes, we name this classifier the ICHMM.

By using an ICHMM instead of a fully coupled HMM (FCHMM) (Figure 5.9), the overall complexity reduces from ($O(TN^{2D})$ or $O(TD^2N^2)$) to $O(DTN^2)$ given that each HMM chain has a complexity of $O(TN^2)$. In addition, because we are using a single HMM chain to train on a

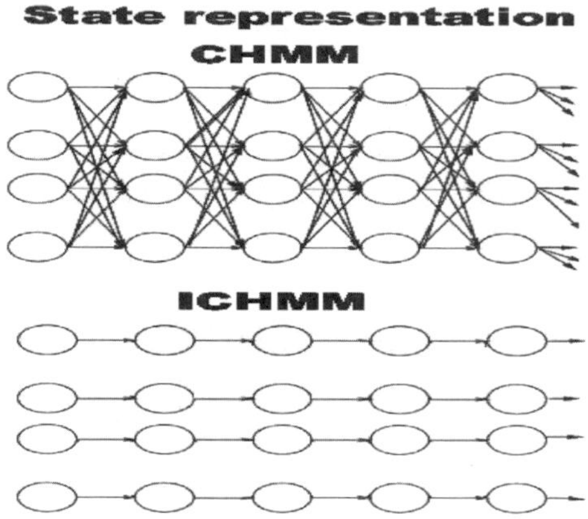

FIGURE 5.9: State representation of ICHMM vs. CHMM.

single neural channel, the number of parameters is greatly reduced and can support the amount of training data. Specifically, the individual HMM chains in the ICHMM contain around 70 parameters for a training set of 10 000 samples as opposed to almost 18 000 parameters necessary for a comparable FCHMM (due to the dependent states).

Each of the individual HMM chains is trained with the Baum–Welch algorithm as before, but now directly from the input data. We compose the likelihood ratio for the decision by

$$l(O) = \frac{\prod_{i=1}^{D} P(O_T^{(i)} \mid \lambda_M^{(i)})}{\prod_{i=1}^{D} P(O_T^{(i)} \mid \lambda_r^{(i)})} > \zeta \tag{5.60}$$

To give a qualitative interpretation of these weak classifiers, we present in Figure 5.10 the probabilistic ratios from 14 single-channel HMM chains (shown between the top and bottom

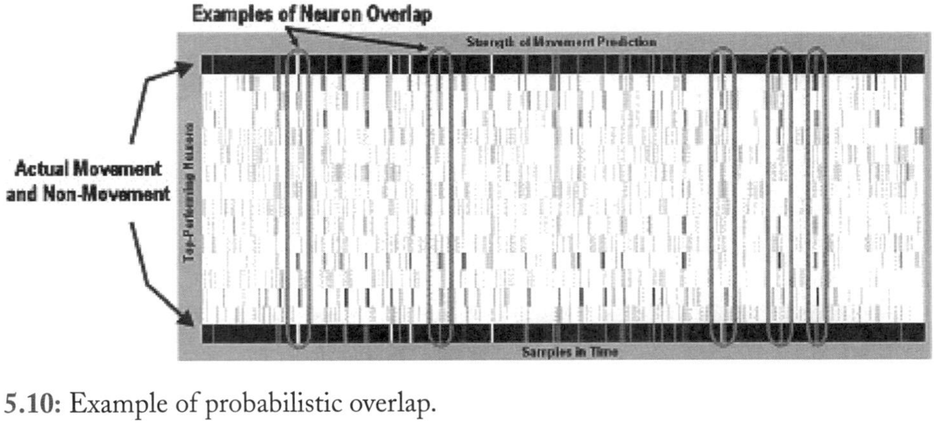

FIGURE 5.10: Example of probabilistic overlap.

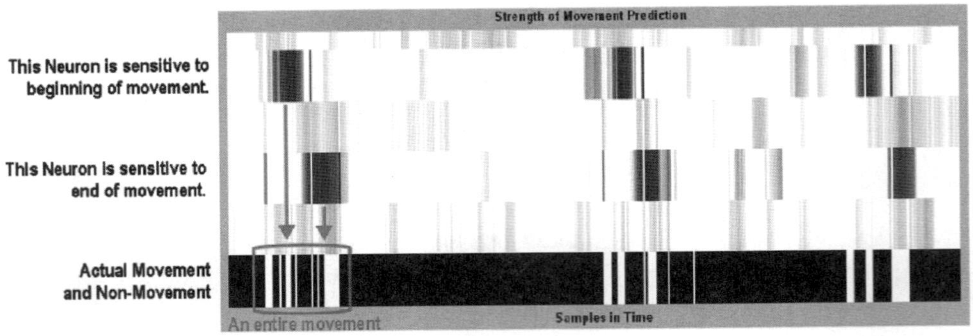

FIGURE 5.11: Zoomed-in example of probabilistic overlap.

movement segmentations) that produced the best classifications individually. Specifically, we present the ratio $\frac{P(O^{(i)}|\lambda_m^{(i)})}{P(O^{(i)}|\lambda_r^{(i)})}$ for each neural channel in a grayscale gradient format. The darker bands represent ratios larger than 1 and correspond to a higher probability for the movement class. Lighter bands represent ratios smaller than 1 and correspond to a higher probability for the rest class. The conditional probabilities nearly equal to one another show up as gray bands, indicating that classification for the movement or rest classes is inconclusive.

Overall, Figure 5.10 illustrates that the single channel HMMs can roughly predict movement and rest segments from the neural recordings. In Figure 5.11, they even seem tuned to certain parts of the trajectory like rest-food, food mouth, and mouth-rest. Specifically, we see there are more white bands $P(O|\lambda_r) \gg P(O|\lambda_m)$ during rest segments and darker bands $P(O|\lambda_m) \gg P(O|\lambda_r)$ during movement segments.

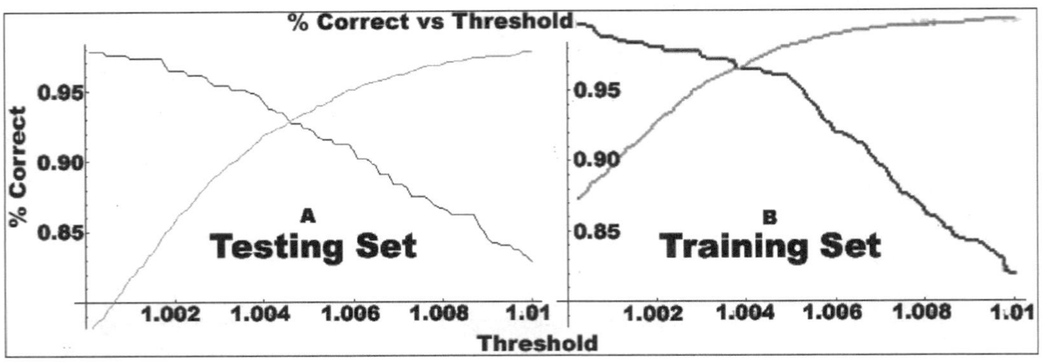

FIGURE 5.12: Training and testing with neural data.

TABLE 5.3: Results compared against different systems on same data

	SINGLE LINEAR MODEL	BIMODEL WITH VQ-HMM	BIMODEL WITH ICHMM
Classification %	–	87%	93%
CC (move)	0.75 ± 0.20	0.79 ± 0.27	0.86 ± 0.11
CC (rest)	0.06 ± 0.22	0.06 ± 0.24	0.03 ± 0.26

Although the standard receiver operator characteristic (ROC) curve can determine the threshold for an appropriate performance (i.e., compromise between number of false alarms and false detections), it fails to convey an explicit relationship between the classes and the threshold when optimizing performance. We use the true positives and true negatives of the respective classes to provide an explicit interclass relationship. For nomenclature simplicity, we label these two curves the likelihood ratio operating characteristic (LROC) curves because they represent the quantities in the likelihood ratio. The "optimal" operating point on the LROC occurs when the two curves of the classes intersect because this intersection represents equal classification for the two classes.

In Figure 5.12, the LROC curves show that the ICHMM is a significant improvement in classification over our previous VQ-HMM classifier. This is evident from the equilibrium point showing that the movement and rest classifications occur around 93% as opposed to 87% in our previous work mentioned in the VQ-HMM section. In Figure 5.12, we also see that the threshold is similar in the training and testing (ζ = 1.0044; ζ = 1.0048, respectively), which show that the threshold can be reliably set from the training data.

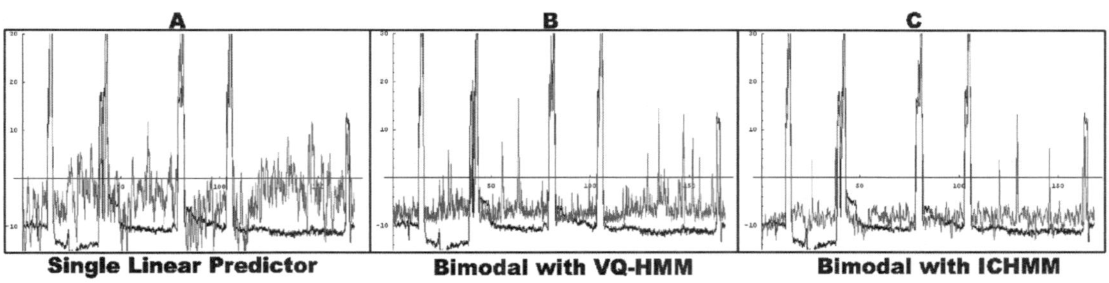

FIGURE 5.13: Trajectory results compared.

Figure 5.13 shows the predicted hand trajectories of each modeling approach, superimposed over the desired (actual) arm trajectories for the test data; for simplicity, we only plot the trajectory along the z coordinate. Qualitatively, we can see that the ICHMM performs better than the others in terms of reaching targets. Overall, prediction performance of the ICHMM classifier is slightly better than the VQ-HMM classifier, and superior to the single linear predictor, as evidenced by the average CCs (across the three coordinates) of 0.64, 0.80, and 0.86 for the single linear predictor, the bimodel system with VQ-HMM and the bimodel system with ICHMM.

5.6 SUMMARY

In brain–machine interfaces, the Kalman filter method can provide a rigorous and well-understood framework to model the encoding of hand movement in motor cortex, and for inferring or decoding this movement from the firing rates of a cell population. When the restrictive Gaussian assumptions and linear system model holds, the Kalman filter algorithm provides an elegant analytic optimal solution to the tracking problem. If one assumes that the observation time-series (neural activity) is generated by a linear system, then the tuning can be optimally estimated by a linear filter. The second assumption of Gaussianity of the posterior density of the kinematic stimulus given the neural spiking activities reduces all the richness of the interactions to second order information (mean and the covariance). These two assumptions may be too restrictive for BMI applications and may be overcome with methods such as particle filtering.

Unfortunately, in the BMI application this particular formulation is also faced with problems of parameter estimation. The generative model is required to find the mapping from the low-dimensional kinematic parameter state space to the high-dimensional output space of neuronal firing patterns (100+ dimensions). Estimating model parameters from the collapsed space to the high-dimensional neural can be difficult and yield multiple solutions. For this modeling approach, our use of physiological knowledge in the framework of the model actually complicates the mapping process. As an alternative, one could disregard any knowledge about the system and use a strictly data-driven methodology to build the model. However, if the distributions are not constrained to be Gaussian, but can be described by unbiased consistent means and covariance, the filter can still be optimally derived using a least squared argument.

For the HMMs, the results presented here show that the final prediction performance of the bimodel system using the ICHMM is much better than using the VQ-HMM, and superior to that of a single linear predictor. Overall, the ICHMM produces good results with few parameters. The one caveat to the ICHMM is the reliance on the z threshold. Fortunately, this threshold is retrievable from the training set. Interestingly, the z threshold can be viewed as global weighting for the two classes in this system. If one frames the ICHMM as mixture of experts (ME) perhaps boosting or bagging could be used to locally weight these simplistic classifiers in future work. The ME generates complex and powerful models by combining simpler models that often map different regions of the input space [47]. With boosting, models are weighted to create a strong ensemble decisions so that a weighted majority "votes" for the appropriate class labeling [47]. This is analogous to what the ICHMM currently does with a global z weighting or biasing.

In the next chapter, we will shift the focus from signal processing methods that use estimates of the instantaneous spike rate through binning and concentrate on the spike trains directly. The trade-offs of computing in BMIs directly with spike trains or with rate codes can have an impact on the ability to resolve modulation, correlation, integration, and coincidence in neuronal representations necessary for accurate reconstruction of the sensory or motor function. The implications and significance of the choice of spike traines in BMI approaches will be reviewed next.

REFERENCES

1. Georgopoulos, A.P., A.B. Schwartz, and R.E. Kettner, *Neuronal population coding of movement direction.* Science, 1986. **233**(4771): pp. 1416–1419.
2. Georgopoulos, A., et al., *On the relations between the direction of two-dimensional arm movements and cell discharge in primate motor cortex.* Journal of Neuroscience, 1982. **2**: pp. 1527–1537.
3. Flament, D. and J. Hore, *Relations of motor cortex neural discharge to kinematics of passive and active elbow movements in the monkey.* Journal of Neurophysiology, 1988. **60**(4): pp. 1268–1284.
4. Wessberg, J., et al., *Real-time prediction of hand trajectory by ensembles of cortical neurons in primates.* Nature, 2000. **408**(6810): pp. 361–365.
5. Taylor, D.M., S.I.H. Tillery, and A.B. Schwartz, *Direct cortical control of 3D neuroprosthetic devices.* Science, 2002. **296**(5574): pp. 1829–1832.
6. Serruya, M.D., et al., *Brain–machine interface: Instant neural control of a movement signal.* Nature, 2002. **416**: pp. 141–142.
7. Sanchez, J.C., et al. *A comparison between nonlinear mappings and linear state estimation to model the relation from motor cortical neuronal firing to hand movements.* in SAB Workshop on Motor Control in Humans and Robots: on the Interplay of Real Brains and Artificial Devices. 2002. University of Edinburgh, Scotland.
8. Sanchez, J.C., et al. *Input–output mapping performance of linear and nonlinear models for estimating hand trajectories from cortical neuronal firing patterns,* in International Work on Neural Networks for Signal Processing. 2002. Martigny, Switzerland.
9. Moran, D.W., and A.B. Schwartz, *Motor cortical representation of speed and direction during reaching.* Journal of Neurophysiology, 1999. **82**(5): pp. 2676–2692.
10. Kalaska, J.F., et al., *A comparison of movement direction-related versus load direction-related activity in primate motor cortex, using a two-dimensional reaching task.* Journal of Neuroscience, 1989. **9**(6): pp. 2080–2102.

11. Georgopoulos, A.P., et al., *Mental rotation of the neuronal population vector.* Science, 1989. **243**(4888): pp. 234–236.
12. Thach, W.T., *Correlation of neural discharge with pattern and force of muscular activity, joint position, and direction of intended next movement in motor cortex and cerebellum.* Journal of Neurophysiology, 1978. **41**: pp. 654–676.
13. Scott, S.H. and J.F. Kalaska, *Changes in motor cortex activity during reaching movements with similar hand paths but different arm postures.* Journal of Neurophysiology, 1995. **73**(6): pp. 2563–2567.
14. Todorov, E., *Direct cortical control of muscle activation in voluntary arm movements: A model.* Nature Neuroscience, 2000. **3**(4): pp. 391–398.
15. Gao, Y., et al. *A quantitative comparison of linear and non-linear models of motor cortical activity for the encoding and decoding of arm motions*, in the 1st International IEEE EMBS Conference on Neural Engineering. 2003. Capri, Italy.
16. Wu, W., et al. *Inferring hand motion from multi-cell recordings in motor cortex using a Kalman filter.* in SAB Workshop on Motor Control in Humans and Robots: on the Interplay of Real Brains and Artificial Devices. 2002. University of Edinburgh, Scotland.
17. Kalman, R.E., *A new approach to linear filtering and prediction problems.* Transactions of the ASME-Journal of Basic Engineering, 1960. **82**(Series D): pp. 35–45.
18. Andrieu, C., et al., *Particle methods for change detection, system identification, and control.* Proceedings of the IEEE, 2004. **92**(3): pp. 423–438.
19. Jordan, M.I., Learning in Graphical Models. 1998, New York: MIT Press.
20. Rabiner, L.R. and B.H. Juang, Fundamentals of Speech Recognition. Prentice-Hall Signal Processing Series. 1993, Englewood Cliffs, NJ: PTR Prentice-Hall.
21. Mussa-Ivaldi, F.A., *Modular features of motor control and learning.* Current Opinion in Neurobiology, 1999. **9**: pp. 713–717.
22. Dayan, P. and L. Abbott, Theoretical Neuroscience: Computational and Mathematical Modeling of Neural Systems. 2001, Cambridge, MA: MIT Press.
23. Scott, S.H., *Population vectors and motor cortex: neural coding or epiphenomenon.* Nature, 2000. **3**(4): pp. 307–308.
24. Moran, D.W. and A.B. Schwartz, *Motor cortical activity during drawing movements: population representation during spiral tracing.* Journal of Neurophysiology, 1999. **82**(5): pp. 2693–2704.
25. Chen, M.-H., Q.-M. Shao, and J.G. Ibrahim, Monte Carlo Methods in Bayesian Computation. Springer Series in Statistics. 2000, New York: Springer.
26. Brockwell, A.E., A.L. Rojas, and R.E. Kaas, *Recursive Bayesian decoding of motor cortical signals by particle filtering.* Journal of Neurophysiology, 2003. **91**: pp. 1899–1907.

27. Erdogmus, D., J.C. Sanchez, and J.C. Principe. *Modified Kalman filter based method for training state-recurrent multilayer perceptrons*. in International Work on Neural Networks for Signal Processing. 2002. Martigny, Switzerland.
28. Wu, W., M.J. Black, D. Munford, Y. Gao, E. Bienenstock, and J.P. Donoghue, *Modeling and decoding motor cortical activity using a switching Kalman filter*. IEEE Transactions on Biomedical Engineering, 2004. **51**: pp. 933–942.
29. Wu, W., et al., *Bayesian population decoding of motor cortical activity using a Kalman filter*. Neural Computation, 2005. **18**: pp. 80–118.
30. Sorenson, H.W., Kalman Filtering: Theory and Application. 1985, New York: IEEE Press.
31. Wan, E.A. and R.v.d. Merwe, *The Unscented Kalman Filter*, in Kalman Filtering and Neural Networks, S. Haykin, ed. 2001, New York: Wiley Publishing.
32. Schwartz, A.B., D.M. Taylor, and S.I.H. Tillery, *Extraction algorithms for cortical control of arm prosthetics*. Current Opinion in Neurobiology, 2001. **11**(6): pp. 701–708.
33. Liu, J.S. and R. Chen, *Sequential Monte Carlo methods for dynamic systems*. Journal of the American Statistical Association, 1998. **93**: pp. 1032–1044.
34. Doucet, A., S. Godsill, and C. Andrieu, *On sequential Monte Carlo methods for Bayesian filtering*. Statistical Computation, 2000. **10**: pp. 197–208.
35. Doucet, A., Monte Carlo Methods for Bayesian Estimation of Hidden Markov Models: Application to Radiation Signals. 1997, Orsay, France: University Paris-Sud.
36. Arulampalam, M.S., et al., *A tutorial on particle filters for online nonlinear/non-Gaussian Bayesian Tracking*. IEEE Transactions on Signal Processing, 2002. **50**(2): pp. 174–188.
37. Andersen, R.A., et al., *Multimodal representation of space in the posterior parietal cortex and its use in planning movements*. Annual Review of Neuroscience, 1997. **20**: pp. 303–330.
38. Shenoy, K.V., et al., *Neural prosthetic control signals from plan activity*. NeuroReport, 2003. **14**: pp. 591–597.
39. Rabiner, L.R., *A tutorial on hidden markov models and selected applications in speech recognition*. Proceedings of the IEEE, 1989. **77**(2): pp. 257–286.
40. Baum, L.E., et al., *A maximization technique occurring in the statistical analysis of probabilistic functions of Markov chains*. Annals of Mathematical Statistics, 1970. **41**(1): p. 164-171.
41. Kim, S.P., et al., *Divide-and-conquer approach for brain machine interfaces: Nonlinear mixture of competitive linear models*. Neural Networks, 2003. **16**(5–6): pp. 865–871.
42. Linde, Y., A. Buzo, and R.M. Gray, *An algorithm for vector quantizer design*. IEEE Transactions in Communication, 1980. **28**(1): pp. 84–95.
43. Darmanjian, S., et al. *Bimodel brain–machine interface for motor control of robotic prosthetic*, in IEEE International Conference on Intelligent Robots and Systems. 2003. Las Vegas, NV.

44. Nicolelis, M.A., et al., *Simultaneous encoding of tactile information by three primate cortical areas.* Nature Neuroscience, 1998. **1**(7): pp. 621–630.
45. Freeman, W.J., Mass Action in the Nervous System: Examination of the Neurophysiological Basis of Adaptive Behavior Through EEG. 1975, New York: Academic Press.
46. Ghahramani, Z. and M.I. Jordan, *Factorial hidden Markov models.* Machine Learning, 1997. **29**: pp. 245–275.
47. Jacobs, R.A., et al., *Adaptive mixtures of local experts.* Neural Computation, 1991. **3**: pp. 79–87.

• • • •

CHAPTER 6

Adaptive Algorithms for Point Processes

Additional Contributor: Yiwen Wang

In this chapter, we review the design of adaptive filters for *point processes* under the Gaussian assumption, and then introduce a Monte Carlo sequential estimation, to probabilistically reconstruct the state from discrete (spiking) observation events. In the previous chapter, we covered a similar topic but the observations were assumed continuous random variables (i.e., firing rates). On the other hand, here they are point processes, that is, 0/1 signals that contain the information solely in their time structure. So the fundamental question is how to adapt the Bayesian sequential estimation models described in Chapter 5 to point processes. The answer is provided by working with the probability of neural firing (which is a continuous random variable). We will use the well-accepted Poisson model of spike generation introduced in Chapter 2, but this time making the firing rate dependent on the system state.

Methods for analyzing spike trains have been applied primarily to understand how neurons encode information [1]. In motor control BMIs, the problem is actually the reverse, where a process called *decoding* [2] identifies how a spike train in motor cortex can explain the movement of a limb. However, the primary methodologies are still inspired by the *encoding* methods. For example, the population vector method of Georgopoulos [3] is a generative model of the spike activity based on the tuning curve concept (preferential firing for a given hand position/speed) that has been extensively utilized in encoding methods. In BMIs, the population vector technique has been championed by Schwartz et al. [4]. All the encoding methods effectively model the PDF of the spike firings. An alternative methodology that effectively bypasses this requirement is the use of maximum likelihood methods, assuming a specific PDF. In neuroscience, the Poisson distribution assumption is very common because it has been validated in numerous experimental setups, but it cannot account for multimodal firing histograms that are often found in neurons. The Poisson model has been improved with a time varying mean to yield what is called the inhomogeneous Poisson model.

A general adaptive filtering paradigm for point processes was recently proposed in [5] to reconstruct the hand position from the discrete observation of neural firings. This algorithm modeled the neural spike train as an inhomogeneous Poisson process feeding a kinematic model through a nonlinear tuning function. The point process counterparts of the Kalman filter, recursive least squares, and steepest descent algorithms were derived and recently compared in the decoding of tuning parameters and states from the ensemble neural spiking activity [6]. The point process analogue of the Kalman filter performs the best, because it provides an adjustable step size to update the state, which is estimated from the covariance information. However, this method still assumes that the posterior density of the state vector given the discrete observation is Gaussian distributed, which is rather unlikely. A Monte Carlo sequential estimation algorithm for point processes has been recently proposed to infer the kinematic information directly from neural spike trains [7]. Given the neural spike train, the posterior density of the kinematic stimulus can be estimated at each time step without the Gaussian assumption. The preliminary simulations have shown a better velocity reconstruction from exponentially tuned neural spike trains.

These spike-based BMI methods require preknowledge of the neuron receptive properties, and an essential stationary assumption is used when the receptive field is estimated from a block of data, which may not account for changes in response of the neural ensemble from open to closed loop experiments [8]. All the encoding methods effectively model the PDF (often called the intensity function) of spike firings. PDF estimation is a difficult problem that is seldom attempted because it requires lots of data and stationary conditions. An alternative methodology that effectively bypasses this requirement is the use of maximum likelihood methods, assuming a specific PDF. Unfortunately, the extension of these methods to multineuron spike trains is still based on the assumption of spike independence, which does not apply when neurons are part of neural assemblies.

6.1 ADAPTIVE FILTERING FOR POINT PROCESSES WITH A GAUSSIAN ASSUMPTION

One can model a point process by using a Bayesian approach to estimate the system state by evaluating the posterior density of the state given the discrete observation [6]. This framework provides a nonlinear time-series probabilistic model between the state and the spiking event [10].

Given an observation interval $(0,T]$, the number $N(t)$ of events (e.g., spikes) can be modeled as a stochastic inhomogeneous Poisson process characterized by its conditional intensity function $\lambda(t \mid \mathbf{x}(t), \boldsymbol{\theta}(t), \boldsymbol{Z}(t))$, that is, the instantaneous rate of events, defined as

$$\lambda(t \mid \mathbf{x}(t), \boldsymbol{\theta}(t), \boldsymbol{Z}(t)) = \lim_{\Delta t \to 0} \frac{\Pr(N(t+\Delta t) - N(t) = 1 \mid \mathbf{x}(t), \boldsymbol{\theta}(t), \boldsymbol{Z}(t))}{\Delta t} \tag{6.1}$$

where $\mathbf{x}(t)$ is the system state, $\boldsymbol{\theta}(t)$ is the parameter of the adaptive filter, and $\boldsymbol{Z}(t)$ is the history of all the states, parameters, and the discrete observations up to time t. The relationship between the single-parameter Poisson process λ, the state $\mathbf{x}(t)$, and the parameter $\boldsymbol{\theta}(t)$ is a nonlinear model represented by

$$\lambda(t \mid \mathbf{x}(t), \boldsymbol{\theta}(t)) = f(\mathbf{x}(t), \boldsymbol{\theta}(t)) \tag{6.2}$$

The nonlinear function $f(\cdot)$ is assumed to be known or specified according to the application. Let us consider hereafter the parameter $\boldsymbol{\theta}(t)$ as part of the state vector $\mathbf{x}(t)$. Given a binary observation event ΔN_k over the time interval $(t_{k-1}, t_k]$, the posterior density of the whole state vector $\mathbf{x}(t)$ at time t_k can be represented by Bayes' rule as

$$p(\mathbf{x}_k \mid \Delta N_k, \boldsymbol{Z}_k) = \frac{p(\Delta N_k \mid \mathbf{x}_k, \boldsymbol{Z}_k)\, p(\mathbf{x}_k \mid \boldsymbol{Z}_k)}{p(\Delta N_k \mid \boldsymbol{Z}_k)} \tag{6.3}$$

where $p(\Delta N_k \mid \mathbf{x}_k, \boldsymbol{Z}_k)$ is the probability of observing spikes in the interval $(t_{k-1}, t_k]$, considering the Poisson process

$$\Pr(\Delta N_k \mid \mathbf{x}_k, \boldsymbol{Z}_k) = (\lambda(t_k \mid \mathbf{x}_k, \boldsymbol{Z}_k)\Delta t)^{\Delta N_k} \exp(-\lambda(t_k \mid \mathbf{x}_k, \boldsymbol{Z}_k)\Delta t) \tag{6.4}$$

and $p(\mathbf{x}_k \mid \boldsymbol{Z}_k)$ is the one-step prediction density given by the Chapman–Kolmogorov equation as

$$p(\mathbf{x}_k \mid \boldsymbol{Z}_k) = \int p(\mathbf{x}_k \mid \mathbf{x}_{k-1}, \boldsymbol{Z}_k)\, p(\mathbf{x}_{k-1} \mid \Delta N_{k-1}, \boldsymbol{Z}_{k-1})\, d\mathbf{x}_{k-1} \tag{6.5}$$

where the state $\mathbf{x}_k$ evolves according to the linear relation

$$\mathbf{x}_k = F_k \mathbf{x}_{k-1} + \eta_k \tag{6.6}$$

F_k establishes the dependence on the previous state and η_k is zero-mean white noise with covariance Q_k. Substituting (6.4) and (6.5) in (6.3), the posterior density of the state $p(\mathbf{x}_k \mid \Delta N_k, \boldsymbol{Z}_k)$ can be recursively estimated from the previous value based on all the spike observations.

Assuming the posterior density given by (6.3) and the noise term η_k in the state evolution equation (6.6) are Gaussian distributed, the Chapman–Kolmogorov equation (6.5) becomes a convolution of two Gaussian functions, from which the estimation of the state at each time has a closed-form expression given by (see [11] for details)

$$\mathbf{x}_{k|k-1} = F_k \mathbf{x}_{k-1|k-1} \tag{6.7a}$$

$$P_{k|k-1} = F_k P_{k-1|k-1} F_k' + Q_k \tag{6.7b}$$

$$(P_{k|k})^{-1} = (P_{k|k-1})^{-1} + \left[\left(\frac{\partial \log \lambda}{\partial \mathbf{x}_k}\right)' [\lambda \Delta t_k] \left(\frac{\partial \log \lambda}{\partial \mathbf{x}_k}\right) - (\Delta N_k - \lambda \Delta t_k) \frac{\partial^2 \log \lambda}{\partial \mathbf{x}_k \partial \mathbf{x}_k'} \right]_{\mathbf{x}_{k|k-1}} \tag{6.7c}$$

$$\mathbf{x}_{k|k} = \mathbf{x}_{k|k-1} + P_{k|k} \left[\left(\frac{\partial \log \lambda}{\partial \mathbf{x}_k}\right)' (\Delta N_k - \lambda \Delta t_k) \right]_{\mathbf{x}_{k|k-1}} \tag{6.7d}$$

As expected, the Gaussian assumption allows an analytical solution for (6.5) and therefore, for a closed-form solution of (6.3) as (6.7).

Although the above set of equations may seem daunting, they can be interpreted quite easily. First, (6.7a) establishes a prediction for the state based on the previous value. Then, (6.7b) and (6.7c) are used in (6.7d) to correct or refine the previous estimate, after which the recursive calculation is repeated.

6.2 MONTE CARLO SEQUENTIAL ESTIMATION FOR POINT PROCESSES

The Gaussian assumption applied to the posterior distribution in the algorithm just described may not be true in general. Therefore, for the discrete observations case, a nonparametric approach is developed here that poses no constraints on the form of the posterior density.

Suppose at time instant k the previous system state is $\mathbf{x}_{k-1}$. Recall that because the parameter $\boldsymbol{\theta}$ was embedded in the state, all we need is the estimation of the state from the conditional intensity function, because the nonlinear relation $f(\cdot)$ is assumed known. Random state samples are generated using Monte Carlo simulations [12] in the neighborhood of the previous state according to (6.6). Then, weighted Parzen windowing [13] can be used with a Gaussian kernel to estimate the posterior density. Because of the linearity of the integral in the Chapman–Kolmogorov equation and the weighted sum of Gaussians centered at the samples we are still able to evaluate directly from samples the integral. The process is recursively repeated for each time instant propagating the estimate of the posterior density, and the state itself, based on the discrete events over time. Notice that because of the recursive approach, the algorithm not only depends on the previous observation, but also on the whole path of the spike observation events.

Let $\{\mathbf{x}_{0:k}^i, w_k^i\}_{i=1}^{N_S}$ denote a random measure [14] in the posterior density $p(\mathbf{x}_{0:k} \mid N_{1:k})$, where $\{\mathbf{x}_{0:k}^i, i=1,\ldots,N_S\}$ is the set of all state samples up to time k with associated normalized weights $\{w_k^i, i=1,\ldots,N_S\}$ and N_S is the number of samples generated at each time index. Then, the posterior density at time k can be approximated by a weighted convolution of the samples with a Gaussian kernel as

$$p(\mathbf{x}_{0:k} \mid N_{1:k}) \approx \sum_{i=1}^{N_S} w_k^i \cdot k(\mathbf{x}_{0:k} - \mathbf{x}_{0:k}^i, \sigma) \tag{6.8}$$

where $N_{1:k}$ is the spike observation events up to time k modeled by an inhomogeneous Poisson process in previous section, and $k(x-\bar{x},\sigma)$ is the Gaussian kernel in term of x with mean $\bar{x}$ and covariance σ. By generating samples from a proposed density $q(\mathbf{x}_{0:k} \mid N_{1:k})$ according to the principle of importance sampling [15, 16], which usually assumes dependence on $\mathbf{x}_{k-1}$ and N_k only, the weights can be defined by (6.9).

$$w_k^i \propto \frac{p(\mathbf{x}_{0:k}^i \mid N_{1:k})}{q(\mathbf{x}_{0:k}^i \mid N_{1:k})} \tag{6.9}$$

Here, we assume the importance density obeys the properties of Markov chain such that

$$q(\mathbf{x}_{0:k} \mid N_{1:k}) = q(\mathbf{x}_k \mid \mathbf{x}_{0:k-1}, N_{1:k}) q(\mathbf{x}_{0:k-1} \mid N_{1:k-1}) = q(\mathbf{x}_k \mid \mathbf{x}_{k-1}, \Delta N_k) q(\mathbf{x}_{0:k-1} \mid N_{1:k-1}) \tag{6.10}$$

At each iteration, the posterior density $p(\mathbf{x}_{0:k} \mid N_{1:k})$ can be estimated from the previous iteration as (6.11):

$$p(\mathbf{x}_{0:k} \mid N_{1:k}) = \frac{p(\Delta N_k \mid \mathbf{x}_{0:k}, N_{1:k-1}) p(\mathbf{x}_{0:k} \mid N_{1:k-1})}{p(\Delta N_k \mid N_{1:k-1})}$$

$$= \frac{p(\Delta N_k \mid \mathbf{x}_{0:k}, N_{1:k-1}) p(\mathbf{x}_k \mid \mathbf{x}_{0:k-1}, N_{1:k-1})}{p(\Delta N_k \mid N_{1:k-1})} \times p(\mathbf{x}_{0:k-1} \mid N_{1:k-1})$$

$$= \frac{p(\Delta N_k \mid \mathbf{x}_k) p(\mathbf{x}_k \mid \mathbf{x}_{k-1})}{p(\Delta N_k \mid N_{1:k-1})} \times p(\mathbf{x}_{0:k-1} \mid N_{1:k-1})$$

$$\propto p(\Delta N_k \mid \mathbf{x}_k) p(\mathbf{x}_k \mid \mathbf{x}_{k-1}) p(\mathbf{x}_{0:k-1} \mid N_{1:k-1}) \tag{6.11}$$

Replacing (6.10) and (6.11) into (6.9), the weight can be updated recursively as (6.12):

$$w_k^i \propto \frac{p(\Delta N_k \mid \mathbf{x}_k^i) p(\mathbf{x}_k^i \mid \mathbf{x}_{k-1}^i) p(\mathbf{x}_{0:k-1}^i \mid N_{1:k-1})}{q(\mathbf{x}_k^i \mid \mathbf{x}_{k-1}^i, \Delta N_k) q(\mathbf{x}_{0:k-1}^i \mid N_{1:k-1})} = \frac{p(\Delta N_k \mid \mathbf{x}_k^i) p(\mathbf{x}_k^i \mid \mathbf{x}_{k-1}^i)}{q(\mathbf{x}_k^i \mid \mathbf{x}_{k-1}^i, \Delta N_k)} w_{k-1}^i \tag{6.12}$$

Usually, the importance density $q(\mathbf{x}_k^i \mid \mathbf{x}_{k-1}^i, \Delta N_k)$ is chosen to be the prior density $p(\mathbf{x}_k^i \mid \mathbf{x}_{k-1}^i)$, requiring the generation of new samples from $p(\mathbf{x}_k^i \mid \mathbf{x}_{k-1}^i)$ by (6.6) as a prediction stage.

Sequential Importance Resampling [17] should also be applied at every time index to avoid degeneration, so that the sample is i.i.d. from the discrete uniform density with weights $w_{k-1}^i = 1/N_S$. The weights then change proportionally given by

$$w_k^i \propto p(\Delta N_k \mid \mathbf{x}_k^i) \tag{6.13}$$

where $p(\Delta N_k \mid \mathbf{x}_k^i)$ is defined by (6.4) in this section. Using (6.6), (6.13), and the resampling step, the posterior density of the state $\mathbf{x}_k$ given the whole path of the observed events up to time t_k can be approximated as

$$p(\mathbf{x}_k \mid N_{1:k}) \approx \sum_{i=1}^{N_S} p(\Delta N_k \mid \mathbf{x}_k^i) \cdot k(\mathbf{x}_k - \mathbf{x}_k^i) \tag{6.14}$$

Equation (6.14) shows that the posterior density of the current state given the observation is modified by the latest probabilistic measurement of observing the spike event $p(\Delta N_k \mid \mathbf{x}_k^i)$, which is the update stage in the adaptive filtering algorithm.

Without a closed form solution for state estimation, we estimate the posterior density of the state given the observed spike event $p(\mathbf{x}_k \mid N_{1:k})$ at every step and apply two methods to get the state estimation $\tilde{\mathbf{x}}_k$. One possible method is maximum likelihood estimation (MLE), which picks out the sample $\mathbf{x}_k^{i^*}$ with maximum posterior density, or alternatively the expectation of the posterior density can be picked as the state. As we smooth the posterior density by convolving with a Gaussian kernel, we can easily obtain the expectation $\tilde{\mathbf{x}}_k$, and its error covariance V_k by a technique called collapse [18]:

$$\tilde{\mathbf{x}}_k = \sum_{i=1}^{N_S} p(\Delta N_k \mid \mathbf{x}_k^i) \cdot \mathbf{x}_k^i \tag{6.15}$$

$$V_k = \sum_{i=1}^{N_S} p(\Delta N_k \mid \mathbf{x}_k^i) \cdot (\sigma + (\mathbf{x}_k^i - \tilde{\mathbf{x}}_k)(\mathbf{x}_k^i - \tilde{\mathbf{x}}_k)^T) \tag{6.16}$$

From (6.15) and (6.16), it is evident that with simple computation one can easily estimate the next state. Hence, the expectation by collapse is simple and elegant.

The major drawback of the algorithm is computational complexity because the quality of the solution requires many particles $\{\mathbf{x}_{0:k}^i, i = 1, \ldots, N_S\}$ to approximate the posterior density. Smoothing the particles with kernels as in (6.14) alleviates the problem in particular when collapsing is utilized, but still the computation is much higher than calculating the mean and covariance of the PDF with a Gaussian assumption.

6.3 SIMULATION OF MONTE CARLO SEQUENTIAL ESTIMATION USING SPIKE TRAINS

Next, we will describe how the framework presented in Section 6.2 can be used to determine neuronal receptive fields. In a conceptually simplified motor cortical neural model [19], the one-dimensional velocity can be reconstructed from the neuron spiking events by Monte Carlo sequential estimation algorithm. This algorithm can provide a probabilistic approach to infer the most probable velocity as one of the components of the state. This decoding simulation updates the state estimation simultaneously and applies this estimation to reconstruct the signal, which assumes the interdependence between the encoding and decoding so that the accuracy of the receptive field estimation and the accuracy of the signal reconstruction are reliable. Notice that dealing with a point process is a more complex problem than when firing rates (continuous observations) are used as traditionally considered in particle filtering [20].

Let us first explain how the simulated data was generated. The tuning function of the receptive field that models the relation between the velocity and the firing rate is assumed exponential and given by (6.17)

$$\lambda(t_k) = \exp(\mu + \beta_k v_k) \tag{6.17}$$

where $\exp(\mu)$ is the background firing rate without any movement and β_k is the modulation in firing rate due to the velocity v_k. In practice in the electrophysiology laboratory, this function is unknown. Therefore, an educated guess needs to be made about the functional form, but the exponential function is widely utilized [21].

The desired velocity was generated as a random walk with a noise variance 2.5×10^{-5} at each 1-msec time step, as shown in Figure 6.1.

The background-firing rate $\exp(\mu)$ and the modulation parameter β_k are set as 1 and 3, respectively, for the whole simulation time, 200 sec. A neuron spike is drawn as a Bernoulli random variable with probability $\lambda(t_k)\Delta t$ within each 1-msec time window [22]. The neuron spike train is shown in Figure 6.2.

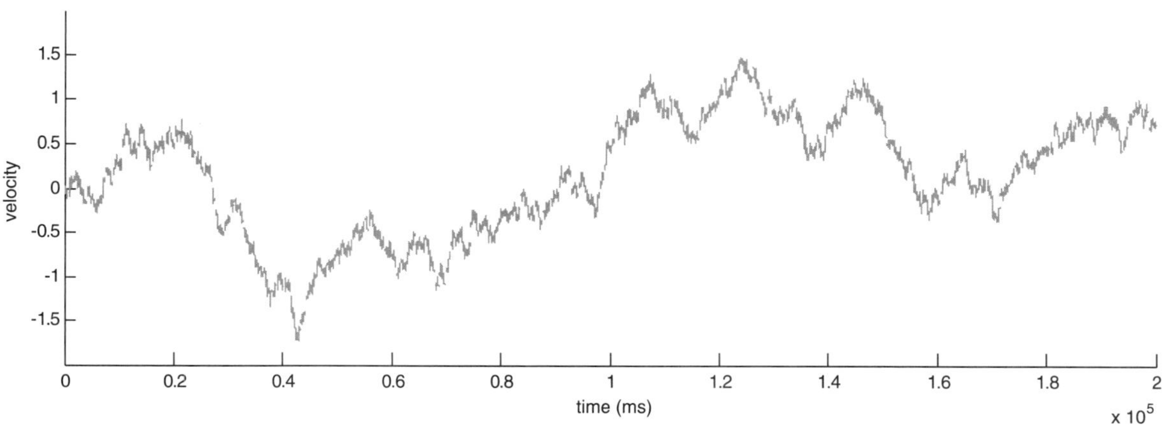

FIGURE 6.1: The desired velocity generated by random walk.

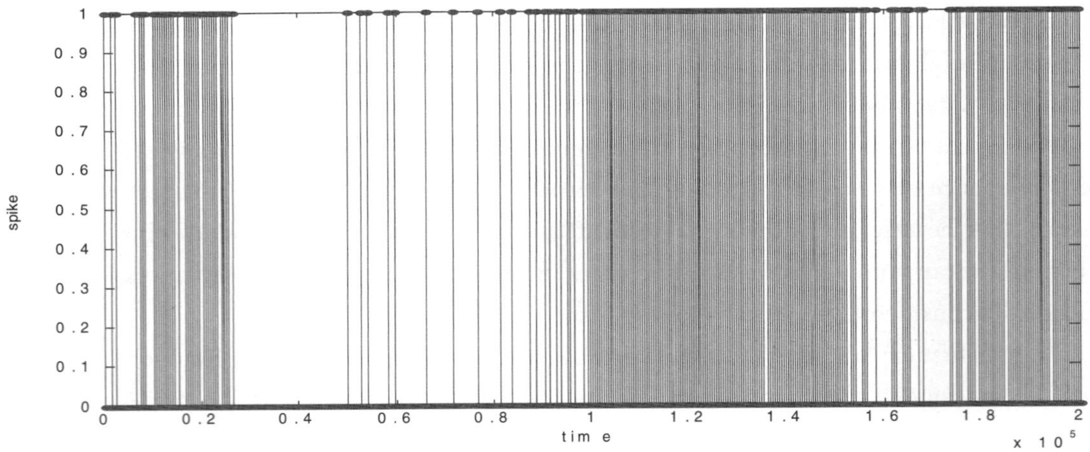

FIGURE 6.2: Simulated neuron spike train generated by an exponential tuning function.

With the exponential tuning function operating on the velocity, we can see that when the velocity is negative, there are few spikes; whereas when the velocity is positive, many spikes appear. The problem lies in obtaining from this spike train the desired velocity of Figure 6.1, assuming the Poisson model of (6.17) and of course using one of the sequential estimation techniques discussed.

To implement the Monte Carlo sequential estimation of the point process, we regard both modulation parameter β_k and velocity v_k as the state $\mathbf{x}_k = [v_k\ \beta_k]^{\mathrm{T}}$. One hundred samples of velocity v_0^i and modulation parameter β_k were initialized, respectively, with a uniform and with a Gaussian distribution. The new samples are generated according to the linear state evolution (6.6), where $F_k = \mathrm{diag}[1\ \ 1]$ and Q_k is the covariance matrix of the i.i.d. noise. The kernel size utilized in (6.14) to estimate the maximum of the posterior density (through MLE) was the average spike interval.

To obtain realistic performance assessments of the different models (MLE and collapse), the state estimations $\tilde{v}_k\ \tilde{\beta}_k$ for the duration of the trajectory are drawn 20 different times for different runs of the noise covariance matrices Q_k for state generation. The MSE between the desired trajectory and the model output is shown in Table 6.1 for the adaptive filtering and sequential estimation with both MLE and collapse. In general, if Q_k is too large, one needs many samples to estimate the PDF of the state appropriately and the state trajectory may not be smooth. If it is too small, the reconstructed velocity may get stuck in the same position, whereas the simulated one moves away by a distance much larger than Q_k.

The best velocity reconstruction, shown in the first row of Table 6.1, by both methods is shown in Figure 6.1.

From Figure 6.3 and Table 6.1, we can see that compared with the desired velocity (dash-dotted line) the best velocity reconstruction was achieved by the sequential estimation with collapse (solid black line). It is more sensitive than the adaptive filtering approach involving the Gaussian

TABLE 6.1: Comparison results of all algorithms with different Q_k

MSE DIAG(Q_K)	ADAPTIVE FILTERING OF POINT PROCESS	SEQUENTIAL ESTIMATION	
		MLE	COLLAPSE
[2e-5, 1e-7]	0.04801	0.1199	0.04522
[2e-5, 1e-6]	0.1081	0.1082	0.0489
[1e-5, 1e-6]	0.1076	0.1013	0.0588

assumption (solid gray line).The performance difference can be attributed to a more accurate estimate of the real posterior density because no further assumptions are made. The MLE estimation does not perform well in this simulation for reasons that are not totally clear, but may be related to the multiple maxima of the posterior. As an example in Figure 6.4, the posterior density at time 45.092 sec is shown (blue solid line) with three ripples and is obviously not Gaussian distributed.

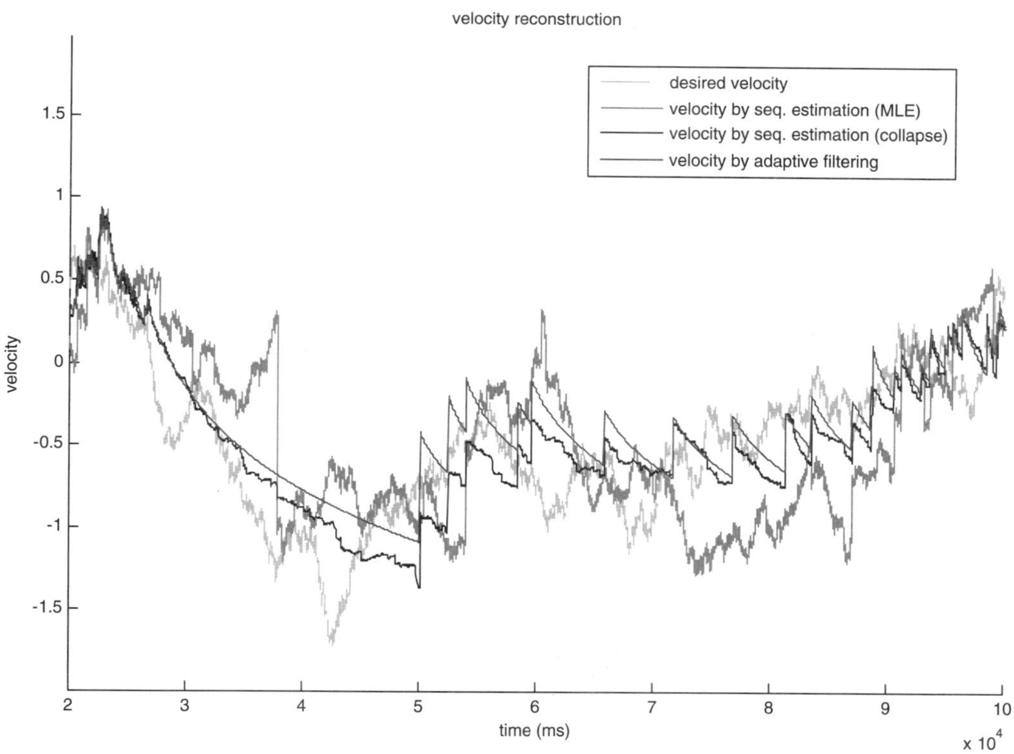

FIGURE 6.3: Velocity reconstruction by different algorithms.

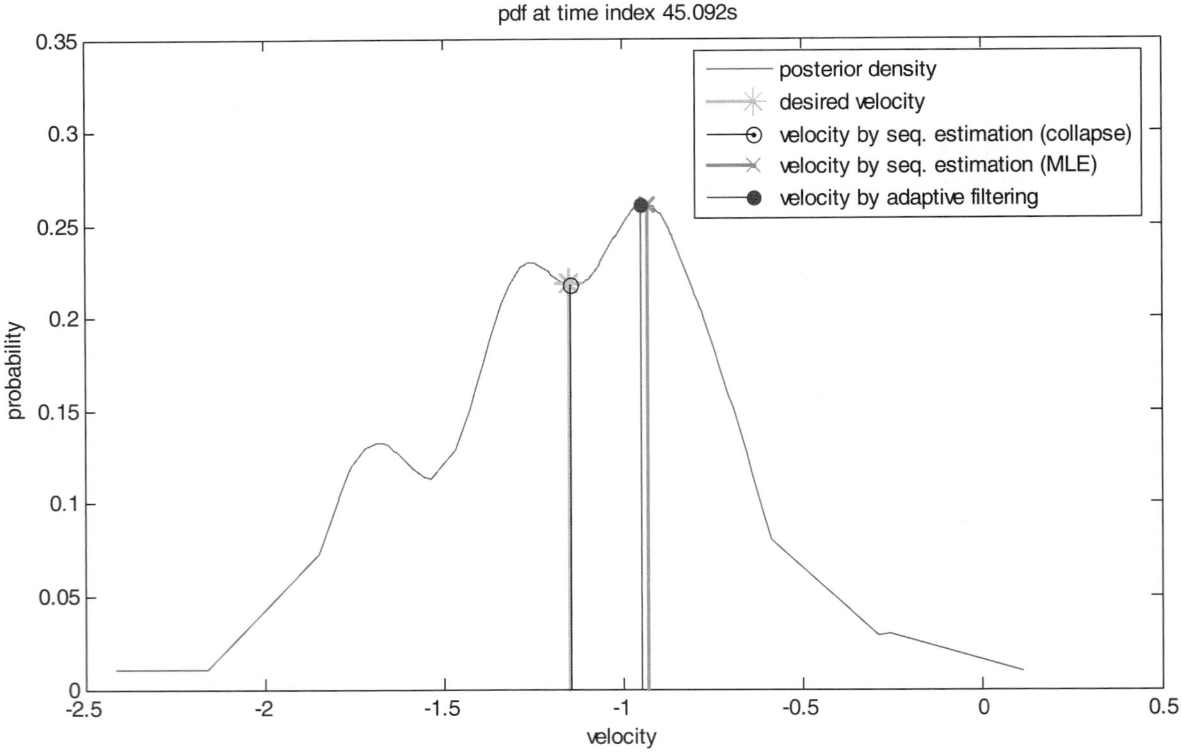

FIGURE 6.4: $p(\Delta N_k \mid v_k^i)$ at time 45.092 sec.

The velocity estimated by the sequential estimation with collapse denoted by the black circle is the closest to the desired velocity (green star).

In summary, the Monte Carlo sequential estimation on point processes shows a good capability to estimate the state from the discrete spiking events.

6.4 ENCODING/DECODING IN MOTOR CONTROL

It is interesting that in black box modeling, the motor BMI is posed as a decoding problem, that is, a transformation from motor neurons to behavior. However, when we use the Bayesian sequential estimation, decoding is insufficient to solve the modeling problem. The full dimension of the difficulty is still partially hidden in the Kalman filter (how to update the state given the observations?), but it becomes crystal clear when we develop the point process approach. To implement *decoding*, it is important to also model how neurons *encode* movement. Therefore, one sees that generative models do in fact require more information about the task and are therefore an opportunity to investigate further neural functionality.

One aspect that underlies the nervous system operation is the different time scales of the neural spike trains (few milliseconds) versus behavior (motor system dynamics are at hundreds of milliseconds to the second scale). Although at the microscale the spike timing is appropriately modeled by a Poisson random process, behavior is exquisitely organized albeit at a much larger time scale. Modeling methods should appropriately exploit this gap in time scales, instead of using simple averaging of spike occurrences in windows, as we do today.

The tuning function is a good example of a model that builds a functional stochastic relationship across time scales. In the estimation of the tuning function, most researchers use an exponential nonlinearity, but for a more realistic assessment, it is necessary to estimate the tuning functions of the neurons in a training set, and also to estimate the delays between the neural firing and the actual movement. This will have to be done in the training data, but once completed—and assuming that the neurons do not change their characteristics—the model is ready to be used for testing (possibly even in other types of movements not found in the training set).

6.4.1 Spike Generation From Neuronal Tuning

Many different functional forms of tuning have been proposed. Most consist of linear projections of the neural modulation on two or three dimensions of kinematic vectors and bias. Moran and Schwartz introduced an exponential velocity and direction tuned motor cortical model [19]. Eden et al have used a Gaussian tuning function for hippocampal pyramidal neurons [6] . These general mathematical models may not be optimal for dealing with the real data because the tuning properties across the ensemble can vary significantly. Moreover, the tuning may change over time. The accuracy of the tuning function estimation can directly affect the modeling in the Bayesian approach and the results of kinematic estimation.

One appropriate methodology is to estimate neural tuning using the training set data obtained in experiments. Marmarelis and Naka [23] developed a statistical method, called white noise analysis, to model the neural responses with stochastic stimuli. This method was improved by Simoncelli et al. [1]. By parametric model identification, the nonlinear property between the neural spikes and the stimuli was directly estimated from data, which is more reliable than just assuming a linear or Gaussian shape. Here we will describe how to use sequential state estimation on a point process algorithm to infer the kinematic vectors from the neural spike train, which is the opposite of sensory representation in spite of the fact that the kinematic vector can be regarded as the outcome of the motor cortex neuromodulation. The tuning function between the kinematic vector and the neural spike train is exactly the observation model between the state and the observed data in our algorithm.

6.4.1.1 Modeling and Assumptions. Recall that in Chapter 2 we show that a tuning function can be modeled as a linear filter coupled to a static nonlinearity followed by a Poisson model, the so-called linear–nonlinear–Poisson (LNP) model (Figure 6.5).

FIGURE 6.5: Block diagram of the linear–nonlinear–Poisson model.

For the BMI, the velocity of the hand movement is measured during a [-300, 500]-msec window with respect to every spike. Because the task is dynamic, a velocity vector at each time step is considered as the input (equal to the size of the space for simplicity). The linear filter projects the velocity vector v into its weight vector k (representing a direction in space), which produces a scalar value that is converted by a nonlinear function f and applied to the Poisson spike-generating model as the instantaneous conditional firing probability for that particular direction in space $p(\text{spike} \mid \vec{k} \cdot \vec{v})$. The filter weights are obtained optimally by least squares $\vec{k} = (E[\vec{v}^T \vec{v}] + \alpha I)^{-1} E_{\vec{v}|spike}[\vec{v}]$, where $E_{v|\text{spike}}$ is the conditional expectation of the velocity data given the spikes, which corresponds to the cross-

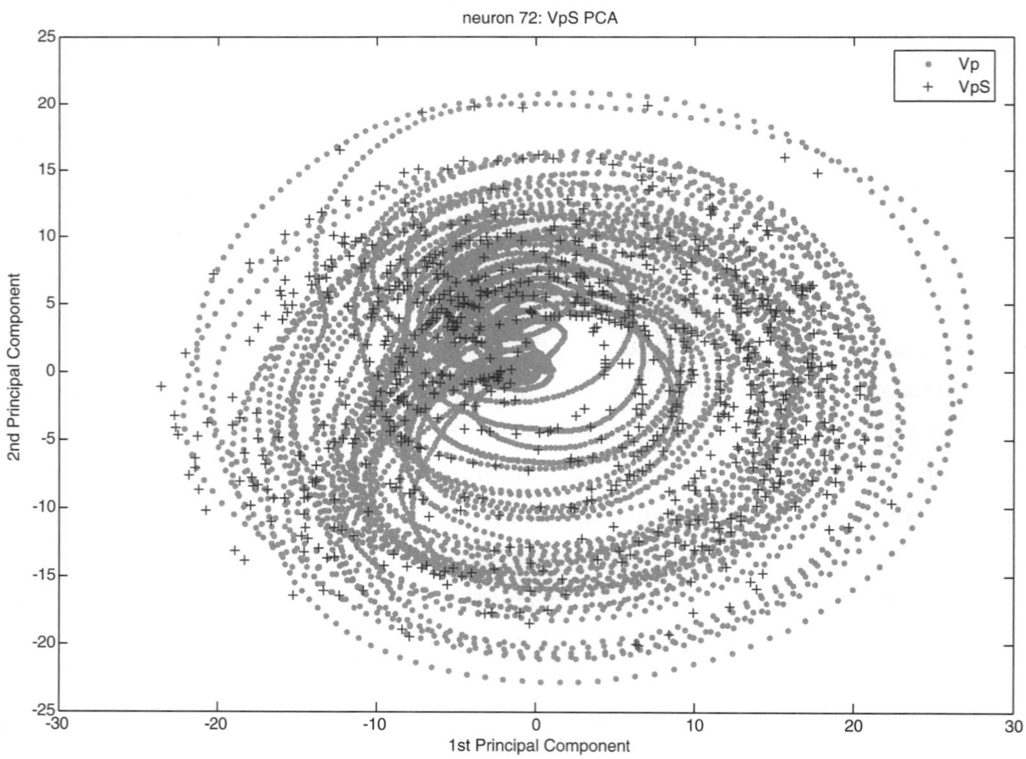

FIGURE 6.6: PCA on spike-triggered velocity vectors compared with the projection of all velocity vectors.

correlation in least squares and for spike trains, but has this form because the binary nature of the data. The parameter α is a regularizing parameter to properly condition the inverse.

This linear filter actually has a geometry interpretation. Figure 6.6 shows the first and the second PCA components of the spike-triggered velocity vectors for neuron 72 (the blue plus symbol in the figure). The velocity vectors for the same segment, which are represented by the green dots, are also projected onto the same PCA component directions. If the neuron has tuning, we expect that the first two components of spike-triggered velocity will be distributed differently from the velocity vectors.

The optimal linear filter actually projects the multidimensional velocity vectors along the direction where they differ the most from the spike-triggered velocity vectors.

For the time interval selected for the spike analysis (i.e., the time interval valid for a Poisson assumption in the collected data), a number is randomly drawn from a normalized uniform distribution (i.e., 0 to 1) and compared with the instantaneous conditional firing probability. If the

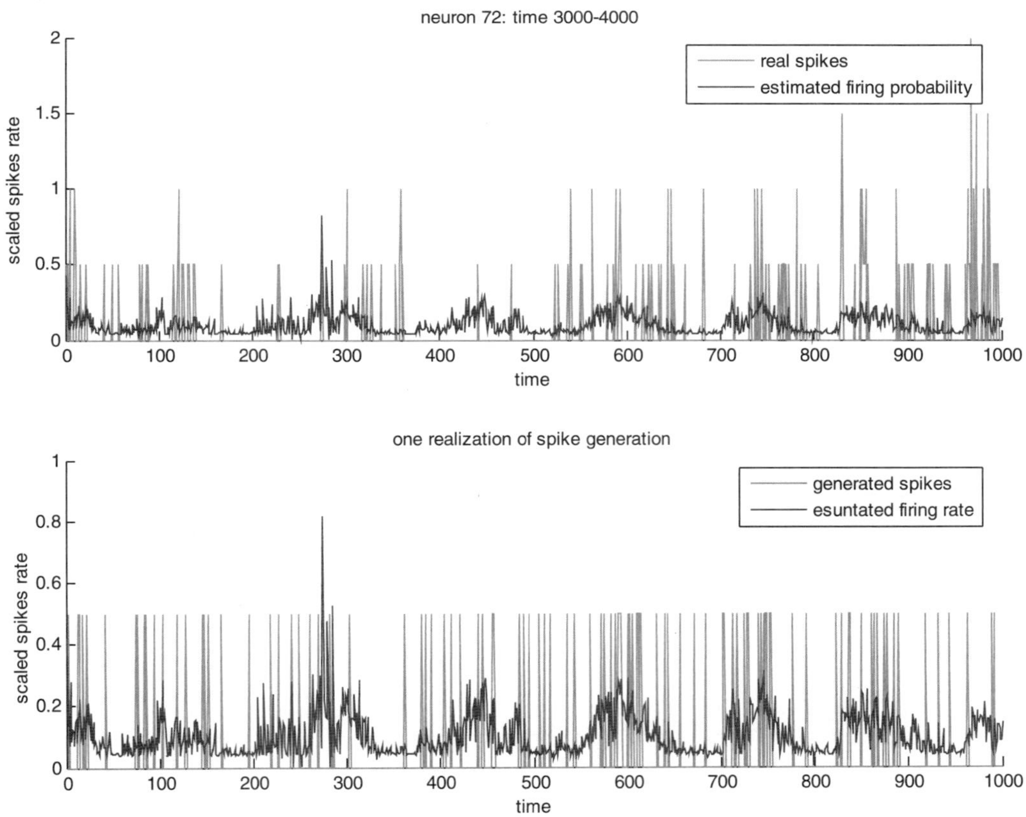

FIGURE 6.7: Spike generation from the estimated tuning function for neuron 72.

number is smaller than the probability, then a spike is generated in this time interval. The tuning function can be shown equivalent to $\lambda_t = f(\vec{k} \cdot \vec{v}_t)$ with spike$_t$ = Poisson (λ_f).

Figure 6.7 shows one realization of a generated spike train using the estimated tuning function for neuron 72. The upper plot depicts spikes scaled by ½ because of the temporal resolution presented in the figure. The blue line is the estimated instantaneous conditional firing probability $p(\text{spike} \mid \vec{k} \cdot \vec{v})$, which corresponds to the mean λ of the inhomogeneous Poisson process. During most of the duration depicted in the figure, the estimated λ can follow the real spike density change accurately. The bottom plot displays one realization of spikes generated from the estimated λ. The total number of spikes 113 is close to the real number of spike 126 during the time length and the correlation coefficient is between the spike trains smoothed by a Gaussian kernel are shown in Table 6.2 for different number of realizations. Note that the realizations by the Poisson spike generator can be quite different.

6.4.2 Estimating the Nonlinearity in the LNP Model

The nonlinear encoding function *f* for each neuron was estimated using an intuitive nonparametric binning technique. Given the linear filter vector $\vec{k}$, we drew the histogram of all the velocity vectors filtered by $\vec{k}$ and smoothed the histogram by convolving with a Gaussian kernel. The same procedure was repeated to draw the smoothed histogram for the outputs of the spike-triggered velocity vectors filtered by $\vec{k}$. The nonlinear function *f*, which gives the conditional instantaneous firing rate to the Poisson spike-generating model, was then estimated as the ratio of the two smoothed histograms.

Statistically speaking, the kernel smoothed histogram of all the velocity vectors filtered by $\vec{k}$ is an approximation to the marginal PDF $p(\vec{k} \cdot \vec{v})$ of the multidimensional velocity vectors projected

TABLE 6.2: Performance as a function of the number of realizations

NO. OF REALIZATIONS	CORRELATION COEFFICIENT
100	0.89
200	0.94
300	0.96
1000	0.98
10 000	0.99

onto the linear filter direction. The kernel-smoothed histogram of spike-triggered velocity vectors filtered by $\vec{k}$ is an approximation of the joint PDF $p(\text{spike}, \vec{k}\cdot\vec{v})$ projected to the linear filter direction. The ratio actually obeys the Bayesian formulation:

$$p(\text{spike} \mid \vec{k}\cdot\vec{v}) = \frac{p(\text{spike}, \vec{k}\cdot\vec{v})}{p(\vec{k}\cdot\vec{v})} \tag{6.18}$$

Figure 6.8 depicts the marginal PDF $p(\vec{k}\cdot\vec{v})$, which is the blue line in the top plot, and the joint PDF $p(\text{spike}, \vec{k}\cdot\vec{v})$, which is represented by the red line in the top plot, computed for neuron 110.

The nonlinear estimation for *f* is shown in the bottom plot of Figure 6.8. The *x* axis is the range of the filtered multidimensional velocity vectors. The *y* axis is the corresponding instantaneous conditional probability of spike given the linear filter output. The accuracy of the nonlinear function estimation is not constant across the range, and this may explain the ripples seen at the extremes of the range, so regularization should be utilized. This *f* is estimated from the real data,

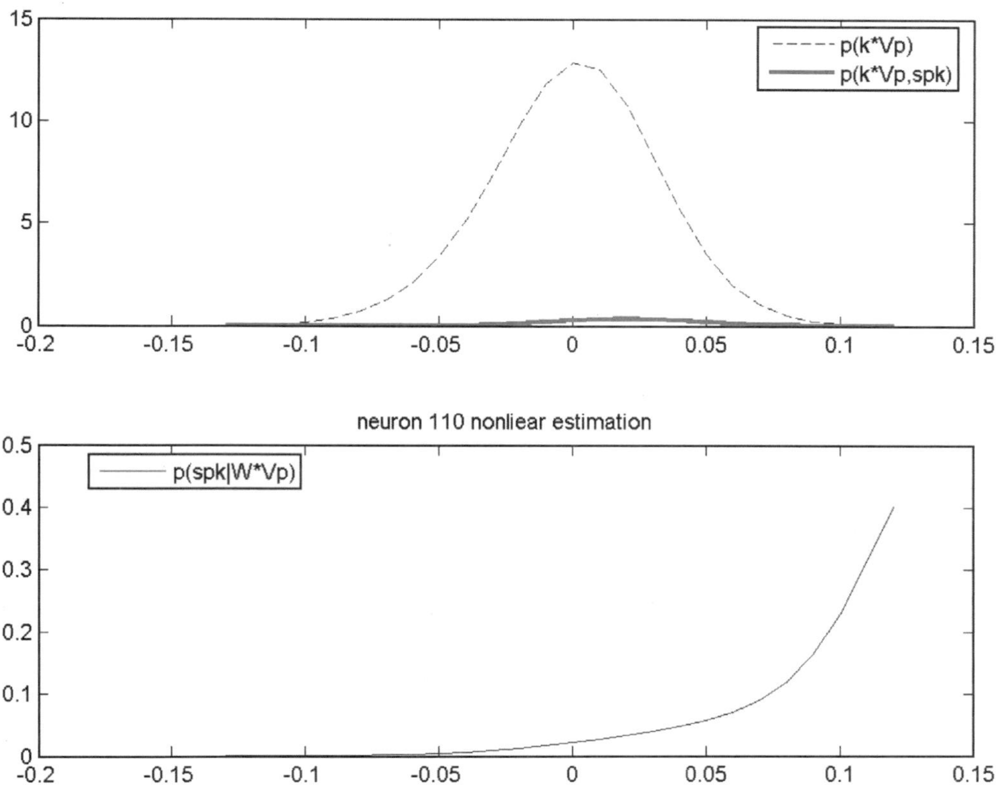

FIGURE 6.8: Nonlinear tuning estimation for neuron 110.

and thus provides more accurate nonlinear properties than just assuming the exponential or Gaussian function. It can actually be implemented as a look up table for the nonlinear tuning function f in testing as

$$p(\text{spike} \mid \vec{k} \cdot \vec{v}^{\,t}_{\text{test}}) = \frac{\sum_j k(\vec{k} \cdot \vec{v}^{\,t}_{\text{test}} - \vec{k} \cdot \vec{v}^{\,j}_{\text{spike,training}})}{\sum_i k(\vec{k} \cdot \vec{v}^{\,t}_{\text{test}} - \vec{k} \cdot \vec{v}^{\,i}_{\text{training}})} \tag{6.19}$$

where k is the Gaussian kernel, $\vec{v}^{\,t}_{\text{test}}$ is a possible sample we generate at time t in the test data. $\vec{v}^{\,j}_{\text{spike,training}}$ is one spike-triggered velocity vector sample in the training data, and $\vec{v}^{\,i}_{\text{training}}$ is one velocity vector sample in the training data.

6.4.3 Estimating the Time Delay From the Motor Cortex to Movement

Before implementing the generative BMI based on spikes, one has to align the spike trains with the corresponding hand movements, which is accomplished by estimating the causal time delays due to propagation effects of signals in the motor and peripheral nervous system, as shown in Figure 6.9.

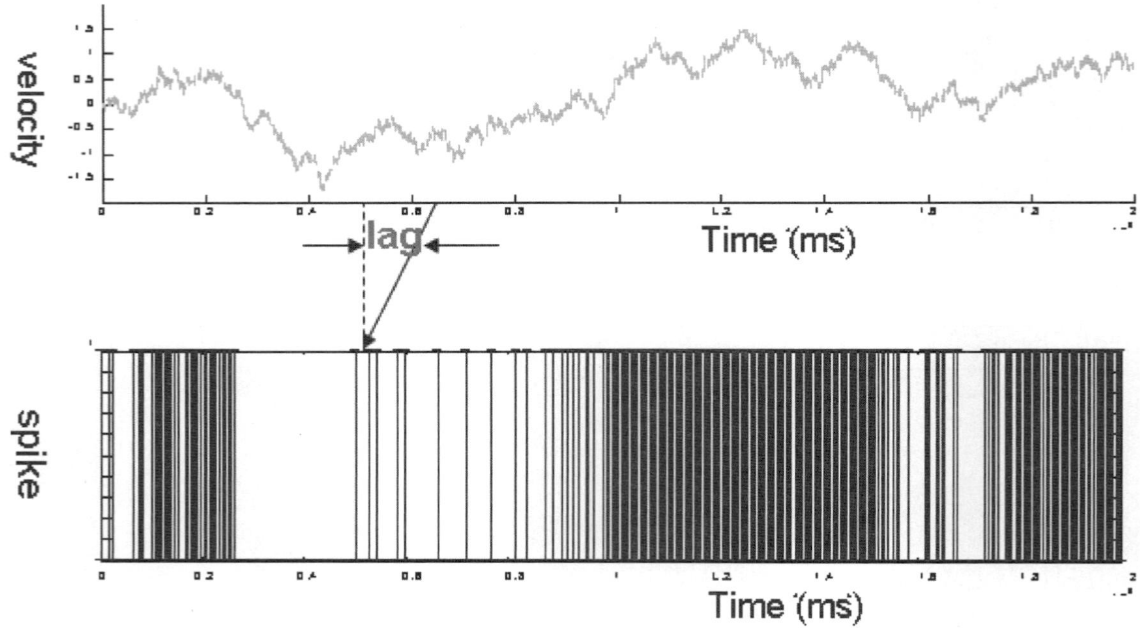

FIGURE 6.9: Illustration of the time delay between neuron spike train (bottom plot) and the kinematics response (upper plot).

To contend with the problem of selecting delays, the mutual information between the spike and the delayed linear filter kinematics vector can be computed as a function of the time lag after a spike arrives. Considering that the response delay ranges from 0 to 500 msec after a neuron fires, the mutual information is determined as a function of lag according to (6.20).

$$I_{(\text{spike};\vec{k}\cdot\vec{\mathbf{x}})}(\text{lag}) = \sum_{\vec{k}\cdot\vec{\mathbf{x}}} p(\vec{k}\cdot\vec{\mathbf{x}}(\text{lag})) \sum_{\text{spike}=0,1} p(\text{spike} \mid \vec{k}\cdot\vec{\mathbf{x}}(\text{lag})) \log_2 \left(\frac{p(\text{spike} \mid \vec{k}\cdot\vec{\mathbf{x}}(\text{lag}))}{p(\text{spike})} \right) \tag{6.20}$$

where to approximate $p(\text{spike} \mid \vec{k}\cdot\vec{\mathbf{x}}(\text{lag}))$ exactly the same method as the estimation of the nonlinear f we discussed in Section 6.4.1. For all 185 neurons, the mutual information as the function of time delay was obtained from 10 000 continuous samples of kinematics. The time delay with most mutual information was assigned as the best time lag for each neuron.

Figure 6.10 shows the mutual information as the function of time delay from 0 to 500 msec after spikes for five neurons: neurons 72, 77, 80, 99, and 108. The cross on each curve marks the best time lag, which was 110, 170, 170, 130, and 250 msec, respectively. The average best time delay for 185 neurons is 220 msec and was chosen for the subsequent analysis to simplify the implementation.

6.4.4 Point Process Monte Carlo Sequential Estimation Framework for BMIs

The decoding problem for BMIs is shown in Figure 6.11. The spike times from multiple neurons are the multichannel point process observations. The process begins by first windowing the data to determine if a spike is present or not (assign 1 if there is a spike; otherwise, assign 0). The interval should be chosen to be small enough so that only few intervals have more than one spike, and therefore still obey largely the Poisson assumption. One must be careful when selecting the kinematic state (position, velocity, or acceleration) in decoding because it is still not clear which one the neuron actually encodes for. The analysis presented here will consider all three kinematic variables. The velocity is derived as the difference between the current and previous recoded positions, and the acceleration is estimated by first differences from the velocity. For fine timing resolution, all of the kinematics are interpolated and time-synchronized with the neural spike trains.

The whole process can be specified in the following steps:

Step 1: Preprocess and analysis

1. Generate spike trains from stored spike times.
2. Synchronize all the kinematics with the spike trains.
3. Assign the kinematic vector $\vec{\mathbf{x}}$ to be reconstructed.

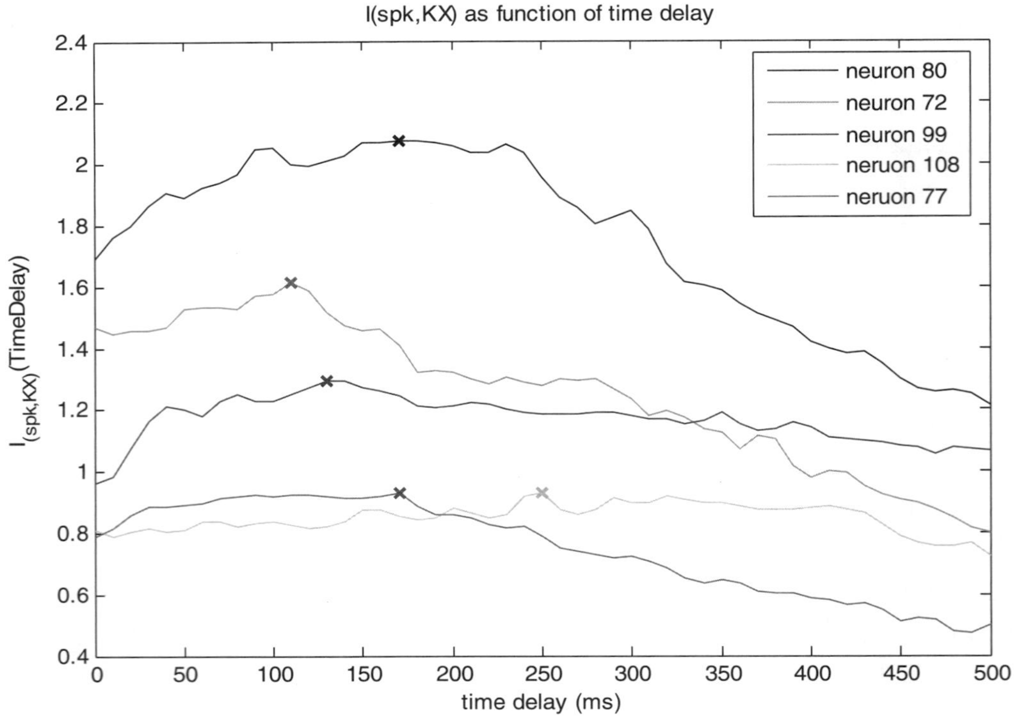

FIGURE 6.10: Mutual information as function of time delay for five neurons.

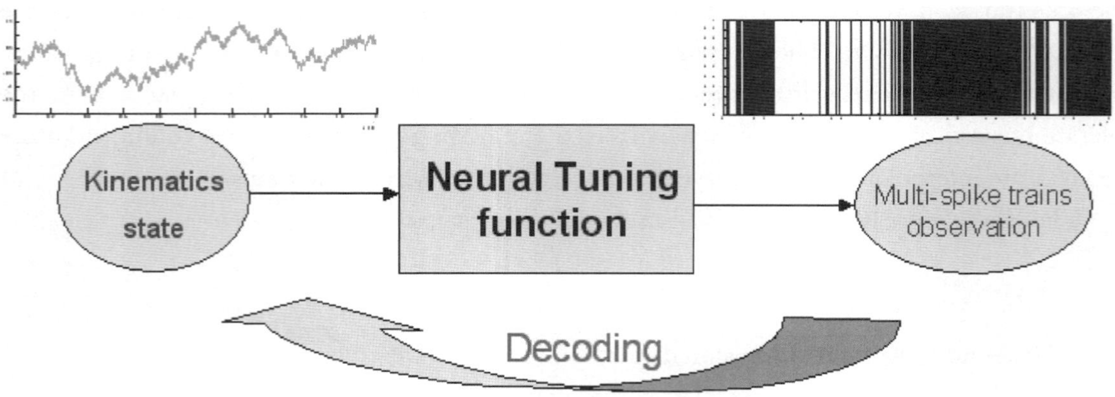

FIGURE 6.11: Schematic of BMI decoding by probabilistic approach.

Step 2: Model estimation

1. Estimate the kinematic dynamics of the system model $F_k = (E[\vec{\mathbf{x}}_{k-1}\vec{\mathbf{x}}_{k-1}^{\mathrm{T}}])^{-1}E[\vec{\mathbf{x}}_{k-1}\vec{\mathbf{x}}_k^{\mathrm{T}}]$
2. For each neuron j, estimate the tuning function
   - Linear model $\vec{k}^j = (E[\vec{\mathbf{x}}^{\mathrm{T}}\vec{\mathbf{x}}])^{-1}E_{\vec{\mathbf{x}}|\mathrm{spike}^j}[\vec{\mathbf{x}}]$
   - Nonlinear function $f^j(\vec{k}^j \cdot \vec{\mathbf{x}}_t) = \dfrac{p(\mathrm{spike}^j, \overrightarrow{k^j \cdot \mathbf{x}})}{p(\vec{k}^j \cdot \vec{\mathbf{x}})}$
   - Build the inhomogeneous Poisson generator.

Step 3: Monte Carlo sequential kinematics estimation

For each time k, a set of samples for state $\vec{\mathbf{x}}_k^i$ are generated, i=1:N

1. Predict new state samples $\vec{\mathbf{x}}_k^i = F_k \vec{\mathbf{x}}_k^i + \eta_k$, i = 1:N
2. For each neuron j,
   - Estimate the conditional firing rate $\lambda_k^{i,j} = f^j(\vec{k}^j \cdot \vec{\mathbf{x}}_k^i)$, i = 1:N
   - Update the weights $w_k^{i,j} \propto p(\Delta N_k^j \mid \lambda_t^{i,j})$, i = 1:N
3. Draw the weight for the joint posterior density $W_k^i = \prod_j w_k^{i,j}$, i=1:N
4. Normalize the weights $W_k^i = \dfrac{W_k^i}{\sum W_k^i}$, i = 1:N
5. Draw the joint posterior density $p(\vec{\mathbf{x}}_k \mid N_{1:k}) \approx \sum_{i=1}^{N} W_k^i \cdot k(\vec{\mathbf{x}}_k - \vec{\mathbf{x}}_k^i)$
6. Estimate the state $\mathbf{x}_k^*$ from the joint posterior density by MLE or expectation.
7. Resample $\vec{\mathbf{x}}_k^i$ according to the weights W_k^i.

6.4.5 Decoding Results Using Monte Carlo Sequential Estimation

The Monte Carlo sequential estimation framework was tested for the 2D control of a computer cursor. In the preprocessing step we consider a dataset consisting of 185 neurons. Here the state vector is chosen as the instantaneous kinematic vector $\vec{\mathbf{x}} = [\vec{\mathbf{p}}_x\ \vec{\mathbf{v}}_x\ \vec{\mathbf{a}}_x\ \vec{\mathbf{p}}_y\ \vec{\mathbf{v}}_y\ \vec{\mathbf{a}}_y]^T$ to be reconstructed directly from the spike trains. The position, velocity, and acceleration are all included in the vector to allow the possibility of sampling different neurons that represent particular aspects of the kinematics. Picking only the velocity, for example, might provide only partial information between the neural spikes and other kinematics. For each neuron in the ensemble, a small time interval of 10 msec was selected to construct the point process observation sequence. With this interval, 99.62% of windows had a spike count less than 2. For each neuron, 1 was assigned when there was one spike or more appearing during the interval, otherwise 0 was assigned.

After preprocessing the data, the kinematics model F_k can be estimated using the least squares solution. A regularization factor should be used to compute the linear filter weights in the tuning function (1e-4 in this analysis). It should also be noted that this Monte Carlo approach is stochastic, that is, it introduces variation between realizations even with fixed parameters because of the Poisson spike generation model. The most significant parameters include the kernel size σ, the number of particles x_n, and the intensity parameter e_k. For example, in the approximation of the noise distribution $p(\eta)$ approximated by the histogram of $\eta_k = \mathbf{x}_k - F_k\mathbf{x}_{k-1}$, the intensity parameter e_k or number of samples used to approximate the noise distribution can influence computational time and performance of the approximation. The kernel size σ is used to smooth the nonlinearity in the tuning function and in this estimation of the nonlinear function f^j it was assigned 0.02. This kernel size should be chosen carefully to avoid losing the characteristics of the tuning curve and also to minimize noise in the curve.

Figure 6.12 shows one realization of the reconstructed kinematics from all 185 neurons for 1000 samples. The left and right column plots display respectively the reconstructed kinematics for

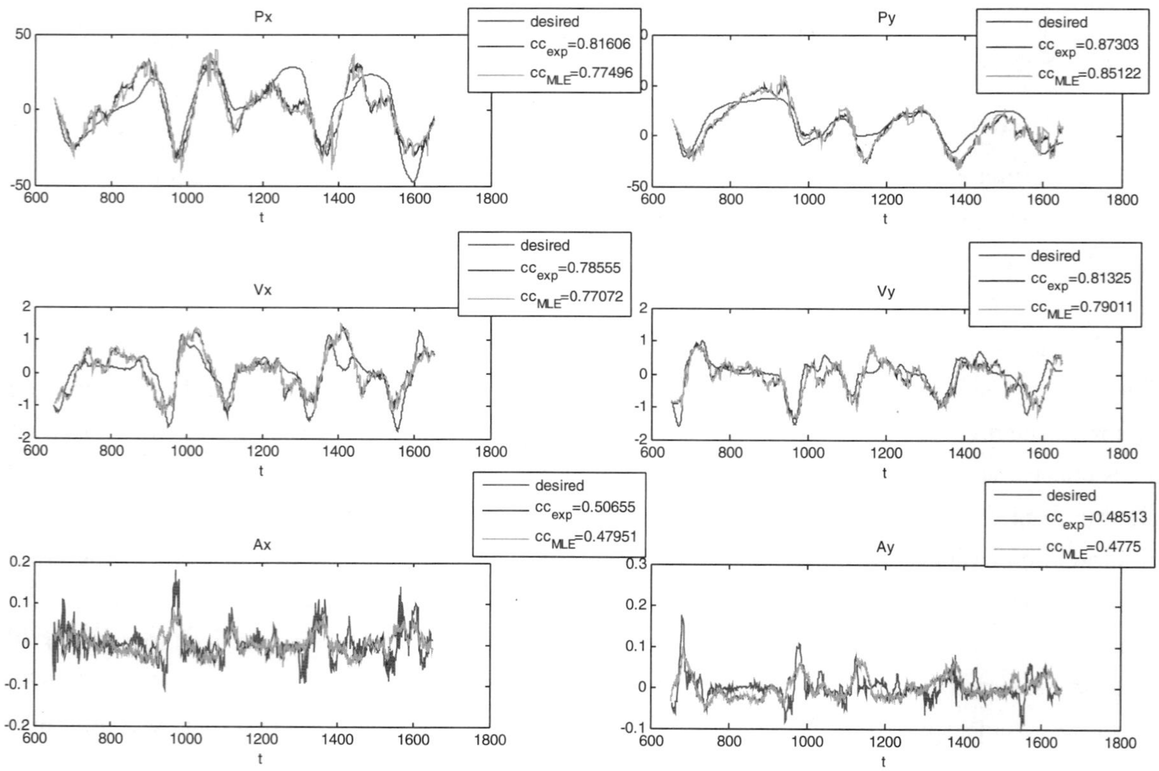

FIGURE 6.12: Reconstructed kinematics for a 2D reaching task.

TABLE 6.3: Average correlation coefficient between the desired kinematics and the reconstructions for 20 realizations

CC	POSITION		VELOCITY		ACCELERATION	
	x	*y*	*x*	*y*	*x*	*y*
Expectation	0.7958 ± 0.0276	0.8600 ± 0.0353	0.7274 ± 0.0278	0.7730 ± 0.0155	0.5046 ± 0.0148	0.4873 ± 0.0080
MLE	0.7636 ± 0.0334	0.8327 ± 0.0417	0.7119 ± 0.0292	0.7672 ± 0.0149	0.4872 ± 0.0200	0.4791 ± 0.0115

x axis and y axis. The three rows of plots from top to bottom display respectively the reconstructed position, the velocity, and the acceleration. In each subplot, the red line indicates the desired signal, the blue line indicates the expectation estimation, and green line indicates the MLE estimation. The Monte Carlo approach offers the best reconstruction for the position and the velocity. However, it should be remembered that the reconstructions of the kinematic states will vary among the realizations. The statistical information of correlation coefficients between the desired signal and the estimations for 20 times Monte Carlo runs is given in Table 6.3.

The reconstructed kinematics estimated from the expectation of the joint posterior density performed better than the one from the noise MLE. The position shows the best results as 0.8161 for x and 0.8730 for y than other two kinematic variables. This result may be because velocity and acceleration are derived as differential variables, where the noise in the estimation might be magnified. Another interesting phenomenon is that the y kinematics is reconstructed always better than x, which was the same situation in previous approaches.

To compare the performance with another generative model, Kalman filter algorithm was applied on the same data [24] to predict the hand positions from the binned neural spiking rate. Compared to the approach shown here, the Kalman filter simply assumes both the kinematic dynamic system model and tuning function are linearly related and the posterior density is Gaussian distributed. The average correlation coefficients for reconstructed position and y are 0.62 ± 0.26 and 0.82 ± 0.11, respectively, with sliding window for 40 samples prediction, which is 4 sec long. Because we have different sampling frequency (100 Hz for kinematics rather than 10 Hz), here the average correlation coefficients are calculated with 50% overlapping window. The average correlations for positions x and y are 0.8401 ± 0.0738 and 0.8945 ± 0.0477, respectively, which is better than the Kalman filter results.

6.5 SUMMARY

Although spike trains are very telling of neuronal processing, they are also very removed from the time and macroscopic scales of behavior. Therefore, the spike train methodology begs an answer to the question of how to optimally bridge the time scale of spikes events (milliseconds) with the time scale of behavior (seconds). Most often, the relatively rudimentary method of binning is used, but much of the resolution of the representation is lost, suggesting that better, model-based methodologies need to be developed.

The fundamental problem in spike-based BMI modeling is how to find more effective ways to work directly with spikes for modeling that overcome the difficulties of the conventional approaches. Here we have shown how a Monte Carlo sequential estimation framework could be used as a probabilistic approach to reconstruct the kinematics directly from the multichannel neural spike trains. We investigated an information theoretical tuning depth to evaluate the neuron tuning properties. To describe the functional relationship between neuron firing and movement, a parametric LNP model was utilized. With the knowledge gained from the neuron physiology tuning analysis, a novel signal processing algorithm based on a Monte-Carlo sequential estimation could be applied directly to point processes to convert the decoding role of a brain–machine interface into a problem of state sequential estimation. The Monte Carlo sequential estimation modifies the amplitude of the observed discrete neural spiking events by the probabilistic measurement contained in the posterior density. The Monte Carlo sequential estimation provided a better approximation of the posterior density than point process adaptive filtering with a Gaussian assumption. Compared with the Kalman filter applied to a cursor control task, the preliminary kinematics reconstruction using the Monte Carlo sequential estimation framework showed better correlation coefficients between the desired and estimated trajectory.

However, the signal processing methodologies of spike train modeling are not yet equipped to handle the stochastic and point process nature of the spike occurrence. Many parameters are assumed and need to be estimated with significant complexity. Conventional signal processing methods on random processes work easily with optimal projections; therefore, it would be very useful to create directly from spike trains a metric space that would allow inner product computations because all the available tools of binned methods could be applied immediately.

REFERENCES

1. Simoncelli, E.P., et al., *Characterization of neural responses with stochastic stimuli.* 3rd ed. The New Cognitive Neuroscience. 2004, Cambridge, MA: MIT Press.
2. Paninski, L., et al., *Superlinear population encoding of dynamic hand trajectory in primary motor cortex.* Journal of Neuroscience, 2004. **24**(39): pp. 8551–8561.
3. Georgopoulos, A.P., A.B. Schwartz, and R.E. Kettner, *Neuronal population coding of movement direction.* Science, 1986. **233**(4771): pp. 1416–1419.
4. Schwartz, A.B., D.M. Taylor, and S.I.H. Tillery, *Extraction algorithms for cortical control of arm prosthetics.* Current Opinion in Neurobiology, 2001. **11**(6): pp. 701–708.
5. Brown, G.D., S. Yamada, and T.J. Sejnowski, *Independent component analysis at the neural cocktail party.* Trends in Neurosciences, 2001. **24**(1): pp. 54–63.
6. Eden, U.T., et al., *Dynamic analysis of neural encoding by point process adaptive filtering.* Neural Computation, 2004. **16**: pp. 971–998.
7. Wang, Y., A.R.C. Paiva, and J.C. Principe. *A Monte Carlo sequential estimation for point process optimum filtering*, in IJCNN. 2006. Vancouver.
8. Helms Tillery, S.I., D.M. Taylor, and A.B. Schwartz, *Training in cortical control of neuroprosthetic devices improves signal extraction from small neuronal ensembles.* Reviews in the Neurosciences, 2003. **14**: pp. 107–119.

9. Tillery, S.I.H., D.M. Taylor, and A.B. Schwartz, *Training in cortical control of neuroprosthetic devices improves signal extraction from small neuronal ensembles*. Reviews in the Neurosciences, 2003. **14**(1–2): pp. 107–119.
10. Brown, E.N., L.M. Frank, and M.A. Wilson. *Statistical approaches to place field estimation and neuronal ensemble decoding*. Society for Neuroscience Abstracts, 1996. **22**: p. 910.
11. Brown, E.N., et al., *An analysis of neural receptive field plasticity by point process adaptive filtering*. Proceedings of the National Academy of Sciences of the United States of America, 2001. **98**(12): pp. 12 261–12 266.
12. Carpenter, J., P. Clifford, and P. Fearnhead, *Improved particle filter for nonlinear problems*. IEEE Proceedings on Radar, Sonar, and Navigation 1999. **146**: pp. 2–7.
13. Parzen, E., *On the estimation of a probability density function and the mode*. Annals of Mathematical Statistics, 1962. **33**(2): pp. 1065–1076.
14. Arulampalam, M.S., et al., *A tutorial on particle filters for online nonlinear/non-Gaussian Bayesian tracking*. IEEE Transactions on Signal Processing, 2002. **50**(2): pp. 174–188.
15. Bergman, N., Recursive Bayesian Estimation: Navigation and Tracking Applications. 1999, Linkoping, Sweden: Linkoping University.
16. Doucet, A., On Sequential Simulation-Based Methods for Bayesian Filtering. 1998, Cambridge, UK: University of Cambridge.
17. Gordon, N.J., D.J. Salmond, and A.F.M. Smith, *Novel approach to nonlinear/non-Gaussian Bayesian state estimation*. IEEE Proceedings on Radar and Signal Processing, 1993. **140**: pp. 107–113.
18. Wu, W., et al., *Modeling and decoding motor cortical activity using a switching Kalman filter*. IEEE Transactions on Biomedical Engineering, 2004. **51**(6): pp. 933–942.
19. Moran, D.W., and A.B. Schwartz, *Motor cortical representation of speed and direction during reaching*. Journal of Neurophysiology, 1999. **82**(5): pp. 2676–2692.
20. Brockwell, A.E., A.L. Rojas, and R.E. Kaas, *Recursive Bayesian decoding of motor cortical signals by particle filtering*. Journal of Neurophysiology, 2003. **91**: pp. 1899–1907.
21. Trappenberg, T.P., *Fundamentals of Computational Neuroscience*. New York: Oxford University Press, 2002.
22. Brown, E.N., et al., *The time-rescaling theorem and its application to neural spike train data analysis*. Neural Computation, 2002. **14**: pp. 325–346.
23. Marmarelis, P.Z., and K.-I. Naka, *White-noise analysis of a neuron chain: An application of the Wiener theory*. Science, 1972. **175**(4027): pp. 1276–1278.
24. Sanchez, J.C., *From cortical neural spike trains to behavior: Modeling and analysis*. 2004, Gainesville, FL: Department of Biomedical Engineering, University of Florida.

• • • •

CHAPTER 7

BMI Systems

Additional Contributors: Erin Patrick, Toshikazu Nishida, David Cheney, Du Chen, Yuan Li, John G. Harris, Karl Gugel, Shalom Darmanjian, and Rizwan Bashirullah

The success of translational neuroprosthetics to aid patients who suffer from neurological disorders hinges on the goal of providing a direct interface for neural rehabilitation, communication, and control over a period of years. Without the assurance that the neural interface will yield reliable recording, preprocessing, and neural encoding/decoding in a small portable (or implantable) package over the patient's lifetime, it is likely that BMIs will be limited to acute clinical interventions. However, the reality of the present systems is far from long life or fully implanted systems.

The present state of BMI technology requires that human or animal subjects be wired to large equipment racks that amplify and process the multidimensional signals collected directly from the brain to control in real time the robotic interface as shown in Figure 7.1. However, the vision for the future is to develop portable BMIs as shown in Figure 7.1c, which incorporates all of the complexity of the laboratory system but in a portable package. Data bandwidths can typically exceed 1Mb/sec when monitoring neural signal activity from multiple electrodes and as a result, many of the systems with wireless links are not fully implantable and require large backpacks [1–7]. To overcome these challenges, several parallel paths have been proposed [7–12] to ultimately achieve a fully implantable wireless neural recording interface. But the truth is that the development of a clinical neural interface requires beyond state-of-the-art electronics, not only DSP algorithms, and models as treated in this book. The purpose of this chapter is to review the basic hardware building blocks that are required for a portable BMI system, establish the bottlenecks, and present developments that are currently underway.

As discussed in Chapter 1, the design space in portable BMIs is controlled by three axes: computation throughput (for real time), power consumption, and size–weight. Obviously, all three depend on the number of recording electrodes, so BMI technology should be specified by *the size–weight, power, and computation per channel*. Each electrode can sense action potentials and LFPs, and creates a digital stream of 300 kbits/sec with conventional analog/digital (A/D) converters

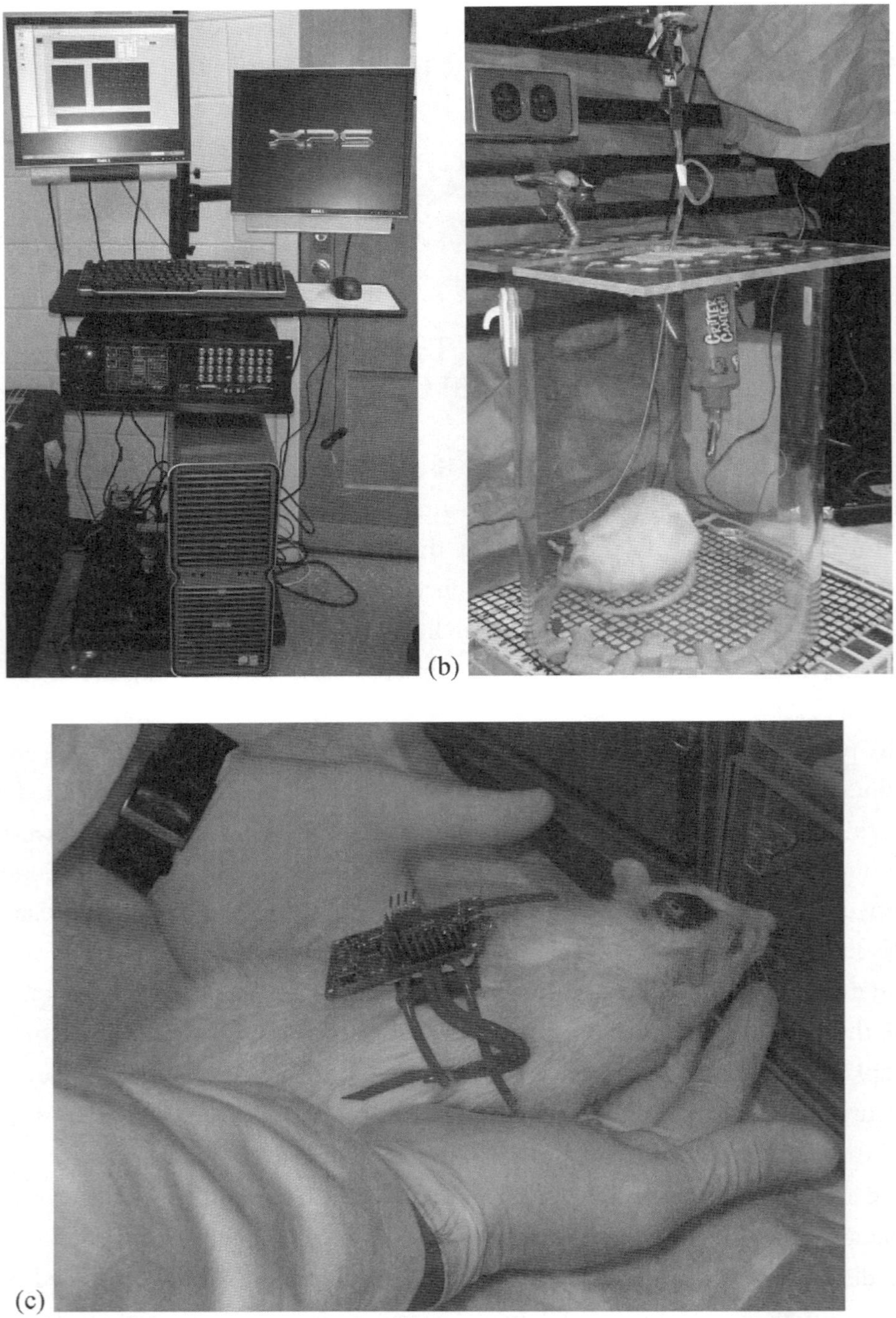

(a) (b) (c)

FIGURE 7.1: Experimental BMI systems. (a) Standard recording system used in the electrophysiology laboratory. (b) Tethered recording of a behaving animal. (c) Wireless "backpack" system for BMIs.

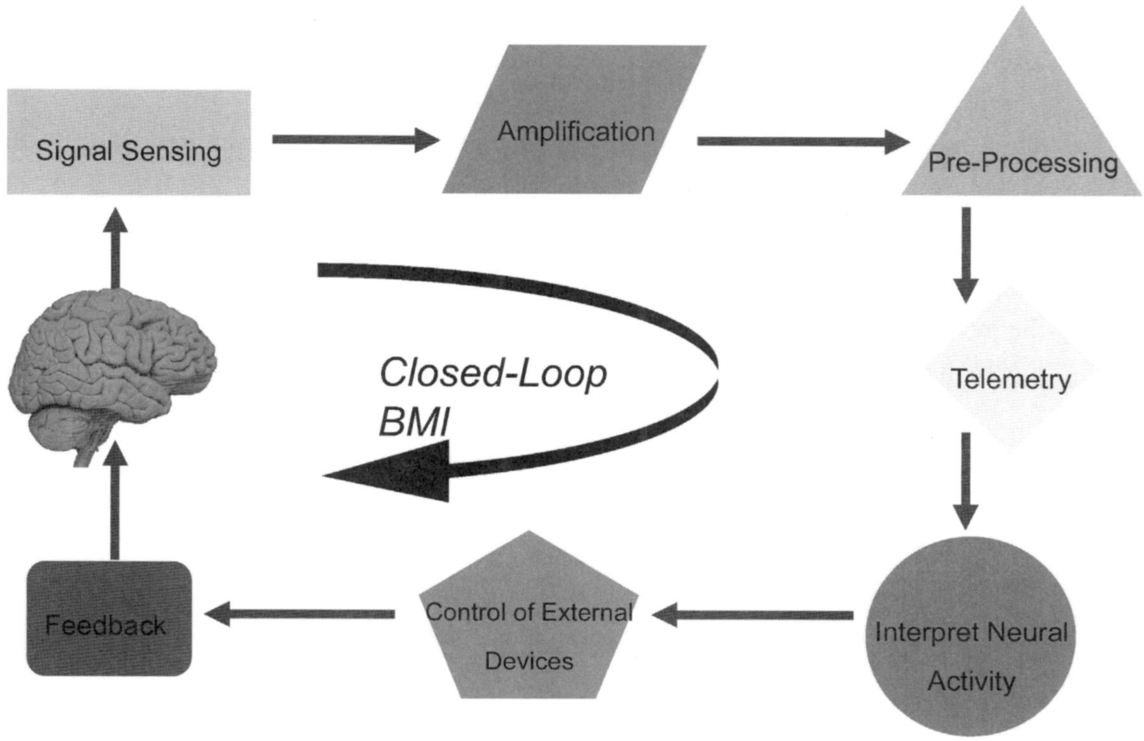

FIGURE 7.2: Hardware block diagram for a closed-loop BMI.

(25 kHz sampling and 12 bits for action potentials, and with 300 Hz with 12 bits for LFPs). Figure 7.2 shows a more detailed block diagram of a closed-loop BMI, where the most important parts are clearly detailed: the electrodes (signal sensing), the amplification, the signal processing (both at the preprocessing and the interpretation), the wireless telemetry, the robotic interface, and the feedback to the user.

We will present two basic approaches for portable BMIs, one short term and another more long term. These approaches will be called first- and second-generation BMI systems.

Generation 1 is a "backpack" approach allowing some mobility to the animal or human patient. Several research groups are working on designs that fall into this category with one or two commercial products in the market. The degrees of freedom (materials, geometries, amplification, A/D, wireless) to accommodate the experimentalist and engineering are enormous, and are visible on the very different engineering solutions proposed and implemented by many research groups [8–23]. Because continuous operation for many hours is a necessity, several modes of operation are possible with the available electrodes and headstages [24, 25]:

Mode 1: Compress the raw neural recordings (action potentials and LFPs) and transmit the full information to a remote processor where it can be recovered with a given precision for visualization and/or BMI model building. The bottleneck in mode 1 is the compression algorithm (i.e., how many channels with which fidelity can be transmitted through the high-speed wireless channel, with a given power budget).

Mode 2: Extract features from raw neural recordings (preprocessing) and transmit only the action potentials or spike counts to a remote station for model building. This is the simplest of the modes, but the issue is to do spike detection and sorting with sufficient precision before binning.

Mode 3: Locally perform the model building and just transmit the commands to the robotic device through a low-bandwidth wireless channel, which requires mode 2. In this case, the wireless channel needs only low bandwidth (send three numbers every 100 msec).

Mode 4: Mix and match among all of the above.

These different possibilities affect the computation and wireless resources, and a flexible hardware solution would be very beneficial.

Generation 2 moves the amplification and preprocessing of neural data into a fully implanted device that transmits the data wirelessly to a base station, where data can be visualized. BMI models are implemented outside the implant and whose outputs control a robotic device through a low-bandwidth wireless channel. Generation 2 systems are at a research stage. The interfaces typically consist of a microelectrode array that can record (motor) or control neural networks chemically (drug delivery) and via stimulation (retinal, cochlear). Because the device is implanted beneath the scalp, it needs to be self-contained; therefore, the electrodes are directly integrated into the electronics, the hardware substrate, and the packaging, and remotely powered, monitored, and controlled via a bidirectional telemetry link [22]. Reports in the literature have provided a wide variety of layouts and physical forms with the entire implant surface area of approximately 100 mm^2 and heights above the skull of no more than several millimeters.

This chapter describes recent progress in each of these areas. Several possible generation 2 architectures are possible and will be briefly reviewed.

7.1 SENSING NEURONAL ACTIVITY: THE ELECTRODES

The ability to quantify neuronal representations has been fueled by advancements in neural recording technology and surgical techniques. The trend since the last century has been to develop acquisition systems with the ability to spatially resolve the activity of large ensembles of single neurons. This goal has been met by advancing electrode design, amplifier design, recording D/A, and surgical technique. For motor control neural interfaces, the functionally representative modulation of activity in cortical columns (i.e., hand, arm, etc., of the homunculus) is the signal of interest. In the rat, cortical columns consist of a dense network of cells with estimates of 100 000 cells/mm3 with pyramidal cells appearing 89% of the time [26]. The ability to resolve sufficient spatial and temporal neuronal representation for the development of BMIs will likely require reliable neural probes that are capable of recording large ensembles of single neurons for long periods of time. The process of extracting signals from the motor, premotor, and parietal cortices of a behaving animal involves the implantation of subdural microwire electrode arrays into the brain tissue (usually layer V) [27]. Current chronic fixed probe technology requires at the surgical stage for the sensor to be placed very close to the neurons of interest for obtaining the highest quality recordings. Moreover, once the probe is placed, the relative distance to the neuron should remain fixed. Dealing with the experimental variability of chronically implantable arrays has been a formidable task because the neural tissue can shift over time with respect to the "fixed" electrode tips, and sensor sites can degrade from immune response and encapsulation. Therefore, the viability of chronic electrodes and recording from single neurons has been limited from months to years. This limitation presents a problem for long-term animal model research and neuroprosthetic development for humans. A variety of approaches have been proposed to improve the electrode tissue interface which include novel electrode material coatings for engineering the glial response and tissue inflamation [28, 29], multisite recording shanks for targeting columnar cortical structure [22, 30], microdrive electrodes for tuning into cellular activity [31], and rapid injectable probes for minimizing implantation injury [32].

Two major types of fixed probes commonly used for neural recording are: 1) conventional wire microelectrodes assembled (most often by hand) from insulated tungsten wires and 2) micromachined electrodes fabricated using integrated circuit microfabrication technologies. Conventional wire microelectrodes are still the "workhorse" of neural recordings because of ease of fabrication and good characteristics in terms of cell yields; however, because they are assembled from discrete components, there is a limit to the electrode array density as well as their functionality. Wire microelectrodes have been extensively used for acute and chronic applications of neural recording [33–35] and have provided the precise firing information of single neurons from cortical and subcortical structures. Typical wire microelectrodes consist of bundles of 50-μm-diameter insulated tungsten wire with sharp electropolished tips. However, the overall bundle size is large for subcutaneous insertion, and the wire spacing is not accurately controlled. For BMIs, the ultimate application of a fully implantable device warrants the need for integration between the amplifiers and electrode arrays. Traditional microwire arrays may not be prepared for this functional integration.

The limitations of mechanically assembled wire electrode arrays can be overcome by using microfabrication and micromachining techniques used in integrated circuits [36, 37]. Moreover, integration of signal processing circuitry onto the substrate of the probe is possible. The recording sites typically consist of exposed metal pads located on rigid shanks that are connected via interconnect traces to output leads or to signal processing circuitry on a monolithic substrate. Some place multiple recording sites along the length of the shank to allow for interrogating cells at varying depths in the neural tissue [36, 37]. An electrode design known as the Utah probe [38–40] uses micromachined square arrays of 100 single contact electrodes bonded to a demultiplexer connector with local amplification. Despite advances in micromachined neural probes, the fixed nature of the probes limit the ability to optimize the recording of the neural signal. If the probe is spatially far from the neuron of interest, the measured action potential is low. The multisite micromachined neural probe provides one approach to resolve this issue. By using a multiplexer to sample the neural signal at different sites at different depths into the neural tissue, a discrete number of locations may be monitored without additional surgery to reposition the probe. An alternative approach is to use moveable probes instead of fixed probes. It is well known that the ability to reposition electrodes after implantation surgery can improve the signal-to-noise ratio (SNR) and yield of neuronal recordings. A variety of microdrives have been proposed that either mechanically or electrically move arrays of electrodes in unison or independently with manual or automated mechanisms. Fabrication of microdrives often requires sophisticated miniature mechanical components and significant skill, time, and cost. For precise targeting of individual units, the probe excursion should traverse 1–2 m in intervals of 5–30 μm. Recently, motorized screw microdrives have been reported to provide a 10-fold increase in yield compared to manual drives [41]. In addition, miniature piezoelectric linear actuators require no gears or screws for electrode movement [31]. Physically, the design challenges also involve specifying small form factors and lightweight components that can fit on the skull of the patient.

Rigid substrate micromachined probes described e may introduce chronic tissue damage due to mechanical forces encountered from strain transferred from the mount to the probes floating in neural tissue [42]. As one way to mitigate this challenge, Rousche et al. [43] designed a flexible substrate microelectrode array, where thin metal electrode sites and subsequent wiring are enclosed between polymers. This flexible electrode array adds needed strain relief, yet cannot be inserted into the brain matter directly. An incision must be created first in order for the electrode to be implanted. Ref. [44] incorporates rigid probes with planar electrode sites on the probe shank hybrid-packaged to a flexible cable that connects electrodes to output ports. This design may be inserted directly into neural tissue, keeping damage to a minimum, and provides greater reliability of the probe assembly.

Here, we report a neural microelectrode array design that leverages the recording properties of conventional microwire electrode arrays with the additional features of precise control of the electrode geometries and flexible micromachined ribbon cable integrated with the rigid probes. The goal is to produce electrode arrays that have high neuronal yield, are highly customizable in terms of geometry/layout, minimize tissue damage, and are easy to mass fabricate.

7.1.1 Microelectrode Design Specifications

An optimal electrode design for BMI chronic in vivo recording requires

- small profile probes that exert the least amount of tissue damage during insertion and chronic recording;
- structurally robust probes that do not buckle during insertion into tissue;
- low probe impedance that is stable during chronic recording;
- recording sites selective to single neuron action potentials;
- adaptability to on-chip processing circuitry.

Using conventional micromachining techniques, one can design small-profile metal traces enclosed between flexible polyimide insulation, making a cable, as shown in Figure 7.3. The actual probes extend from the cable 2 mm and include 20 × 50 μm electrode sites on the tips. The electrode area is chosen for sufficient compromise between signal selectivity and noise performance. The corresponding probe dimensions assure adequate structural integrity according to calculation using the Euler–Bernoulli beam theory. The metal traces and corresponding bond sites can be made to any size specification and spacing distance via photolithography. Therefore, custom application specific

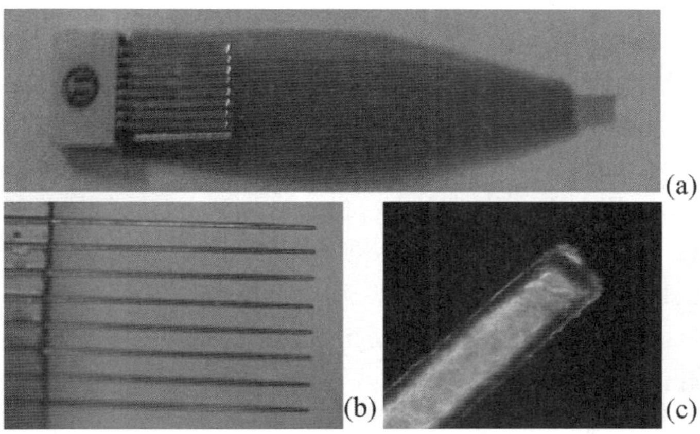

FIGURE 7.3: (a) Flexible substrate microelectrode array with Omnetics connector. (b) Microelectrode array. (c) Probe tip showing insulation along shank and gold plating on tip.

integrated circuits for signal amplification and spike detection may be packaged on the flexible substrate using flip-chip bonding techniques in future designs.

7.1.2 Process Flow

All processing for building the electrode array is performed on the surface of a 100-mm-diameter silicon wafer covered with Kapton tape (which provides adequate adhesion for the subsequent polyimide layers). The polyimide bottom insulation layer (PI 2611; HD Microsystems, Cupertino, CA) is spin deposited and cured to a final thickness of 20 μm. Sputtered nickel, 100 Å, is patterned to define the probe, wiring, and bond pad dimensions. Then, Ni is electrodeposited on the Ni seed to an average thickness of 20 μm via a 10-mA direct current for 4 hours in a nickel sulfamate bath (Nickel S; Technic Inc.). Adhesion promoter (VM9611, HD Microsystems) is next applied followed by three spin coatings of the PI 2611 to achieve the final 20-μm top layer of insulation. Al (1000 Å) is patterned as a hard mask for the subsequent oxygen plasma etch. The etching process includes an O_2 reactive ion etch that removes the polyimide from the top of the bond pads and the probe tips. The remaining polyimide under the probe tips is isotropically etched in a plasma barrel asher. Then the probe-cable assembly is removed from the substrate wafer and primed for parylene-C deposition with a silane adhesion promoter (Acros Organcis, Geel, Belgium). The parylene-C vapor deposition step insulates the shank of the metal probes to a thickness of 2–4 μm. Then the probe ends are manually cut with a blade to expose bare metal for the electrode sites. Finally, the probes are immersed in an electroless gold plating solution (TechniIMGold AT, 600, Technic Inc., Cranston, RI) that covers the electrode sites as well as the bond pad sites with 0.1 μm of gold. An Omnetics connector is then fixed to the bond pads with silver epoxy.

7.1.3 In Vivo Testing

Adult male 250-g Sprague–Dawley rats were used to test the recording performance of the flexible electrode arrays. All procedures have been approved by the University of Florida Institutional Animal Care and Use Committee Board and were performed in the University of Florida McKnight Brain Institute. Before surgery, the rats were anesthetized and the surgical site was thoroughly sterilized. The top of the skull was then exposed by a midsaggital incision from between the eyes, and the landmarks bregma and lambda are located on the skull [45]. The microwire array was implanted to a depth of 1.66 mm into the forelimb region of the primary motor cortex. The electrodes are stereotaxically moved to the appropriate site and lowered to the appropriate depth using a micropositioner (1 mm/h) to minimize distress to the brain tissue (FHC, Bowdowinham, ME) as shown in Figure 7.4. The array was then grounded using a 1/16-in.-diameter stainless steel screw. A low-profile Omnetics connector was used to attach the recording wire.

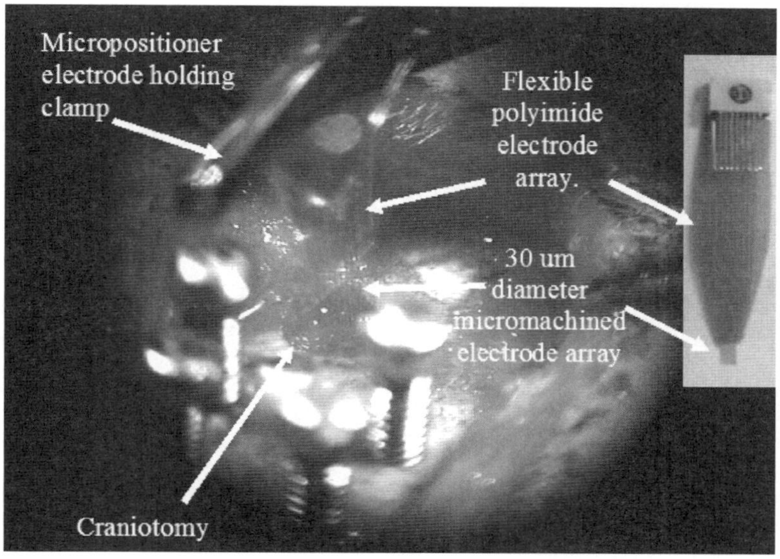

FIGURE 7.4: Implantation of the flexible substrate microwire array.

Extracellular potentials recorded at 12 207 Hz during surgery were analyzed and spike sorted using Spike2 (CED, Cambridge, UK) software package. Recordings were analyzed over a period of 130 sec. To detect and sort neural activity within each channel, an automated waveform matching system within Spike2 was used to construct templates using threshold detection.

Once a set of waveform templates were generated for a data stream, all templates (noise) that did not match characteristic neural depolarization behavior were removed [46–48]. The remaining waveform templates were sorted according to amplitude and shape, and any waveform templates that were significantly similar to each other were combined into a single template. Clustering of waveform variance within templates was verified through PCA. Each waveform template was statistically unique and representative of a distinct neuron within the channel.

Once neuron waveforms were isolated and sorted, peak-to-peak amplitude was evaluated by computing the average waveform of all spikes within the neuron template and measuring the potential difference from the apex of the repolarization peak to the apex of the depolarization peak. The noise floor of each channel was evaluated by computing the root mean square value of a 5-sec period of noise. Using these two values, the SNR for each neuron template was calculated. To ensure proper reporting, all spike waveform templates that possessed peak-to-peak amplitude of magnitude less than three times the value of the noise floor were considered too close to the noise to be reliably and consistently distinguished, and were removed from the study. Values of neural yield, noise floor, amplitude, and SNR are reported for each channel within Table 7.1. Action potential amplitudes as

TABLE 7.1: Neuronal yield for eight channel microelectrode array

Electrode	1	2	3	4	5	6	7	8
Yield (neurons)	2	2	2	3	6	5	3	4
Noise floor (μV, RMS)	4.1	5.0	5.3	4.4	5.2	3.8	3.7	4.3
Neuron amplitude (μV, PtP)	20.1	23.3	32.6	26.1	114.7	90.4	31.4	45.1
	13.2	15.5	24.7	18.3	56.8	52.3	13.4	29.7
				14.2	34.6	35.7	11.7	21.0
					21.3	21.0		16.0
					18.8	13.8		
					17.4			
SNR (dB)	13.8	13.4	15.8	15.5	26.9	27.6	18.6	20.4
	10.2	9.9	13.4	12.4	20.8	22.8	11.2	16.8
				10.2	16.5	19.5	10.0	13.8
					12.2	14.8		11.4
					11.2	11.2		
					10.5			

large as 115 μV and as small as 13 μV are discriminated by the electrode and recording system. The average RMS noise floor is 4 μV. Figure 7.5 shows recorded data from electrode number 6.

Acute electrophysiological recordings show excellent yield of recordable neurons and SNRs ranging from 10 to 27 dB. The benefits of the microfabrication design allows for tailoring the electrode geometry for neuronal structures of interest. Moreover, high channel count arrays can be constructed by layering the proposed design. The flexible cable additionally provides strain relief from the fixed external connection and can be used to minimize tissue damage in chronic applications. Additional studies on the reliability of the electrode array as a chronic recording device is needed to

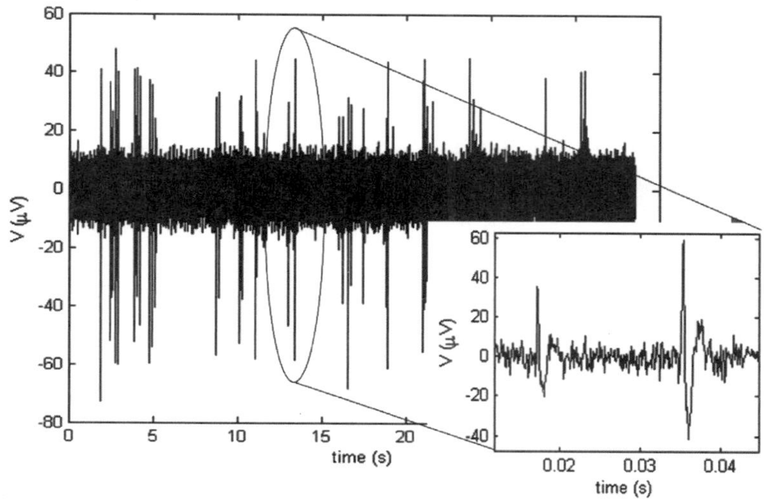

FIGURE 7.5: Data from neural recording in the rat motor cortex at a depth of 1.66 μm during implantation surgery. Inset shows two distinct neurons recorded by single probe.

confirm the biocompatibility of the electrode material. However, because of the adaptability of the fabrication process, other metals such as platinum may be readily incorporated.

7.2 AMPLIFICATION

The next step in the signal processing chain is amplification. Because of the very small signal levels (microvolts) the amplification needs to be done very close to the frontend. The design of the headstage, that is, the connector that links the electrodes and the cables to the amplifier is critical. Impedance matching must be carefully controlled, and most designs use separate ground and reference.

The amplifiers have gains of 40–70 dB, a frequency response between 0.1 Hz and 10 kHz, and are therefore AC coupled. One of the most important requirements is high common mode rejection because of the large electrical artifacts normally present (60 Hz interference and other electrical noises). In rack-mounted instrumentation, the amplifiers are well within the technology envelope and several manufacturers provide excellent instruments (Tucker-Davis Technologies, Plexon, Grass, Biosemi, Axon, Alpha-Omega). The difficulty is in portable designs. We present briefly below an analog amplifier we have developed at the University of Florida (Harris).

7.2.1 Integrated Bioamplifier

The custom amplifier consists of two stages: a low-noise 40-dB preamplifier and a second-stage 40-dB amplifier that were fabricated in the 0.6-μm AMI complementary metal oxide semiconductor

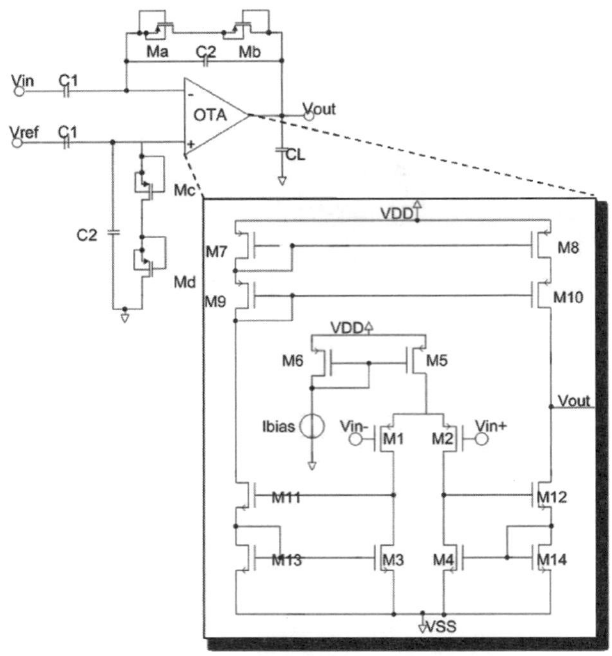

FIGURE 7.6: First-stage amplifier.

(CMOS) process. Figure 7.6 shows the schematic of the preamplifier and the second stage is in Figure 7.7. The amplifier designs are based on the work of Reid Harrison [49].

The first-stage midband gain, $A_{M,}$ is $-C_1/C_2$, the lower corner frequency is at $\omega_1 \approx 1/(RC_2)$, and the higher corner frequency is at $\omega_2 \approx g_m C_2/(C_L C_1)$, where g_m is the transconductance of the operational transconductance amplifier (OTA) shown in the lower portion of Figure 7.5 [50]. To obtain a low cutoff frequency, a large resistance is needed in the feedback loop provided by two diode-connected transistors, Ma and Mb, acting as "pseudo-resistors" without sacrificing die area.

Noise performance is important in biological preamplifier designs, it is minimized by carefully choosing the width and length of the transistors. However, there is a trade-off between stability and low noise. Because low noise is critical in this application, decreasing the transconductance of M1, M2, M7, M8, and the four transistors of the two Wilson current mirrors can minimize the thermal noise. Flicker noise, which is important in low-frequency applications, is minimized by increasing the device area. Because p type metal oxide semiconductor devices exhibit less flick noise than n type, p type is used in the input pair.

However, the typical neural signal amplitude is on the order of tens to hundreds of microvolts, the output voltage of the preamplifier is still too small for signal processing by an A/D converter. Therefore, a second-stage amplifier is required [51], in which C_1/C_3 determines the midband gain,

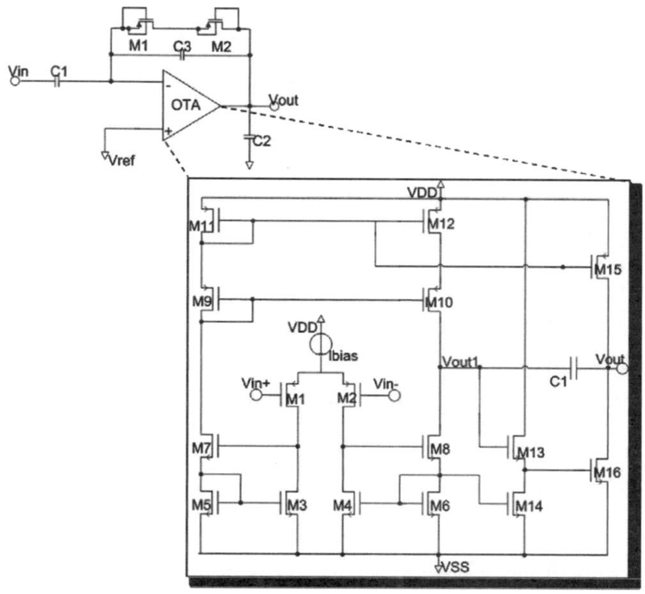

FIGURE 7.7: Second-stage amplifier.

TABLE 7.2: Amplifier specifications		
PARAMETER	PREAMPLIFIER	SECOND-STAGE AMPLIFIER
Supply voltage (V)	5	5
Power consumption (μW)	80	~120
Gain (dB)	39.46	~40
Bandwidth (kHz)	5.4	19
Lower cutoff frequency (Hz)	~0.3	<1
Die area (μm^2)	91 800	66 400
Input referred noise (μV rms)	5.94	Not available

whereas V_{ref} and diode-connected resistors provide the DC operational bias. The OTA is a typical two-stage OTA with a P-type input differential pair loaded with Wilson current mirrors followed by a class A output stage.

Table 7.2 shows the performance of our current designs per electrode using 0.5-μm technology. Better performance can be obtained with newer technologies. Dies can also be wired bound to the circuit board to minimize the physical size and avoid noise.

7.3 THE PICO SYSTEM

We have developed and tested a data acquisition/preprocessor/wireless system to implement mode 2. We called it the PICO system, and its goal is to acquire neural recordings from the headstage, amplify the signals, perform spike detection (feature extraction/compression), spike sorting, and wirelessly transmit binned data (rate codes) to a remote station. Alternatively, this system can simply amplify the raw neural data, do spike detection with a simple threshold, and send just these windows to a remote system. As is, it already plays an important role in BMIs because it implements a first-generation backpack. However, it does not allow the neurophysiologist to analyze and store the continuous raw voltage traces being collected because of bandwidth limitations (mode 1), nor it is capable of performing the high speed computations of mode 3. This is where a neural signal processor (NSP) (local decoding algorithm implementation) will add capabilities in the future by performing data reduction at the frontend such that the data can be reconstructed in the backend for visualization and archiving purposes. In the particular path described here, the previously reported Pico Recording system [52] is capable of replacing a multiconductor commutator cable used in animal experimental paradigms with a high-bandwidth, low-power wireless interface that allows the subject more mobility and a more natural response to a given experiment. This new wireless interface presents a compact architecture as well as new challenges related to the power consumption of the remote device.

To optimize the trade-off between power consumption and processing power, the low-power, low-noise, high-gain amplifier shown in Section 7.2 was designed in conjunction with off-the-shelf components, specifically MSP430F1611 as the microcontroller and the Nordic nRF2401 wireless transceiver. The processor consumes less than 2 mW of power with 8 million instructions per second of processing and has the added benefit of conserving space by integrating two universal asynchronous receiver/transmitters, three direct memory access channels, and an eight-channel, 12-bit A/D converter all in one small package. In this work, we demonstrate the integration of the low-power amplifier with the Pico system and collect data from a live subject.

The Pico Neural Data Collection System consists of the Pico Remote (battery-powered neural amplifiers, a neural data processor, and a wireless transceiver), the Pico Receiver (a wireless-transceiver to USB converter), and PC-based software (for spike sorting and firing rate quantification). Figure 7.8

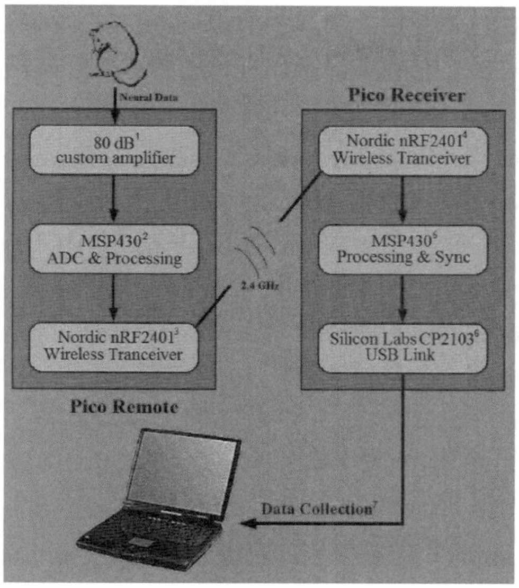

FIGURE 7.8: Block diagram for the PICO wireless neural recording system.

provides a block diagram for signal flow throughout the system. The Pico Remote senses single- and multiunit neuronal activity from 16 chronically implanted microelectrodes from a laboratory rat. The signals are amplified (80 dB) with a custom amplifier1 and fed to one of eight 12-bit A/D channels2 sampling at 20 kHz on the MSP430. The MSP430 has two modes: spike sorting and spike reading. While in the spike sorting mode, raw sampled data from a single probe are sent over the Nordic transceiver for reception by the Pico Receiver. This mode allows the user to analyze in detail the raw recording from the microelectrodes to set thresholds for detecting and sorting spikes (action potentials). In the spike reading mode, the MSP430 executes the spike detection through user defined thresholds and sorting. The firing rate for each neuron is computed locally and sent over the Nordic transceiver for reception by the Pico Receiver.

The Nordic transceiver is capable of up to 250 kbit/sec sustained data transfer of packet data. Data sent over the Nordic transceivers are fed to another MSP430 processor for buffering and formatting and for the eventual pass through to a Silicon Laboratories CP2103 USB bridge to software running on a PC.

The MSP430F1611 processor samples each analog channel at 20 kHz with an on-board 12-bit A/D converter. Depending on the mode in which the processor is functioning, the samples are handled differently. While in the spike sorting mode, no signal processing occurs and the raw samples are sent directly to the transceiver for processing by software running on the PC. The data are sent as 12-bit samples at rate of 20k samples/sec for a data rate of 240k samples/sec. In the spike counting mode, each sample is compared against thresholds that define a spike using well established spike-sorting techniques. If all the threshold conditions are met, a counter is incremented. Each probe can have as many as three rate counters corresponding to three neurons per electrode. Every 100 msec, the bin counts are sent to the wireless transceiver for processing on a PC.

On the receiving side of the RF link, data are collected via a Nordic nRF2401 transceiver on the Pico Receiver. The RF link is controlled by another MSP430F1611 microcontroller and results are passed on to a PC via a USB bridge (CP2103 from Silicon Laboratory). The USB bridge is compliant with the USB 2.0 specification and has a maximum throughput of 12 Mbits/sec.

The Pico Remote is configured with software running on Windows-based PCs. The testing software has three separate applications: Spike Capture, Spike Detection Parameter Initialization, and Binned Spike Data (firing rate) Display. The Spike Capture application is a Matlab graphical user interface that captures and displays raw sample data sent from the Pico Remote. This application does not configure the Pico Remote but is used solely for viewing and storing a single channel's raw sample data for analysis. The Spike Detection Parameter Initialization application is in development and will accommodate configuring three counters for each probe. Figure 7.9 shows screenshots of the user interface.

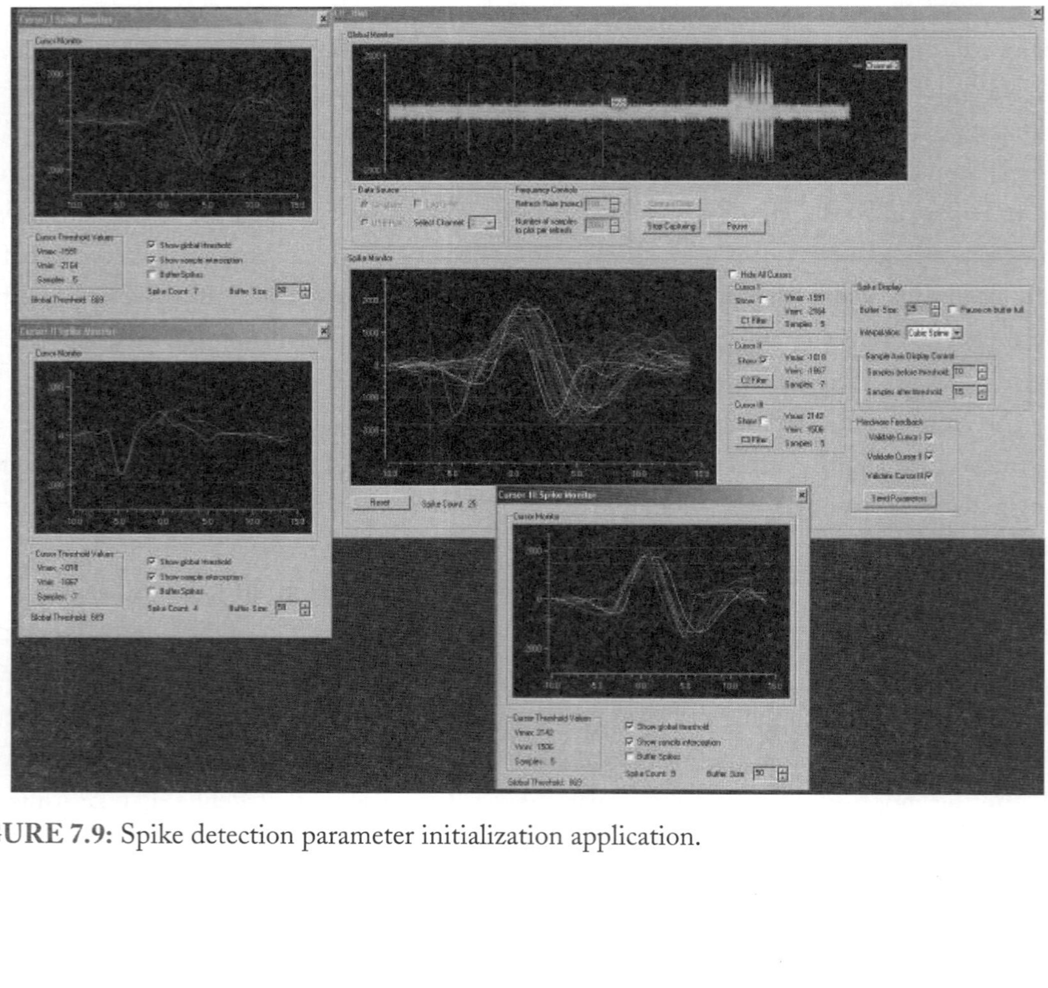

FIGURE 7.9: Spike detection parameter initialization application.

The Spike Rate Data Display application is used for demonstration purposes. In an actual laboratory environment, a Windows service reads the spike counts from the USB port and stores them in shared memory so any Windows application can process them.

7.3.1 In Vivo Testing and Results

Because invasive microwire recording technologies are targeting single-unit neuronal activity, we will use the action potential (spike) as our standard for evaluation of data collected from the neural probes [48]. The Pico system and custom amplifier have been previously tested extensively using a Bionic Neural Simulator to generate stereotypical physiological signals on the electronics workbench. In vivo tests used adult male 250-g Sprague–Dawley rats to benchmark the recording performance of the custom amplifier and Pico system. All procedures have been approved by the University of Florida IACUC Board. A 16-electrode (2 × 8) microwire array was implanted to a depth of 1.66 mm into the forelimb region of the primary motor cortex. The array was then grounded by using a 1/16-in.-diameter stainless steel screw implanted in the skull.

Figure 7.10 shows a 30-sec trace of neural recordings captured with the Pico system and gives a global perspective into the peak-to-peak amplitudes and occurrences of spiking neurons. The average of the data over the 30 sec is expected to be about zero. Instead, there is an offset of nearly 9 μV, which has been determined to be attributed to not precisely biasing the amplifiers.

In Figure 7.11, we zoom in on two individual action potentials where the range of the noise floor can take values of ±25 μV with the spikes reaching −75 μV.

CED's Spike 2 analysis software was used to extract neural data from the dataset. Figure 7.12 shows the average waveshape of two neurons firing on the probe from which we sampled.

It is difficult to calculate meaningful SNRs on neurophysiological data, because the spike signal is not continuous like those used in traditional communication's calculations [53]. Table 7.3 presents an attempt to derive meaningful SNR calculations from the average spike value. The RMS voltage of the noise floor was calculated by taking the square root of the average of the square of

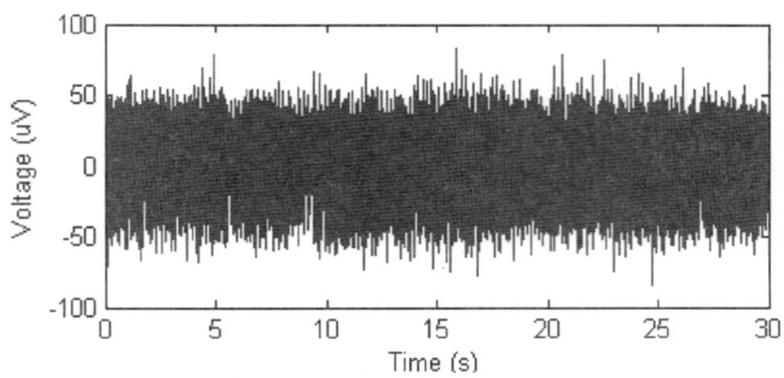

FIGURE 7.10: Thirty seconds of extracellular recordings from an anesthetized rat.

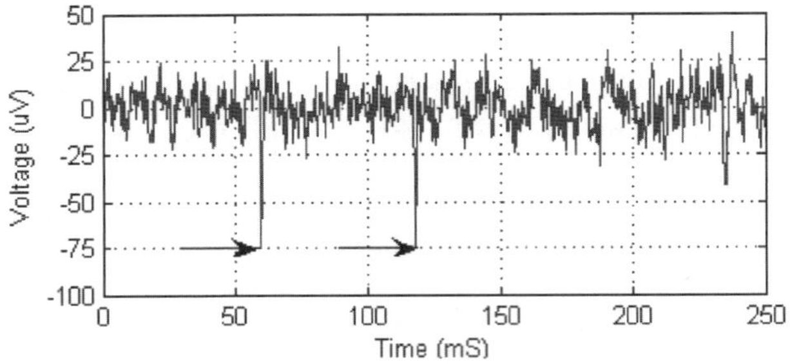

FIGURE 7.11: Two hundred fifty microseconds of extracellular recordings with two spikes.

samples in the noise floor. The samples used in the calculation were extracted by looking for periods of data where no spikes appeared.

To provide a baseline for quantifying the noise attributed by the system, the inputs to the probes were tied to the reference and the corresponding trace is given in Figure 7.13. These results suggest that the source of the dc offset is inherent in the system, which supports the need to carefully adjust the bias of the amplifiers. When these data are normalized, the RMS voltage of the noise using the same process described earlier yields 0.66 μV, which is much lower when compared to the 8.24 μV in the actual test.

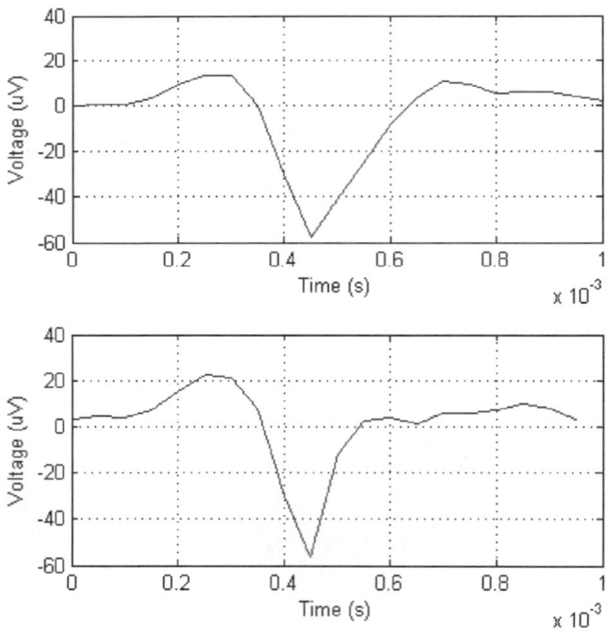

FIGURE 7.12: Average waveform spikes of two different neurons.

TABLE 7.3: SNR calculations

	NEURON 1	NEURON 2
Noise V_{rms} (μV)	8.24	8.24
Neuron V_{pp} (μV)	71.13	79.33
SNR (dB)	18.72	19.67

For all of the neural recordings presented here, the system power consumption for the Pico Remote is quantified in Table 7.4. The Pico remote consumes around 130 mW, whereas bin counting, on a 3.7-V, 150 mA h LiON battery, would provide about 4.25 h of operation.

Now that the initial goal of bringing actual neural data from a live subject over the wireless link to a PC has been achieved, we can move forward in processing the data on the Pico remote by applying our spike counting algorithm. There are still additional tests that have to be completed, such as, comparing in a parallel test, raw signal data and spike counts with a commercial system, such as those from TDT Systems.

To ensure data integrity over the wireless link, we have created specifications for a robust communications protocol that utilizes packet acknowledgements and resends. Any data sent over the wireless network will be acknowledged by the receiving device. If the sending device does not receive an acknowledgement in a few milliseconds, the data are resent. A new version of the Nordic part allows for data resends in hardware reducing power and firmware complexity. Once the latest Nordic parts and the protocol are implemented, we expect data loss to be very close to zero.

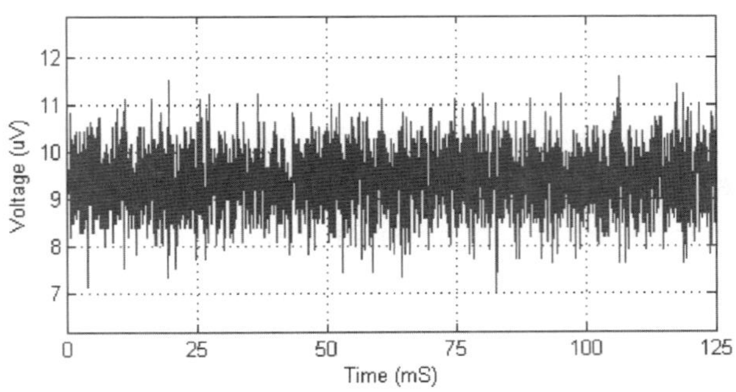

FIGURE 7.13: Noise introduced by the Pico system.

TABLE 7.4: Pico Remote power consumption

SYSTEM	SPIKE SORTING (mW)	BIN COUNTING (mW)
Pico Remote	100	87
Pico Remote with TDT with headstage	143	130

Once the tests are completed and the design is stable, the system will be repackaged in a small rat backpack not larger than 1.5 × 3 in. in size and weighing a few ounces.

7.4 PORTABLE DSP DESIGNS: THE NEURAL SIGNAL PROCESSOR

Because the modeling is in the critical path of BMIs, an effort to create low-power flexible DSP boards was pursued in the scope of generation 1 implementations. The system described here serves as the digital portion of the overall BMI structure and is responsible for mapping the neural firings into action in the external world. To do this mapping, the system first acquires digitized neural data from the analog module. After acquisition and spike detection, the system then computes a linear or nonlinear model to map the neural data to a trajectory. Once a trajectory is determined, the system finally transmits commands wirelessly to an off-board robotic arm as described in mode 3. Mode 2 probably can be implemented without the use of a DSP. Mode 1 also requires a DSP for data compression. Recent results with a specially designed self-organizing map [54] or a vector quantizer tuned to action potentials [54] show that compression ratios of at least 100 are possible preserving spike shape.

Because the system needs to be carried by a subject (human, primate, or rat) without wires, there are portability and wireless constraints that will affect the design choices. First, the size and weight must be small enough to accommodate the various subject species. This involves the actual size of the components and the printed circuit board layout. Second, power consumption must be low to allow the use of lighter batteries and yield longer experimental trials. With power and size constraints placed on the system, the choice of processor and wireless transceiver becomes a concern. Because most of the prediction models require the use of floating point arithmetic, the processor needs to be floating point with enough speed to attain real-time model computations yet low power enough to appeal to the system constraints. In addition, the wireless transceiver must not only be small with a sufficient transmission bandwidth for current and near-future needs, it must also be low power. In the following subsections, we detail the NSP component choices that address these system requirements. Although these choices address the system requirements, the modular design allows for replacement of the individual components as new technologies are developed with better specifications (Figure 2).

7.4.1 Digital Signal Processor

The central component to any computational system is the processor. The processor has the ability to determine the speed, computational throughput, and power consumption of the entire system. Moreover, it dictates what additional support devices will be used for the overall system design. For the NSP, we chose the Texas Instruments (Dallas, TX) TMS320VC33 (C33) DSP. We chose this processor because it is suitable for on-board BMI modeling in terms of computational needs, power consumption, and versatility (Figure 7.14).

Specifically, the C33 meets high speed and floating-point requirements because it is a floating-point DSP capable of up to 200 mega-floating point operations per second (with over clocking). It achieves such high speeds by taking advantage of its dedicated floating point/integer multiplier. In addition, it works in tandem with the arithmetic logic unit, so that it has the ability to compute two

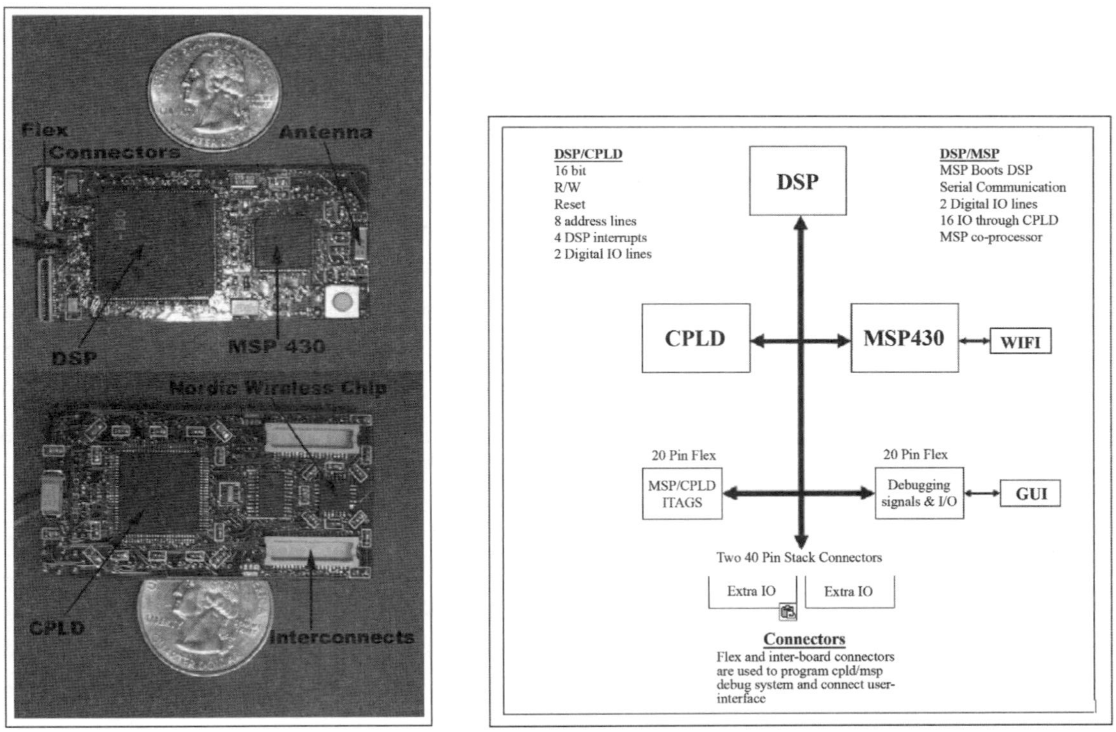

FIGURE 7.14: Modules of BMI DSP system.

mathematical operations in a single cycle. With support for both assembler and C/C++, code for the DSP can be optimized at low and high programming levels for additional speed support. Even at 200 MFLOPS, the C33 is also well suited to low power needs because it uses less than 200 mW. It achieves such power savings due in part to its 1.8-V core and other power-saving measures built into the processor. Using adjustable clock speeds, the DSP can exchange processing speed for lower power consumption. This may be desirable because some BMI experiments may require increased battery life in exchange for less computational speed (i.e., less neural data). Finally, the C33 is able to fulfill other expandability requirements of the system. First, it supports a large address space by providing a 24-bit address bus to read/write 16 million different locations. This address space is used to map the different hardware components of the system and future hardware interfaces. It includes a 32-bit parallel data bus and serial port for multiple communication methods and hardware multiplexing. This processor can also provide expansion because it has four hardware interrupts, two 32-bit timers, and a direct memory access controller that can be used for future requirements or hardware interfaces.

Wireless Communication. As with the PICO system, the wireless connection is the second most important hardware module in the NSP. This module, similar to the processor, has the ability to define the size and power consumption of the full system. Early on, we determined that 802.11B was the most convenient protocol for our group and collaborators [18]; unfortunately, it required a PCMCIA card that was very large and consumed the majority of the power from the system. Because the digital module does not need a lot of bandwidth to transmit 100-msec trajectory coordinates (20 bytes), and the possible connection to the analog frontend if wireless uses at most 500 kbits/sec, we sacrifice the bandwidth of the PCMCIA card and 802.11Bs universal protocol for a tiny low power 2.4-GHz wireless solution. This wireless chip (also used in the PICO), Nordic nRF2401, is a single chip transceiver capable of 1 Mbp/sec at 10.5 mA (peak). It also has the ability to reduce its power further to 0.8 mA if we use ShockBurst mode or any variant in between. Compared to the 100- to 300-mA PCMCIA card (depending on throughput), this is a significant savings in power. In addition, the package size of the Nordic chip is only 5 × 5 mm compared to the 4.5 × 2.25-in. PCMCIA card used in the previous design. Although there is an additional antenna, it is only a 6.5 × 2.2 mm surface mount SMD, still making the Nordic chip a significant savings in size and power.

MSP Coprocessor/Boot-Loader. The NSP contains a coprocessor to alleviate the computational strain on the DSP. In turn, this coprocessor supports the current peripheral functions of the digital module and any future functions. Consequently, this allows the DSP to solely concentrate on the algorithms for neural-to-trajectory mappings. We chose the MSP430F1611x to serve as the coprocessor because it is a very low power yet highly functional microcontroller. This 16-bit RISC microcontroller consumes 1 μA of power at 1 MHz and linearly increases power to its maximum at 8 MHz. In addition, there are other power modes that allow it to use infinitesimally small amounts of power with voltage ranges from 1.8 to 3.6 V. For the NSP, we use 3.3 V to supply the microcontroller. Furthermore, the microcontroller has 48 KB of flash RAM that can store data when power is off and 10 KB when the power is on. Combined with a 12A/D-D/A and multiple asynchronous/synchronous serial ports and SPI/I2C ports, this microcontroller is very well endowed for such a small package (10 × 10 mm).

Handling the wireless protocol is one of the main functions this coprocessor is in charge of. It provides the DSP communication transparency when sending and receiving wireless data. Underneath this transparency the multifunctional component (MSP) is controlling the flow and control of the wireless chip. We also use the MSP as a boot loader for the DSP by using its serial port for booting. To boot the DSP serially, we store the DSP program in the flash because the MSP has 48 KB of flash. Essentially, by using the MSP to boot the DSP, we trade one single functioned component for an MSP that is lower power. In terms of expandability, the MSP is fast enough (8 million instructions per second) to handle preprocessing of the neural data for the prediction models running on the DSP. Through a serial port or a digital IO (connected to the flex and interconnects), we can also expand the system to include USB communications for the uploading or downloading of data in real time (from multiple paths MSP/DSP/analog module). We can also use these ports to create a test bed for debugging the system or testing future analog modules. Finally, similar to the wireless chips, we have the ability to scale the NSP system and add multiple coprocessors because they are very low power and small, yet highly functional.

Reconfigurability. 1) Complex Programmable Logic Device (CPLD). By using an Altera C3000, the NSP is able to expand and accommodate future needs. There are options for using on-board clocks as well as off-board and even clocks from the analog module. This makes for a very flexible system for future analog modules that require or provide clock signals. Communication is also reconfigurable because it is critical between the components on-board and modules off board. This includes the serial ports from the DSP and MSP, and the parallel data bus from both processors. Through the CPLD we have the ability to use a USB connection and other serial ports. These in turn can be redirected to the DSP or the MSP. The NSP even has the ability to redirect the wireless connection directly to the external IO for debugging and support. Essentially, the CPLD is very useful for debugging the system and add future features that may become required. It has access to all the connectors and components. Because it is reprogrammable through very high speed integrated circuit hardware (VHDL), it makes for a very versatile hardware component.

2) Connector Interface. To complement the CPLD and allow for functionality of unknown future needs it is necessary to provide as much connectivity between the internal components to the

external. Coupled with the desire to be portable we need connecters that have many pins and yet small enough to not enlarge the system. In addition, we need to be able to provide support functionality such as debugging and test beds.

The first connection type must be able to mate with the analog modules and provide a suitable amount of connections, and yet be strong and compact enough to support another attached printed circuit board. For the NSP, we use two connectors that allow for 80 different connections between the boards. With this amount of connections, we can not only support current needs from an analog module [duke], we can also support other daughter boards that may be designed for testing or future boards.

The second connection type is more inline with debugging and programming the DSP, CPLD, and MSP. In addition, to retain the ability to communicate to systems that may be at great distances we chose to design a flex connecter interface. On these connectors we provide access to the CPLD jtag, MSP serial ports, wireless serial port, and other general CPLD IO for multiplexing communication between the components of the NSP. Finally, the connector layout is such that it allows for stackable modules, providing flexibility for all the three modes of operations described above.

3) Power Subsystem. The two different voltage requirements for the NSP are 1.8 and 3.3 V. These voltages are not only used by the DSP, they are also used for the other hardware modules in the system. In trying to supply the NSP components, the voltages can be supplied from multiple pathways to the NSP. This includes an on-board option and off-board. The main option for power supplying is from on-board. This includes a Texas Instruments TPS70351 Dual-Output LDO Voltage Regulator that provides both 1.8 and 3.3 V voltages on a single chip and only requires 5 V to operate. This chip also provides the power-up sequence required by the DSP once it is initialized. In addition, requiring a single supply is very desirable for a portable system because only one 5-V battery supply is necessary. With the off-board option, we can power the regulator through the flex connectors, interconnect connector or a simple header; this way, a battery or external supply can power the regulator. In addition, we can have the analog module or debugging system provide the individual power requirements. This allows for the isolation of certain components for being powered. With such variability, future hardware can take advantage of the different powering options.

System Software. There are six layers of software in the NSP system environment: 1) PC software, 2) DSP operating system (OS), 3) DSP algorithms, 4)VHDL code 5) MSP430 code, and 6) client/server code. We wrote a PC console program to interface the DSP through the USB. The console program calls functions within the DSP OS to initiate and control the USB communication functions. The DSP OS is also responsible for reading/writing memory locations and various program control functions. This is the main pathway for debugging the system. In tandem with the OS layer of the DSP, there is low-level MSP code for initializing and controlling the wireless controller. This code must interact with the DSP OS and any client code or algorithms that are running simultaneously within the NSP. Once the prediction model completes an epoch or computation cycle, the program must interrupt the MSP and transfer any required data. This process also involves creating the correct packets for transmission to the appropriate server (off-board).

The final layer of code resides in the on-board CPLD. This hardware-based VHDL code is responsible for correctly shuttling data between all of the hardware modules. It achieves this processing through a series of interrupts and control lines that are provided by the individual hardware components. Depending on the functionality of the system, this code is responsible for all the internal and external communication of the NSP system.

We use a high-speed USB connection to transmit 104 binned neural channels and 3D trajectory data to compute an optimal linear combiner using 10-tap FIRs per channel trained with the normalized LMS algorithm for a total of 3120 parameters [55]. The average amount of time for a single prediction when computed over 90 predicted outputs is 211 μsec. This test achieves the same timing results on the computation of predicted output because this was the same DSP used in our previous systems [56, 57]. On a 600-MHz PIII laptop running Matlab 6, the average prediction takes 23 msec. The factor of improvement or speed gain is about 100× for the DSP over the laptop running Matlab. The LMS output results collected at the receiving computer were directly compared to Matlab computed outputs and they agree within seven decimal places. The NSP uses approximately 90 to 120 mA, which equates to 450–600 mW. Overall, this is much lower than the 4 and 1.75 W previously attained by other acquisition hardware [56, 57]. In addition, the board size is 2.5 × 1 in. board size and weighs 9 g.

7.5 FLORIDA WIRELESS IMPLANTABLE RECORDING ELECTRODES

The second-generation systems will have to deal with an immensely more restrictive set of requirements because they will be subcutaneous. We envision a modular design with tens to hundreds of channels. The present requirements that we are working with is an implantable rechargeable device called the Florida wireless implantable recording electrodes (FWIRE) that will measure 1.5 × 1 × 0.5 cm, weigh 2 g, and can collect data from 16 microelectrodes for 100 hours, amplify and send them wirelessly within 500 Kbits/sec to a remote system within 1 m of the subject. A conceptual building block is presented in Figure 7.15.

As illustrated in Figure 7.15, the FWIRE microsystem consists of a flexible substrate that serves as the platform for the signal processing integrated and fine integrated circuit and wireless telemetry radio frequency integrated circuit chips, transmit antenna, receive and power coil, and microwire electrode array. A low-profile rechargeable battery is located below the flexible substrate and is used to power the implant electronics during recording sessions. The external coil wraps

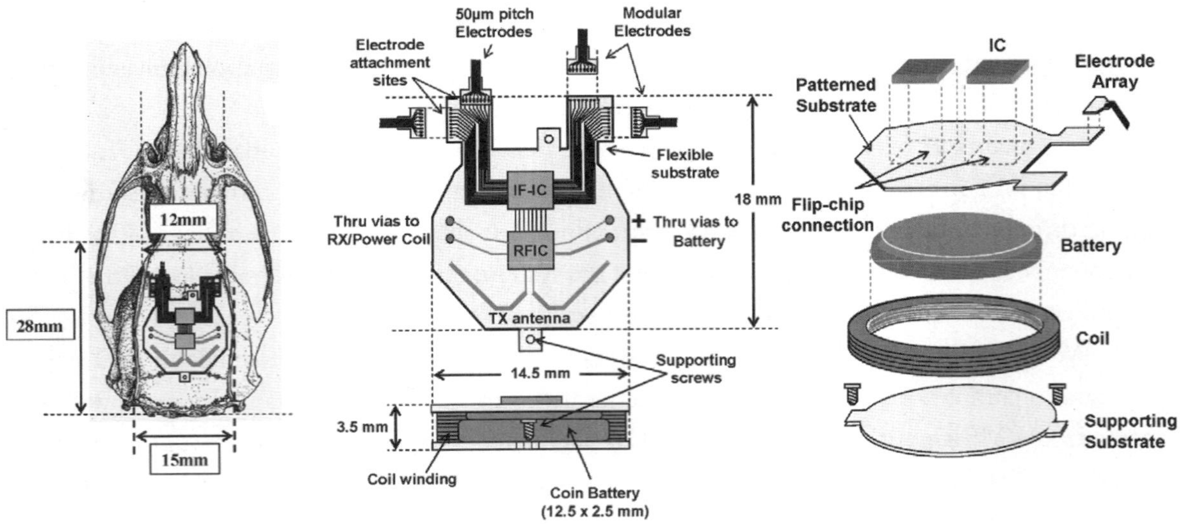

FIGURE 7.15: Component diagram and assembly of the Florida wireless implantable recording electrode system.

around the battery for a very compact package assembly and enables wireless battery recharging. The CMOS ICs are flipchip-bonded onto the flexible substrate, which is encased in medical grade silicone for isolation from fluids. Mechanical stability of the patterned flexible substrate is provided by the underlying battery/coil, supporting substrate, and screws, which attaches the platform to the skull of the behaving animal and provides the required ground reference for the on-chip electronics. Multiple electrode attachment sites onto the flexible substrate using micromachined flexible polyimide ribbon cable provide additional flexibility for experimentation. The metal traces and corresponding bond sites can be made to any size specification and spacing distance via photolithography. The initial prototypes of the electrode array shown in Section 7.1 consists of a row of eight (30 µm diameter with 50 µm pitch) gold-plated nickel electrodes with parylene-C insulated shanks that extend 4 mm from the edge of the flexible polyimide cable [58]. In vivo studies produced a high neuronal yield with SNR ranging from 10 to 27 dB. The resulting FWIRE platform is a highly modular architecture that takes into consideration the mechanical and physical constraints of the implant site, and facilitates independent development and testing of its individual components.

7.5.1 Neural Signal Processing and Representation

To achieve these objectives, very fundamental aspects of both the electronics and signal processing must be modified. We believe that a hybrid analog–digital signal processing approach is necessary, and started some preliminary steps in that direction. The first is to rethink Nyquist sampling, which is lopsided towards a complex frontend (accurate sample-and-hold that uses large chip areas and consumes a lot of power) versus a very easy backend (lowpass filtering). Here we require exactly the opposite, that is, a sampling scheme that is implemented in a simple frontend circuit eventually with a more difficult reconstruction in the backend. Our group at Florida has recently proposed the integrate-and-fire (IF) representation, which accomplishes these goals. The signal is integrated up to a threshold and a pulse is fired when the voltage crosses a threshold. Hence, the time between the pulses is proportional to the integral of the voltage, providing an aperiodic sampling scheme. Provided that there are sufficient pulses on an interval, one can show that this scheme obeys the definition of sampling (a one to one mapping with a unique inverse). What is interesting is that the IF representation is very easy to build in hardware and is immune to noise in the chip on during transmission because of the pulse nature of the representation. As long as delays are constants, they do not affect the accuracy of the representation. The quantization noise is associated with how precise is the pulse timing, so data rates will not increase with precision as in conventional A/Ds. Moreover, this representation naturally improves the fidelity of the high-amplitude portions of the signals (where there are more spikes), so in many applications (as in spike trains) considerably lower data rates are necessary to represent the signals when compared with Nyquist samplers. The power–size–bandwidth specification of this sampler seems better than conventional Nyquist samplers.

The availability of the IF representation also opens the door to blend analog and signal processing in a much more interesting way. Figure 7.16 shows our present thinking on three approaches to decrease data rates through processing at the frontend.

Strategy 1 is sends out the data directly after the IF representation. The data rates depend on the input structure and the threshold. By selecting higher thresholds the data rates decrease, but aliasing starts cropping up in the low amplitude portions of the signal.

Strategy 2 does signal compression in the pulse domain before sending the data out. The idea is to use the time between the pulses and their polarity to recognize the events of importance that in our case will be neural spikes. The objective is to decrease the data rates and only send out events that are of interest.

The third strategy is to ultimately do the processing on the chip by creating signal processing algorithms that work with point processes. The output will be directly connected to an external device for brain control.

Integrate and Fire Representation. The voltage output of the amplifier is first converted into current and, by integrating this current, the amplitude information is encoded into an asynchronous digital pulse train. This IF module efficiently encodes the analog information as an asynchronous train of pulses [59]. The principle is to encode amplitude in the time difference between events. The advantage is that a single event represents a high-resolution amplitude measurement, with

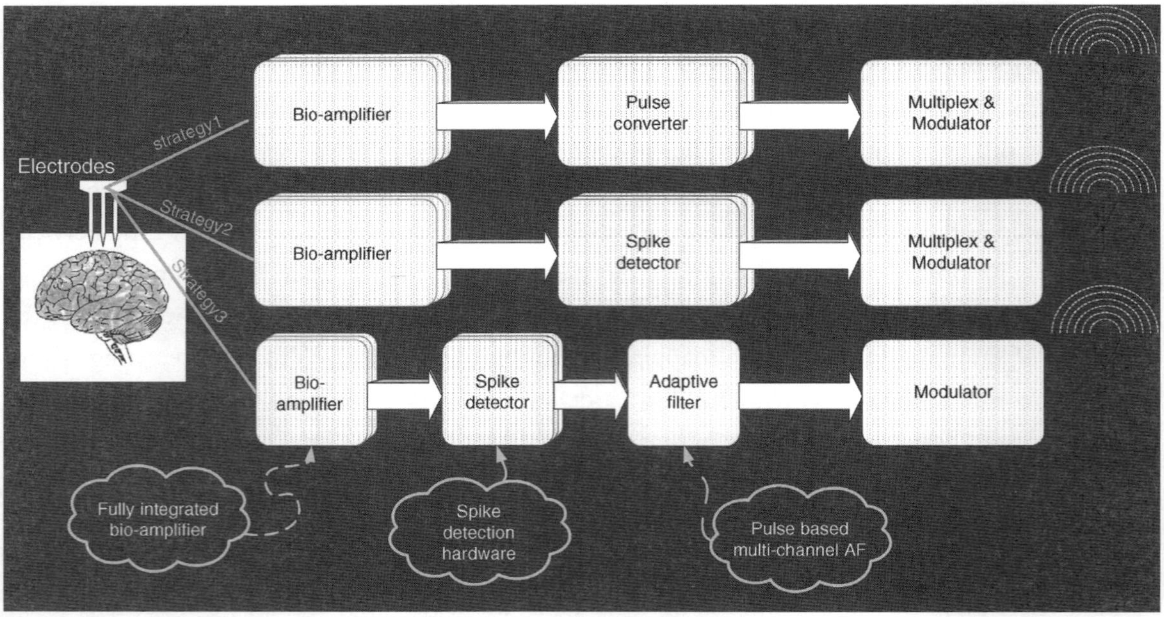

FIGURE 7.16: Three strategies for contending with bandwidth limitations of wireless implantable neural recording systems.

substantial bandwidth and power savings. These pulses have better noise immunity than conventional analog signals in transmission and also eliminates the need for a traditional ADC. The latest version using a delta–sigma modulation style to produce timing events resulted in 52 μW of power dissipation with 8-bit resolution [60]. Given a band-limited continuous signal $y(t)$, we create an alternate representation

$$\hat{y}(t) = h(t) * \sum_k w_k \delta(t - t_k) = \sum_k w_k h(t - t_k) \tag{7.1}$$

where $|y(t) - \hat{y}(t)| <_{\varepsilon}$ for all time, w_k are appropriate weights, and $h(t)$ is the sinc function. Sampling times t_i are given by: $f(t_k) = k_\theta$, where

$$\int_{t_k}^{t_{k+1}} y(t)\,\mathrm{d}t = \theta \text{ and } f(t) = \int_{t_0}^{t} y(\tau)\,\mathrm{d}\tau \tag{7.2}$$

Previously, simple linear low-pass filters were used as crude reconstruction filters from IF outputs, computing something similar to rate coding from the spike train, and the performance was poor. Pulse times are determined by the local area under the signal input $x(t)$, and they can be used for perfect reconstruction. Let us assume that $x(t)$ is band-limited to Ω_s and the maximum interspike interval T_{max} satisfies $T_{max} < \pi/\Omega_s$. From mathematical frame theory, it is known that any band-limited signal can be expressed as a lowpass filtered version of an appropriately weighted sum of delayed impulse functions, which is expressed in equation (1). So the signal recovery problem is how to calculate the appropriate weights. We can define the timing of pulses: $T = [t_1, t_2, \ldots, t_k, t_{k+1}, \ldots]$. Let $(t) = v_{in}A_{in} - V_{ref-supply} + V_{mid}$. Substituting equation (7.2) into equation (7.1), we obtain

$$\theta = \int_{t_i}^{t_{i+1}} y(t)\,dt = \int_{t_i}^{t_{i+1}} \sum_j w_j h(t - t_j)\,dt = \sum_j w_j \int_{t_i}^{t_{i+1}} w_j h(t - t_j)\,dt \tag{7.3}$$

$$c_{ij} = \int_{t_i}^{t_{i+1}} h(t - t_j)\,dt \tag{7.4}$$

which can be rewritten in matrix notation as $CW = \theta$. Unfortunately, C is usually ill-conditioned, necessitating the use of a SVD-based pseudoinverse to calculate the weight vector **W**. We have extended the reconstruction algorithm to work for a biphasic pulse train with positive and negative pulse indicating when a positive or negative threshold is surpassed [59].

7.5.2 Wireless

A wireless interface for the FWIRE microsystem is currently being developed to facilitate in vivo neural recording in behaving animals. As shown in Figure 7.17, the system includes a power recovery, regulation, and battery management module; an ASK clock and data recovery (CDR) circuit to download external system commands; a transmitter and signal processing circuits to upload neural data from the various recording channels; and a small controller with register bank to oversee system functionality, decode/encode data packets, and store on-chip settings. All components are fully integrated with the exception of the power/downlink coil, the uplink antenna, and the battery.

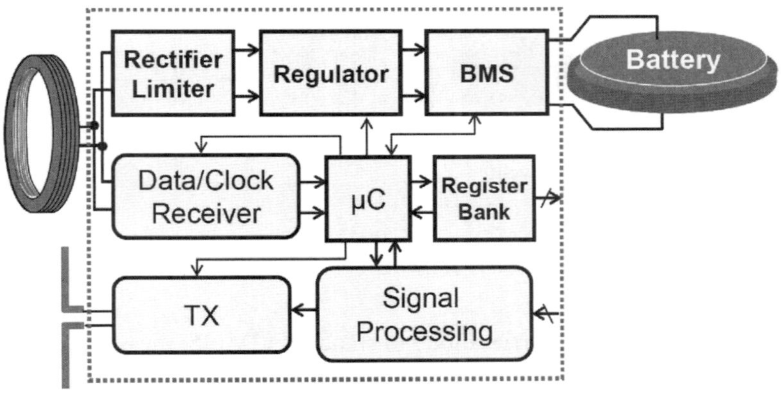

FIGURE 7.17: Wireless interface system block diagram.

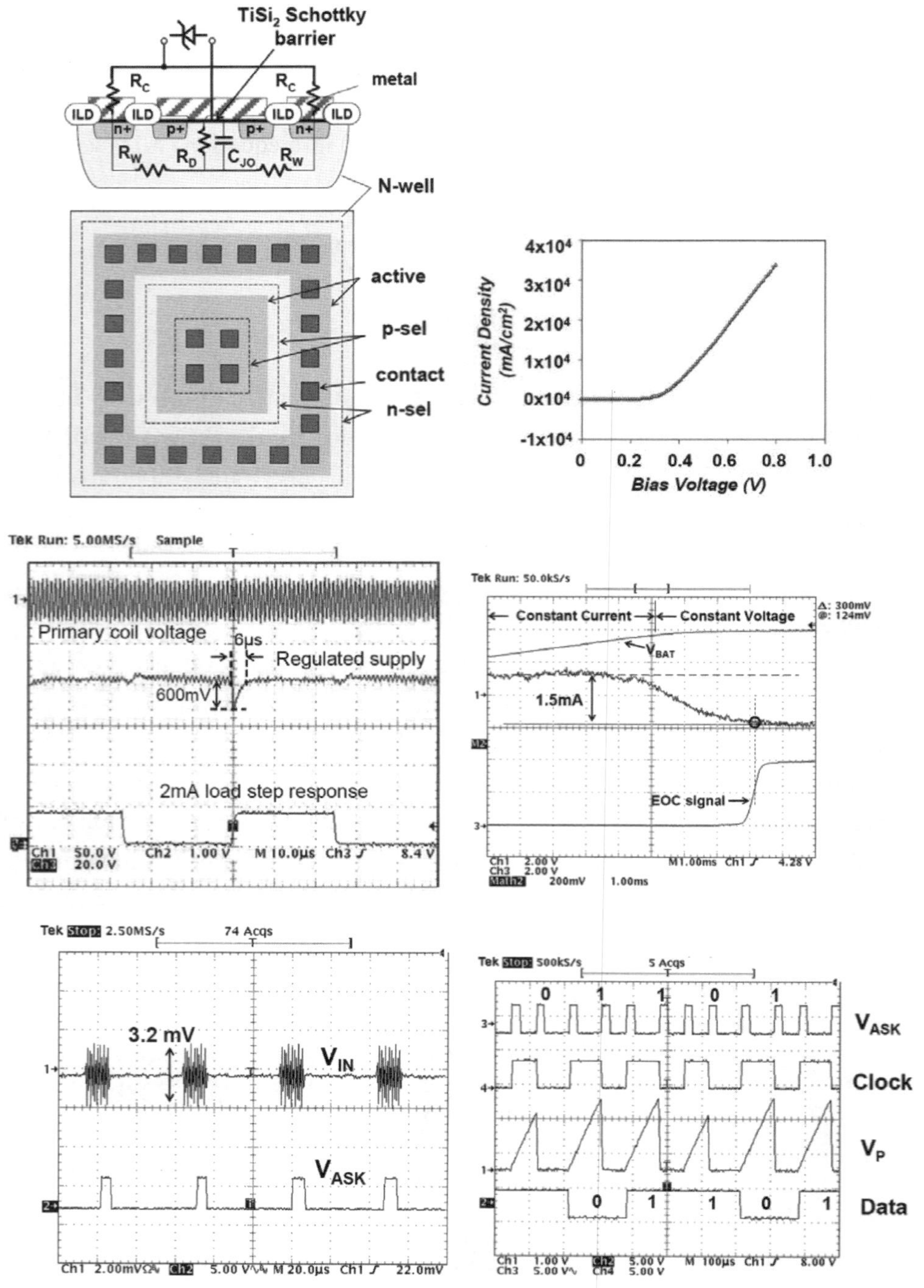

FIGURE 7.18: (a) Schottky contact barrier diode in standard CMOS; measured (b) diode current density versus bias voltage; (c) regulator transient response; (d) battery control loop charging profiles; (e) ASK detector; (f) CDR waveforms.

7.5.2.1 Wireless Power and Data Interface

A 4-MHz LC tank with on-chip resonant capacitor, full-wave rectifier, storage capacitor, low drop-out voltage regulator, and RF limiting circuit provide a stable supply and overvoltage protection to the implant electronics. Schottky contact barrier diodes are used in the bridge rectifier circuit to improve the frequency response and lower the turn-on voltage. As shown in Figure 7.18a, Ti-Si_2 Schottky barrier diodes are fabricated in standard 0.6-μm CMOS process by selectively blocking the n+/p+ implants in desired diffusion areas [61]. The measured I–J characteristic of the Schottky barrier devices is shown in Figure 7.18b. The devices have an ideality factor of ~1.16 and barrier height of ~0.5 eV, which was computed using the Richardson–Dushman equation for the thermionic current [62]. The measured turn-on voltage is ~300 mV, which is lower than standard pn junction based implementations [63]. The diodes are fabricated with small schottky contact area as this yields a higher cutoff frequency [64], and multiple cells are placed in parallel to improve the current handling capability. Figure 7.18c shows the transient regulator response when an externally generated 0- to 2-mA load step is applied as the link is powered by the primary coil voltage. The measured response is within 15% (or 600 mV/4.1 V) of the target 4.1-V supply. The regulator exhibits a load regulation of 2 mV/mA (or 240 ppm/1 mA), a line regulation 2 mV/V, and a low dropout voltage of 50 mV. At 4 MHz, the worst-case measured peak-to-peak ripple voltage at the output is within 100 mV. A battery management circuit has been implemented to charge and monitor the external Li ion cell battery [61]. Charging profile experiments for the battery control loop (Figure 7.18d) shows that during the constant-current phase, the circuit delivers 1.5 mA, resulting in a linear increase in battery voltage. The end-of-charge (EOC) during the constant-voltage phase is detected once the battery current reaches 5% of the nominal constant charging current of 1.5 mA, triggering the EOC signal. An on-chip CDR circuit is used to download external system commands. Synchronization of clock and data is achieved via a modulation scheme based on amplitude shift keying and pulse position modulation facilitates CDR using an ASK demodulator and a charge-pump with latched comparator. The CDR is operational for an input data range of 4–18 kb/sec, and exhibits a sensitivity of 3.2 mV p-p at 1 MHz.

7.6 SUMMARY

The breadth of challenges that stand between concept generation and realization of the next generation of BMIs calls for tandem and synergistic development of the many facets of neurotechnology. Here, we have provided several examples of how the goal of system development (as opposed to component development) of a fully integrated BMI implant is leading to beyond-state-of-the art technologies that might not be realized readily in a piecewise fashion. To develop the necessary signal processing, neural interfaces, or electronics independently, as has been largely conducted in the past, may not yield optimal BMI systems that are capable of chronically interfacing with large neural assemblies and delivering meaningful function to the patient. The complete specifications of future BMI systems will be driven by the technical challenges encountered in a particular specialization of BMI neurotechnology and transferred bidirectionally to the others. For example, the choice of the appropriate physiological scale to derive the most robust control hinges upon the electrode design and polymer science to sense signals throughout the life of the patient. However, without feedback on the performance of neural decoding, one may have the nonoptimal but biocompatible electrodes that do not provide the most relevant information. It is clear that the next generation of BMI technologies cannot be built solely from existing engineering principles. So the question then becomes, "What is the best strategy to deal with technology bottlenecks?" Fortunately, neural engineering approaches can benefit from the intersection of top–down signal processing techniques with bottom–up theories of neurophysiology. By reverse engineering the neural interface problem using modified versions of standard signal processing techniques, one can overcome the bottlenecks.

REFERENCES

1. Xu, S., et al., *A multi-channel telemetry system for brain microstimulation in freely roaming animals.* Journal of Neuroscience Methods, 2004. **133**: pp. 57–63.
2. Beechey, P., and D.W. Lincoln, *A miniature FM transmitter for radio-telemetry of unit activity.* Journal of Physiology (London), 1969. **203**(1): p. 5.
3. Edge, G.M., G. Horn, and G. Stechler, *Telemetry of single unit activity from a remotely controlled microelectrode.* Journal of Physiology (London), 1969. **204**(1): p. 2.
4. Eichenbaum, H., et al., *Compact miniature microelectrode-telemetry system.* Physiology & Behavior, 1977. **18**(6): pp. 1175–1178.
5. Grohrock, P., U. Hausler, and U. Jurgens, *Dual-channel telemetry system for recording vocalization-correlated neuronal activity in freely moving squirrel monkeys.* Journal of Neuroscience Methods, 1997. **76**(1): pp. 7–13.
6. Strumwasser, F., *Long-term recording from single neurons in brain of unrestrained mammals.* Science, 1958. **127**(3296): pp. 469–470.
7. Warner, H., et al., *A remote control brain telestimulator with solar cell power supply.* IEEE Transactions on Biomedical Engineering, 1968. **BM15**(2): p. 94.
8. Akin, T., et al., *A modular micromachined high-density connector system for biomedical applications.* IEEE Transactions on Biomedical Engineering, 1999. **46**(4): pp. 471–480.
9. Bai, Q., and K.D. Wise, *Single-unit neural recording with active microelectrode arrays.* IEEE Transactions on Biomedical Engineering, 2001. **48**(8): pp. 911–920.

10. Berger, T.W., et al., *Brain-implantable biomimetic electronics as the next era in neural prosthetics.* Proceedings of the IEEE, 2001. **89**(7): pp. 993–1012.
11. Blum, N.A., et al., *Multisite microprobes for neural recordings.* IEEE Transactions on Biomedical Engineering, 1991. **38**: pp. 68–74.
12. Campbell, P.K., et al., *A silicon-based, 3-dimensional neural interface—manufacturing processes for an intracortical electrode array.* IEEE Transactions on Biomedical Engineering, 1991. **38**(8): pp. 758–768.
13. Chang, J.C., G.J. Brewer, and B.C. Wheeler, *Modulation of neural network activity by patterning.* Biosensors and Bioelectronics, 2001. 16(7–8): pp. 527–533.
14. Jones, K.E., and R.A. Normann, *An advanced demultiplexing system for physiological stimulation.* IEEE Transactions on Biomedical Engineering, 1997. **44**(12): pp. 1210–1220.
15. Kennedy, P.R., *The cone electrode—a long-term electrode that records from neurites grown onto its recording surface.* Journal of Neuroscience Methods, 1989. 29(3): pp. 181–193.
16. Kruger, J., *A 12-fold microelectrode for recording from vertically aligned cortical-neurons.* Journal of Neuroscience Methods, 1982. **6**(4): pp. 347–350.
17. Moxon, K.A., et al., *Ceramic-based multisite electrode arrays for chronic single-neuron recording.* IEEE Transactions on Biomedical Engineering, 2004. **51**(4): pp. 647–656.
18. Obeid, I., M.A.L. Nicolelis, and P.D. Wolf, *A low power multichannel analog front end for portable neural signal recordings.* Journal of Neuroscience Methods, 2004. **133**(1–2): pp. 27–32.
19. Palmer, C.I., *Long-term recordings in the cat motor cortex unit-activity and field potentials from sensory and brain-stem stimulation as a means of identifying electrode position.* Journal of Neuroscience Methods, 1990. **31**(2): pp. 163–181.
20. Salcman, M., *Trauma after electrode implantation.* Archives of Neurology, 1976. **33**(3): pp. 215–215.
21. Tsai, M.L., and C.T. Yen, *A simple method for fabricating horizontal and vertical microwire arrays.* Journal of Neuroscience Methods, 2003. **131**(1–2): pp. 107–110.
22. Wise, K.D., et al., *Wireless implantable microsystems: High-density electronic interfaces to the nervous system.* Proceedings of the IEEE, 2004. **92**(1): pp. 76–97.
23. Buzsaki, G., *Multisite recording of brain field potentials and unit activity in freely moving rats.* Journal of Neuroscience Methods, 1989. **28**: pp. 209–217.
24. Farshchi, S., et al., *A TinyOS-enabled MICA2-based wireless neural interface.* IEEE Transactions on Biomedical Engineering, 2006. **57**(3): pp. 1416–1424.
25. Mavoori, J., et al., *An autonomous implantable computer for neural recording and stimulation in unrestricted primates.* Journal of Neuroscience Methods, 2005. **148**(1): pp. 71–77.
26. Abeles, M., Corticonics: Neural Circuits of the Cerebral Cortex. 1991, New York: Cambridge University Press.

27. Wessberg, J., et al., *Real-time prediction of hand trajectory by ensembles of cortical neurons in primates.* Nature, 2000. **408**(6810): pp. 361–365.
28. Moxon, K.A., S.C. Leiser, and G.A. Gerhardt, *Ceramic-based multisite electrode arrays for chronic single-neuron recording.* IEEE Transactions on Biomedical Engineering, 2004. **51**(4): pp. 647–656.
29. Chapin, J.K., and K.A. Moxon, eds. *Neural prostheses for restoration of sensory and motor function.* Methods and New Frontiers in Neuroscience. 2001, Boca Raton, FL: CRC Press.
30. Kipke, D.R., R.J. Vetter, and J.C. Williams, *Silicon-substrate intracortical microelectrode arrays for long-term recording of neuronal spike activity in cerebral cortex.* IEEE Transactions on Rehabilitation Engineering, 2003. **11**(2): pp. 151–155.
31. Cham, J.G., et al., *Semi-chronic motorized microdrive and control algorithm for autonomously isolating and maintaining optimal extracellular action potentials.* Journal of Neurophysiology, 2005. **93**(1): pp. 570–579.
32. Rousche, P.J., and R.A. Normann, *A method for pneumatically inserting an array of penetrating electrodes into cortical tissue.* Annals of Biomedical Engineering, 1992. **20**: pp. 413–422.
33. Jaeger, D., S. Gilman, and J.W. Aldridge, *A multiwire microelectrode for single unit recording in deep brain structures.* Journal of Neuroscience Methods, 1990. **32**(2): pp. 143–148.
34. Jellema, T., and J.A.W. Weijnen, *A slim needle-shaped multiwire microelectrode for intracerebral recording.* Journal of Neuroscience Methods, 1991. **40**(2–3): pp. 203–209.
35. Williams, J.C., R.L. Rennaker, and D.R. Kipke, *Long-term neural recording characteristics of wire microelectrode arrays implanted in cerebral cortex.* Brain Research Protocols, 1999. **4**(3): p. 303.
36. Ji, J., and K.D. Wise, *An implantable CMOS circuit interface for multiplexed microelectrode recording arrays.* IEEE Journal of Solid-State Circuits, 1992. **27**(3): pp. 433–443.
37. Najafi, K., and K.D. Wise, *An implantable multielectrode array with on-chip signal processing.* IEEE Journal of Solid State Circuits, 1986. **21**(6): pp. 1035–1044.
38. Maynard, E.M., C.T. Nordhausen, and R.A. Normann, *The Utah Intracortical Electrode Array: A recording structure for potential brain–computer interfaces.* Electroencephalography and Clinical Neurophysiology, 1997. **102**(3): pp. 228–239.
39. Nordhausen, C.T., P.J. Rousche, and R.A. Normann, *Optimizing recording capabilities of the Utah-Intracortical-Electrode-Array.* Brain Research, 1994. **637**(1–2): pp. 27–36.
40. Rousche, P.J., and R.A. Normann, *Chronic recording capability of the Utah Intracortical Electrode Array in cat sensory cortex.* Journal of Neuroscience Methods, 1998. **82**(1): p. 1.
41. Fee, M.S., and A. Leonardo, Miniature motorized microdrive and commutator system for chronic neural recording in small animals. Journal of Neuroscience Methods, 2001. **112**(2): p. 83.
42. Subbaroyan, J., D.C. Martin, and D.R. Kipke, *A finite-element model of the mechanical effects of implantable microelectrodes in the cerebral cortex.* Journal of Neural Engineering, 2005. **2**: pp. 103–113.

43. Rousche, P.J., et al., *'Flexible' polyimide-based intracortical electrode arrays with bioactive capability.* IEEE Transactions on Biomedical Engineering, 2001. **48**(3): pp. 361–371.
44. Vetter, R.J., et al., *Chronic neural recording using silicon-substrate microelectrode arrays implanted in cerebral cortex.* IEEE Transactions on Biomedical Engineering, 2004. **51**(6): pp. 896–904.
45. Paxinos, G., The Rat Brain in Stereotaxic Coordinates. 1997, Sydney: Academic Press.
46. Lewicki, M.S., *A review of methods for spike sorting: The detection and classification of neural action potentials.* Network Computation in Neural Systems, 1998. **9**(4). pp. R53–R78.
47. Wood, F., et al., *On the variability of manual spike sorting.* IEEE Transactions on Biomedical Engineering, 2004. **51**(6): pp. 912–918.
48. Nicolelis, M.A.L., Methods for Neural Ensemble Recordings. 1999, Boca Raton, FL: CRC Press.
49. Harrison, R.R., and C. Charles, *A low-power low-noise CMOS amplifier for neural recording applications.* IEEE Journal of Solid-State Circuits, 2003. **38**: pp. 958–965.
50. Chen, D., An Ultra-Low Power Neural Recording System Using Pulse Representations. 2006, Gainesville, FL: Department of Electrical and Computer Engineering, University of Florida.
51. Li, Y., An Integrated Multichannel Neural Recording System With Spike Outputs. 2006, Gainesville, FL: Department of Electrical and Computer Engineering, University of Florida.
52. Cieslewski, G., et al. *Neural signal sampling via the low power wireless Pico system*, in IEEE International Conference of the Engineering in Medicine and Biology Society. 2006. New York.
53. Nenadic, Z., and J. Burdick, *Spike detection using the continuous wavelet transform.* IEEE Transactions on Biomedical Engineering, 2005. **52**(1): pp. 74–87.
54. Cho, J., et al., *Self-organizing maps with dynamic learning for signal reconstruction.* Neural Networks, 2007. **20**(2): pp. 274–284.
55. Sanchez, J.C., et al. *Interpreting neural activity through linear and nonlinear models for brain machine interfaces*, in International Conference of Engineering in Medicine and Biology Society. 2003. Cancun, Mexico.
56. Shalom, D., et al. *A reconfigurable neural signal processor (NSP) for brain machine interfaces*, in Engineering in Medicine and Biology Society, 2006. EMBS '06. 28th Annual International Conference of the IEEE. 2006.
57. Darmanjian, S., et al. *A portable wireless DSP system for a brain machine interface*, in Neural Engineering, 2005. Proceedings of the 2nd International IEEE EMBS Conference. 2005.
58. Patrick, E., et al. *design and fabrication of a flexible substrate microelectrode array for brain machine interfaces*, in IEEE International Conference of the Engineering in Medicine and Biology Society. 2006. New York.
59. Chen, D., et al. *Asynchronous biphasic pulse signal coding and its CMOS realization*, in Proceedings of IEEE International Symposium on Circuits and Systems (ISCAS). 2006. Kos, Greece.

60. Wei, D., V. Garg, and J.G. Harris. *An asynchronous delta–sigma converter*, in Proceedings of IEEE International Symposium on Circuits and Systems (ISCAS). 2006. Kos, Greece.
61. Li, P., J. Principe, and R. Bashirullah, *A wireless power interface for rechargeable battery operated neural recording implants*, in IEEE Engineering in Medicine and Biology Conference. 2006. New York.
62. Shur, M., Physics of Semiconductor Devices. 1990, New York: Prentice-Hall.
63. Ghovanloo, M., and K. Najafi, *A modular 32-site wireless neural stimulation microsystem.* IEEE Journal on Solid-State Circuits, 2004. **39**(12): pp. 2457–2466.
64. Sankaran, S., and K.K. O, *Schottky barrier diodes for millimeter wave detection in a foundry CMOS process.* IEEE Electron Device Letters, 2005. **26**(7): pp. 492–494.

• • • •

Author Biography

Justin C. Sanchez Dr. Sanchez's research interests are in Neural Engineering and neural assistive technologies. Topics include the analysis of neural ensemble recordings, adaptive signal processing, Brain-Machine Interfaces, motor system electrophysiology, treatment of movement disorders, and the neurophysiology of epilepsy. He is an Assistant Professor of Pediatrics, Neuroscience, and Biomedical Engineering at the University of Florida College of Medicine, Engineering, and McKnight Brain Institute in Gainesville, Florida. He received his Ph.D. (2004) and M.E. degrees in Biomedical Engineering and B.S. degree in Engineering Science (Highest Honors—2000) with a minor in Biomechanics from the University of Florida. The goal of his research is to develop state-of-the-art novel medical treatments by operating at the interface between basic neural engineering research and clinical care. This direction of research is motivated by the potential of direct neural interfaces for delivering therapy and restoring functionality to disabled individuals using engineering principles. In 2005, he won two prestigious awards for his work including Excellence in Neuroengineering and more recently an American Epilepsy Society Young Investigator Award. In 2006 he founded the Gainesville Engineering in Medicine and Biology/Communications Joint Societies Chapter and serves as the IEEE Gainesville Section Director for membership development. His neural engineering electrophysiology laboratory is currently developing direct neural interfaces for use in the research and clinical settings and has published over 35 peer review papers and holds 3 patents in neuroprosthetic design. He is the founding member of the Neuroprosthetics Group (NRG) at the University of Florida (http://nrg.mbi.ufl.edu).

Dr. José C. Príncipe is Distinguished Professor of Electrical and Biomedical Engineering at the University of Florida since 2002. He joined the University of Florida in 1987, after an eight year appointment as Professor at the University of Aveiro, in Portugal. Dr. Principe holds degrees in electrical engineering from the University of Porto (Bachelor), Portugal, University of Florida (Master and Ph.D.), USA and a Laurea Honoris Causa degree from the Universita Mediterranea in Reggio Calabria, Italy. Dr. Principe's interests lie in nonlinear non-Gaussian optimal signal processing and modeling and in biomedical engineering. He created in 1991 the Computational NeuroEngineering Laboratory to synergistically focus the research in biological information processing models. He recently received the Gabor Award from the International Neural Network Society for his contributions.

Dr. Principe is a Fellow of the IEEE, Fellow of the AIMBE, past President of the International Neural Network Society, and past Editor in Chief of the Transactions of Biomedical Engineering, as well as a former member of the Advisory Science Board of the FDA. He holds 5 patents and has submitted seven more. Dr. Principe was supervisory committee chair of 50 Ph.D. and 61 Master students, and he is author of more than 400 refereed publications (3 books, 4 edited books, 14 book chapters, 116 journal papers and 276 conference proceedings).

Introduction to Statistics for Biomedical Engineers

Kristina M. Ropella
Department of Biomedical Engineering
Marquette University

MORGAN & CLAYPOOL PUBLISHERS

This text is dedicated to all the students who have completed my BIEN 084 statistics course for biomedical engineers and have taught me how to be more effective in communicating the subject matter and making statistics come alive for them. I also thank J. Claypool for his patience and for encouraging me to finally put this text together. Finally, I thank my family for tolerating my time at home on the laptop.

ABSTRACT

There are many books written about statistics, some brief, some detailed, some humorous, some colorful, and some quite dry. Each of these texts is designed for a specific audience. Too often, texts about statistics have been rather theoretical and intimidating for those not practicing statistical analysis on a routine basis. Thus, many engineers and scientists, who need to use statistics much more frequently than calculus or differential equations, lack sufficient knowledge of the use of statistics. The audience that is addressed in this text is the university-level biomedical engineering student who needs a bare-bones coverage of the most basic statistical analysis frequently used in biomedical engineering practice. The text introduces students to the essential vocabulary and basic concepts of probability and statistics that are required to perform the numerical summary and statistical analysis used in the biomedical field. This text is considered a starting point for important issues to consider when designing experiments, summarizing data, assuming a probability model for the data, testing hypotheses, and drawing conclusions from sampled data.

A student who has completed this text should have sufficient vocabulary to read more advanced texts on statistics and further their knowledge about additional numerical analyses that are used in the biomedical engineering field but are beyond the scope of this text. This book is designed to supplement an undergraduate-level course in applied statistics, specifically in biomedical engineering. Practicing engineers who have not had formal instruction in statistics may also use this text as a simple, brief introduction to statistics used in biomedical engineering. The emphasis is on the application of statistics, the assumptions made in applying the statistical tests, the limitations of these elementary statistical methods, and the errors often committed in using statistical analysis. A number of examples from biomedical engineering research and industry practice are provided to assist the reader in understanding concepts and application. It is beneficial for the reader to have some background in the life sciences and physiology and to be familiar with basic biomedical instrumentation used in the clinical environment.

KEYWORDS

probability model, hypothesis testing, physiology, ANOVA, normal distribution, confidence interval, power test

Contents

CHAPTER 1

Introduction

Biomedical engineers typically collect all sorts of data, from patients, animals, cell counters, microassays, imaging systems, pressure transducers, bedside monitors, manufacturing processes, material testing systems, and other measurement systems that support a broad spectrum of research, design, and manufacturing environments. Ultimately, the reason for collecting data is to make a decision. That decision may concern differentiating biological characteristics among different populations of people, determining whether a pharmacological treatment is effective, determining whether it is cost-effective to invest in multimillion-dollar medical imaging technology, determining whether a manufacturing process is under control, or selecting the best rehabilitative therapy for an individual patient.

The challenge in making such decisions often lies in the fact that all real-world data contains some element of uncertainty because of random processes that underlie most physical phenomenon. These random elements prevent us from predicting the exact value of any physical quantity at any moment of time. In other words, when we collect a sample or data point, we usually cannot predict the exact value of that sample or experimental outcome. For example, although the average resting heart rate of normal adults is about 70 beats per minute, we cannot predict the exact arrival time of our next heartbeat. However, we can approximate the likelihood that the arrival time of the next heartbeat will fall in a specific time interval if we have a good probability model to describe the random phenomenon contributing to the time interval between heartbeats. The timing of heartbeats is influenced by a number of physiological variables [1], including the refractory period of the individual cells that make up the heart muscle, the leakiness of the cell membranes in the sinus node (the heart's natural pacemaker), and the activity of the autonomic nervous system, which may speed up or slow down the heart rate in response to the body's need for increased blood flow, oxygen, and nutrients. The sum of these biological processes produces a pattern of heartbeats that we may measure by counting the pulse rate from our wrist or carotid artery or by searching for specific QRS waveforms in the ECG [2]. Although this sum of events makes it difficult for us to predict exactly when the new heartbeat will arrive, we can guess, with a certainty amount of confidence when the next beat will arrive. In other words, we can assign a probability to the likelihood that the next heartbeat will arrive in a specified time interval. If we were to consider all possible arrival times and assigned a probability to those arrival times, we would have a probability model for the heartbeat intervals. If we can find a probability model to describe the likelihood of occurrence of a certain event or experimental outcome, we can use statistical methods to make decisions. The probability models describe characteristics of the population or phenomenon being studied. Statistical analysis then makes use of these models to help us make decisions about the population(s) or processes.

The conclusions that one may draw from using statistical analysis are only as good as the underlying model that is used to describe the real-world phenomenon, such as the time interval between heartbeats. For example, a normally functioning heart exhibits considerable variability in beat-to-beat intervals (Figure 1.1). This variability reflects the body's continual effort to maintain homeostasis so that the body may continue to perform its most essential functions and supply the body with the oxygen and nutrients required to function normally. It has been demonstrated through biomedical research that there is a loss of heart rate variability associated with some diseases, such as diabetes and ischemic heart disease. Researchers seek to determine if this difference in variability between normal subjects and subjects with heart disease is significant (meaning, it is due to some underlying change in biology and not simply a result of chance) and whether it might be used to predict the progression of the disease [1]. One will note that the probability model changes as a consequence of changes in the underlying biological function or process. In the case of manufacturing, the probability model used to describe the output of the manufacturing process may change as

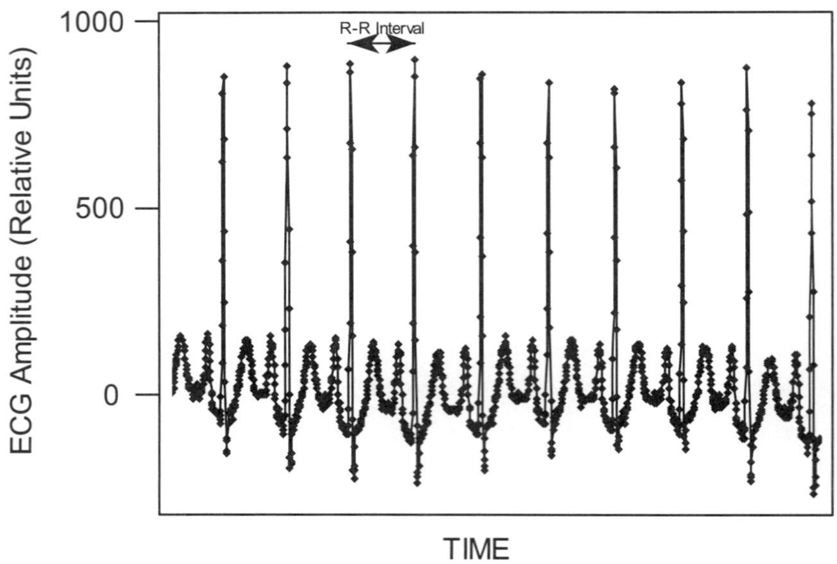

FIGURE 1.1: Example of an ECG recording, where R-R interval is defined as the time interval between successive R waves of the QRS complex, the most prominent waveform of the ECG.

a function of machine operation or changes in the surrounding manufacturing environment, such as temperature, humidity, or human operator.

Besides helping us to describe the probability model associated with real-world phenomenon, statistics help us to make decisions by giving us quantitative tools for testing hypotheses. We call this *inferential statistics*, whereby the outcome of a statistical test allows us to draw conclusions or make inferences about one or more populations from which samples are drawn. Most often, scientists and engineers are interested in comparing data from two or more different populations or from two or more different processes. Typically, the default hypothesis is that there is no difference in the distributions of two or more populations or processes, and we use statistical analysis to determine whether there are true differences in the distributions of the underlying populations to warrant different probability models be assigned to the individual processes.

In summary, biomedical engineers typically collect data or samples from various phenomena, which contain some element of randomness or unpredictable variability, for the purposes of making decisions. To make sound decisions in the context of the uncertainty with some level of confidence, we need to assume some probability model for the populations from which the samples have been collected. Once we have assumed an underlying model, we can select the appropriate statistical tests for comparing two or more populations and then use these tests to draw conclusions about

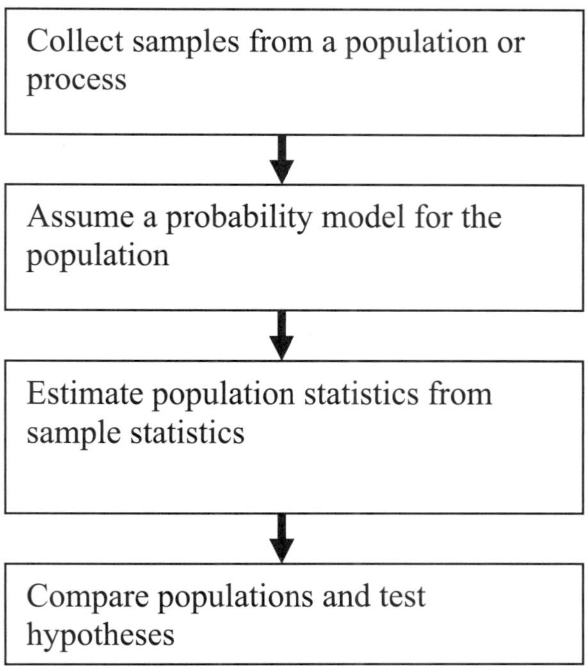

FIGURE 1.2: Steps in statistical analysis.

our hypotheses for which we collected the data in the first place. Figure 1.2 outlines the steps for performing statistical analysis of data.

In the following chapters, we will describe methods for graphically and numerically summarizing collected data. We will then talk about fitting a probability model to the collected data by briefly describing a number of well-known probability models that are used to describe biological phenomenon. Finally, once we have assumed a model for the populations from which we have collected our sample data, we will discuss the types of statistical tests that may be used to compare data from multiple populations and allow us to test hypotheses about the underlying populations.

• • • •

CHAPTER 2

Collecting Data and Experimental Design

Before we discuss any type of data summary and statistical analysis, it is important to recognize that the value of any statistical analysis is only as good as the data collected. Because we are using data or samples to draw conclusions about entire populations or processes, it is critical that the data collected (or samples collected) are representative of the larger, underlying population. In other words, if we are trying to determine whether men between the ages of 20 and 50 years respond positively to a drug that reduces cholesterol level, we need to carefully select the population of subjects for whom we administer the drug and take measurements. In other words, we have to have enough samples to represent the variability of the underlying population. There is a great deal of variety in the weight, height, genetic makeup, diet, exercise habits, and drug use in all men ages 20 to 50 years who may also have high cholesterol. If we are to test the effectiveness of a new drug in lowering cholesterol, we must collect enough data or samples to capture the variability of biological makeup and environment of the population that we are interested in treating with the new drug. Capturing this variability is often the greatest challenge that biomedical engineers face in collecting data and using statistics to draw meaningful conclusions. The experimentalist must ask questions such as the following:

- What type of person, object, or phenomenon do I sample?
- What variables that impact the measure or data can I control?
- How many samples do I require to capture the population variability to apply the appropriate statistics and draw meaningful conclusions?
- How do I avoid biasing the data with the experimental design?

Experimental design, although not the primary focus of this book, is the most critical step to support the statistical analysis that will lead to meaningful conclusions and hence sound decisions.

One of the most fundamental questions asked by biomedical researchers is, "What size sample do I need?" or "How many subjects will I need to make decisions with any level of confidence?"

We will address these important questions at the end of this book when concepts such as variability, probability models, and hypothesis testing have already been covered. For example, power tests will be described as a means for predicting the sample size required to detect significant differences in two population means using a *t* test.

Two elements of experimental design that are critical to prevent biasing the data or selecting samples that do not fairly represent the underlying population are randomization and blocking.

Randomization refers to the process by which we randomly select samples or experimental units from the larger underlying population such that we maximize our chance of capturing the variability in the underlying population. In other words, we do not limit our samples such that only a fraction of the characteristics or behaviors of the underlying population are captured in the samples. More importantly, we do not bias the results by artificially limiting the variability in the samples such that we alter the probability model of the sample population with respect to the probability model of the underlying population.

In addition to randomizing our selection of experimental units from which to take samples, we might also randomize our assignment of treatments to our experimental units. Or, we may randomize the order in which we take data from the experimental units. For example, if we are testing the effectiveness of two different medical imaging methods in detecting brain tumor, we will randomly assign all subjects suspect of having brain tumor to one of the two imaging methods. Thus, if we have a mix of sex, age, and type of brain tumor participating in the study, we reduce the chance of having all one sex or one age group assigned to one imaging method and a very different type of population assigned to the second imaging method. If a difference is noted in the outcome of the two imaging methods, we will not artificially introduce sex or age as a factor influencing the imaging results.

As another example, if one are testing the strength of three different materials for use in hip implants using several strength measures from a materials testing machine, one might randomize the order in which samples of the three different test materials are submitted to the machine. Machine performance can vary with time because of wear, temperature, humidity, deformation, stress, and user characteristics. If the biomedical engineer were asked to find the strongest material for an artificial hip using specific strength criteria, he or she may conduct an experiment. Let us assume that the engineer is given three boxes, with each box containing five artificial hip implants made from one of three materials: titanium, steel, and plastic. For any one box, all five implant samples are made from the same material. To test the 15 different implants for material strength, the engineer might randomize the order in which each of the 15 implants is tested in the materials testing machine so that time-dependent changes in machine performance or machine-material interactions or time-varying environmental condition do not bias the results for one or more of the materials. Thus, to fully randomize the implant testing, an engineer may literally place the numbers 1–15 in a hat and also assign the numbers 1–15 to each of the implants to be tested. The engineer will then blindly draw one of the 15 numbers from a hat and test the implant that corresponds to that number. This way the engineer is not testing all of one material in any particular order, and we avoid introducing order effects into the data.

The second aspect of experimental design is blocking. In many experiments, we are interested in one or two specific factors or variables that may impact our measure or sample. However, there may be other factors that also influence our measure and confound our statistics. In good experimental design, we try to collect samples such that different treatments within the factor of interest are not biased by the differing values of the confounding factors. In other words, we should be certain that every treatment within our factor of interest is tested within each value of the confounding factor. We refer to this design as blocking by the confounding factor. For example, we may want to study weight loss as a function of three different diet pills. One confounding factor may be a person's starting weight. Thus, in testing the effectiveness of the three pills in reducing weight, we may want to block the subjects by starting weight. Thus, we may first group the subjects by their starting weight and then test each of the diet pills within each group of starting weights.

In biomedical research, we often block by experimental unit. When this type of blocking is part of the experimental design, the experimentalist collects multiple samples of data, with each sample representing different experimental conditions, from each of the experimental units. Figure 2.1 provides a diagram of an experiment in which data are collected before and after patients receives therapy, and the experimental design uses blocking (left) or no blocking (right) by experimental unit. In the case of blocking, data are collected before and after therapy from the same set of human subjects. Thus, within an individual, the same biological factors that influence the biological response to the therapy are present before and after therapy. Each subject serves as his or her own control for factors that may randomly vary from subject to subject both before and after therapy. In essence, with blocking, we are eliminating biases in the differences between the two populations

Block (Repeated Measures)		
Subject	Measure before treatment	Measure after treatment
1	M11	12
2	M21	M22
3	M31	M32
.		
.		
K	MK1	MK2

No Block (No repeated measures)			
Subject	Measure before treatment	Subject	Measure after treatment
1	M1	K+1	M(K+1)
2	M2	K+2	M(K+2)
3	M3	K+3	M(K+3)
.		.	
.		.	
K	MK	K+K	M(K+K)

FIGURE 2.1: Samples are drawn from two populations (before and after treatment), and the experimental design uses block (left) or no block (right). In this case, the block is the experimental unit (subject) from which the measures are made.

(before and after) that may result because we are using two different sets of experimental units. For example, if we used one set of subjects before therapy and then an entirely different set of subjects after therapy (Figure 2.1, right), there is a chance that the two sets of subjects may vary enough in sex, age, weight, race, or genetic makeup, which would lead to a difference in response to the therapy that has little to do with the underlying therapy. In other words, there may be confounding factors that contribute to the difference in the experimental outcome before and after therapy that are not only a factor of the therapy but really an artifact of differences in the distributions of the two different groups of subjects from which the two samples sets were chosen. Blocking will help to eliminate the effect of intersubject variability.

However, blocking is not always possible, given the nature of some biomedical research studies. For example, if one wanted to study the effectiveness of two different chemotherapy drugs in reducing tumor size, it is impractical to test both drugs on the same tumor mass. Thus, the two drugs are tested on different groups of individuals. The same type of design would be necessary for testing the effectiveness of weight-loss regimens.

Thus, some important concepts and definitions to keep in mind when designing experiments include the following:

- *experimental unit*: the item, object, or subject to which we apply the treatment and from which we take sample measurements;
- *randomization*: allocate the treatments randomly to the experimental units;
- *blocking*: assigning all treatments within a factor to every level of the blocking factor. Often, the blocking factor is the experimental unit. Note that in using blocking, we still randomize the order in which treatments are applied to each experimental unit to avoid ordering bias.

Finally, the experimentalist must always think about how representative the sample population is with respect to the greater underlying population. Because it is virtually impossible to test every member of a population or every product rolling down an assembly line, especially when destructive testing methods are used, the biomedical engineer must often collect data from a much smaller sample drawn from the larger population. It is important, if the statistics are going to lead to useful conclusions, that the sample population captures the variability of the underlying population. What is even more challenging is that we often do not have a good grasp of the variability of the underlying population, and because of expense and respect for life, we are typically limited in the number of samples we may collect in biomedical research and manufacturing. These limitations are not easy to address and require that the engineer always consider how fair the sample and data analysis is and how well it represents the underlying population(s) from which the samples are drawn.

• • • •

CHAPTER 3
Data Summary and Descriptive Statistics

We assume now that we have collected our data through the use of good experimental design. We now have a collection of numbers, observations, or descriptions to describe our data, and we would like to summarize the data to make decisions, test a hypothesis, or draw a conclusion.

3.1 WHY DO WE COLLECT DATA?

The world is full of uncertainty, in the sense that there are random or unpredictable factors that influence every experimental measure we make. The unpredictable aspects of the experimental outcomes also arise from the variability in biological systems (due to genetic and environmental factors) and manufacturing processes, human error in making measurements, and other underlying processes that influence the measures being made.

Despite the uncertainty regarding the exact outcome of an experiment or occurrence of a future event, we collect data to try to better understand the processes or populations that influence an experimental outcome so that we can make some predictions. Data provide information to reduce uncertainty and allow for decision making. When properly collected and analyzed, data help us solve problems. It cannot be stressed enough that the data must be properly collected and analyzed if the data analysis and subsequent conclusions are to have any value.

3.2 WHY DO WE NEED STATISTICS?

We have three major reasons for using statistical data summary and analysis:

1. The real world is full of random events that cannot be described by exact mathematical expressions.
2. Variability is a natural and normal characteristic of the natural world.
3. We like to make decisions with some confidence. This means that we need to find trends within the variability.

3.3 WHAT QUESTIONS DO WE HOPE TO ADDRESS WITH OUR STATISTICAL ANALYSIS?

There are several basic questions we hope to address when using numerical and graphical summary of data:

1. Can we differentiate between groups or populations?
2. Are there correlations between variables or populations?
3. Are processes under control?

Finding physiological differences between populations is probably the most frequent aim of biomedical research. For example, researchers may want to know if there is a difference in life expectancy between overweight and underweight people. Or, a pharmaceutical company may want to determine if one type of antibiotic is more effective in combating bacteria than another. Or, a physician wonders if diastolic blood pressure is reduced in a group of hypertensive subjects after the consumption of a pressure-reducing drug. Most often, biomedical researchers are comparing populations of people or animals that have been exposed to two or more different treatments or diagnostic tests, and they want to know if there is difference between the responses of the populations that have received different treatments or tests. Sometimes, we are drawing multiple samples from the same group of subjects or experimental units. A common example is when the physiological data are taken before and after some treatment, such as drug intake or electronic therapy, from one group of patients. We call this type of data collection *blocking* in the experimental design. This concept of blocking is discussed more fully in Chapter 2.

Another question that is frequently the target of biomedical research is whether there is a correlation between two physiological variables. For example, is there a correlation between body build and mortality? Or, is there a correlation between fat intake and the occurrence of cancerous tumors. Or, is there a correlation between the size of the ventricular muscle of the heart and the frequency of abnormal heart rhythms? These type of questions involve collecting two set of data and performing a correlation analysis to determine how well one set of data may be predicted from another. When we speak of correlation analysis, we are referring to the linear relation between two variables and the ability to predict one set of data by modeling the data as a linear function of the second set of data. Because correlation analysis only quantifies the linear relation between two processes or data sets, nonlinear relations between the two processes may not be evident. A more detailed description of correlation analysis may be found in Chapter 7.

Finally, a biomedical engineer, particularly the engineer involved in manufacturing, may be interested in knowing whether a manufacturing process is under control. Such a question may arise if there are tight controls on the manufacturing specifications for a medical device. For example, if the engineer is trying to ensure quality in producing intravascular catheters that must have diameters between 1 and 2 cm, the engineer may randomly collect samples of catheters from the assembly line at random intervals during the day, measure their diameters, determine how many of the catheters meet specifications, and determine whether there is a sudden change in the number of catheters that fail to meet specifications. If there is such a change, the engineers may look for elements of the manufacturing process that change over time, changes in environmental factors, or user errors. The engineer can use control charts to assess whether the processes are under control. These methods of statistical analysis are not covered in this text, but may be found in a number of references, including [3].

3.4 HOW DO WE GRAPHICALLY SUMMARIZE DATA?

We can summarize data in graphical or numerical form. The numerical form is what we refer to as statistics. Before blindly applying the statistical analysis, it is always good to look at the raw data, usually in a graphical form, and then use graphical methods to summarize the data in an easy to interpret format.

The types of graphical displays that are most frequently used by biomedical engineers include the following: scatterplots, time series, box-and-whisker plots, and histograms.

Details for creating these graphical summaries are described in [3–6], but we will briefly describe them here.

3.4.1 Scatterplots

The scatterplot simply graphs the occurrence of one variable with respect to another. In most cases, one of the variables may be considered the independent variable (such as time or subject number), and the second variable is considered the dependent variable. Figure 3.1 illustrates an example of a scatterplot for two sets of data. In general, we are interested in whether there is a predictable relationship that maps our independent variable (such as respiratory rate) into our dependent variable (such a heart rate). If there is a linear relationship between the two variables, the data points should fall close to a straight line.

3.4.2 Time Series

A time series is used to plot the changes in a variable as a function of time. The variable is usually a physiological measure, such as electrical activation in the brain or hormone concentration in the blood stream, that changes with time. Figure 3.2 illustrates an example of a time series plot. In this figure, we are looking at a simple sinusoid function as it changes with time.

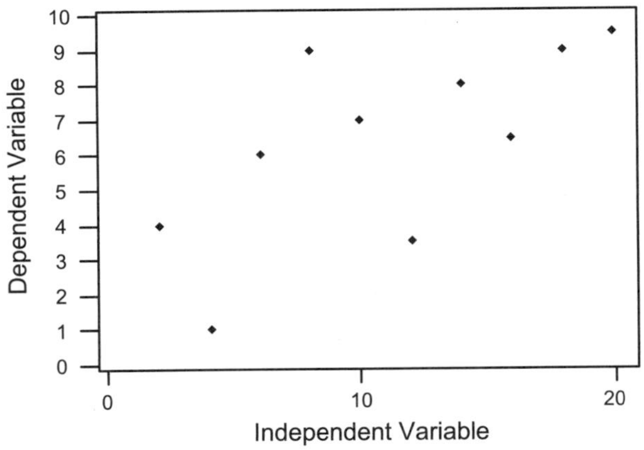

FIGURE 3.1: Example of a scatterplot.

3.4.3 Box-and-Whisker Plots

These plots illustrate the first, second, and third quartiles as well as the minimum and maximum values of the data collected. The second quartile (Q2) is also known as the median of the data. This quantity, as defined later in this text, is the middle data point or sample value when the samples are listed in descending order. The first quartile (Q1) can be thought of as the median value of the samples that fall below the second quartile. Similarly, the third quartile (Q3) can be thought of as the median value of the samples that fall above the second quartile. Box-and-whisker plots are useful in that they highlight whether there is skew to the data or any unusual outliers in the samples (Figure 3.3).

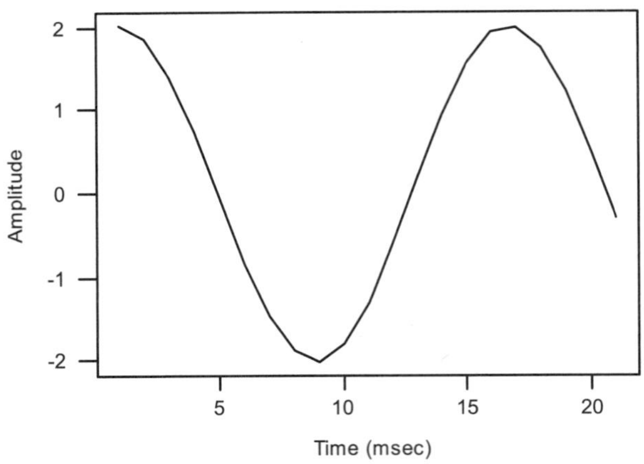

FIGURE 3.2: Example of a time series plot. The amplitude of the samples is plotted as a function of time.

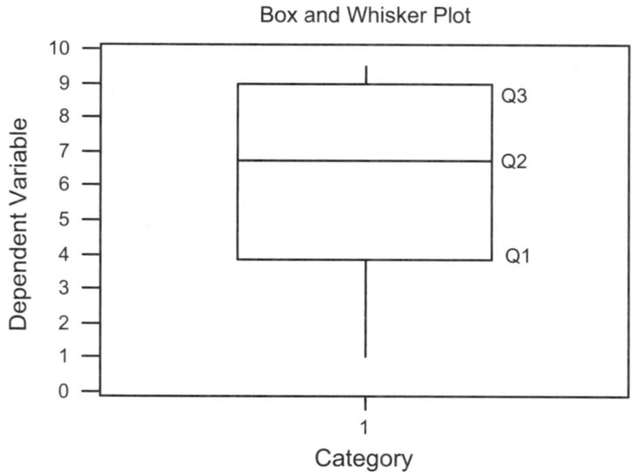

FIGURE 3.3: Illustration of a box-and-whisker plot for the data set listed. The first (Q1), second (Q2), and third (Q3) quartiles are shown. In addition, the whiskers extend to the minimum and maximum values of the sample set.

3.4.4 Histogram

The histogram is defined as a frequency distribution. Given N samples or measurements, x_i, which range from X_{min} to X_{max}, the samples are grouped into nonoverlapping intervals (bins), usually of equal width (Figure 3.4). Typically, the number of bins is on the order of 7–14, depending on the nature of the data. In addition, we typically expect to have at least three samples per bin [7]. Sturgess' rule [6] may also be used to estimate the number of bins and is given by

$$k = 1 + 3.3 \log(n).$$

where k is the number of bins and n is the number of samples.

Each bin of the histogram has a lower boundary, upper boundary, and midpoint. The histogram is constructed by plotting the number of samples in each bin. Figure 3.5 illustrates a histogram for 1000 samples drawn from a normal distribution with mean (μ) = 0 and standard deviation (σ) = 1.0. On the horizontal axis, we have the sample value, and on the vertical axis, we have the number of occurrences of samples that fall within a bin.

Two measures that we find useful in describing a histogram are the absolute frequency and relative frequency in one or more bins. These quantities are defined as

a) f_i = absolute frequency in ith bin;
b) f_i/n = relative frequency in ith bin, where n is the total number of samples being summarized in the histogram.

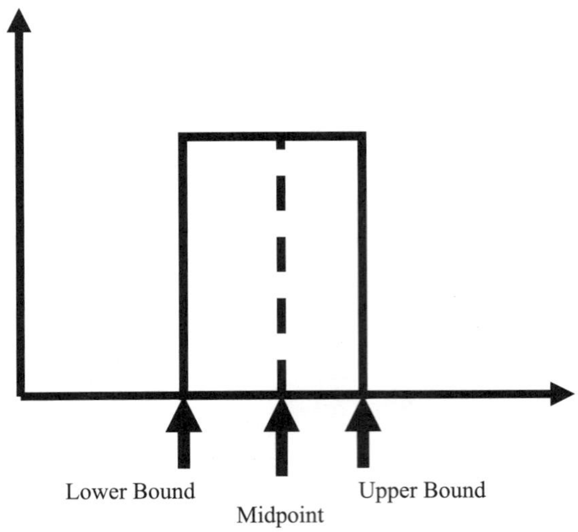

FIGURE 3.4: One bin of a histogram plot. The bin is defined by a lower bound, a midpoint, and an upper bound.

A number of algorithms used by biomedical instruments for diagnosing or detecting abnormalities in biological function make use of the histogram of collected data and the associated relative frequencies of selected bins [8]. Often times, normal and abnormal physiological functions (breath sounds, heart rate variability, frequency content of electrophysiological signals) may be differentiated by comparing the relative frequencies in targeted bins of the histograms of data representing these biological processes.

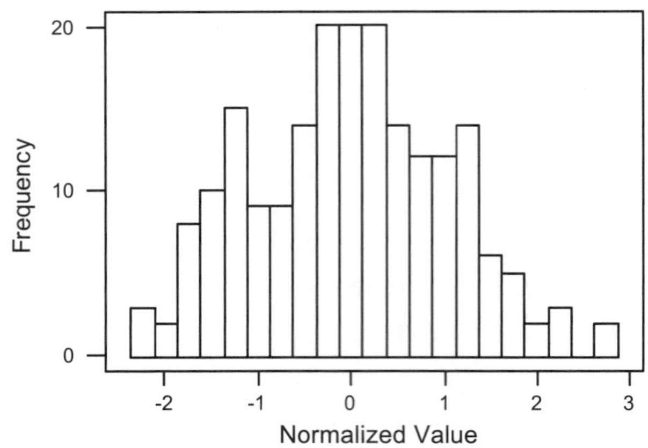

FIGURE 3.5: Example of a histogram plot. The value of the measure or sample is plotted on the horizontal axis, whereas the frequency of occurrence of that measure or sample is plotted along the vertical axis.

The histogram can exhibit several shapes. The shapes, illustrated in Figure 3.6, are referred to as symmetric, skewed, or bimodal.

A skewed histogram may be attributed to the following [9]:

1. mechanisms of interest that generate the data (e.g., the physiological mechanisms that determine the beat-to-beat intervals in the heart);
2. an artifact of the measurement process or a shift in the underlying mechanism over time (e.g., there may be time-varying changes in a manufacturing process that lead to a change in the statistics of the manufacturing process over time);
3. a mixing of populations from which samples are drawn (this is typically the source of a bimodal histogram).

The histogram is important because it serves as a rough estimate of the true probability density function or probability distribution of the underlying random process from which the samples are being collected.

The probability density function or probability distribution is a function that quantifies the probability of a random event, x, occurring. When the underlying random event is discrete in nature, we refer to the probability density function as the probability mass function [10]. In either case, the function describes the probabilistic nature of the underlying random variable or event and allows us to predict the probability of observing a specific outcome, x (represented by the random variable), of an experiment. The cumulative distribution function is simply the sum of the probabilities for a group of outcomes, where the outcome is less than or equal to some value, x.

Let us consider a random variable for which the probability density function is well defined (for most real-world phenomenon, such a probability model is not known.) The random variable is the outcome of a single toss of a dice. Given a single fair dice with six sides, the probability of rolling a six on the throw of a dice is 1 of 6. In fact, the probability of throwing a one is also 1 of 6. If we consider all possible outcomes of the toss of a dice and plot the probability of observing any one of those six outcomes in a single toss, we would have a plot such as that shown in Figure 3.7.

This plot shows the probability density or probability mass function for the toss of a dice. This type of probability model is known as a uniform distribution because each outcome has the exact same probability of occurring (1/6 in this case).

For the toss of a dice, we know the true probability distribution. However, for most real-world random processes, especially biological processes, we do not know what the true probability density or mass function looks like. As a consequence, we have to use the histogram, created from a small sample, to try to estimate the best probability distribution or probability model to describe the real-world phenomenon. If we return to the example of the toss of a dice, we can actually toss the dice a number of times and see how close the histogram, obtained from experimental data, matches

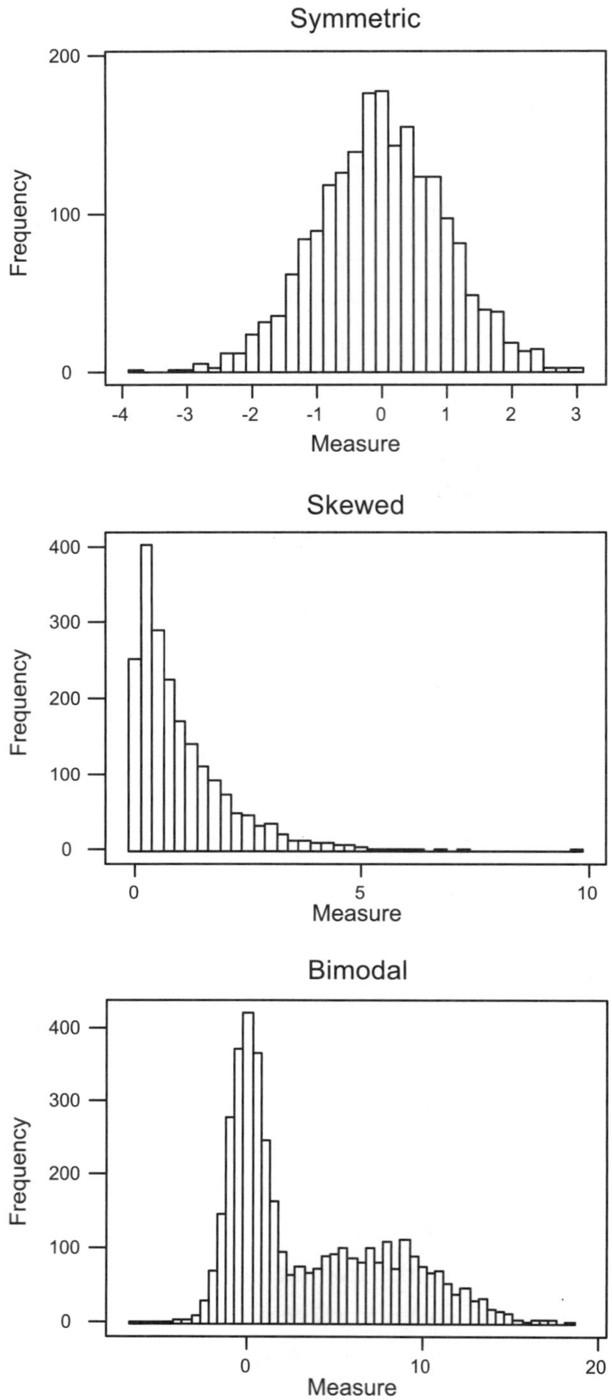

FIGURE 3.6: Examples of a symmetric (top), skewed (middle), and bimodal (bottom) histogram. In each case, 2000 sampled were drawn from the underlying populations.

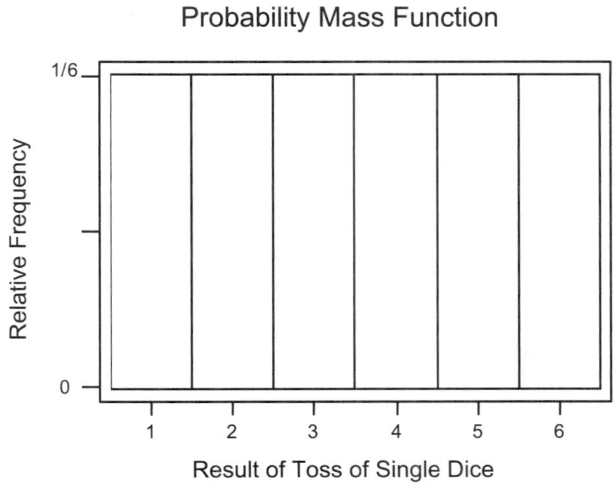

FIGURE 3.7: The probability density function for a discrete random variable (probability mass function). In this case, the random variable is the value of a toss of a single dice. Note that each of the six possible outcomes has a probability of occurrence of 1 of 6. This probability density function is also known as a uniform probability distribution.

the true probability mass function for the ideal six-sided dice. Figure 3.8 illustrates the histograms for the outcomes of 50 and 1000 tosses of a single dice. Note that even with 50 tosses or samples, it is difficult to determine what the true probability distribution might look like. However, as we approach 1000 samples, the histogram is approaching the true probability mass function (the uniform distribution) for the toss of a dice. But, there is still some variability from bin to bin that does not look as uniform as the ideal probability distribution illustrated in Figure 3.7. The message to take away from this illustration is that most biomedical research reports the outcomes of a small number of samples. It is clear from the dice example that the statistics of the underlying random process are very difficult to discern from a small sample, yet most biomedical research relies on data from small samples.

3.5 GENERAL APPROACH TO STATISTICAL ANALYSIS

We have now collected our data and looked at some graphical summaries of the data. Now we will use numerical summary, also known as statistics, to try to describe the nature of the underlying population or process from which we have taken our samples. From these descriptive statistics, we assume a probability model or probability distribution for the underlying population or process and then select the appropriate statistical tests to test hypotheses or make decisions. It is important to note that the conclusions one may draw from a statistical test depends on how well the assumed probability model fits the underlying population or process.

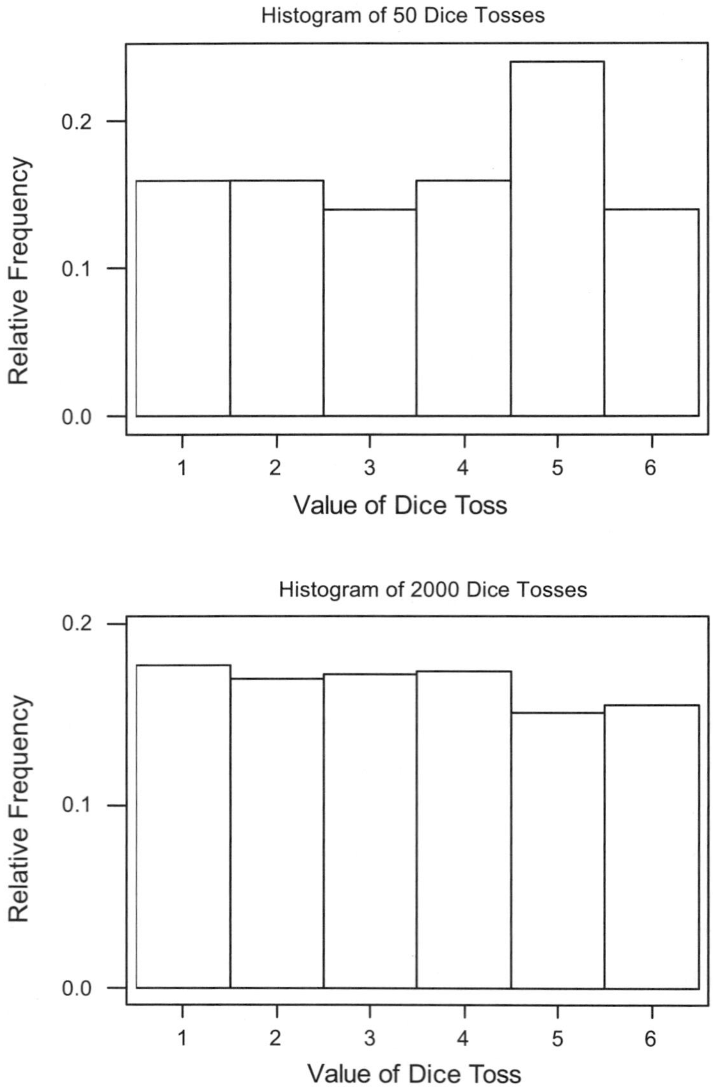

FIGURE 3.8: Histograms representing the outcomes of experiments in which a single dice is tossed 50 (top) and 2000 times (lower), respectively. Note that as the sample size increases, the histogram approaches the true probability distribution illustrated in Figure 3.7.

As stated in the Introduction, biomedical engineers are trying to make decisions about populations or processes to which they have limited access. Thus, they design experiments and collect samples that they think will fairly represent the underlying population or process. Regardless of what type of statistical analysis will result from the investigation or study, all statistical analysis should follow the same general approach:

1. Measure a limited number of representative samples from a larger population.
2. Estimate the true statistics of larger population from the sample statistics.

Some important concepts need to be addressed here. The first concept is somewhat obvious. It is often impossible or impractical to take measurements or observations from an entire population. Thus, the biomedical engineer will typically select a smaller, more practical sample that represents the underlying population and the extent of variability in the larger population. For example, we cannot possibly measure the resting body temperature of every person on earth to get an estimate of normal body temperature and normal range. We are interested in knowing what the normal body temperature is, on average, of a healthy human being and the normal range of resting temperatures as well as the likelihood or probability of measuring a specific body temperature under healthy, resting conditions. In trying to determine the characteristics or underlying probability model for body temperature for healthy, resting individuals, the researcher will select, at random, a sample of healthy, resting individuals and measure their individual resting body temperatures with a thermometer. The researchers will have to consider the composition and size of the sample population to adequately represent the variability in the overall population. The researcher will have to define what characterizes a normal, healthy individual, such as age, size, race, sex, and other traits. If a researcher were to collect body temperature data from such a sample of 3000 individuals, he or she may plot a histogram of temperatures measured from the 3000 subjects and end up with the following histogram (Figure 3.9). The researcher may also calculate some basic descriptive statistics for the 3000 samples, such as sample average (mean), median, and standard deviation.

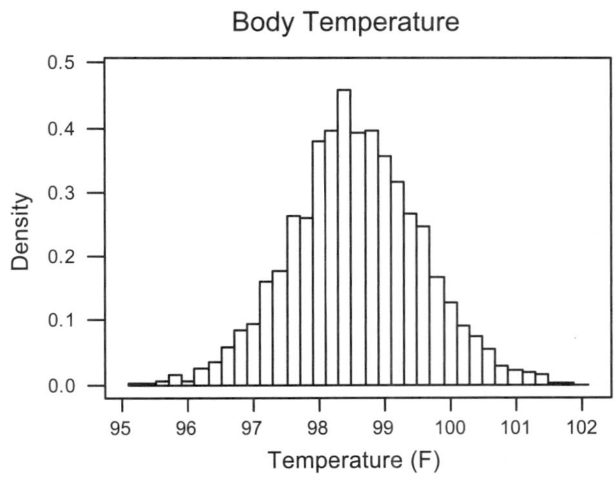

FIGURE 3.9: Histogram for 2000 internal body temperatures collected from a normally distributed population.

Once the researcher has estimated the sample statistics from the sample population, he or she will try to draw conclusions about the larger (true) population. The most important question to ask when reviewing the statistics and conclusions drawn from the sample population is how well the sample population represents the larger, underlying population.

Once the data have been collected, we use some basic descriptive statistics to summarize the data. These basic descriptive statistics include the following general measures: central tendency, variability, and correlation.

3.6 DESCRIPTIVE STATISTICS

There are a number of descriptive statistics that help us to picture the distribution of the underlying population. In other words, our ultimate goal is to assume an underlying probability model for the population and then select the statistical analyses that are appropriate for that probability model.

When we try to draw conclusions about the larger underlying population or process from our smaller sample of data, we assume that the underlying model for any sample, "event," or measure (the outcome of the experiment) is as follows:

$$X = \mu \pm \text{individual differences} \pm \text{situational factors} \pm \text{unknown variables},$$

where X is our measure or sample value and is influenced by μ, which is the true population mean; individual differences such as genetics, training, motivation, and physical condition; situation factors, such as environmental factors; and unknown variables such as unidentified/nonquantified factors that behave in an unpredictable fashion from moment to moment.

In other words, when we make a measurement or observation, the measured value represents or is influenced by not only the statistics of the underlying population, such as the population mean, but factors such as biological variability from individual to individual, environmental factors (time, temperature, humidity, lighting, drugs, etc.), and random factors that cannot be predicted exactly from moment to moment. All of these factors will give rise to a histogram for the sample data, which may or may not reflect the true probability density function of the underlying population. If we have done a good job with our experimental design and collected a sufficient number of samples, the histogram and descriptive statistics for the sample population should closely reflect the true probability density function and descriptive statistics for the true or underlying population. If this is the case, then we can make conclusions about the larger population from the smaller sample population. If the sample population does not reflect variability of the true population, then the conclusions we draw from statistical analysis of the sample data may be of little value.

There are a number of probability models that are useful for describing biological and manufacturing processes. These include the normal, Poisson, exponential, and gamma distributions [10]. In this book, we will focus on populations that follow a normal distribution because this is the most frequently encountered probability distribution used in describing populations. Moreover, the most frequently used methods of statistical analysis assume that the data are well modeled by a normal ("bell-curve") distribution. It is important to note that many biological processes are not well modeled by a normal distribution (such as heart rate variability), and the statistics associated with the normal distribution are not appropriate for such processes. In such cases, nonparametric statistics, which do not assume a specific type of distribution for the data, may serve the researcher better in understanding processes and making decisions. However, using the normal distribution and its associated statistics are often adequate given the central limit theorem, which simply states that the sum of random processes with arbitrary distributions will result in a random variable with a normal distribution. One can assume that most biological phenomena result from a sum of random processes.

3.6.1 Measures of Central Tendency

There are several measures that reflect the central tendency or concentration of a sample population: sample mean (arithmetic average), sample median, and sample mode.

The sample mean may be estimated from a group of samples, x_i, where i is sample number, using the formula below.

Given n data points, $x_1, x_2, \ldots, x_n$:

$$\overline{x} = \frac{1}{n}\sum_{i=1}^{n} x_i .$$

In practice, we typically do not know the true mean, μ, of the underlying population, instead we try to estimate true mean, μ, of the larger population. As the sample size becomes large, the sample mean, $\overline{x}$, should approach the true mean, μ, assuming that the statistics of the underlying population or process do not change over time or space.

One of the problems with using the sample mean to represent the central tendency of a population is that the sample mean is susceptible to outliers. This can be problematic and often deceiving when reporting the average of a population that is heavily skewed. For example, when reporting income for a group of new college graduates for which one is an NBA player who has just signed a multimillion-dollar contract, the estimated mean income will be much greater than what most graduates earns. The same misrepresentation is often evident when reporting mean value for homes in a specific geographic region where a few homes valued on the order of a million can hide the fact that several hundred other homes are valued at less than $200,000.

Another useful measure for summarizing the central tendency of a population is the sample median. The median value of a group of observations or samples, x_i, is the middle observation when samples, x_i, are listed in descending order.

For example, if we have the following values for tidal volume of the lung:

2, 1.5, 1.3, 1.8, 2.2, 2.5, 1.4, 1.3,

we can find the median value by first ordering the data in descending order:

2.5, 2.2, 2.0, 1.8, 1.5, 1.4, 1.3, 1.3,

and then we cross of values on each end until we reach a middle value:

~~2.5~~, ~~2.2~~, ~~2.0~~, 1.8, 1.5, ~~1.4~~, ~~1.3~~, ~~1.3~~.

In this case, there are two middle values; thus, the median is the average of those two values, which is 1.65.

Note that if the number of samples, n, is odd, the median will be the middle observation. If the sample size, n, is even, then the median equals the average of two middle observations. Compared with the sample mean, the sample median is less susceptible to outliers. It ignores the skew in a group of samples or in the probability density function of the underlying population. In general, to fairly represent the central tendency of a collection of samples or the underlying population, we use the following rule of thumb:

1. If the sample histogram or probability density function of the underlying population is symmetric, use mean as a central measure. For such populations, the mean and median are about equal, and the mean estimate makes use of all the data.
2. If the sample histogram or probability density function of the underlying population is skewed, median is a more fair measure of center of distribution.

Another measure of central tendency is mode, which is simply the most frequent observation in a collection of samples. In the tidal volume example given above, 1.3 is the most frequently occurring sample value. Mode is not used as frequently as mean or median in representing central tendency.

3.6.2 Measures of Variability

Measures of central tendency alone are insufficient for representing the statistics of a population or process. In fact, it is usually the variability in the population that makes things interesting and leads to uncertainty in decision making. The variability from subject to subject, especially in physiological function, is what makes finding fool-proof diagnosis and treatment often so difficult. What works for one person often fails for another, and, it is not the mean or median that picks up on those subject-to-subject differences, but rather the variability, which is reflected in differences in the probability models underlying those different populations.

When summarizing the variability of a population or process, we typically ask, "How far from the center (sample mean) do the samples (data) lie?" To answer this question, we typically use the following estimates that represent the spread of the sample data: interquartile ranges, sample variance, and sample standard deviation.

The interquartile range is the difference between the first and third quartiles of the sample data. For sampled data, the median is also known as the second quartile, Q2. Given Q2, we can find the first quartile, Q1, by simply taking the median value of those samples that lie below the second quartile. We can find the third quartile, Q3, by taking the median value of those samples that lie above the second quartile. As an illustration, we have the following samples:

$$1, 3, 3, 2, 5, 1, 1, 4, 3, 2.$$

If we list these samples in descending order,

$$5, 4, 3, 3, 3, 2, 2, 1, 1, 1,$$

the median value and second quartile for these samples is 2.5. The first quartile, Q1, can be found by taking the median of the following samples,

$$2.5, 2, 2, 1, 1, 1,$$

which is 1.5. In addition, the third quartile, Q3, may be found by taking the median value of the following samples:

$$5, 4, 3, 3, 3, 2.5,$$

which is 3. Thus, the interquartile range, Q3 - Q1 = 3 - 1.5 = 2.

Sample variance, s^2, is defined as the "average distance of data from the mean" and the formula for estimating s^2 from a collection of samples, x_i, is

$$s^2 = \frac{1}{n-1}\sum_{i-1}^{n}(x_i - \bar{x})^2.$$

Sample standard deviation, s, which is more commonly referred to in describing the variability of the data is

$$s = \sqrt{s^2} \text{ (same units as original samples).}$$

It is important to note that for normal distributions (symmetrical histograms), sample mean and sample deviation are the only parameters needed to describe the statistics of the underlying phenomenon. Thus, if one were to compare two or more normally distributed populations, one only need to test the equivalence of the means and variances of those populations.

• • • •

CHAPTER 4

Assuming a Probability Model From the Sample Data

Now that we have collected the data, graphed the histogram, estimated measures of central tendency and variability, such as mean, median, and standard deviation, we are ready to assume a probability model for the underlying population or process from which we have obtained samples. At this point, we will make a rough assumption using simple measures of mean, median, standard deviation and the histogram. But it is important to note that there are more rigorous tests, such as the χ^2 test for normality [7] to determine whether a particular probability model is appropriate to assume from a collection of sample data.

Once we have assumed an appropriate probability model, we may select the appropriate statistical tests that will allow us to test hypotheses and draw conclusions with some level of confidence. The probability model will dictate what level of confidence we have when accepting or rejecting a hypothesis.

There are two fundamental questions that we are trying to address when assuming a probability model for our underlying population:

1. How confident are we that the sample statistics are representative of the entire population?
2. Are the differences in the statistics between two populations significant, resulting from factors other than chance alone?

To declare any level of "confidence" in making statistical inference, we need a mathematical model that describes the probability that any data value might occur. These models are called probability distributions.

There are a number of probability models that are frequently assumed to describe biological processes. For example, when describing heart rate variability, the probability of observing a specific time interval between consecutive heartbeats might be described by an exponential distribution [1, 8]. Figure 3.6 in Chapter 3 illustrates a histogram for samples drawn from an exponential distribution.

Note that this distribution is highly skewed to the right. For R-R intervals, such a probability function makes sense physiologically because the individual heart cells have a refractory period that prevents them from contracting in less that a minimum time interval. Yet, a very prolonged time interval may occur between beats, giving rise to some long time intervals that occur infrequently.

The most frequently assumed probability model for most scientific and engineering applications is the normal or Gaussian distribution. This distribution is illustrated by the solid black line in Figure 4.1 and often referred to as the bell curve because it looks like a musical bell.

The equation that gives the probability, $f(x)$, of observing a specific value of x from the underlying *normal* population is

$$f(x) = \frac{1}{\sigma\sqrt{2\pi}} e^{-\frac{1}{2}\left(\frac{x-\mu}{\sigma}\right)^2}, \quad \infty < x < \infty$$

where μ is the true mean of the underlying population or process and σ is the standard deviation of the same population or process. A graph of this equation is given illustrated by the solid, smooth curve in Figure 4.1. The area under the curve equals one.

Note that the normal distribution is

1. a symmetric, bell-shaped curve completely described by its mean, μ, and standard deviation, σ.
2. by changing μ and σ, we stretch and slide the distribution.

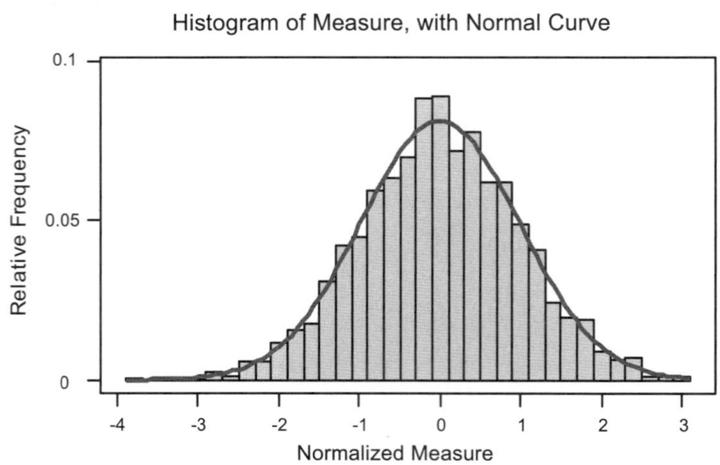

FIGURE 4.1: A histogram of 1000 samples drawn from a normal distribution is illustrated. Superimposed on the histogram is the ideal normal curve representing the normal probability distribution function.

Figure 4.1 also illustrates a histogram that is obtained when we randomly select 1000 samples from a population that is normally distributed and has a mean of 0 and a variance of 1. It is important to recognize that as we increase the sample size n, the histogram approaches the ideal normal distribution shown with the solid, smooth line. But, at small sample sizes, the histogram may look very different from the normal curve. Thus, from small sample sizes, it may be difficult to determine if the assumed model is appropriate for the underlying population or process, and any statistical tests that we perform may not allow us to test hypotheses and draw conclusions with any real level of confidence.

We can perform linear operations on our normally distributed random variable, x, to produce another normally distributed random variable, y. These operations include multiplication of x by a constant and addition of a constant (offset) to x. Figure 4.2 illustrates histograms for samples drawn from each of populations x and y. We note that the distribution for y is shifted (the mean is now equal to 5) and the variance has increased with respect to x.

One test that we may use to determine how well a normal probability model fits our data is to count how many samples fall within ±1 and ±2 standard deviations of the mean. If the data and underlying population or process is well modeled by a normal distribution, 68% of the samples should lie within ±1 standard deviation from the mean and 95% of the samples should lie within

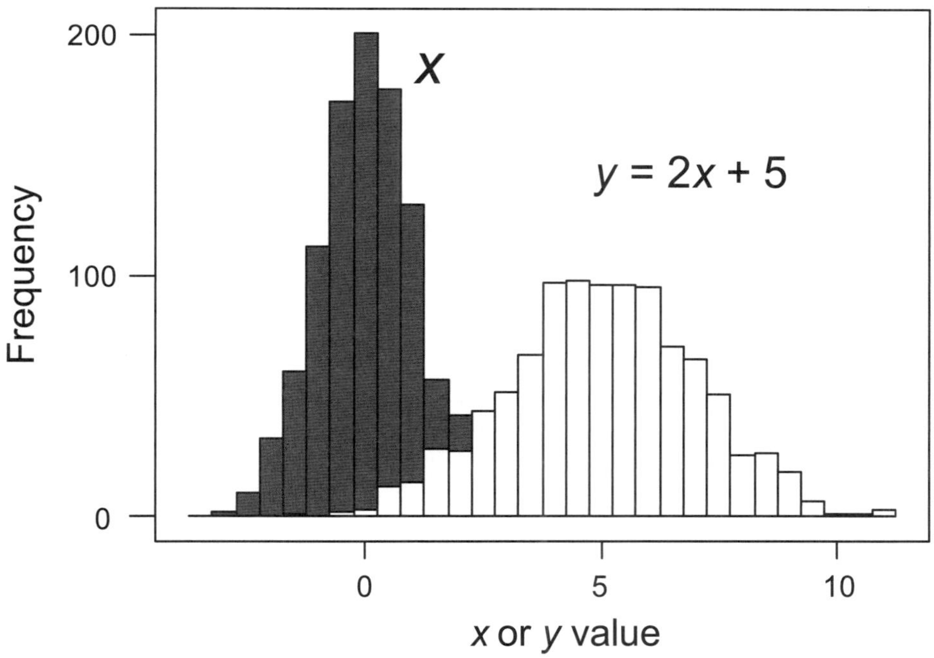

FIGURE 4.2: Histograms are shown for samples drawn from populations x and y, where y is simply a linear function of x. Note that the mean and variance of y differ from x, yet both are normal distributions.

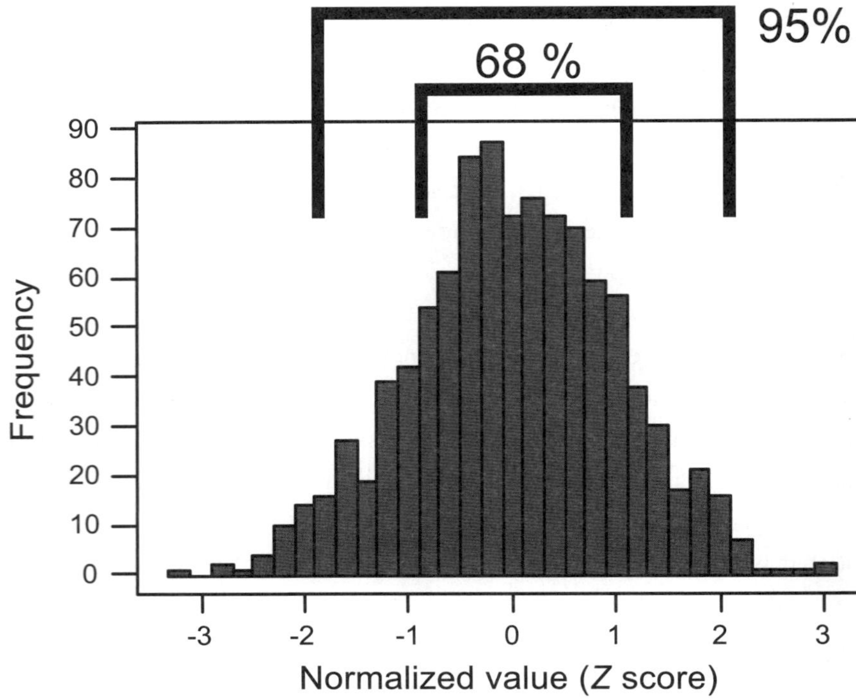

FIGURE 4.3: Histogram for samples drawn from a normally distributed population. For a normal distribution, 68% of the samples should lie within ±1 standard deviation from the mean (0 in this case) and 95% of the samples should lie within ±2 standard deviations (1.96 to be precise) of the mean.

±2 standard deviations from the mean. These percentages are illustrated in Figure 4.3. It is important to remember these few numbers, because we will frequently use this 95% interval when drawing conclusions from our statistical analysis.

Another means for determining how well our sampled data, x, represent a normal distribution is the estimate Pearson's coefficient of skew (PCS) [5]. The coefficient of skew is given by

$$\text{PCS} = \frac{3\left|\bar{x} - x_{\text{median}}\right|}{s}.$$

If the PCS > 0.5, we assume that our samples were not drawn from a normally distributed population.

When we collect data, the data are typically collected in many different types of physical units (volts, celsius, newtons, centimeters, grams, etc.). For us to use tables that have been developed for probability models, we need to normalize the data so that the normalized data will have a mean of 0 and a standard deviation of 1. Such a normal distribution is called a standard normal distribution and is illustrated in Figure 4.1.

The standard normal distribution has a bell-shaped, symmetric distribution with $\mu = 0$ and $\sigma = 1$.

To convert normally distributed data to the standard normal value, we use the following formulas,

$$z = (x - \mu)/\sigma \quad \text{or} \quad z = (x - \bar{x})/s,$$

depending on if we know the true mean, μ, and standard deviation, α, or we only have the sample estimates, $\bar{x}$ or s.

For any individual sample or data point, x_i, from a sample with mean, $\bar{x}$, and standard deviation, s, we can determine its z score from the following formula:

$$z_i = \frac{x_i - \bar{x}}{s}.$$

For an individual sample, the z score is a "normalized" or "standardized" value. We can use this value with our equations for probability density function or our standardized probability tables [3] to determine the probability of observing such a sample value from the underlying population.

The z score can also be thought of as a measure of the distance of the individual sample, x_i, from the sample average, $\bar{x}$, in units of standard deviation. For example, if a sample point, x_i has a z score of $z_i = 2$, it means that the data point, x_i, is 2 standard deviations from the sample mean.

We use normalized z scores instead of the original data when performing statistical analysis because the tables for the normalized data are already worked out and available in most statistics texts or statistical software packages. In addition, by using normalized values, we need not worry about the absolute amplitude of the data or the units used to measure the data.

4.1 THE STANDARD NORMAL DISTRIBUTION

The standard normal distribution is illustrated in Table 4.1.

The z table associated with this figure provides table entries that give the probability that $z \leq a$, which equals the area under the normal curve to the left of $z = a$. If our data come from a normal distribution, the table tells us the probability or "chance" of our sample value or experimental outcomes having a value less than or equal to a.

Thus, we can take any sample and compute its z score as described above and then use the z table to find the probability of observing a z value that is less than or equal to some normalized value, a. For example, the probability of observing a z value that is less than or equal to 1.96 is 97.5%. Thus, the probability of observing a z value greater than 1.96 is 2.5%. In addition, because of symmetry in the distribution, we know that the probability of observing a z value greater than −1.96 is also 97.5%, and the probability of observing a z value less than or equal to −1.96 is 2.5%. Finally,

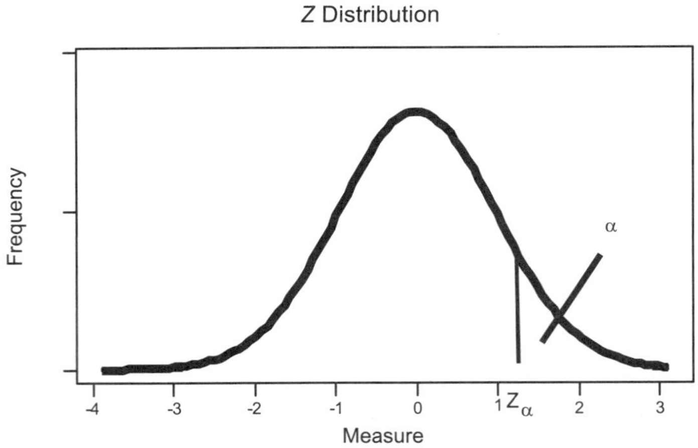

Area to left of z_α equals the $\Pr(z < z_\alpha) = 1 - \alpha$; thus, the area in the tail to the right of z_α equals α.

TABLE 4.1: Standard z distribution function: areas under standardized normal density function

z	0.00	0.01	0.02	0.03	0.04	0.05	0.06	0.07	0.08	0.09
0.0	0.5000	0.5040	0.5080	0.5120	0.5160	0.5199	0.5239	0.15279	0.5319	0.5359
0.1	0.5398	0.5438	0.5478	0.5517	0.5557	0.5596	0.5636	0.5675	0.5714	0.5753
0.2	0.5793	0.5832	0.5871	0.5910	0.5948	0.5987	0.6026	0.6064	0.6103	0.6141
…										
1.7	0.9554	0.9564	0.9573	0.9582	0.9591	0.9599	0.9608	0.9616	0.9625	0.9633
1.8	0.9641	0.9649	0.9656	0.9664	0.9671	0.9678	0.9686	0.9693	0.9699	0.9706
1.9	0.9713	0.9719	0.9726	0.9732	0.9738	0.9744	0.9750	0.9756	0.9761	0.9767
2.0	0.9772	0.9778	0.9783	0.9788	0.9793	0.9798	0.9803	0.9808	0.9812	0.9817
…										
2.4	0.9918	0.9920	0.9922	0.9925	0.9927	0.9929	0.9931	0.9932	0.9934	0.9936
2.5	0.9938	0.9940	0.9941	0.9943	0.9945	0.9946	0.9948	0.9949	0.9951	0.9952
2.6	0.9953	0.9955	0.9956	0.9957	0.9959	0.9960	0.9961	0.9962	0.9963	0.9964
…										
3.0	0.9987	0.9987	0.9987	0.9988	0.9988	0.9989	0.9989	0.9989	0.9990	0.9990

Commonly used z values:

AREA IN RIGHT TAIL, α	Z_α
0.10	1.282
0.05	1.645
0.025	1.96
0.010	2.326
0.005	2.576

the probability of observing a z value between −1.96 and 1.96 is 95%. The reader should study the z table and associated graph of the z distribution to verify that the probabilities (or areas under the probability density function) described above are correct.

Often, we need to determine the probability that an experimental outcome falls between two values or that the outcome is greater than some value a or less or greater than some value b. To find these areas, we can use the following important formulas, where Pr is the probability:

$$\Pr(a \leq z \leq b) = \Pr(z \leq b) - \Pr(z \leq a)$$
$$= \text{area between } z = a \text{ and } z = b.$$

$$\Pr(z \leq a) = 1 - \Pr(z < a)$$

$$= \text{area to right of } z = a$$
$$= \text{area in the right “tail”}.$$

Thus, for any observation or measurement, x, from any normal distribution:

$$\Pr(a \leq x \leq b) = \Pr\left(\frac{a-\mu}{\sigma} \leq z \leq \frac{b-\mu}{\sigma}\right),$$

where μ is the mean of normal distribution and σ is the standard deviation of normal distribution.

In other words, we need to normalize or find the z values for each of our parameters, a and b, to find the area under the standard normal curve (z distribution) that represents the expression on the left side of the above equation.

Example 4.1 The mean intake of fat for males 6 to 9 years old is 28 g, with a standard deviation of 13.2 g. Assume that the intake is normally distributed. Steve's intake is 42 g and Ben's intake is 25 g.

1. What is the proportion of area between Steve's daily intake and Ben's daily intake?
2. If we were to randomly select a male between the ages of 6 and 9 years, what is the probability that his fat intake would be 50 g or more?

Solution: x = fat intake

1. The problem may be stated as: what is $\Pr(25 \leq x \leq 42)$?
   Assuming a normal distribution, we convert to z scores:
   What is $\Pr(((25 - 28)/13.2) < z < ((42 - 28)/13.2)))$?
   $= \Pr(-0.227 \leq z \leq 1.06) = \Pr(z \leq 1.06) - \Pr(z \leq -0.227)$ (using formula for $\Pr(a \leq z \leq b)$)
   $= \Pr(z \leq 1.06) - [1 - \Pr(z \leq 0.227)] = 0.8554 - [1 - 0.5910] = 0.4464$ or 44.6% of area under the z curve.
2. The problem may be stated as, "What is $\Pr(x > 50)$?"
   Normalizing to z score, what is $\Pr(z > (50 - 28)/13.2)$?
   $= \Pr(z > 1.67)$
   $= 1 - \Pr(z \leq 1.67) = 1 - 0.9525 = 0.0475$, or 4.75% of the area under the z curve.

Example 4.2 Suppose that the specifications on the range of displacement for a limb indentor are 0.5 ± 0.001 mm. If these displacements are normally distributed, with mean = 0.47 and standard deviation = 0.002, what percentage of indentors are within specifications?

Solution: x = displacement.

The problem may be stated as, "What is $\Pr(0.499 \leq x \leq 0.501)$?"

Using z scores, $\Pr(0.499 \leq x \leq 0.501) = \Pr((0.499 - 0.47)/0.002 \leq z \leq (0.501 - 0.47/0.002))$

$= \Pr(14.5 \leq z \leq 15.5) = \Pr(z \leq 15.5) - \Pr(z \leq 14.5) = 1 - 1 = 0$

It is useful to note that if the distribution of the underlying population and the associated sample data are not normal (i.e. skewed), transformations may often be used to make the data normal, and the statistics covered in this text may then be used to perform statistical analysis on the transformed data. These transformations on the raw data include logs, square root, and reciprocal.

4.2 THE NORMAL DISTRIBUTION AND SAMPLE MEAN

All statistical analysis follows the same general procedure:

1. Assume an underlying distribution for the data and associated parameters (e.g., the sample mean).

2. Scale the data or parameter to a "standard" distribution.
3. Estimate confidence intervals using a standard table for the assumed distribution. (The question we ask is, "What is the probability of observing the experimental outcome by chance alone?")
4. Perform hypothesis test (e.g., Student's *t* test).

We are beginning with and focusing most of this text on the normal distribution or probability model because of its prevalence in the biomedical field and something called the central limit theorem. One of the most basic statistical tests we perform is a comparison of the means from two or more populations. The sample mean is it itself an estimate made from a finite number of samples. Thus, the sample mean, $\bar{x}$, is itself a random variable that is modeled with a normal distribution [4].

Is this model for $\bar{x}$ legitimate? The answer is yes, for large samples, because of the central limit theorem, which states [4, 10]:

If the individual data points or samples (each sample is a random variable), x_i, *come from any arbitrary probability distribution, the sum (and hence, average) of those data points is normally distributed as the sample size,* n, *becomes large.*

Thus, even if each sample, such as the toss of a dice, comes from a nonnormal distribution (e.g., a uniform distribution, such as the toss of a dice), the sum of those individual samples (such as the sum we use to estimate the sample mean, $\bar{x}$) will have a normal distribution. One can easily assume that many of the biological or physiological processes that we measure are the sum of a number of random processes with various probability distributions; thus, the assumption that our samples come from a normal distribution is not unreasonable.

4.3 CONFIDENCE INTERVAL FOR THE SAMPLE MEAN

Every sample statistic is in itself a random variable with some sort of probability distribution. Thus, when we use samples to estimate the true statistics of a population (which in practice are usually not known and not obtainable), we want to have some level of confidence that our sample estimates are close to the true population statistics or are representative of the underlying population or process.

In estimating a confidence interval for the sample mean, we are asking the question: "How close is our sample mean (estimated from a finite number of samples) to the true mean of the population?"

To assign a level of confidence to our statistical estimates or statistical conclusions, we need to first assume a probability distribution or model for our samples and underlying population and then we need to estimate a confidence interval using the assumed distribution.

The simplest confidence interval that we can begin with regarding our descriptive statistics is a confidence interval for the sample mean, $\bar{x}$. Our question is how close is the sample mean to the true population or process mean, μ?

Before we can answer this, we need to assume an underlying probability model for the sample mean, $\bar{x}$, and true mean, μ. As stated earlier, it may be shown that for a large samples size, the sample mean, $\bar{x}$, is well modeled by a normal distribution or probability model. Thus, we will use this model when estimating a confidence interval for our sample mean, $\bar{x}$.

Thus, $\bar{x}$ is estimated from the sample. We then ask, how close is $\bar{x}$ (sample mean) to the true population mean, μ?

It may be shown that if we took many groups of n samples and estimated $\bar{x}$ for each group,

1. the average or sample mean of $\bar{x} = \mu$, and
2. the standard deviation of $\bar{x} = s/\sqrt{n}$.

Thus, as our sample size, n, gets large, the distribution for $\bar{x}$ approaches a normal distribution.

For large n, $\bar{x}$ follows a normal distribution, and the z score for $\bar{x}$ may be used to estimate the following:

$$\Pr(a \le \bar{x} \le b) = \Pr\left(\frac{a-\mu}{\sigma/\sqrt{n}} \le z \le \frac{b-\mu}{\sigma/\sqrt{n}}\right).$$

This expression assumes a large n and that we know σ.

Now we look at the case where we might have a large n, but we do not know σ. In such cases, we replace σ with s to get the following expression:

$$\Pr(a \le \bar{x} \le b) = \Pr\left(\frac{a-\mu}{s/\sqrt{n}} \le z \le \frac{b-\mu}{s/\sqrt{n}}\right),$$

where $s/\sqrt{n}$ is called the sample standard error and represents the standard deviation for $\bar{x}$.

Let us assume now for large n, we want to estimate the 95% confidence interval for $\bar{x}$. We first scale the sample mean, $\bar{x}$, to a z value (because the central limit theorem says that $\bar{x}$ is normally distributed)

$$z = \frac{\bar{x} - \mu}{s/\sqrt{n}}.$$

We recall that 95% of z values fall between ±1.96 (approximately 2σ) of the mean, and for the z distribution,

$$\Pr(-1.96 \le z \le 1.96) = 0.95.$$

Substituting for z,

$$z = \frac{\bar{x} - \mu}{s / \sqrt{n}}.$$

we get

$$0.95 = \Pr\left(-1.96 \le \frac{\bar{x} - \mu}{s / \sqrt{n}} \le 1.96\right).$$

If we use the following notation in terms of the sample standard error:

$$\mathrm{SE}(\bar{x}) = \frac{s}{\sqrt{n}}.$$

Rearranging terms for the expression above, we note that the probability that μ lies between ±1.96 (or 2) standard deviations of $\bar{x}$ is 95%:

$$0.95 = \Pr\left(\bar{x} - 1.96\mathrm{SE}(\bar{x}) \le \mu \le \bar{x} + 1.96\mathrm{SE}(\bar{x}).\right)$$

Note that 1.96 is referred to as $z_{\alpha/2}$. This z value is the value of z for which the area in the right tail of the normal distribution is $\alpha/2$. If we were to estimate the 99% confidence interval, we would substitute $z_{0.01/2}$, which is 2.576, into the 1.96 position above.

Thus, For large n and any confidence level, $1 - \alpha$, the $1 - \alpha$ confidence interval for the true population mean, μ, is given by:

$$\mu = \bar{x} \pm z_{\alpha/2} SE(\bar{x}).$$

This means that there is a $(1 - \alpha)$percent probability that the true mean lies within the above interval centered about $\bar{x}$.

Example 4.3 Estimate of confidence intervals

Problem: Given a collection of data with, $\bar{x}$ = 505 and s = 100. If the number of samples was 1000, what is the 95% confidence interval for the population mean, μ?

Solution: If we assume a large sample size, we may use the z distribution to estimate the confidence interval for the sample mean using the following equation:

$$\mu = \bar{x} \pm z_{\alpha/2}\,\mathrm{SE}(\bar{x}).$$

We plug in the following values:

$$\bar{x} = 505;$$
$$\mathrm{SE}(\bar{x}) = s/\sqrt{n} = 100/\sqrt{1000}.$$

For 95% confidence interval, $\alpha = 0.05$.

Using a z table to locate $z(0.05/2)$, we find that the value of z that gives an area of 0.025 in the right tail is 1.96

Plugging in $\bar{x}$, $\mathrm{SE}(\bar{x})$, and $z(\alpha/2)$ into the estimate for the confidence interval above, we find that the 95% confidence interval for μ = [498.80, 511.20].

Note that if we wanted to estimate the 99% confidence interval, we would simply use a different z value, z (0.01/2) in the same equation. The z value associated with an area of 0.005 in the right tail is 2.576. If we use this z value, we estimate a confidence interval for μ of [496.86, 515.14]. We note that the confidence interval has widened as we increased our confidence level.

4.4 THE *t* DISTRIBUTION

For small samples, $\bar{x}$ is no longer normally distributed. Therefore, we use Student's t distribution to estimate the true statistics of the population. The t distribution, as illustrated in Table 4.2 looks like a z distribution but with slower taper at the tails and flatter central region.

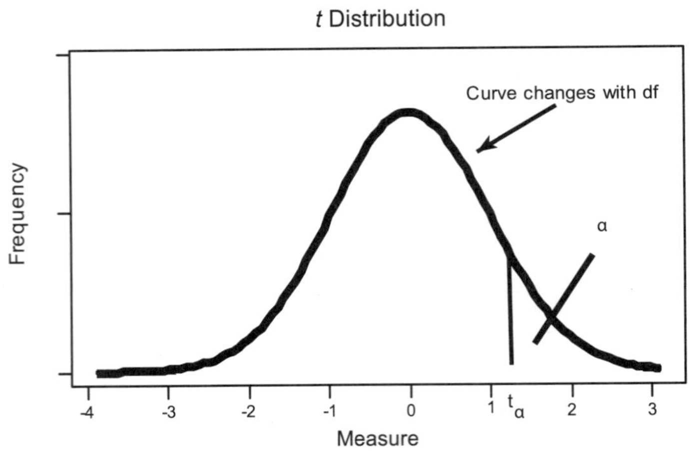

Table entry = $t(\alpha; df)$, where α is the area in the tail to the right of $t(\alpha; df)$ and df is degrees of freedom.

TABLE 4.2: Percentage points for Student's t distribution

df	α = area to right of $t(\alpha; df)$				
	0.10	0.05	0.025	0.01	0.005
1	3.078	6.314	12.706	31.821	63.657
2	1.886	2.920	4.303	6.965	9.925
3	1.638	2.353	3.182	4.541	5.841
4	1.533	2.132	2.776	3.747	4.604
5	1.476	2.015	2.571	3.365	4.032
...					
10	1.372	1.812	2.228	2.764	3.169
11	1.363	1.796	2.201	2.718	3.106
12	1.356	1.782	2.179	2.681	3.055
13	1.350	1.771	2.160	2.650	3.012
14	1.345	1.761	2.145	2.624	2.977
15	1.341	1.753	2.131	2.602	2.947
...					
30	1.310	1.697	2.042	2.457	2.750
40	1.303	1.684	2.021	2.423	2.704
60	1.296	1.671	2.000	2.390	2.660
120	1.289	1.658	1.980	2.358	2.617
∞	1.282	1.645	1.960	2.326	2.576
z_α(large sample)	1.282	1.645	1.960	2.326	2.576

We use t distribution, with slower tapered tails, because with so few samples, we have less certainty about our underlying distribution (probability model).

Now our normalized value for $\bar{x}$ is given by

$$\frac{\bar{x} - \mu}{s / \sqrt{n}},$$

which is known to have a t distribution rather than the z distribution that we have discussed thus far. The t distribution was first invented by W.S. Gosset [4, 11], a chemist who worked for a brewery. Gosset decided to publish his t distribution under the alias of Student. Hence, we often refer to this distribution as Student's t distribution.

The t distribution is symmetric like the z distribution and generally has a bell shape. But the amount of spread of the distribution to the tails, or the width of the bell, depends on the sample size, n. Unlike the z distribution, which assumes an infinite sample size, the t distribution changes shape with sample size. The result is that the confidence intervals estimated with t values are more "spread out" than for z distribution, especially for small sample sizes, because with such samples sizes, we are penalized for not having sufficient samples to represent the extent of variability of the underlying population or process. Thus, when we are estimating confidence intervals for the sample mean, $\bar{x}$, we do not have as much confidence in our estimate. Thus, the interval widens to reflect this decreased certainty with smaller sample sizes. In the next section, we estimate the confidence interval for the same example given previously, but using the t distribution instead of the z distribution.

4.5 CONFIDENCE INTERVAL USING t DISTRIBUTION

Like the z tables, there are t tables where the values of t that are associated with different areas under the probability curve are already calculated and may be used for statistical analysis without the need to recalculate the t values. The difference between the z table and t table is that now the t values are a function of the samples size or degrees of freedom. Table 4.2 gives a few lines of the t table from [3].

To use the table, one simply looks for the intersection of degrees of freedom, df, (related to sample size) and α value that one desires in the right-hand tail. The intersection provides the t value for which the area under the t curve in the right tail is α. In other words, the probability that t will be less than or equal to a specific entry in the table is $1 - \alpha$. For a specific sample size, n, the degrees of freedom, $df = n - 1$.

Now we simply substitute t for z to find our confidence intervals. So, the confidence interval for the sample mean, $\bar{x}$, using the t distribution now becomes

$$\mu = \bar{x} \pm t\left(\frac{\alpha}{2}; n-1\right) \mathrm{SE}(\bar{x})$$

where μ is the true mean of the underlying population or process from which we are drawing samples, $\mathrm{SE}(\bar{x})$ is the standard error of the underlying population or process, t is the t value for which there is an area of $\alpha/2$ in the right tail, and n is the sample size.

Example 4.4 Confidence interval using *t* distribution

Problem: We consider the same example used previously for estimating confidence intervals using z values. In this case, the sample size is small ($n = 20$), so we now use a t distribution.

Solution: Now, our estimate for confidence interval for $\mu = \bar{x} \pm t\left(\frac{\alpha}{2}; n-1\right) \mathrm{SE}(\bar{x})$.

Again, we plug in the following values:

$$\bar{x} = 505;$$
$$\mathrm{SE}(\bar{x}) = s/\sqrt{n} = 100/\sqrt{20}.$$

For 95% confidence interval, $\alpha = 0.05$.

Using a t table to locate $t(0.05/2, 20 - 1)$, we find that for 19 *df*, the value of t that gives an area of 0.025 in the right tail is 2.093.

Plugging in $\bar{x}$, $\mathrm{SE}(\bar{x})$, and $t(\alpha, n - 1)$ into the estimate for the confidence interval above, we find that the 95% confidence interval for $\mu = [458.20, 551.80]$ We note that this confidence interval is widened compared with that previously estimated using the z values. This is expected because the t distribution is wider than the z distribution at small sample sizes, reflecting the fact that we have less confidence in our estimate of $\bar{x}$ and, hence, μ when the sample size is small.

Confidence intervals may be estimated for most descriptive statistics, such as the sample mean, sample variance, and even the line of best fit determined through linear regression [3]. As noted above, the confidence interval reflects the variability in the parameter estimate and is influenced by the sample size, the variability of the population, and the level on confidence that we desire. The greater the desired confidence, the wider the interval. Likewise the greater the variability in the samples, the wider the confidence interval.

Example 4.5 Review of probability concepts

Problem: Assembly times were measured for a sample of 15 glucose infusion pumps. The mean time to assemble a glucose infusion pump was 15.8 minutes, with a standard deviation of 2.4 minutes. Assuming a relatively symmetric distribution for assembly times,

1. What percentage of infusion pumps require more than 17 seconds to assemble?
2. What is the 99% confidence interval for the true mean assembly time (μ)?
3. What is the 99% confidence interval for mean assembly time if the sample size is 2500?

Solution:

1. x = assembly time
   What is the Pr (x > 17)?
   Pr (x > 17) = Pr (z > (17 – 15.8)/2.4) = Pr (z > 0.5) = 1 – Pr ($z \leq 0.5$) = 1 – 0.6915) = 0.3085, or 38.05% of the infusion pumps.
2. $= \bar{x} \pm t(\alpha/2, n - 1)\mathrm{SE}(x)$
   $= 15.8 \pm t((0.01)/2; 15 - 1)\ (2.4\sqrt{15})$
   $= 15.8 \pm t(0.005, 14)\ (0.6196)$
   $= 15.8 \pm 2.977\ (0.6196)$
   $= [13.96, 17.64]$
   Because the samples size of 2500 is now large, we use a z value for estimating the confidence interval, the $\mu = \bar{x} \pm z\ (\alpha/2)\mathrm{SE}(x)$
3. $= \bar{x} \pm z\ (\alpha/2)\mathrm{SE}(x)$
   $= 15.8 \pm z\ (0.01/2)\ (2.4\sqrt{2500})$
   $= 15.8 \pm 2.576\ (0.048)$
   $= [15.68, 15.92]$

• • • •

CHAPTER 5

Statistical Inference

Now that we have collected our data, estimated some basic descriptive statistics and assumed a probability model for the underlying population or process, we are prepared to perform some statistical analysis that will allow us to compare populations and test hypotheses. For this text, we are only going to discuss statistical analysis for which the normal distribution is a good probability model for the underlying population(s). If the data or samples suggest that the underlying population or process is not well modeled by a normal distribution, then we will need to resort to other types of statistical analysis, such a nonparametric techniques [6], which do not assume an underlying probability model. Some of these tests include the Wilcoxon rank sum test, Mann–Whitney *U* test, Kruskal–Wallis test, and the runs test [6, 12]. More advanced texts provide details for administering these nonparametric tests. Keep in mind that if the statistical analysis presented in this text is used for populations or processes that are not normally distributed, the results may be of little value, and the investigator may miss important findings from the data.

Assuming our data represents a normally distributed population or process, we are now prepared to perform a variety of statistical tests that allow us to test hypotheses about the equality of means and variances across two or more populations. Remember that for normal distributions, only the mean and standard deviation are required to completely characterize the probabilistic nature of the population or process. Thus, if we are to compare two normally distributed populations, we need only compare the means and variances of the populations. If the populations or processes are not normally distributed, there may be other parameters, such as skew and kurtosis, which differentiate two or more populations or processes.

5.1 COMPARISON OF POPULATION MEANS

One of the most fundamental questions asked by scientists or engineers performing experiments is whether two populations, methods, or treatments are really different in central tendency. More specifically, are the two population means, reflected in the sample means collected under two different experimental conditions, significantly different? Or, is the observed difference between two means simply because of chance alone?

Some examples or questions asked by biomedical engineers that require comparing two population means include the following:

1. Is one chemotherapy drug more effective than a second chemotherapy drug in reducing the size of a cancerous tumor?
2. Is there a difference in postural stability between young and elderly populations?
3. Is there a difference in bone density for women who are premenopausal versus those who are postmenopausal?
4. Is titanium a stronger material for bone implants than steel?
5. For MRI, does one type of pulse sequence perform better than another in detecting white matter tracts in the brain?
6. Do drug-eluting intravascular stents prevent restenosis more effectively than noncoated stents?

In answering these type of questions, biomedical engineers often collect samples from two different groups of subjects or from one group of subjects but under two different conditions. For a specific measure that represents the samples, a sample mean is typically estimated for each group or under each of two conditions. Comparing the means from two populations reflected in two sets of data is probably the most reported statistical analysis in the scientific and engineering literature and may be accomplished using something known as a *t* test.

5.1.1 The *t* Test

When performing the *t* test, we are asking the question, "Are the means of two populations really different?" or "Would we see the observed differences simply because of random chance?" The two populations are reflected in data that have been collected under two different conditions. These conditions may include two different treatments or processes.

To address this question, we use one of the following two tests, depending on whether the two data sets are independent or dependent on one another:

1. Unpaired *t* test for two sets of independent data
2. Paired *t* test for two sets of dependent data

Before we describe each type of *t* test, we need to discuss the notion of hypothesis testing.

5.1.1.1 Hypothesis Testing

Whenever we perform statistical analysis, we are testing some hypothesis. In fact, before we even collect our data, we formulate a hypothesis and then carefully design our experiment and collect and analyze the data to test the hypothesis. The outcome of the statistical test will allow us, if the assumed probability model is valid, to accept or reject the hypothesis and do so with some level of confidence.

There are basically two forms of hypotheses that we test when performing statistical analysis. There is the null hypothesis and the alternative hypothesis.

The null hypothesis, denoted as H_0, is expressed as follows for the *t* test comparing two population means, μ_1 and μ_2:

$$H_0: \mu_1 = \mu_2.$$

The alternative hypothesis, denoted as H_1, is expressed as one of the following for the *t* test comparing two population means, μ_1 and μ_2:

$$H_1: \mu_1 \neq \mu_2 \qquad \text{(two-tailed t test)},$$

$$H_1: \mu_1 < \mu_2 \qquad \text{(one tailed t test)},$$

or

$$H_1: \mu_1 > \mu_2 \qquad \text{(one-tailed t test)}.$$

If the H_1 is of the first form, where we do not know in advance of the data collection which mean will be greater than the other, we will perform a two-tailed test, which simply means that the level of significance for which we accept or reject the null hypothesis will be double that of the case in which we predict in advance of the experiment that one mean will be lower than the second based on the physiology or the engineering process or other conditions affecting experimental outcome. Another way to express this is that with a one-tailed test, we will have greater confidence in rejecting or accepting the null hypothesis than with the two-tailed condition.

5.1.1.2 Applying the *t* Test

Now that we have stated our hypotheses, we are prepared to perform the *t* test. Given two populations with n_1 and n_2 samples, we may compare the two population means using a *t* test. It is important to remember the underlying assumptions made in using the *t* test to compare two population means:

1. The underlying distributions for both populations are normal.
2. The variances of the two populations are approximately equal: $s_1^2 = s_2^2$.

These are big assumptions we make when n_1 and n_2 are small. If these assumptions are poor for the data being analyzed, we need to find different statistics to compare the two populations.

Given two sets of sampled data, x_i and y_i, the means for the two populations or processes reflected in the sampled data can be compared using the following t statistic:

5.1.1.3 Unpaired *t* Test

The unpaired t statistic may be estimated using:

$$T = \frac{\bar{x} - \bar{y}}{\sqrt{\frac{(n_1 - 1)S_x^{\,2} + (n_2 - 1)S_y^{\,2}}{n_1 + n_2 - 2}\left(\frac{1}{n_1} + \frac{1}{n_2}\right)}},$$

where n_1 is the number of x_i observations, n_2 is the number of y_i observations, S_x^2 is the sample variance of x_i, S_y^2 = sample variance of y_i, $\bar{x}$ is the sample average for x_i, and $\bar{y}$ is the sample average for y_i.

Once the T statistic has been computed, we can compare our estimated T value to t values given in a table for the t distribution. More specifically, if we want to reject the null hypothesis with $1 - \alpha$ level of confidence, then we need to determine whether our estimated T value is greater than the t value entry in the t table associated with a significance level of α (one-sided t test) or $\alpha/2$ (two-sided t test). In other words, we need to know the t value from the t distribution for which the area under the t curve to the right of t is α. Our estimated T must be greater than this t to reject the null hypothesis with $1 - \alpha$ level of confidence.

Thus, we compare our T value to the t distribution table entry for

$$t(\alpha, n_1 + n_2 - 2) \qquad \text{(one-sided)}$$

or

$$t(\alpha/2, n_1 + n_2 - 2) \qquad \text{(two-sided)},$$

where α is the level of significance (equal to 1 − level of confidence), and n_1 and n_2 are the number of samples from each of the two populations being compared.

Note that α is the level of significance at which we want to accept or reject our hypothesis. For most research, we reject the null hypothesis when $\alpha \leq 0.05$. This corresponds to a confidence level of 95%.

For example, if we want to reject our null hypothesis with a confidence level of 95%, then for a one-sided t test, our estimated T must be greater than $t(0.05, n_1 + n_2 - 2)$ to reject H_0 and accept H_1 with 95% confidence or a significance level of 0.05 or less. Remember that our confidence = $(1 - 0.05) \times 100\%$. This test was for H_1: $\mu_1 < \mu_2$ or $\mu_1 > \mu_2$ (one-sided). If our H_1 was H_1:

$\mu_1 \neq \mu_2$, then our measured T must be greater than $t(0.05/2, n_1 + n_2 - 2)$ to reject the null hypothesis with the same 95% confidence.

Thus, to test the following alternative hypotheses:

1. For H_1: $\mu_1 \neq \mu_2$: Use two-tail t test. For ($\alpha < 0.05$), the estimated $T > t(0.025, n_1 + n_2 - 2)$ to reject H_0 with 95% confidence.
2. For H_1: $\mu_1 < \mu_2$ or H_1: $\mu_1 < \mu_2$: Use one-tail t test. For ($\alpha < 0.05$), estimated $T > t(0.05, n_1 + n_2 - 2)$ to reject H_0 with 95% confidence.

α is also referred to as the type I error. For the t test, α is the probability of observing a measured T value greater than the table entry, $t(\alpha; df)$ if the true means of two underlying populations, x and y, were actually equal. In other words, there is no significant ($\alpha < 0.05$) difference in the two population means, but the sampled data analysis led us to conclude that there is a difference and thus reject the null hypothesis. This is referred to as a type I error.

Example 5.1 An example of a biomedical engineering challenge where we might use the t test to improve the function of medical devices is in the development of implantable defibrillators for the detection and treatment of abnormal heart rhythms [13]. Implantable defibrillators are small electronic devices that are placed in the chest and have thin wires that are placed in the chambers of the heart. These wires have electrodes that detect the small electrical current traversing the heart muscle. Under normal conditions, these currents follow a very orderly pattern of conduction through the heart muscle. An example of an electrogram, or record of electrical activity that is sensed by an electrode, under normal conditions is given in Figure 5.1. When the heart does not function normally and enters a state of fibrillation (Figure 5.1), whereby the heart no longer contracts normally or pumps blood to the rest of the body, the device should shock the heart with a large electrical current from the device in an attempt to convert the heart rhythm back to normal. For a number of reasons, it is important that the device accurately detect the onset of life-threatening arrhythmias, such as fibrillation, and administer an appropriate shock. To administer a shock, the device must use some sort of signal processing algorithms to automatically determine that the electrogram is abnormal and characteristic of fibrillation.

One algorithm that is used in most devices for differentiating normal heart rhythms from fibrillation is a rate algorithm. This algorithm is basically an amplitude threshold crossing algorithm whereby the device determines how often the electrogram exceeds an amplitude threshold in a specified period of time and then estimates a rate from the detected threshold crossings. Figure 5.2 illustrates how this algorithm works.

Before such an algorithm was put into implantable defibrillators, researchers and developers had to demonstrate whether the rate estimated by such a device truly differed between normal and

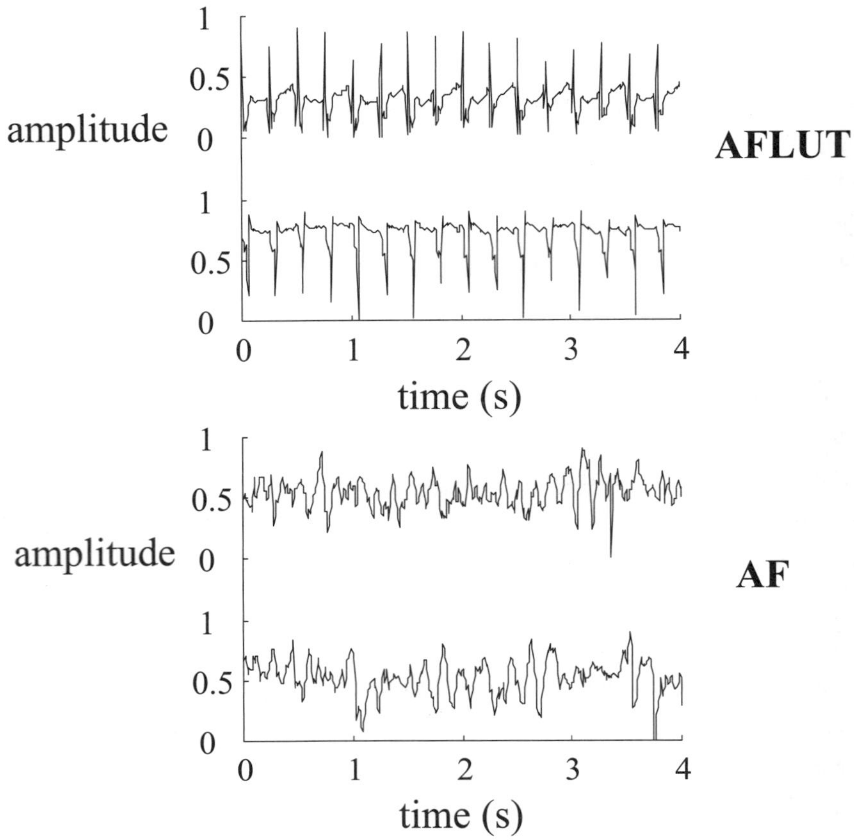

FIGURE 5.1: Electrogram recordings measured from electrodes placed inside the left atrium of the heart. For each rhythm, there are two electrogram recordings taken from two different locations in the atrium: atrial flutter (AFLUT) and atrial fibrillation (AF).

fibrillatory rhythms. It was important to have little overlap between the normal and fibrillatory rhythms so that a rate threshold could be established, whereby rates that exceeded the threshold would lead to the device administering a shock. For this algorithm to work, there must be a significant difference in rates between normal and fibrillatory rhythms and little overlap that would lead to shocks being administered inappropriately and causing great pain or risk the induction of fibrillation. Overlap in rates between normal and fibrillatory rhythms could also result in the device missing detection of fibrillation because of low rates.

To determine if rate is a good algorithm for detecting fibrillatory rhythms, investigators might actually record electrogram signals from actual patients who have demonstrated normal or fibrillatory rhythms and use the rate algorithm to estimate a rate and then compare the mean rates for normal rhythms against mean rates for fibrillatory rhythms. For example, let us assume

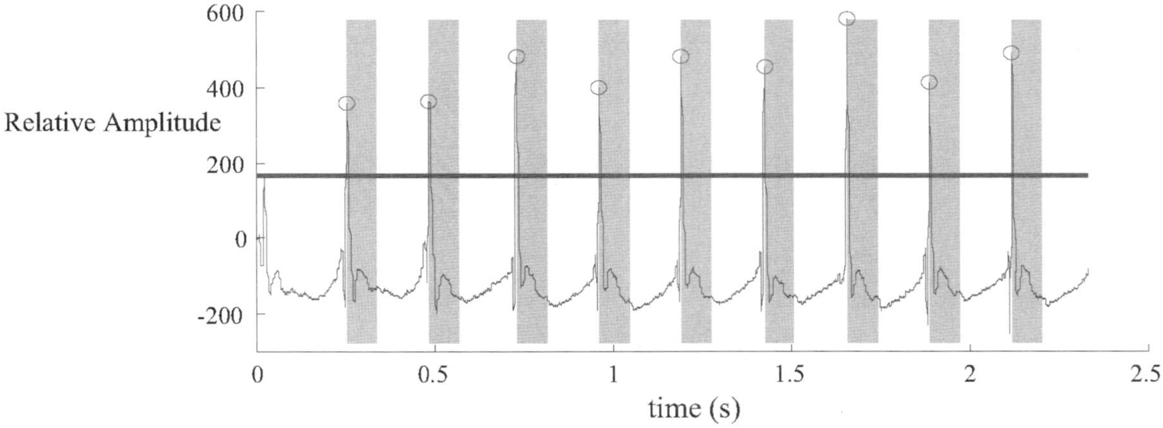

FIGURE 5.2: A heart rate is estimated from an electrogram using an amplitude threshold algorithm. Whenever the amplitude of the electrogram signal (solid waveform) exceeds an amplitude threshold (solid gray horizontal line) within a specific time interval (shaded vertical bars), an "event" is detected. A rate is calculated by counting the number of events in a certain time period.

investigators collected 15-second electrogram recordings for examples of fibrillatory (n_1 = 10) and nonfibrillatory rhythms (n_2 = 11) in 21 patients. We note that the fibrillatory data were collected from different subjects than the normal data. Thus, we do not have blocking in this experimental design.

A rate was estimated, using the device algorithm, for each individual 15-second recording. Figure 5.3 shows a box-and-whisker plot for each of the two data sets.

The descriptive statistics for the two data sets are given by the following table:

	NONFIBRILLATORY ELECTROGRAMS (N_1 = 11)	FIBRILLATORY ELECTROGRAMS (N_2 = 10)
Mean	96.82	239.0
Standard deviation	22.25	55.3

There are several things to note from the plots and descriptive statistics. First, the sample size is small; thus, it is likely that the variability of electrogram recordings from normal and fibrillatory rhythms is not adequately represented. Moreover, the data are skewed and do not appear to be well modeled by a normal distribution. Finally, the variability in rates appears to be much greater for fibrillatory rhythms than for normal rhythms. Thus, some of the assumptions we make regarding

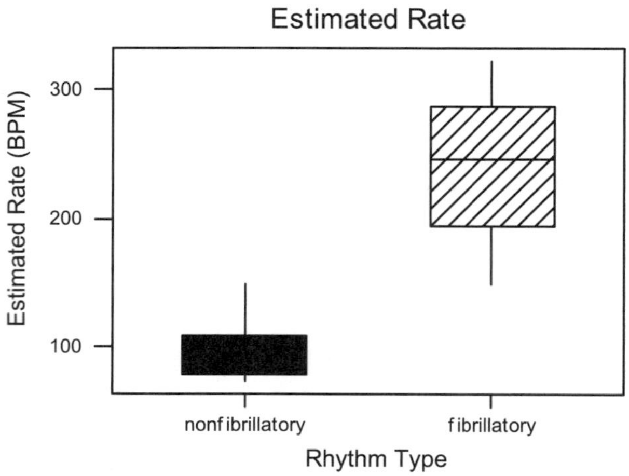

FIGURE 5.3: Box-and-whisker plot for rate data estimated from samples of nonfibrillatory and fibrillatory heart rhythms.

the normality of the populations and the equality of variances are probably violated in applying the *t* test. However for the purpose of illustration, we will perform a *t* test using the unpaired *t* statistic.

To find an estimated *T* statistic from our sample data, we plug in the means, variances, and sample sizes into the equation for the unpaired *T* statistic. We turn the crank and find that for this example, the estimated *t* value is 7.59. If we look at our *t* tables for [11 + 10 − 2] degrees of freedom, we can reject H_0 at $\alpha < 0.005$ because our estimated *T* value exceeds the *t* table entries for $\alpha = 0.05$ ($t = 1.729$), $\alpha = 0.01$ ($t = 2.539$), and $\alpha = 0.005$ ($t = 2.861$).

Thus, we reject the null hypothesis with $(1 - 0.005) \times 100\%$ confidence and state that mean heart rate for fibrillatory rhythms is greater than mean heart rate for normal rhythms. Thus, the rate algorithm should perform fairly well in differentiating normal from fibrillatory rhythms. However, we have only tested the population means. As one may see in Figure 5.3, there is some overlap in individual samples between normal and fibrillatory rhythms. Thus, we might expect the device to make errors in administering shock inappropriately when the heart is in normal but accelerated rhythms (as in exercise), or the device may fail to shock when the heart is fibrillating but at a slow rate or with low amplitude.

In applying the *t* test, it is important to note that you can never prove that two means are equal. The statistics can only show that a specific test cannot find a difference in the population means, and, not finding a difference is not the same as proving equality. The null or default hypothesis is that there is no difference in the means, regardless of what the true difference is between the two means. Not finding a difference with the collected data and appropriate statistical test does not mean that the means are proved equal. Thus, we do not accept the null hypothesis with a level of significance. We simply accept the null hypothesis and do not accept with a confidence or significance level. We only assign a level of confidence when we reject the null hypothesis.

5.1.1.4 Paired *t* Test

In the previous example, we used an unpaired *t* test because the two data sets came from two different, unrelated, groups of patients. The problem with such an experimental design is that differences in the two patient populations may lead to differences in mean heart rate that have nothing to do with the actual heart rhythm but rather differences in the size or age of the hearts between the two groups of patients or some other difference between patient groups. A better way to conduct the previous study is to collect our normal and fibrillatory heart data from one group of patients. A single defibrillator will only need to differentiate normal and fibrillatory rhythms within a single patient. By blocking on subjects we can eliminate the intersubject variability in electrogram characteristics that may plague the rate algorithm from separating populations. It may be more reasonable to assume that rates for normal and fibrillatory rhythms differ more within a patient than across patients. In other words, for each patient, we collect an electrogram during normal heart function and fibrillation. In such experimental design, we would compare the means of the two data sets using a paired *t* test.

We use the paired *t* test when there is a natural pairing between each observation in data set, *X*, with each observation in data set, *Y*. For example, we might have the following scenarios that warrant a paired *t* test:

1. Collect blood pressure measurements from 8 patients before and after exercise.
2. Collect both MR and CT data from the same 20 patients to compare quality of blood vessel images.
3. Collect computing time for an image processing algorithm before and after a change is made in the software.

In such cases, the *X* and *Y* data sets are no longer independent. In the first example above, because we are collecting the before and after data from the same experimental units, the physiology and biological environment that impacts blood pressure before exercise in each patients also affects blood pressure after exercise. Thus, there are a number of variables that impact blood pressure that we cannot directly control but whose effects on blood pressure (beside the exercise effect) can be controlled by using the same experimental units (human subjects in this case) for each data set. In such cases, the experimental units serve as their own control. This is typically the preferred experimental design for comparing means.

For the paired t test, we again have a null hypothesis and an alternative hypothesis as stated above for the unpaired t test. However, in a paired t test, we use a t test on the difference, $W_i = X_i - Y_i$, between the paired data points from each of the two populations.

We now calculate the paired T statistic: (n = number of pairs)

$$T = \frac{\overline{W}}{S_w / \sqrt{n}},$$

where $\overline{W}$ is the average difference of the differences, W_i, and S_w is the standard deviation of the differences, W_i.

As for the unpaired t test, we now have an estimated T value that we can compare with the t values in the table for the t distribution to determine if the estimated T value lies in the extreme values (greater than 95%) of the t distribution.

To reject the null hypothesis at a significance level of α (confidence level of $1 - \alpha$), our estimated T value must be greater than $t(\alpha, n-1)$, where n is the number of pairs of data.

If the estimated T value exceeds $t(\alpha, n-1)$, we reject H_0 and accept H_1 at a significance level of α or a confidence level of $1 - \alpha$.

Once again, if H_1: $\mu_1 < \mu_2$ or H_1: $\mu_1 < \mu_2$, we perform a one-sided test where our T statistic must be greater than $t(\alpha, n-1)$ to reject the null hypothesis at a level of α. If the null hypothesis that we begin with before the experiment is H_1: ($\mu_1 \neq \mu_2$), then we perform a two-sided test, and the T statistic must be greater than $t(\alpha/2, n-1)$ to reject the null hypothesis at a significance level of α.

5.1.1.5 Example of a Biomedical Engineering Challenge

In relation to abnormal heart rhythms discussed previously, antiarrhythmic drugs may be used to slow or terminate an abnormal rhythm, such as fibrillation. For example, a drug such as procainamide may be used to terminate atrial fibrillation. The exact mechanism whereby the drug leads to termination of the rhythm is not exactly known, but it is thought that the drug changes the refractory period and conduction velocity of the heart cells [8]. Biomedical engineers often use signal processing on the electrical signals generated by the heart as an indirect means for studying the underlying physiology. More specifically, engineers may use spectral analysis or Fourier analysis to look at changes in the spectral content of the electrical signal, such as the electrogram, over time. Such changes may tell us something about the underlying electrophysiology.

For example, spectral analysis has been used to look at changes in the frequency spectrum of atrial fibrillation with drug administration. In one such study [8], biomedical engineers were interested in looking at changes in median frequency of atrial electrograms after drug administration. Figure 5.4 shows an example of the frequency spectrum for atrial fibrillation and the location

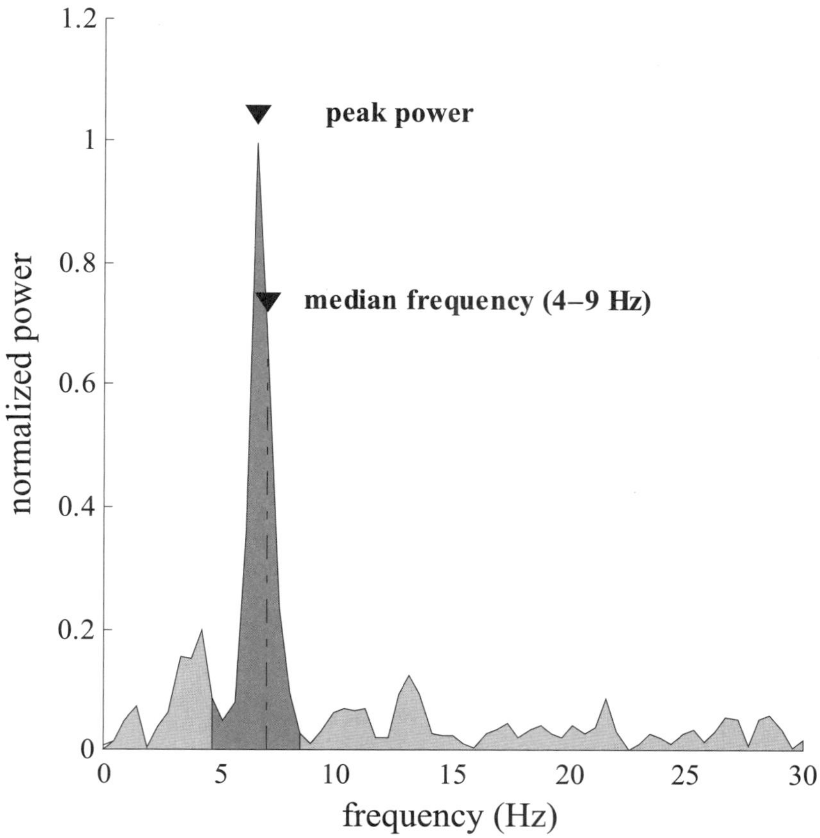

FIGURE 5.4: Frequency spectrum for an example of atrial fibrillation. Median frequency is defined as that frequency that divides the area under the power curve in the 4- to 9-Hz band in half.

of the median frequency, which is the frequency that divides the power of the spectrum (area under the spectral curve between 4 and 9 Hz) in half. One question posed by the investigators is whether median frequency decreases after administration of a drug such procainamide, which is thought to slow the electrical activity of the heart cells.

Thus, an experiment was conducted to determine whether there was a significant difference in mean median frequency between fibrillation before drug administration and fibrillation after drug administration. Electrograms were collected in the right atrium in 11 patients before and after the drug was administered. Fifteen-second recordings were evaluated for the frequency spectrum, and the median frequency in the 4- to 9-Hz frequency band was estimated before and after drug administration.

Figure 5.5 illustrates the summary statistics for median frequency before and after drug administration. The question is whether there was a significant decrease in median frequency after

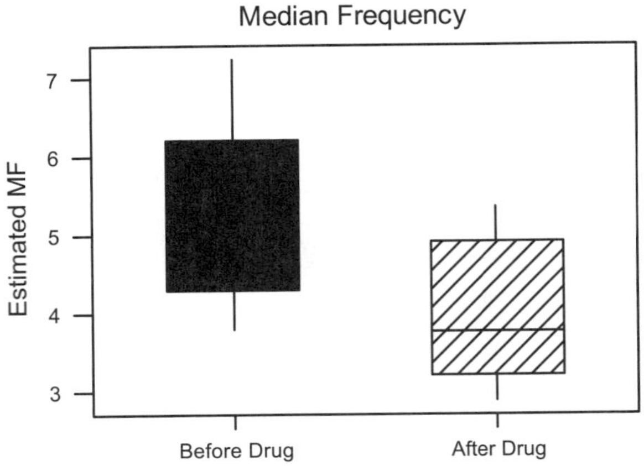

FIGURE 5.5: Box-and-whisker plot for median frequency estimated from samples of atrial fibrillation recorded before, and after a drug was administered.

administration of procainamide. In this case, the null hypothesis is that there is no change in mean median frequency after drug administration. The alternative hypothesis is that mean median frequency decreases after drug administration. Thus, we are comparing two means, and the two data sets have been collected from one set of patients. We will require a paired t test to reject or accept the null hypothesis.

BEFORE DRUG	AFTER DRUG	WI
4.30	2.90	1.4
4.15	2.97	1.18
3.80	3.20	0.60
5.10	3.30	1.80
4.30	3.75	0.55
7.20	5.35	1.85
6.40	5.10	1.30
6.20	4.90	1.30
6.10	4.80	1.30
5.00	3.70	1.30
5.80	4.50	1.30

To perform a paired t test, we create another column, W_i, as noted in the table above. We find the following for W_i: $\overline{W} = 1.262$ and $S_w = 0.402$. If we use these estimates for $\overline{W}$ and S_w in our equation for the paired T statistic, we find that $T = 10.42$ for $n = 11$ pairs of data. If we compare our estimated T value to the t distribution, we find that our T value is greater than the table entry for $t\,(0.005, 11 - 1)$;therefore, we may reject H_0 at a significance less than 0.005. In other words, we reject the null hypothesis at the $[1 - 0.005] \times 100\%$ confidence level.

Errors in Drawing Conclusions From Statistical Tests

When we perform statistical analysis, such as the t test, there is a chance that we are mistaken in rejecting or accepting the null hypothesis. When we draw conclusions from a statistical analysis, we typically assign a confidence level to our conclusion. This means that there is always some chance that our conclusions are incorrect.

There are two types of errors that may occur when we draw conclusions from a statistical analysis. These errors are referred to as types I and II.

Type I errors:

1. also referred to as a false-positive error;
2. occurs when we accept H_1 when H_0 is true;
3. may result in a false diagnosis of disease.

Type II errors:

1. also referred to as a false-negative error;
2. occurs when we accept H_0 when H_1 is true;
3. may result in a missed diagnosis (often more serious than a type I error).

If we think about the medical environment, a type I error might occur when a person is given a diagnostic test to detect streptococcus bacteria, and the test indicates that the person has the streptococcus bacteria when in fact, the person does not have the bacteria. The result of such an error means that the person spends money for an antibiotic that is serving no purpose.

A type II error would occur when the same person actually has the streptococcus bacteria, but the diagnostic test results in a negative outcome, and we conclude that the person does not have the streptococcus bacteria. In this example, this type of error is more serious than the type I error because the streptococcus bacteria left untreated can lead to a number of serious complications for the body.

The α value that we refer to as the level of significance is also the probability of a type I error. Thus, the smaller the level of significance at which we may reject the null hypothesis, the smaller the type I error and the lower the probability of making a type I error.

We typically use β to denote the type II error. We will discuss this error further at the end of chapter seven when we discuss power tests.

5.2 COMPARISON OF TWO VARIANCES

We used the t test to compare the means for two populations or processes. Two populations may also be compared for differences in variance. As discussed earlier, populations that are normally distributed are completely characterized by their mean and variance. Thus, if we want to test for differences between two normal populations, we need only compare their two means and their two variances.

Figure 5.6 illustrates the probability density functions for two normal populations (black and red traces). The four diagrams illustrate how two different normally distributed populations may compare with each other. The right two panels differ from the left two panels in the means of the populations. The top two panels differ from the bottom two panels in variance of the populations.

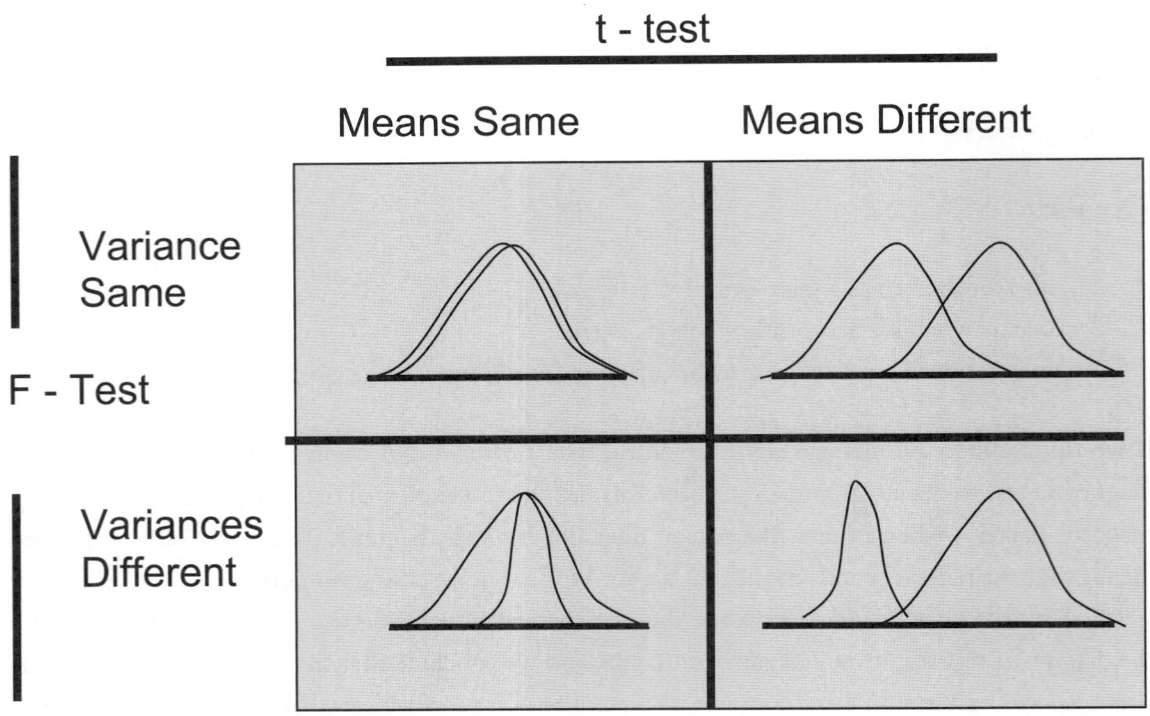

FIGURE 5.6: Two normal populations may differ in their means (top row), their variances (left half), or both (bottom right corner). t and F tests may be used to test for significant differences in the population means and population variances, respectively.

As indicated across the top of the tracings, a *t* test is used to test for differences in mean between the two populations. As indicated along the vertical direction, an *F* test is used to test for significant differences in the variances of the populations. Note that two normal populations may differ significantly in both mean and variance.

To compare the variances of two populations, we use what is referred to as an *F* test. As for the *t* test, the *F* test assumes that the data consist of independent random samples from each of two normal populations. If the two populations are not normally distributed, the results of the *F* test may be meaningless.

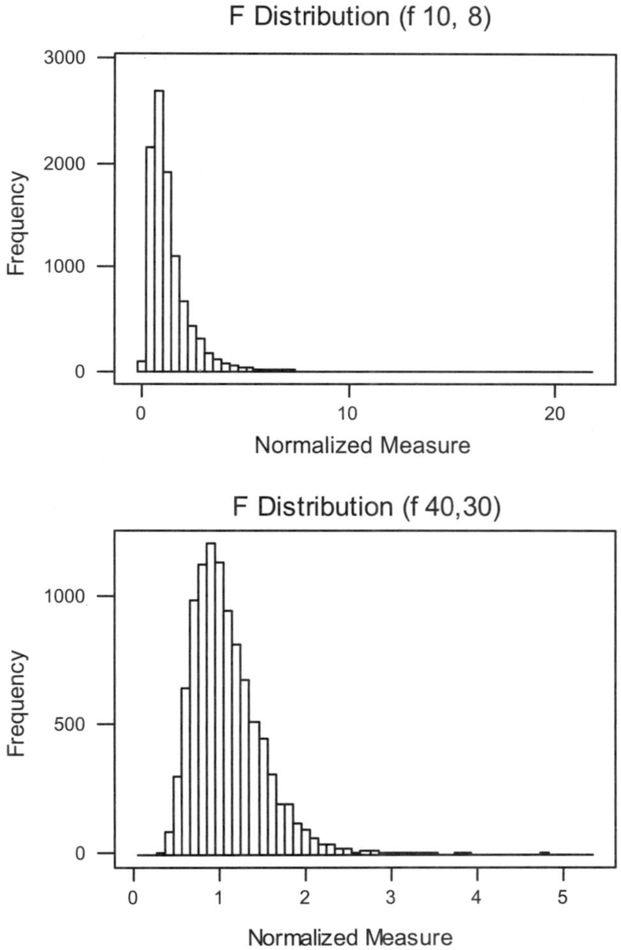

FIGURE 5.7: Histograms of samples drawn from two different *F* distributions. In the top panel, the two degrees of freedom are 10 and 8. In the lower panel, the two degrees of freedom are 40 and 30.

TABLE 5.1: Values from the *F* distribution for areas of α in the tail to the right of $F(dn, dd, \alpha)$

	dn							
dd	**1**	**2**	**3**	**4**	**5**	**…**	**10**	**11**
1	161, ***4052***	200, ***4999***	216, ***5403***	225, ***5625***	230, ***5764***		242, ***6056***	243, ***6082***
2	18.51, ***98.49***	19.00, ***99.01***	19.16, ***99.17***	19.25, ***99.25***	19.30, ***99.30***		19.39, ***99.40***	19.40, ***99.41***
3	10.13, ***34.12***	9.55, ***30.81***	9.28, ***29.46***	9.12, ***28.71***	9.01, ***28.24***		8.78, ***27.23***	8.76, ***27.13***
4	7.71, ***21.20***	6.94, ***18.00***	6.59, ***16.69***	6.39, ***15.98***	6.26, ***15.52***		5.96, ***14.54***	5.93, ***14.45***
5	6.61, ***16.26***	5.79, ***13.27***	5.41, ***12.06***	5.19, ***11.39***	5.05, ***10.97***		4.74, ***10.05***	4.70, ***9.96***
⋮								
10	4.96, ***10.04***	4.10, ***7.56***	3.71, ***6.55***	3.48, ***5.99***	3.33, ***5.64***		2.97, ***4.85***	2.94, ***4.78***
12	4.75, ***9.33***	3.88, ***6.93***	3.49, ***5.95***	3.26, ***5.41***	3.11, ***5.06***		2.76, ***4.30***	2.72, ***4.22***
15	4.54, ***8.68***	3.68, ***6.36***	3.29, ***5.42***	3.06, ***4.89***	2.90, ***4.56***		2.55, ***3.80***	2.51, ***3.73***
⋮								
20	4.35, ***8.10***	3.49, ***5.85***	3.10, ***4.94***	2.87, ***4.43***	2.71, ***4.10***		2.35, ***3.37***	2.31, ***3.30***
50	4.03, ***7.17***	3.18, ***5.06***	2.79, ***4.20***	2.56, ***3.72***	2.40, ***3.41***		2.02, ***2.70***	1.98, ***2.62***
100	3.94, ***6.90***	3.09, ***4.82***	2.70, ***3.98***	2.46, ***3.51***	2.30, ***3.20***		1.92, ***2.51***	1.88, ***2.43***
200	3.89, ***6.76***	3.04, ***4.71***	2.65, ***3.38***	2.41, ***3.41***	2.26, ***3.11***		1.87, ***2.41***	1.83, ***2.34***
∞	3.84, ***6.64***	2.99, ***4.60***	2.60, ***3.78***	2.37, ***3.32***	2.21, ***3.02***		1.83, ***2.32***	1.79, ***2.24***

dn = degrees of freedom for numerator; *dd* = degrees of freedom for denominator; α = the area in distribution tail to right of $F(dn, dd, \alpha)$ = 0.05 or ***0.01***.

					dn				
12	**14**	**…**	**20**	**30**	**40**	**50**	**100**	**200**	**∞**
244, ***6106***	245, ***6142***		248, ***6208***	250, ***6258***	251, ***6286***	252, ***6302***	253, *6334*	254, ***6352***	254, ***6366***
19.41, ***99.42***	19.42, ***99.43***		19.44, ***99.45***	19.46, ***99.47***	19.47, ***99.48***	19.47, ***99.48***	19.49, ***99.49***	19.49, ***99.49***	19.50, ***99.50***
8.74, ***27.05***	8.71, ***26.92***		8.66, ***26.69***	8.62, ***26.50***	8.60, ***26.41***	8.58, ***26.30***	8.56, ***26.23***	8.54, ***26.18***	8.53, ***26.12***
5.91, ***14.37***	5.87, ***14.24***		5.80, ***14.02***	5.74, ***13.83***	5.71, ***13.74***	5.70, ***13.69***	5.66, ***13.57***	5.65, ***13.52***	5.63, ***13.46***
4.68, ***9.89***	4.64, ***9.77***		4.56, ***9.55***	4.50, ***9.38***	4.46, ***9.29***	4.44, ***9.24***	4.40, ***9.13***	4.38, ***9.07***	4.36, ***9.02***
2.91, ***4.71***	2.86, ***4.60***		2.77, ***4.41***	2.70, ***4.25***	2.67, ***4.17***	2.64, ***4.12***	2.59, ***4.01***	2.56, ***3.96***	2.54, ***3.91***
2.69, ***4.16***	2.64, ***4.05***		2.54, ***3.86***	2.46, ***3.70***	2.42, ***3.61***	2.40, ***3.56***	2.35, ***3.46***	2.32, ***3.41***	2.30, ***3.36***
2.48, ***3.67***	2.43, ***3.56***		2.33, ***3.36***	2.25, ***3.20***	2.21, ***3.12***	2.18, ***3.07***	2.12, ***2.97***	2.10, ***2.92***	2.07, ***2.87***
2.28, ***3.23***	2.23, ***3.13***		2.12, ***2.94***	2.04, ***2.77***	1.99, ***2.69***	1.96, ***2.63***	1.90, ***2.53***	1.87, ***2.47***	1.84, ***2.42***
1.95, ***2.56***	1.90, ***2.46***		1.78, ***2.26***	1.69, ***2.10***	1.63, ***2.00***	1.60, ***1.94***	1.52, ***1.82***	1.48, ***1.76***	1.44, ***1.68***
1.85, ***2.36***	1.79, ***2.26***		1.68, ***2.06***	1.57, ***1.89***	1.51, ***1.79***	1.48, ***1.73***	1.39, ***1.59***	1.34, ***1.51***	1.28, ***1.43***
1.80, ***2.28***	1.74, ***1.17***		1.62, ***1.97***	1.52, ***1.79***	1.45, ***1.69***	1.42, ***1.62***	1.32, ***1.48***	1.26, ***1.39***	1.19, ***1.28***
1.75, ***2.18***	1.69, ***2.07***		1.57, ***1.87***	1.46, ***1.69***	1.40, ***1.59***	1.35, ***1.52***	1.24, ***1.36***	1.17, ***1.25***	1.00, ***1.00***

As with the *t* test, the *F* test is used to test the following hypotheses:

1. null hypothesis: H_0: $\sigma_1^2 = \sigma_2^2$ and
2. alternative hypothesis: H_1: $\sigma_1^2 > \sigma_2^2$,

where σ_1^2 and σ_2^2 are the variances of the two populations.

To reject or accept the null hypothesis, we compute the following *F* statistic:

$$F = \frac{s_1^2}{s_2^2},$$

where s_1^2 and s_2^2 are the sample variance estimates of the two populations. The ratio of two variances from two normal populations is also a random variable that follows an *F* distribution. The *F* distribution is illustrated in Figure 5.7. As with the *t* distribution, the *F* distribution varies with two parameters, such as the samples sizes of the two populations. Table 5.1 shows a fraction of an *F* table, where two degrees of freedom, *dn* and *dd*, are required to locate an *F* value in the *F* table. In this figure, the table entries are given for significance levels (α values) of 0.05 and 0.01. The *F* values associated with 0.05 and 0.01 significance levels are given in regular type and boldface-italics, respectively. Thus, for any two degrees of freedom, there are two *F* values provided, one for the 95% confidence level and one for the 99% confidence level.

To make use of the *F* table with the *F* test, we estimate an *F* statistic using the sample variance estimates from each of the two populations we are trying to compare. Note that for the use of this *F* table, the larger of the two variances should be put in the numerator of the equation above.

We now compare our estimated *F* statistic to the entries in the *F* table associated with *dn* and *dd* degrees of freedom and appropriate confidence level (only the 95% and 99% *F* values are provided in the table provided). $dn = n_1$ and $dd = n_2$ are the number of samples in each population, with n_1 being the number of samples in the population with variance placed in the numerator.

If we are to reject the null hypothesis outlined above, our calculated *F* statistic must be > $F(\alpha, dn, dd)$ in table to reject H_0 with confidence $(1 - \alpha) \times 100\%$. The degrees of freedom, $dn = n_1$, is the value used to locate the table entry in the horizontal direction (numerator), and $dd = n_2$ is the degrees of freedom used to locate the table entry in the vertical direction (denominator).

EXAMPLE 5.2 *F* test

	POPULATION A ($N_1 = 9$)	POPULATION B ($N_2 = 9$)
Mean	0.026	0.027
Variance	2.0E – 5	7.4E – 5

Given the data listed in table above, where we have collected 9 samples from each of two populations, A and B, we can estimate our *F* statistic to be

$$F = (7.4E-5)/(2.0E-5) = 3.7.$$

Using an *F* table that has table entries for both the 95% and 99% confidence levels, we find that the table entry for α equal to 0.05 and 9 *df* for both populations is

$$F(0.05, 9, 9) = 3.18.$$

Our estimated value of 3.7 exceeds the table entry of 3.18; thus, we may reject H_0 with 95% confidence and accept that population B has significantly greater variance than population A. However, we note that the table entry for $F(0.01, 9, 9)$ equals 5.35 and is greater than our estimated value. Thus, we cannot reject the null hypothesis with 99% confidence.

5.3 COMPARISON OF THREE OR MORE POPULATION MEANS

Thus far, we have discussed statistical analysis for comparing the means and variances (or standard deviations) for two populations based on collecting two sets of data. These two sets of data may be independent of each other or they may be dependent on each other, which we refer to as paired or blocked data. Another way to think of this pairing is to call it blocking on the experimental unit, meaning that both sets of data were collected from the same experimental units but under different conditions. For example, the experimental units may be human subjects, animals, medical instruments, manufacturing lines, cell cultures, electronic circuits, and more. If, for example, the experimental unit is human beings, the two sets of data may be derived before and after a drug has been administered or before and after the subjects have engaged in exercise. In another example, we may measure two different types of data, such as heart rate and blood pressure, each from the same group of subjects. This is referred to as repeated measures because we are taking multiple sets of measures from the same experimental units. In repeated measures, something in the experimental conditions, experimental unit, or measure being collected has changed from one set of measures to the next, but the sampling is blocked by experimental unit.

In many biomedical engineering applications, we need to compare the means from three or more populations, processes, or conditions. In such cases, we use a method of statistical analysis called analysis of variance, or ANOVA. Although the name implies that one is analyzing variances, the conclusions that stem from such analysis are in regard to significant difference in the means of three or more populations or processes.

5.3.1 One-Factor Experiments

We begin our discussions of ANOVA by discussing the design and analysis of one-factor experiments. In such experiments, we are drawing samples from three or more populations for which one factor has been varied from one population to the next.

As stated in Chapter 2, experimental design should include randomization and blocking. Randomization ensures that we do not introduce bias into the data because of ordering effects in collecting our data. Moreover, blocking helps us to reduce the effect of intersubject variability on the differences between populations.

Some biomedical engineering challenges that may require the use of ANOVA include the following:

1. Compare wear time between hip implants of various materials: In this example, the treatment = implant material (e.g., titanium, steel, polymer resins).
2. Compare MR pulse sequences in the ability to image tissue damage after stroke: In this example, the treatment = MR pulse sequence.
3. Compare ability of various drugs to reduce high blood pressure: In this example, the treatment = drug.

5.3.1.1 Example of Biomedical Engineering Challenge

An example of a biomedical engineering problem that might use ANOVA to test a hypothesis is in the study of reflex mechanisms in spinal cord injured patients [14]. One of the interests of investigators is how hip flexion (torque) is dependent on ankle movement in patients with spinal chord injury. In this example, there are actually two experimental factors that may affect hip flexion:

1. Range of ankle motion: in this case, the treatment = range of ankle extension.
2. Speed of ankle flexion: in this case, the treatment = speed of ankle extension.

We might evaluate one factor at a time using a one-factor or one-way ANOVA. Or, we might evaluate the impact of both factors on mean hip flexion using two-factor or two-way ANOVA. In either case, we have a null hypothesis and alternative hypotheses for the effect of each factor, such as range of ankle motion, on the population mean, such as hip flexion. Our null hypothesis is that there is no significant difference in mean hip flexion across ankle motion,

$$H_0: \mu_1 = \mu_2 = \mu_3 = \dots = \mu_n.$$

The alternative hypothesis, H_1, states that at least two of the population means, such as hip flexion for two different ankle motions, differ significantly from each other.

TABLE 5.2: One-factor experiment

TREATMENT	OBSERVATIONS	SAMPLE MEAN	SAMPLE VARIANCE
1	$y_{11}, y_{22}, \ldots, y_{1n}$	$\bar{y}_1$	$\sigma^2 y_1$
2	$y_{21}, y_{22}, \ldots, y_{2m}$	$\bar{y}_2$	$\sigma^2 y_2$
$\vdots$			
i	$y_{i1}, y_{i2}, \ldots, y_p$	$\bar{y}_k$	$\sigma^2 y_i$
$\vdots$			
k	$y_{k1}, y_{k2}, \ldots y_{kq}$	$\bar{y}_k$	$\sigma^2 y_k$

The basic design of the one-factor experiment is given in Table 5.2. In this case, there are k treatments in one factor. k is the number of treatments, y_{ij} is the individual j sample or data point for the ith treatment, $\bar{y}_i$ is the sample mean for the ith treatment, and σ^2_{yi} is the sample variance for the ith treatment (adapted from [3]).

For example, we might be interested in comparing weight loss for three different types of diet pills. In this case, k = 3. Under the weight-loss column, one will find the weight loss of individual subjects who used one of the three diet pills. Note that in this example, the number of samples varies with each drug type. In a balanced experiment, there should be an equal number of samples for each type of treatment. In the third and fourth columns, we have the sample mean and standard deviation for weight loss for each of the diet pills. The question we are trying to answer is whether there is a significant difference in mean weight loss as a function of diet pill type.

DIET PILL	WEIGHT LOSS	MEAN	STANDARD DEVIATION
Placebo	10, 12, 5, 8, 5, 20	10.0	5.62
Drug B	30, 5, 12, 20	16.75	10.75
Drug C	2, 10, 5, 5, 10, 20, 25, 40	14.63	12.92

We make a number of assumptions when using ANOVA to compare differences in means between three or more populations or processes:

1. Subjects are randomly assigned to a specific treatment (in this case diet pill).
2. The populations or processes (such as weight loss) are approximately normally distributed.
3. Variance is approximately equal across treatment groups.

In this specific weight-loss example, no blocking is used. In other words, in no case did the same subject receive more than one treatment or diet pill. This may end up being problematic and lead to erroneous conclusions because we are not accounting for intersubject variability in response to the diet drug. For example, we might assume that the amount of weight loss is related to starting weight. It is possible that the subjects assigned to drug A all had starting weights less than the subjects given drug B. Thus, differences in the weight loss may be significant but may have little to do with the actual diet pill. The difference in weight loss may be because of the differences in starting weight rather than the diet pill. Because the above experiment does not use blocking, intersubject variability that is not accounted for may confound our conclusions regarding the effectiveness of diet pill.

The question we are attempting to address in using ANOVA is, "Are the means of the k treatments equal, or have our samples been drawn from treatments (populations) with different means?"

Within each treatment, i (or population, i), we assume that the variability across observed samples, Y_{ij}, is influenced by the population (treatment) mean, μ_i, and independent random variables, e_{ij}, that are normally distributed, with a mean of zero and variance σ^2. In other words, the samples, Y_{ij}, collected for each trial, j, within each treatment, i, can be expressed as

$$Y_{ij} = \mu_i + e_{ij}.$$

With our model, we are trying to determine how much of the variability in Y is because of the factor or treatment (population with mean, μ_i), and how much is because of random effects, e_{ij}, which we cannot control for or have not captured in the model provided above.

When we perform an ANOVA for a one-factor experiment, we can organize the analysis and results in the following table (Table 5.3):

TABLE 5.3: One-factor ANOVA with k treatments (no blocking)

SOURCE	*df*	SS	MS	*F*
Treatment	$k-1$			MS_{treat}/MS_{error}
Error	$N-k$			

N = total number of samples across all treatments; k = number of treatments within the one factor; F = the statistic that we will compare with our F tables (F distribution) to either reject or accept the null hypothesis; $SS_{treatment}$ = between-treatment sum of squares, a measure of the variability among treatment means; SS_{error} = within-treatment sum of squares, a measure of the sum of variances across all k treatments; $MS_{treatment} = SS_{treatment}/k-1$; $MS_{error} = SS_{error}/N-k$.

For k treatments, the specific equations for the SS elements are the following:

$$SS_{\text{treatment}} = \sum_{i=1}^{k} n_i(\bar{y}_i - \bar{y}_{\text{grand}})^2$$

and

$$SS_{\text{error}} = \sum_{i=1}^{k}\sum_{j=1}^{n_i}(Y_{ij} - \bar{Y}_i)^2 = \sum_{i=1}^{k}\sigma_{yi}^{\;2}.$$

$\bar{Y}_{\text{grand}}$ is the sample mean for all samples across all treatments combined.

Finally, the F statistic that we are most interested in estimating to reject or accept our null hypothesis is

$$F = MS_{\text{treatment}}/MS_{\text{error}}.$$

To test our hypothesis at a significance level of α, we now compare our estimated F statistic to the F distribution provided in the F tables for the table entry, $F(\alpha; k-1, N-k)$. If our estimated F value is greater than the table entry, we may reject the null hypothesis with $(1-\alpha) \times 100\%$ confidence.

Example 5.3 A heart valve manufacturer has three different processes for producing a leaf valve. Random samples of 50 valves were selected five times from each type of manufacturing process. Each valve was tested for defective opening mechanisms. The number of defective valves in each sample of 50 valves is summarized in the following table:

PROCESS A	PROCESS B	PROCESS C
1	5	3
4	8	1
3	6	1
7	9	4
5	10	0

Using an $\alpha = 0.05$, we want to determine whether the mean number of defects differs between processes (treatment). (Our null hypothesis, H_0, is that mean number of defects is the same across processes.)

To answer the question, we need to complete the following ANOVA table.

One-way ANOVA: processes A, B, and C				
SOURCE	***df***	**SS**	**MS**	***f***
Factor	2			
Error	12			
Total	14			

We first find SS_{factor} (= $SS_{treatment}$).

We know that

$$SS_{treatment} = \sum_{i=1}^{k} n_i(\overline{y}_i - \overline{y}_{grand})^2.$$

From the data given, we know that there are three treatments, A, B, and C; thus, $k = 3$. We will assign $A = 1$, $B = 2$, and $C = 3$. We also know that $n_1 = n_2 = n_3 = 5$. We can estimate the sample mean for each treatment or process to obtain, $\overline{y}_1 = 4$, $\overline{y}_2 = 7.6$, $\overline{y}_3 = 1.8$. We can also use all 15 samples to find $\overline{y}_{grand} = 4.47$.

Now, we can use these estimates in the equation for $SS_{treatment}$:

$$SS_{treatment} = 5(4 - 4.47)^2 + 5(7.6 - 4.47)^2 + 5(1.8 - 4.47)^2 = 85.70.$$

Now we solve for SS_{error}:

Given that

$$SS_{error} = \sum_{i=1}^{k}\sum_{j=1}^{n_i}(Y_{ij} - \overline{Y}_i)^2,$$

we need to estimate the inner summation for each treatment, i, noted by the outer summation. Thus, for $i = 1$,

$$SS_1 = (1 - 4)^2 + (4 - 4)^2 + (3 - 4)^2 + (7 - 4)^2 + (5 - 4)^2 = 20;$$

for $i = 2$, $SS_2 = (5 - 7.6)^2 + (8 - 7.6)^2 + (6 - 7.6)^2 + (9 - 7.6)^2 + (10 - 7.6)^2 = 17.2;$

for $i = 3$, $SS_3 = (3 - 1.8)^2 + (1 - 1.8)^2 + (1 - 1.8)^2 + (4 - 1.8)^2 + (0 - 1.8)^2 = 10.8.$

Now, summing over all k treatments, where $k = 3$, we find

$$SS_{error} = SS_1 + SS_2 + SS_3 = 48.0.$$

Now, we can find the mean squared errors:

$$MS_{treatment} = SS_{treatment}/DF_{treatment} = 87.5/2 = 43.7.$$

$$MS_{error} = SS_{error}/DF_{error} = 48.0/12.0 = 4.$$

Finally, our F statistic is given by

$$F = MS_{treatment}/MS_{error} = 43.7/4 = 10.92.$$

We now complete our ANOVA table.

One-way ANOVA: processes A, B, and C				
SOURCE	***df***	**SS**	**MS**	***f***
Factor	2	87.5	43.7	10.92
Error	12	48.0	4.0	

Now, we want to determine if the mean number of defects differ among processes A, B, and C. Our null hypothesis is that the mean number of defects does not differ among processes. To reject this hypothesis at the 95% confidence level ($\alpha = 0.05$), our estimated F statistic must be greater than $F(0.05, 2, 12) = 3.88$, found in the F distribution tables. Our F statistic = $10.92 > 3.88$; thus, we can reject the null hypothesis with 95% confidence and conclude that the number of defects does vary with manufacturing process and that our samples for treatments A, B, and C were drawn from populations with different means. In fact, the F distribution table list a value for $F(0.01, 2, 12) = 6.93$. Thus, we can reject the null hypothesis at the 99% confidence level as well.

Example 5.4 Four types of MR scanners are being evaluated for speed of image acquisition. The following table summarizes the speeds measured (in minutes) for three samples of each scanner type.

SCANNER A	SCANNER B	SCANNER C	SCANNER D
2.0	4.0	3.0	6.0
1.8	4.5	2.5	5.5
2.7	5.5	2.0	3.5

Does the mean speed of image acquisition differ amongst the four scanner types? (Use α = 0.01.)

To answer this question, we may work through the calculations that we performed in Example 5.3. However, as one discovered in the previous example, as the sample size grows, the number of calculations quickly grows. Thus, most investigators will use a statistical software package to perform the ANOVA calculations. For this example, we used the Minitab software package (Minitab Statistical Software, Release 13.32, Minitab, 2000) to perform the ANOVA. ANOVA produces the following table:

ANOVA for speed					
SOURCE	***df***	**SS**	**MS**	***f***	α
Factor	3	19.083	6.361	9.07	0.006
Error	8	5.613	0.702		
Total	11	24.697			

The F statistic of 9.07 results in an area with α = 0.006 in the right tail of the *F* distribution. Because α < 0.01, the value of α at which we were testing our hypothesis, we can reject the null hypothesis and accept that alternative hypothesis that at least one of the scanners differs from the remaining scanners in the mean speed of acquisition.

In the examples given previously, we did not use blocking in the experimental design. In other words, the populations from which we collected samples differed from treatment to treatment. In some experiments, such as testing weight loss for diet pills, it is not practical or possible to test more than one type of treatment on the same experimental unit.

However, when blocking may be used, it should be used to compare treatments to reduce intersubject variability or differences (that cannot be controlled for) in the experimental outcome. Table 5.4 outlines the experimental design for a one-factor experiment that makes use of blocking. In this experimental design, all experimental units are subject to the every treatment. The result is

TABLE 5.4: One factor with k treatment and blocking

TREATMENT (ANKLE SPEED)	SUBJECT NUMBER				TREATMENT MEAN
	1	2	3	4	
10	5	2	5	6	
20	10	2	8	4	
30	15	8	15	10	
Block mean					

that we now have treatment means and block means. Thus, when we perform the ANOVA analysis, we may test two different sets of hypotheses. The first null hypothesis is that there is not significance difference in treatment means. The second null hypothesis is that there is no significant difference in means across subjects or experimental units (intersubject variability is not significant).

In this example, treatment could be ankle speed, and the block would be the patient, such that each ankle speed is tested on each and every patient subject. In such a design, it is important to randomize the order in which different ankle speeds are tested so that there is no bias in hip flexion due to ordering effects of ankle speed or machine wear.

Note that the example above is also a balanced experimental design because there are same numbers of data points in every cell of the table.

Within each treatment, i (or population, i), we now assume that the variability across observed samples, Y_{ij}, is influenced by the population (treatment) mean, μ_i, the block effects, β_j, and independent random variables, e_{ij}, which are normally distributed with a mean of zero and variance σ^2. In other words, the samples, Y_{ij}, collected for each trial, j, within each treatment, i, can be expressed as

$$Y_{ij} = \mu_i + \beta_j + e_{ij}.$$

In other words, we are trying to determine how much of the variability in Y is because of the factor or treatment (population with mean, μ_i) and how much is due to block effects, β_j, and random effects, e_{ij}, that we cannot control for or have not captured in the model provided above. In such a model, we assume that treatment effects and block effects are additive. This is not a good assumption when there are interaction effects between treatment and block. Interaction effects mean

TABLE 5.5: One-factor ANOVA with block

SOURCE	*df*	SS	MS	*F*
Treatment	$k - 1$			MS_{treat}/MS_{error}
Block	$b - 1$			MS_{block}/MS_{error}
Error	$(b - 1)(k - 1)$			

that the effect of a specific treatment may depend on a specific block. When we address two-factor experiments in the next section, we will discuss further these interaction effects.

We can summarize the ANOVA for one factor with block with the following ANOVA table (assuming no interaction effects) (Table 5.5; adapted from [3]).

Note that we now have two estimated F values to consider. $F_{treatment}$ is compared with the table entry for $F(\alpha; k - 1, (b - 1)(k - 1))$, and the null hypothesis stating that there is no difference in mean hip flexion across treatments is rejected if the estimated F value is greater than the table entry.

F_{block} is compared with the table entry for $F(\alpha; b - 1, (b - 1)(k - 1))$, and the null hypothesis stating that there is no difference in mean hip flexion across subjects is rejected if the estimated F statistic is greater than the table entry.

Example 5.5 One-factor experiment with block

HeartSync manufactures four types of defibrillators that differ in the strength of the electrical shock given for an episode of fibrillation. A total of 280,000 patients were divided into four groups of 70,000 patients each. Each group was assigned to one of the four defibrillators, and the number of shocks that failed to defibrillate was recorded for four consecutive years. The results were as follows:

YEAR AFTER IMPLANT	DEVICE A	DEVICE B	DEVICE C	DEVICE D
1	6	1	9	2
2	8	1	10	2
3	5	3	8	0
4	10	2	11	5

There are two questions we wish to address with these data:

1. Using $\alpha = 0.01$, does the mean number of failures differ significantly as a function of device type?
2. Using $\alpha = 0.01$, does the mean number of failures differ significantly as a function of year after implant? In other words, is year after implant a major source of variability between populations?

Again, rather than estimate the calculations by hand, we may use statistical software, such as Minitab, to obtain the following results using ANOVA with block:

ANOVA for number of failures				
SOURCE	*df*	SS	MS	*F*
Device	3	173.188	57.729	35.67
Year	3	20.688	6.896	4.26
Error	9	14.563	1.618	
Total	15	208.438		

The F statistic for device, $F = 35.67$, is greater than the critical F value, $F(0.01, 3, 9) = 6.99$, given in the F table. Thus, we can reject the null hypothesis and accept that alternative hypothesis that at least one of the devices differs from the remaining devices in the mean number of failure. The second part of the questions tests the hypothesis that the mean number of failures differs significantly as a function of year after implant (the blocking factor). The F statistic for year, $F = 4.26$, is less than the critical F value, $F(0.01, 3, 9) = 6.99$, given in the F table; thus, we accept our null hypothesis, which means failure rate does not differ between years after implant.

5.3.2 Two-Factor Experiments

In the original biomedical engineering challenge describing hip flexion reflexes, we discussed two factors that influence hip flexion: ankle speed and range of ankle extension.

In a two-factor experiment leading to a two-way ANOVA, there are two factors being varied, A and B, where A has a treatments and B has b treatments, and there are n samples at every combination of A and B.

The two-factor experiment is said to be completely crossed if there are samples collected for every combination of factors A and B. In addition, the experiment is said to be balanced if we have same number of samples for every combination of factors A and B.

Table 5.6 below illustrates a two-factor experiment that is completely balanced and crossed. Each of the two treatments within factor A is crossed with each of the three treatments within factor B. Also note that for each combination of A and B, we have three samples.

The question that we are trying to address with a two-factor ANOVA is whether there are differences in treatment means for each of the two factors. In addition, we wish to know if there are interaction effects such that there are significant differences in means as a function of the cross-interaction between factors. In other words, there are significant differences in sample means, and hence, the means of the underlying populations, when specific combinations of factors A and B occur together.

Once the data have been collected, as illustrated in Table 5.6 above, we may perform a two-factor ANOVA to test the following three null hypotheses:

$$H_0: \mu_{A1} = \mu_{A2} = \mu_{A3} = \ldots. = \mu_{Aa};$$

$$H_0: \mu_{B1} = \mu_{B2} = \mu_{B3} = \ldots. = \mu_{Bb};$$

$$H_0: \mu_{A1B1} = \mu_{A1B2} = \mu_{A1B3} = \mu_{A2B1} = \mu_{A2B2} = \ldots = \mu_{AaBb}.$$

For each of the three null hypotheses, the associated alternative hypothesis is that there is a significant difference in at least two of the population means for a given factor or combination of factors.

The analysis and results of a two-factor ANOVA may be organized as in Table 5.7 [3].

The equations for the SS and MS for each of the factors and interaction factors are beyond the scope of this text but may be found in [3]. In practice, the investigator will use a popular statistical software package such as Minitab, SPSS, or SAS to estimate these SS and MS values (because of computational burden) and simply refer to the F statistics to reject or accept the null hypotheses.

TABLE 5.6: Two-factor experiment

	FACTOR B		
Factor A	1	2	3
1	1.2, 1.4, 2.1	2.3, 2.2, 2.6	6.4, 5.8, 3.2
2	3.2, 4.1, 3.6	4.1, 4.3, 4.0	8.2, 7.8, 8.3

TABLE 5.7: Two-factor ANOVA table (each with multiple treatments)

SOURCE	*df*	SS	MS	*F*
A	$a - 1$			MS_A/MS_{error}
B	$b - 1$			MS_B/MS_{error}
AB	$(a - 1)(b - 1)$			MS_{AB}/MS_{error}
Error	$ab(n - 1)$			

We note that there are three *F* statistics to test each of the three null hypotheses described earlier. In each case, we compare our estimated *F* statistic with the *F* values in our *F* table representing the *F* distribution. More specifically, we compare our estimates *F* values to the following table entries:

To test H_0: $\mu_{A1} = \mu_{A2} = \mu_{A3} = \ldots. = \mu_{Aa}$, compare F_A with $F(\alpha; a - 1, ab(n - 1))$;

To test H_0: $\mu_{B1} = \mu_{B2} = \mu_{B3} = \ldots. = \mu_{Bb}$, compare F_B with $F(\alpha; b - 1, ab(n - 1))$;

To test H_0: $\mu_{A1B1} = \mu_{A1B2} = \mu_{A1B3} = \mu_{A2B1} = \mu_{A2B2} = \ldots = \mu_{AaBb}$,

compare F_{AB} with $F(\alpha; (a - 1)(b - 1), ab(n - 1))$.

In each of the three tests, if the estimated *F* statistic is greater than the table entries for the *F* distribution, one may reject the null hypothesis and accept the alternative hypothesis with $(1 - \alpha) \times$ 100% confidence.

Example 5.6 An example of a two-factor experiment that will be evaluated using two-factor ANOVA occurs when biomedical engineers are looking at the effectiveness of rehabilitative therapy and pharmacological therapy on the recovery of movement in a limb after stroke. Details of the experimental design include the following:

1. Factor T: therapy used (there are three types of therapies, T1, T2 and T3);
2. Factor D: drug used (there are three types of Drugs, D1, D2 and D3);
3. 36 patients are randomly assigned to each combination of T and D;
4. measure: number of days to meet recovery criteria.

Experimental design for two factors

T1			T2			T3		
D1	D2	D3	D1	D2	D3	D1	D2	D3
20	25	13	22	8	16	9	15	7
15	16	12	16	10	19	12	10	10
18	10	22	17	9	11	8	9	9
24	20	10	12	11	21	8	10	9

The questions we are trying to address is whether there is a significant difference in mean days of recovery for the three types of rehabilitative therapy, the mean days of recovery for the three types of drug therapy, and differences in mean days of recovery for the nine combinations of rehabilitative therapy and drug therapy (interaction effects).

ANOVA analysis performed using statistical software, known as Minitab, produces the following table summarizing the ANOVA analysis. Note that there are three estimated *F* statistics. We can use the three *F* values to test our hypotheses:

$$H_0: \mu_{T1} = \mu_{T2} = \mu_{T3};$$

$$H_0: \mu_{D1} = \mu_{D2} = \mu_{D3};$$

$$H_0: \mu_{T1D1} = \mu_{T1D2} = \mu_{T1D3} = \mu_{T2B1} = \mu_{T2B2} = \ldots = \mu_{T3D3}.$$

Two-way ANOVA for days of recovery versus T and D

SOURCE	*df*	SS	MS	*F*	α
T	2	337.4	168.7	11.44	0.000
D	2	36.2	18.1	1.23	0.309
Interaction	4	167.8	41.9	2.84	0.043
Error	27	398.3	14.8		

Note that the estimated F statistic for rehabilitative therapy, $F = 11.44$, does exceed the critical value for $\alpha < 0.05$ (actually, α is <0.0001); thus, we conclude that the samples for the three different rehabilitative therapies represent populations with different means. In other words, the different rehabilitative therapies produce different days of recovery. However, the estimated F statistic for the drug therapy does not exceed the critical table F value for $\alpha = 0.05$; thus, we accept the null hypothesis that drug therapy does not significantly affect the days of recovery and that the samples are not drawn from populations with different means. Moreover, the third F statistic, $F = 2.84$, is greater than the table F value for $\alpha < 0.05$, which suggests that there may be a significant difference in days of recovery due to an interaction between rehabilitative therapy and drug therapy.

5.3.3 Tukey's Multiple Comparison Procedure

Once we have established that there is a significant difference in means across treatments within a factor, we may use post hoc tests, such as the Tukey's HSD multicomparison pairwise test [3, 9]. ANOVA simply shows that there is at least one treatment mean that differs from the others. However, ANOVA does not provide information on specifically which treatment mean(s) differs from which treatment mean(s). The Tukey's HSD test allows us to compare the statistical differences in means between all pairs of treatments. For a one-factor experiment with k treatments, there are $k(k - 1)/2$ pairwise comparisons to test. The important point to note is that if α is the probability of a type I error for one comparison, the probability of making at least a type I error for multiple comparisons is much greater. So, if we want $(1 - \alpha) \times 100\%$ confidence for all possible pairwise comparisons, we must start with a much smaller α. Tukey's multiple comparison procedure allows for such an adjustment in significance when performing pairwise comparisons.

• • • •

CHAPTER 6

Linear Regression and Correlation Analysis

Oftentimes, in biomedical engineering research or design, we are interested in whether there is a correlation between two variables, populations, or processes. These correlations may give us information about the underlying biological processes in normal and pathologic states and ultimately help us to model the processes, allowing us to predict the behavior of one process given the state of another correlated process.

Given two sets of samples, X and Y, we ask the question, "Are two variables or random processes, X and Y, correlated?" In other words, can y be modeled as a linear function of x such that

$$\hat{y} = mx + b.$$

Look at the following graph of experimental data (Figure 6.1), where we have plotted the data set, y_i, against the data set, x_i:

We note that the data tend to fall on a straight line. There tends to be a trend such that y increases in proportion to increases in x. Our goal is to determine the line (the linear model) that best fits these data and how close the measured data points lie with respect to the fitted line (generated by the model). In other words, if the modeled line is a good fit to the data, we are demonstrating that y may be accurately modeled as a linear function of x, and thus, we may predict y given x using the linear model.

The key to fitting a line that best predicts process y from process x, is to find the parameters m and b, which minimize the error between model and actual data in a least-squares sense:

$$\min\,[(y - \hat{y})^2].$$

In other words, as illustrated in Figure 6.2, for each measured value of the independent variable, x, there will be the measured value of the dependent variable, y, as well as the predicted or

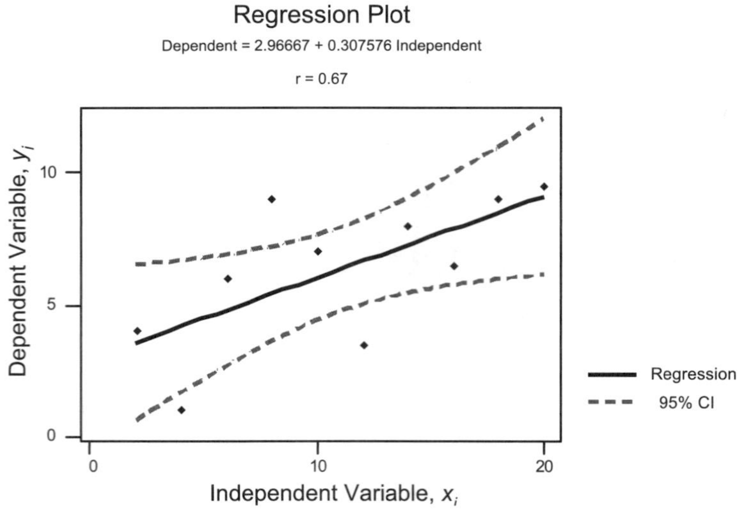

FIGURE 6.1: Results of linear regression applied to the samples illustrated in the scatterplot (black dots). The solid black line illustrates the line of best fit (model parameters listed above the graph) as determined by linear regression. The red dotted curves illustrate the confidence interval for the slope. Finally, the r value is the correlation coefficient.

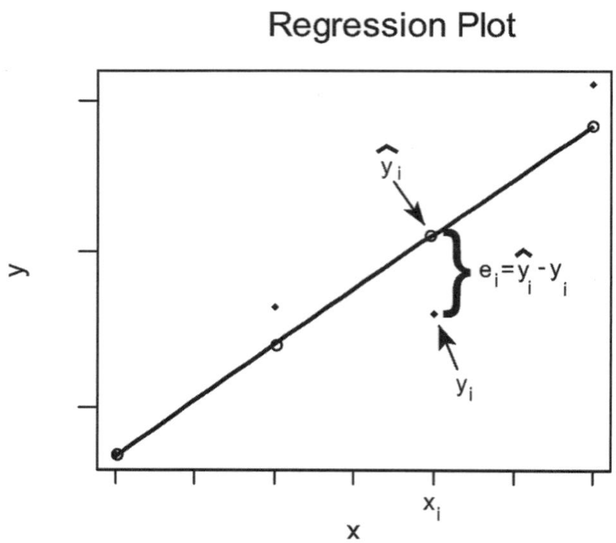

FIGURE 6.2: Linear regression can be used to estimate a straight line that best "fits" the measured data points (filled circles). In this illustration, x_i and y_i represent the measured independent and dependent variables, respectively. Linear regression is used to model the dependent variable, y, as a linear function of the independent variable, x. The straight line passing through the measured data points is the result of linear regression whereby the error, e_i, between the predicted value (open circles) of the dependent variable, $\hat{y}_i$, and the measured value of the dependent variable, y_i, is minimized over all data points.

modeled value of y, denoted as $\hat{y}$, that one would obtain if the equation $\hat{y} = mx + b$ is used to predict y. The equation $e_i = y_i - \hat{y}_i$ denotes the errors that occur at each pair of (x_i, y_i) when the modeled value does not exactly align with the predicted value because of factors not accounted for in the model (noise, random effects, and nonlinearities).

In trying to fit a line to the experimental data, our goal is to minimize these e_i between the measured and predicted values of the dependent variable, y. The method used in linear regression and many other biomedical modeling techniques is to find model parameters, such as m and b, that minimize the sum of squared errors, e_i^2.

For linear regression, we seek a least-squares estimate of m and using the following approach:

Suppose we have N samples each of processes x and y. We try to predict y from measured x using the following model:

$$\hat{y} = mx + b.$$

The error in prediction at each data point, x_i, is

$$\text{error}_i = y_i - \hat{y}_i.$$

In the least-squares method, we choose m and b to minimize the sum of squared errors:

$$(y_i - \hat{y}_i)^2 \qquad \text{for } i = 1 \text{ to } N.$$

To find a closed form solution for m and b, we can write an expression for the sum of squared errors:

$$\sum_{i=1}^{N} e_i^2 = \sum_{i=1}^{N} (y_i - \hat{y}_i)^2.$$

We then replace $\hat{y}_i$ with $(mx_i + b)$, our model, and carry out the squaring operations [3, 5].

We can then take derivatives of the above expression with respect to m and then again with respect to b. If we set the derivative expressions to zero to find the minimums, we will have two equations in two unknowns, m and b, and we can simply use algebra to solve for the unknown parameters, m and b. We will get the following expressions for m and b in terms of the measured x_i and y_i:

$$m = \frac{\sum_{i=0}^{N-1} x_i y_i - \left(\sum_{i=0}^{N-1} x_i\right)\left(\sum_{i=0}^{N-1} y_i\right) / N}{\sum_{i=0}^{N-1} x_i^2 - \left(\sum_{i=0}^{N-1} x_i\right)^2 / N}$$

and

$$b = \bar{y} - m\bar{x},$$

where

$$\bar{x} = \frac{1}{N}\sum_{i=0}^{N-1} x_i \quad \text{and} \quad \bar{y} = \frac{1}{N}\sum_{i=0}^{N-1} y_i.$$

Hence, once we have our measured data, we can simply use our equations for m and b to find the line, or linear model, of best fit.

The Correlation Coefficient

It is important to realize that linear regression will fit a line to any two sets of data regardless of how well the data are modeled by a linear model. Even if the data, when plotted as a scatterplot, look nothing like a line, linear regression will fit a line to the data. As biomedical engineers, we have to ask, "How well does the measured data "fit" the line estimated through linear regression?"

One measure of how well the experimental data fit the linear model is the correlation coefficient. The correlation coefficient, r, has a value between −1 and 1 and indicates how well the linear model fits to the data.

The correlation coefficient, r, may be estimated from the experimental data, x_i and y_i, using the following equation:

$$r = \frac{\sum_{i=0}^{N-1}(x_i - \bar{x})(y_i - \bar{y})}{\left[\sum_{i=0}^{N-1}(x_i - \bar{x})^2 \sum_{i=0}^{N-1}(y_i - \bar{y})^2\right]^{1/2}},$$

where

$$\bar{x} = \frac{1}{N}\sum_{i=0}^{N-1} x_i \quad \text{and} \quad \bar{y} = \frac{1}{N}\sum_{i=0}^{N-1} y_i.$$

It is important to note that an $r = 0$ does not mean that the two processes, x and y, are independent. It simply indicates that any dependency between x and y is not well described or modeled by a linear relation. There could be a nonlinear relation between x and y. An $r = 0$ simply means that x and y are uncorrelated in a linear sense. That is, one may not predict y from x using a linear model, $y = mx + b$.

A measure related to the correlation coefficient, r, is the coefficient of determination, R^2, which is a summary statistic that tells us how well our regression model fits our data. R^2 can be used as measure of goodness of fit for any regression model, not just linear regression. For linear regression, R^2 is the square of the correlation coefficient and has a value between 0 and 1. The coefficient of determination tells us how much of the variability in the data may be explained by the model parameters as a fraction of total variability in the data.

It is important to realize that the estimated slope of best fit and the correlation coefficient are statistics that may or may not be significant. Thus, t tests may be performed to test if the slope estimated through linear is significantly different from zero [3]. Likewise, t tests may be performed to test if the correlation coefficient is significantly different from zero. Finally, we may also compute confidence intervals for the estimated slope [3].

• • • •

CHAPTER 7

Power Analysis and Sample Size

Up to this point, we have discussed important aspects of experimental design, data summary, and statistical analysis that will allow us to test hypotheses and draw conclusions with some level of confidence.

However, we have not yet addressed a very important question. The question we ask now is, "how large should my sample be to capture the variability in my underlying population so that my types I and II error rates will be small?" In other words, how large of a sample is required such that the probability of making a type I or II error in rejecting or accepting a null hypothesis will be acceptable under the circumstances. Different situations call for different error rates. An error such as diagnosing streptococcal throat infection when streptococcus bacteria are not present is likely not as serious as missing the diagnosis of cancer. Another way of phrasing the question is, "How powerful is my statistical analysis in accepting or rejecting the null hypothesis?"

If the sample size is too small, the consequence may be that we miss an effect (type II error) because we do not have enough power in the test to demonstrate, with confidence, an effect.

However, when choosing a sample size, it is too easy to simply say that the sample size should be as large as possible. Even if an investigator had access to as many samples as he or she desired, there are practical considerations and constraints that limit the sample size. If the sample size is too large, there are economic and ethical problems to consider. First, there are expenses associated with running an experiment, such as a clinical trial. There are costs associated with the personnel who run the experiments, the experimental units (animals, cell cultures, compensation for human time), perhaps drugs and other medical procedures that are administered, and others. Thus, the greater the number of samples, the greater the expense. Clinical trials are typically very expensive to run.

The second consideration for limiting sample size is an ethical concern. Many biomedical-related experiments or trials involve human or animal subjects. These subjects may be exposed to experimental drugs or therapies that involve some risk, and in the case of animal studies, the animal may be sacrificed at the end of an experiment. Bottom line, we do not wish to use human or animal subjects for no good reason, especially if we gain nothing in terms of the power of our statistical analysis by increasing the sample size.

We recall again that there are two types of errors associated with performing a statistical analysis:

1. Type I: rejecting H_0 when H_0 is true (probability of type I error = α);
2. Type II: accepting H_0 when H_0 is false; missing an effect (probability of type II error = β).

7.1 POWER OF A TEST

When we refer to the power of a given statistical analysis or test, we are quantifying the likelihood, for a given α value (e.g., 0.05), that a statistical test (*t* test, correlation coefficient, ANOVA, etc.) will detect a real effect. For example, if there truly is a difference between the means of two populations, what is the likelihood that we will detect that difference with our *t* test? If the chance of a type II error is β, then the likelihood of detecting a true difference is simply $1 - \beta$, which we refer to as the power of the test. Some practical points about the power of a statistical test are the following:

1. power of test = $1 - \beta$;
2. $0 < \text{power} < 1$
3. In general, we desire power ≥ 0.8 for practical applications.

When trying to establish a sample size for an experiment, we have to decide in advance how much power we want for our statistical analysis, which, in turn, is determined by the amount of types I and II errors we are willing to risk. For some biomedical engineering applications, we can risk greater error than others. Missing a diagnosis may not be problematic in one case or may result in a life-threatening situation. If an implantable defibrillator fails to detect and electrically shock a heart rhythm, such as ventricular fibrillation, it could mean the loss of life. On the other hand, shocking the heart when there is no fibrillation present, which may happen when the device senses electrical activity generated by nearby skeletal muscle, can result in pain and may even initiate a dangerous heart rhythm.

It is often difficult to minimize both α and β at the same time. One is usually minimized at the expense of another. For example, we would rather error on the side of an implantable defibrillator overdetecting fibrillation so as not to miss an occurrence of fibrillation. In the case of detecting breast cancer using mammography, overdetection or underdetection can both be problematic. Underdetection can mean increased chance of death. However, false-positives can lead to unnecessary removal of healthy tissue. One of the biggest challenges for researchers in the biomedical field is to try to find real differences between populations that will improve diagnosis and detection of disease or the performance of medical instruments.

7.2 POWER TESTS TO DETERMINE SAMPLE SIZE

Power tests are used to determine sample size and take into account the effect that we wish to detect with our statistical analysis, the types I and II error rates we will tolerate, and the variability of the populations being sampled in our experiments.

To perform a power test, we can use equations that express power in terms of the factors stated above, or we can make use of power curve and power tables that have already estimated the sample sizes for us. Power curves and tables show the relationship between power ($1 - \beta$) and the effect we are trying to detect. These effects can be difference in two means, difference in two variances, a correlation coefficient, difference in treatment means, and other population differences that we are trying to detect through experiment and data analysis.

It is important to note that there are different equations and, thus, different power curves for different effects, which in turn are detected through different statistical tests. A difference in two means (the effect) is detected with a t test, whereas the difference in two variances (the effect) is detected using an F test. The power curves for the t test are different from the power curves for the F test.

Power curves most frequently used in estimating a sample size for a biomedical experiment in include the following:

1. unpaired or paired t test (difference in one or two population means);
2. Pearson's correlation coefficient (correlation between two populations);
3. ANOVA (difference in three or more population means).

To perform a power test using the power curves, we need the following:

1. the size of effect we want to detect (i.e. difference in two means);
2. an estimate of population parameters (i.e. standard deviation for the population(s) based on pilot data);
3. α level (probability of type I error);
4. power level = ($1 - \beta$) (probability of type II error).

The α and β levels are selected by the investigator before designing the experiment. The size of effect to detect is also chosen by the investigator. The investigator needs to determine in advance how large the effect has to be to be significant for drawing conclusions. In other words, how different do two populations means need to be to signal something different in the underlying physiology, drug effects, or a change in the manufacturing process. Or, how strong does the correlation have to be to mean something to the investigator, in terms of underlying biology or processes? In general, the smaller the difference to be detected with a *t* test or the greater the correlation value to be detected, the greater the sample size.

One might argue that the most difficult piece of information to obtain is an estimate of the standard deviation, or variance, of the underlying populations. It is the assumed variance of the underlying population that plays a large role in determining sample size. In general, the larger the variance in the population, the greater the sample size needed to capture the statistics of the population or process and, consequently, the greater the sample size needed to give power to a statistical analysis.

Table 7.1 shows an example of a power table for the unpaired *t* test. This table was estimated for $\alpha = 0.05$. But similar tables exist for other α levels. These curves and tables allow us to estimate a sample size if we first select a normalized difference in the two means that we wish to detect and we select a power at which to perform the *t* test. The normalized difference is obtained when we take the absolute difference that we wish to detect and divide that difference by the estimated standard deviation of the underlying population. Again, we have a guess for the standard deviation based on pilot data. By normalizing the difference to detect, we need not worry about units of measure and can make used of standardized tables.

TABLE 7.1: Table for power for Student's unpaired (one-tail) *t* test ($\alpha = 0.05$)

	Difference in means (expressed as z score)									
n	0.10	0.20	0.30	0.40	0.50	0.60	0.70	0.80	1.0	1.2
10	0.08	0.11	0.16	0.22	0.29	0.36	0.45	0.53	0.70	0.83
13	0.08	0.13	0.18	0.26	0.34	0.44	0.54	0.63	0.80	0.91
20	0.09	0.15	0.24	0.34	0.46	0.59	0.70	0.80	0.93	0.98
30	0.10	0.19	0.31	0.46	0.61	0.74	0.85	0.92	0.99	
40	0.11	0.22	0.38	0.55	0.72	0.84	0.93	0.97		
60	0.13	0.29	0.50	0.70	0.86	0.95	0.98			
80	0.15	0.35	0.60	0.81	0.93	0.98				
100	0.17	0.41	0.68	0.88	0.97					
200	0.26	0.64	0.91	0.99						

The best way to illustrate the use of power tests is to work through an example. We will use the table in Figure 7.1 to find the sample size for the following example:

Aim: To determine if there is there a difference in mean height for college-age men and women living in the city of make-believe?

Experiment: Measure the heights in random samples of college-age men and women in the city of make-believe.

Question: How many samples do we need to reject our null hypothesis that there is no difference in mean height between college-age men and women in the city of make-believe if indeed there is a true difference in the two populations?

We need to perform a power test to determine how many samples we need to collect from the college-age men and women if we are to detect a true difference in the underlying populations. We are assuming that the two underlying populations are normally distributed. The effect that we are trying to detect is whether there is a significant difference in mean heights; thus, we will be using an unpaired *t* test to analyze the data once they are collected.

Before performing the power test, we need to decide on the magnitude of effect we wish to detect as being significant when we perform our *t* test. For this example, we are going to choose a 3-in. difference in mean heights as being significant. For argument sake, we will claim that a 3-in. difference in population means is of biological significance.

In this example, the effect being tested ($\mu_1 - \mu_2$) is 3 in. To use the standardized curves, we need to normalize this difference by an estimate of the standard deviation, σ, of the underlying populations. This is basically a *z* score for the difference in the two means. Note that we assume the two populations to have roughly equal variance if we are to use a *t* test. To obtain an estimate of σ, let us assume we have collected some pilot data and estimated the sample standard deviation, $s = 5$ in. Now, the normalized difference we wish to detect when we apply our *t* test is 3/5.

We next choose $\alpha = 0.05$ and power = 0.8. Again, these error rates are our choice.

Now, using the magnitude of our effect (3/5), power value (0.8), and α value (0.05), we used the tables or curves in Figure 7.1 to look up the number of required samples from the power tables. For this example, we find the sample size to be approximately 35 samples.

Some general observations may be made about the impact of population variance, types I and II error rates, and the magnitude of the effect on the sample size:

1. For a given power and variance, the smaller the effect to detect, the greater the sample size.
2. The greater the power, the greater the sample size.
3. The greater the population variance, the greater the sample size.

If we use the same example above and the Minitab software to calculate sample size for an unpaired *t* test as we vary several of the inputs to the power test, we find the following sample sizes (Table 7.2):

TABLE 7.2: Relations among sample size, statistical power, standard deviation, and magnitude of effect to detect ($\alpha = 0.05$)

Size of difference in means (in.)	Standard deviation (in)	Power of test	Size of sample
1	5	0.8	394
2	5	0.8	100
3	5	0.8	45
4	5	0.8	26
3	1	0.8	4
3	2	0.8	9
3	10	0.8	176
3	5	0.9	60

Thus, we can estimate, in advance of performing an experiment, a minimal sample size that would allow us to draw conclusions from our data with a certain level of confidence and power.

• • • •

CHAPTER 8

Just the Beginning...

Compared with most textbooks about statistics, the previous seven chapters have provided a brief "elevator pitch" about the fundamentals of statistics and their use in biomedical applications. The reader should be familiar with the basic notion of probability models, the use of such models to describe real-world data, and the use of statistics to compare populations. The reader should be aware of the impact of randomization and blocking on the outcome of statistical analysis. In addition, the reader should have an appreciation for the importance of the normal distribution in describing populations, the use of standardized tables, and the notion that statistical analysis is aimed at testing a hypothesis with some level of confidence. Finally, the reader should know that we can find confidence intervals for any estimated statistic.

In this text, we only covered those statistical analyses that are valid for populations or processes that are well modeled by a normal distribution. Of course, there are many biological processes that are not well modeled by a normal distribution. For other types of distributions, one should read more advanced texts to learn more about nonparametric statistics and statistics for nonnormal distributions or populations. These nonparametric tests do not assume an underlying distribution for the data and are often useful for small size samples.

How do we determine whether our data, and hence underlying population, is well modeled by a normal distribution? We can begin by simply looking at the histogram of the sampled data for symmetry and the proportion of samples that lie within one, two, and three standard deviations of the mean. We can also compare the sample mean to the sample median. A more formal means of quantifying the normality of a sample is to use a χ^2 test [7]. The χ^2 test involves comparing the actual sample distribution to the distribution that would be expected if the sample were drawn from a normal distribution. The differences between the expected frequencies of sample occurrence and the true frequencies of sample occurrence are used to estimate a χ^2 test statistic. This test statistic is then compared with the critical values of the χ^2 distribution to determine the level of significance or confidence in rejecting the null hypothesis. In this case, the null hypothesis is that the frequency distribution (or probability model) for the underlying distribution is no different from a normal distribution. The χ^2 test is also referred to as the goodness-of-fit test. We note that the χ^2 test can also be used to compare a sample distribution with other probability models besides the normal distribution.

We have covered one-and two-factor ANOVA. But one will also encounter multivariate ANOVA, in which there is more that one dependent variable (more than one outcome or measure being sampled.) Such analysis is referred to as MANOVA and is used in a more generalized multiple regression analysis in which there may be more than one independent variable and more than one dependent variable. The dependent variables are multiple measures (or repeated measures) drawn from the same experimental units being subjected to one or more factors (the independent variables). In linear regression, we sought to determine how well we could predict the behavior of one variable given the behavior of a second variable. In other words, we had one independent variable and one dependent variable. In the case of multiple regression analysis, we may assume more than one dependent variable and more than one independent variable to explain the variance in the dependent variable(s). For example, we may have one dependent variable, such as body weight, which we model as a linear function of three independent variables: calorie intake, exercise, and age. In performing a multiple regression analysis, we are trying to predict how much of the variability in body weight is because of each of the three independent variables. Oftentimes, one independent variable is not enough to predict the outcome of the dependent variable. However, we need to keep in mind that one should only add additional predictors (independent variables) that contribute to the dependent variable in a manner that the first predictor does not. In other words, the two or more predictors (independent variables) together must predict the outcome, or dependent variable, better than either independent variable can predict alone. Note that a full, generalized linear model allows for multiple dependent and independent variables. In essence, the multiple regression model is looking for simple correlations between several input and output variables.

Although we have focused on linear regression, the reader should also be aware that there are models for nonlinear regression that may be quite powerful in describing biological phenomena. Also, there is a whole literature on the analysis of errors that result when a model, used to predict the data, is compared with the actual measured data [3]. One can actually look at the residual errors between the modeled or predicted data and the actually measured data. Significant pattern or trends in the magnitude and ordering of residuals typically indicate a poor fit of the model to the data. Such patterns suggest that not all of the predictable variability in the measured data has been accounted for by the model.

Other multivariate analyses include cluster, discriminant, and factor analyses. These topics are covered in numerous statistical texts. These analyses allow one to group a population into subpopulations and explain complex data of seemingly many dimensions of factors into a smaller set of significant factors.

In this text, we have also not covered receiver operator characteristic curves. These are statistical analyses used frequently in the design and assessment of the medical devices or diagnostic tests used to detect disease or abnormalities. These curves, often summarizing terms such as *sen-*

sitivity, *specificity*, and *accuracy*, provide a means for determining how accurately a diagnostic tool, test, or algorithm is in detecting the disease or abnormal physiological function. We have previously discussed types I and II errors, which can also be used to estimate the sensitivity, specificity, and accuracy of a diagnostic test. There is often a trade-off between sensitivity and specificity, which can raise frustration for the biomedical engineer trying to develop safe, accurate, practical, and inexpensive diagnostic tests. The receiver operator characteristic curve is a graph that plots the sensitivity of the test (probability of a true positive result) against the probability of a false-positive test. The operator usually chooses to operate at the point on the receiver operator characteristic curve where sensitivity and specificity are both maximized. In some cases, reducing specificity at the expense of sensitivity may be preferred. For example, when administering a test to detect streptococcal throat infection, we may prefer to maximize sensitivity in order not to miss a diagnosis of strep throat. The trade-off is that we may reduce specificity, and a person without streptococcal infection may be misdiagnosed as having the streptococcus bacteria. The result is that the person ends up paying for and consuming antibiotics that are serving no purpose. In other clinical situations, the trade-off between sensitivity and specificity may be much more complex. For example, we often wish to use a noninvasive imaging tool to detect cancerous breast lesions. High sensitivity is desired so that the test does not miss an occurrence of a cancerous lesion. However, the specificity must also be high so that a patient will not have unnecessary surgery to remove healthy or noncancerous tissue.

Finally, there is a vast area of statistical analysis related to time series, that is, data collected over time. Time is the independent variable being used to predict the dependent variable, which in biomedical applications, is often a biological measure. Statistical time series analysis is nicely introduced in Bendat and Piersol [7] and plays an important role in most biomedical research. Such analysis is often used to develop automated detection algorithms for patient monitoring systems, implantable device, and medical imaging systems.

This text alludes to, but does not explicitly cover, the use of statistical software packages for the analysis of data. There are a number of statistical software programs that are commercially available for the engineer who needs to perform statistical analysis. Some of the software packages are freely available, whereas others can be quite expensive. Many of the software packages offer some tutorial assistance in the use and interpretation of statistical and graphical analysis. Some have user-friendly interfaces that allow the user to quickly load the data and analyze the data without much instruction. Other packages require considerable training and practice. Some of the more popular software packages for statistical analysis used in the biomedical field include SPSS, Minitab, Excel, StatView, and Matlab.

Finally, an important takeaway message for the reader is that statistics is neither cookbook nor cut-and-dry. Even the most experienced biostatisticians may debate over the best analyses to use for biomedical data that are collected under complex clinical conditions. Useful statistical analysis requires that the user first frame a testable hypothesis, that the data be collected from a representative sample, and that the experimental design controls for confounding factors as much as is practical. Once the data are collected, the user needs to be as fair as possible in summarizing and analyzing the data. This requires that the user have a good appreciation for the assumptions made about the underlying populations when using a statistical test as well as the limitations of the statistical test in drawing specific conclusions. When used properly, statistics certainly help us to make good decisions and useful predictions, even in the context of uncertainty and random factors over which we have no control. Biomedical engineers need to embrace statistics and learn to be as comfortable with the application of statistics as they are with the application of algebra, calculus, and differential equations.

• • • •

Bibliography

[1] Task Force of the European Society of Cardiology and the North American Society of Pacing and Electrophysiology, "Heart rate variability. Standards of measurement, physiologic interpretation and clinical use," *Circulation*, vol. 93, pp. 1043–1065, 1996.

[2] Wagner, G.S., *Marriott's Practical Electrocardiology*, 10th ed., Lippincott Williams & Wilkins, Philadelphia, PA, 2001.

[3] Hogg, R.V., and Ledolter, J., *Engineering Statistics*, Macmillan Publishing, New York, 1987.

[4] Gonick, L., and Smith, W., *The Cartoon Guide to Statistics*, HarperPerennial, New York, 1993.

[5] Salkind, N.J., *Statistics for People Who (Think They) Hate Statistics*, 2nd ed., Sage Publications, Thousand Oaks, CA, 2004.

[6] Runyon, R.P., *Fundamentals of Statistics in the Biological, Medical and Health Sciences*, PWS Publishers, Boston, MA, 1985.

[7] Bendat, J.S., and Piersol, A.G., *Random Data. Analysis and Measurement Procedures*, 2nd ed., Wiley-Interscience, John Wiley & Sons, New York, 1986.

[8] Ropella, K.M., Sahakian, A.V., Baerman, J.M., and Swiryn, S., "Effects of procainamide on intra-atrial electrograms during atrial fibrillation: Implications for detection algorithms," *Circulation*, vol. 77, pp. 1047–1054, 1988.

[9] Fung, C.A., *Descriptive Statistics. Module 1 of a Three-Module Series on Basic Statistics for Abbott Labs*, Abbott Labs, Abbott Park, IL, December 11, 1995.

[10] Olkin, I., Gleser, L.J., and Derman, C., *Probability Models and Applications*, Macmillan Publishing, New York, 1980.

[11] Fung, C.A., *Comparing Two or More Populations. Module 2 of a Three-Module Series on Basic Statistics for Abbott Labs*, Abbott Labs, Abbott Park, IL, December 12, 1995.

[12] *Minitab StatGuide. Minitab Statistical Software*, release 13.32, Minitab, State College, PA, 2000.

[13] Ropella, K.M. Baerman, J.M., Sahakian, A.V., and Swiryn, S., "Differentiation of ventricular tachyarrhythmias," *Circulation*, vol. 82, pp. 2035–2043, 1990.

[14] Schmit, B.D., Benz, E., and Rymer, W.Z., "Afferent mechanisms of flexor reflexes in spinal cord injury triggered by imposed ankle movements," *Experimental Brain Research*, vol. 145, pp. 40–49, 2002.

Author Biography

Kristina M. Ropella is currently a professor of biomedical engineering at Marquette University, where she has been a member of the faculty since 1990. She received her Doctor of Philosophy degree in biomedical engineering from Northwestern University (Evanston, IL). Her research interests are in physiological signal processing, electrophysiology of heart and brain, functional magnetic resonance imaging, and biocomputing. She teaches undergraduate and graduate courses in biomedical signal processing, statistical time series analysis, biostatistics, biomedical computing, and biomedical instrumentation design. She is a past recipient of the Marquette University Robert and Mary Gettel Faculty Award for Teaching Excellence and the College of Engineering Outstanding Teacher award. She also directs the Functional Imaging Program, a joint doctoral degree program offered by Marquette University and the Medical College of Wisconsin. She is a fellow of the American Institute of Medical and Biological Engineering, a senior member of the IEEE Engineering in Medicine and Biology Society, and a member of the Biomedical Engineering Society and the American Society for Engineering Education. She serves on the editorial board for IEEE Book Press, biomedical engineering series, and she has chaired a number of education- and university industry-related sessions at the EMBS, BMES, and ASEE society meetings. Her personal interests include raising three children, racquetball, dance, travel, and lay ministry for the church.